Arms Makers of Pennsylvania

Arms Makers of Pennsylvania

JAMES B. WHISKER

Selinsgrove: Susquehanna University Press
London and Toronto: Associated University Presses

Associated University Presses
440 Forsgate Drive
Cranbury, NJ 08512

Associated University Presses
25 Sicilian Avenue
London WC1A 2QH, England

Associated University Presses
P.O. Box 488, Port Credit
Mississauga, Ontario
Canada L5G 4M2

The paper used in this publication meets the requirements of the American National Standard for Permanence of Paper for Printed Library Materials Z39.48-1984.

Library of Congress Cataloging-in-Publication Data

Whisker, James B., 1939–
Arms makers of Pennsylvania.

Bibliography: p.
1. Firearms industry and trade—Pennsylvania—History. 2. Gunsmiths—Pennsylvania—Biography.
I. Title
TS533.3.P4W4 1990 683.4′009748 86-43215
ISBN 0-941664-81-3 (alk. paper)

Printed in the United States of America

Contents

Arms Makers of Pennsylvania

Introduction

Arms makers in Pennsylvania included artisans and craftsmen of various types who performed many functions and services. They made guns used in the earliest skirmishes with the natives and a portion of those used in all subsequent wars. They initially did only repair work on arms made in Europe and imported for use in the various colonies. Eventually, however, they made their own guns, at first assembling these arms from imported parts, and later, when they became mature in their craft, making all parts in their shops. When the needs of the nation and the colony or state were made known, the gunsmiths responded.

Our primary interest, here, will be the long rifles made in individually owned shops by craftsmen working either alone, with a journeyman or apprentice, or with their sons, daughters, nieces, or nephews. Many of these craftsmen were simultaneously arms contractors or repairmen for the government—in wartime for the armies and in peacetime for the militia. But, overwhelmingly, the gunsmiths listed in this book made, and according to the best research published to date, developed, the long, or Kentucky, rifle.

Why call the long rifle developed in Pennsylvania the Kentucky rifle? This question has been discussed at great length and it bears consideration here. The first study of the long rifle was completed by Captain J. G. W. Dillin and published by the National Rifle Association in 1924 under the title, *The Kentucky Rifle*.[1] Ned Roberts wrote the second major book, *The Muzzle Loading Cap Lock Rifle*,[2] in 1944 without giving a specific state identification to the rifle, and this has been interpreted to mean that Roberts did not want to repeat Dillin's title, but that he was reluctant to use the Pennsylvania state association. Professor Henry J. Kauffman entitled his first study *Early American Gunsmiths*,[3] and thus avoided any state association; but when he published his study of the long rifle as made primarily in Pennsylvania, he chose the hyphenated term, *The Pennsylvania-Kentucky Rifle*.[4] In the same year that Kauffman's second book was published, Joe Kindig, Jr. published his monumental tribute to the long rifle makers, *Thoughts on the Kentucky Rifle in Its Golden Age*,[5] although nearly all guns shown from Kindig's own extensive collection had been made in Pennsylvania.

It was not until Samuel E. Dyke wrote his *Pennsylvania Rifle*[6] in 1974 that the birthplace of the long rifle was acknowledged in print. Regrettably, although Dyke had studied the gunsmiths in several northeastern counties, his book covered only his home county, Lancaster. His extensive research in old tax lists existed only in Kentucky Rifle Association newsletters and in mimeographed form.

Many national listings of gunsmiths have appeared in the past half century. These have included, generally, all classes of arms makers, ranging from cottage-industry makers of the Kentucky rifle to inventors of large arms to arms contractors in all American wars to those who held patents, even on arms never actually made except as prototypes. Charles W. Sawyer was the pioneer in such study. His books contain much information, but are now wholly superseded by other studies.[7] In 1940 two important works on arms makers appeared: Stephen Van-Rensselaer's *American Firearms*[8] and Arcadi Gluckman and L. D. Satterlee's *American Gun Makers*.[9] Robert Gardner had been publishing information on arms makers since the early 1930s, and his research culminated in *Small Arms Makers*,[10] but certain information contained in his earlier books, *American Arms and Arms Makers*[11] and *Five Centuries of Gunsmiths, Swordsmiths and Ar-*

mourers,[12] was not repeated in the later book. Of the early students of the American arms industry, Gardner was, by far, the most accurate, and his earlier works contain good documentation. Such was not always the case with works by other authors, although their entries have often been repeated by later compilers of listings. Some authors compiled their listings by examining a gun and making an educated guess about where and when it was made. This led to interesting errors on occasion, as in the case of "Frederick Tell" (Frederick Sell).

Most of the books with listings examined the small, independent gunmakers, while specialized studies covered the major arms manufactories. Articles continue to appear on large arms makers, although one suspects that little new information of major consequence remains to be related. There is still much work to be done on the small arms industry, that is, on those who made the Kentucky rifle. Dr. George Shumway of York, Pennsylvania has been a major publisher of specialized studies, and his contributions are significant. In addition to maintaining Dillin, Roberts, and Kindig in print, Dr. Shumway has published several important volumes of his own. Donald Hutslar's study of Ohio gunsmiths[13] is a major and important contribution to the study of arms makers of that state. Projected as a five-volume study, only the biographical volume has appeared to date, with the remaining four volumes illustrating the guns only in the planning stage. Daniel Hartzler's *Arms Makers of Maryland*[14] is of similarly high quality, and offers a look at the products as well as at the men responsible for arms making in that state. Colonel Russell Harriger's study of the gunsmiths of Elk, Jefferson, and Clarion counties in Pennsylvania[15] set a very high standard of research. It is lavishly illustrated with Dr. Shumway's excellent photographs. Dr. Shumway anticipates publishing approximately twenty-five additional volumes containing other high-quality studies of counties or clusters of interrelated counties.

Old tax records in most Pennsylvania counties gave the occupations of inhabitants until well after the Civil War. Some early tax lists were published in the third series of the Pennsylvania Archives. Unfortunately, the records have been lost in some counties, and misplaced or stored in inaccessible areas in other counties. In some cases obsolete public records have been given over to the care of historical societies. This is a mixed blessing, as some of these societies refuse access to records or charge to have members use them for the researcher. In some cases, courthouse officials are very cooperative, but in other cases, they view the historical researcher as an uninvited intruder into what is supposed to be a lawyer's domain. Tax records are one of the best sources of reliable information available, but studying these in musty basements or dirty lofts can be both time consuming and unpleasant. By 1850 the U.S. census lists began to give occupations, although in many cases we found craftsmen listed as farmers because most of their valuation was established by farmland and most of their income was earned in that way. Records of early land transactions gave occupations. Estate inventories helped establish professions. Early newspapers often contained useful advertisements and notices.

County histories became especially popular in the 1880s, and they seem to have followed a basic formula in most cases. Early inhabitants were discussed in town, city, and township sketches, and more than a few arms makers were located in this way. Many of the biographies were prepared with little attention to the truth because the subscriber had paid to have his name included, and he may even have written the entry himself. Other sketches contained legends and myths and tales that are, to say the least, improbable. A good example of the perpetuation of myth can be found in our entry on Daniel McKinney, below.

By exchanging information with researchers in other states we have been able to offer a more complete picture of the arms-making trade than was heretofore possible. Notable among the studies in progress are Holman J. Swinney's book[16] on New York and Curtis Johnson's research in Illinois.[17] A. W. Lindert[18] is continuing to update his Indiana list. John Bivens[19] is doing the same on North Carolina, and several men are cooperating on Virginia.[20] Lawrence Shelton[21] has completed an excellent study of California gunsmiths and continues to add material. William S. Bowers[22] has done excellent research on areas of Maryland, Virginia, and Pennsylvania. We have integrated appropriate aspects of the books named above in this study.

In the spring of 1961 George N. Hyatt, a prominent collector and publisher of the Kindig and some editions of the Dillin and Roberts books, suggested that an association be formed that would be dedicated to the preservation of the long rifle and the study of its history. In April 1962 the organization was formed as a result of a meeting held at Greensburg, Pennsylvania. The name chosen for it was the Kentucky Rifle Association. In its membership handbook it is extremely defensive of the use of the term Kentucky in its name, and, while admitting that the gun was probably born in Pennsylvania, insists that "Kentucky" ought to be the correct title.[23] The KRA refers those who question the origin and authenticity of the term to the song that started it all, "The Hunters of Kentucky."

In May 1822, just seven years after the Battle of New Orleans, a minstrel named Noah Ludlow excited his audiences in New Orleans with a new song, "The Hunters of Kentucky." This is, to date, the first reference we have found to the term "Kentucky rifle." Ludlow left a vivid account of the reaction of his audience to the song in his memoirs.[24] His account follows:

> As soon as the comedy of the night was over I dressed myself in a buckskin hunting-shirt and leggings, which I had borrowed off a riverman, and with **moccasins** on my feet, and an old slouch hat on my head, and a rifle on my shoulder, I presented myself before the audience. I was saluted with loud applause of hands and feet, and a prolonged whoop, or howl, such as Indians give when they are especially pleased. I sang the first verse, and these manifestations of delight were louder and longer than before: but when I came to the following lines:—
>
> But Jackson he was wide awake, and
> wasn't scared with trifles,
> For well he knew what aim we take
> with our Kentucky rifles;
> So he marched us down to 'Cypress
> Swamp'; the ground was low and mucky;
> There stood John 'Bull,' in martial pomp
> but here was old Kentucky.
>
> As I delivered the last five words I took my old hat off my head, threw it upon the ground, and brought my rifle to the position of taking aim. At that instant came a shout and an Indian yell from the inmates of the pit, and a tremendous applause from the other portions of the house, the whole lasting for nearly a minute and . . . 'the house rose to me!' The whole pit was standing up and shouting. I had to sing the song three times that night before they would let me off."

Ludlow took credit only for the music, for the words had been a poem written by Samuel Woodworth. The text of the song is as follows:[25]

Ye gentlemen and ladies fair.
Who grace this famous city,
Just listen if you've time to spare,
While I rehearse a ditty;
And for an opportunity,
Conceive yourselves quite lucky,
For 'tis not often here you see,
A hunter from Kentucky, O Kentucky & chorus

CHORUS:
The HUNTERS OF Kentucky.
O Kentucky, the hunters of Kentucky.

We are a hardy free-born race,
Each man to fear a stranger,
Whate'er the game, we join in chase,
Despising toil and danger;
And if a daring foe annoys,
Whate'er his strength and forces,
We'll show him that Kentucky boys
Are "alligator horses."
Oh! Kentucky, the hunters of Kentucky,
The hunters of Kentucky.

I s'pose you've read it in the prints.
How Packenham attempted
To make Old Hickory Jackson wince,
But soon his scheme repented;
For we with rifles ready cock'd,
Thought such occasion lucky,
And soon around the General flock'd
The hunters of Kentucky.
Oh! Kentucky, & Chorus

You've heard, I s'pose, how New-Orleans
Is famed for wealth and beauty—
There's girls of every hue, it seems,
From snowy white to sooty;
So Packenham he made his brags,
If he in fight was lucky,
He'd have their girls and cotton bags.
In spite of Old Kentucky.
Oh! Kentucky, & Chorus

But JACKSON, he was wide awake.
And wasn't scared at trifles;
For well he knew what aim we take.
With our Kentucky rifles:
So he led us down to Cypress swamp.
The ground was low and mucky;
There stood John Bull, in martial pomp,
And here was Old Kentucky.
Oh! Kentucky, & Chorus

A bank was raised to hide our breast.
Not that we thought of dying,
But then we always like to rest,
Unless the game is flying:
Behind it stood our little force—
None wished it to be greater,
For every man was half a horse,
And half an alligator.
Oh! Kentucky, & Chorus

They did not let our patience tire.
Before they showed their faces—
We did not choose to waste our fire.
So snugly kept our places:
But when so near we saw them wink,
We thought it time to stop them;
And 'twould have done you good, I think,
To see Kentucky pop them
Oh! Kentucky, & Chorus

They found at last 'twas vain to fight
Where lead was all their booty.
And so they wisely took to flight.
And left us all the beauty.
And now, if danger e'er annoys,
Remember what our trade is.
Just send for us Kentucky boys,
And we'll protect you, Ladies
Oh! Kentucky, & Chorus

To date, no better explanation has been offered for the naming of the Pennsylvania-born long rifle after a state in which it was used by many early pioneers, but that did nothing to develop it. Doubtless, the long rifle was used against more foes in Pennsylvania and Ohio than in Kentucky; and it took more wild game in those states than in

Kentucky as well. But the name seems to be a part of folklore which, unlike history, cannot be corrected. As the KRA handbook notes, "Say Kentucky Rifle and people know what you are talking about. Say anything else and you have to explain."[26]

The first phase of gunsmithing in Pennsylvania began with the arrival of the European colonists and traders. The guns were those common to Europe in the seventeenth century: the wheel lock, matchlock, and flintlock. These guns had fragile mechanisms and often needed repairs. Most professional soldiers and men with military experience knew how to effect fundamental repairs. Presumably, any skilled carpenter or cabinet maker could fabricate a stock. Parts were imported and could have been assembled at will. Native hardwoods would have supplied the stocks as they did generations later. Barrels might have been made from the earliest times from native iron, which was being extracted before 1700. The most complicated parts would have been the locks. While native craftsmen did make locks, at least after 1740, most locks used throughout American history have been imported, primarily from England and Germany. The earliest gunsmiths were almost certainly repairmen and gun assemblers, and not gun makers in the sense that they made all or most gun parts.

We find several armorers listed before 1700. These men may have performed many services, such as cleaning and storing arms and parts and making the minor repairs guns constantly needed. They would not have confined their services to small arms. Many armorers were involved principally with cannon and wall guns. Many armorers did little more than store powder, lead, flints, and other supplies. On rare occasions we find gunsmiths listing themselves as armorers.

The early colonists had a grave obligation to keep and to bear arms in defense of homes, selves, families, and state. The early laws of most of the colonies made this obligation clear by providing stiff penalties for the failure to keep arms, powder, flints, and balls. Nearly any gun would qualify as a militia weapon, ranging from a fowling piece (shotgun) to a rifle. The larger supplies were kept in community storehouses, such as those the English sought to capture at Lexington and Concord at the beginning of the American Revolution. Blacksmiths, whitesmiths (tinsmiths) braiziers and other skilled craftsmen were often impressed into service to maintain the community arms.

The colonists required arms for several reasons in addition to the general military obligation. They hunted wild game, often as the only means of providing food for the family. Sport or recreational hunting may have appealed to city dwellers early on, but for most, hunting was a matter of necessity. Smooth rifles were popular, as they could be used with a single ball for large game as well as for militia duty, and with shot for smaller game. Larger shot, commonly called swan shot (about the same as buckshot today) was used for general purposes, especially against Indians in the wilderness, and for shot a smooth gun was preferable to a rifled one. At distances up to about fifty yards a patched ball shoots about equally well from a smooth and a rifled gun. At longer distances the rifle has a distinct advantage in accuracy.

Guns were used to repel Indian attacks on the frontier. As the colonists found out many times, one need not live on the exact limits of the frontier to be subject to Indian attacks. On occasion, the Indians became bold and attacked small-to-medium-sized towns. Militiamen from the towns were called out on occasion to repel Indian attacks along the frontier.

The Quakers of Pennsylvania eschewed violence in any form, and they would not act as gunsmiths or sanction the use of firearms against the aborigines as long as they retained control of the legislature. Frontiersmen frequently demanded action that the Quakers refused. Eventually, just before the Quakers lost control of the Assembly in 1765, the people on the frontier brought mutilated corpses into Lancaster and Philadelphia and showed the townspeople the results of Indian attacks. When the Assembly finally acted, it did so with a vengeance, placing a bounty on Indian scalps. Guns proved to be most useful in collecting the bounty money.

Favored weapons among the more affluent in Pennsylvania were the German *Jaeger* rifles, used in the forests of Germany and named after the hearty hunters who used them. They had sliding wooden patchboxes and rifled barrels. The patchboxes carried the greased patches in which the user wrapped the lead ball, to assure a close fit between barrel and ball. The patch also helped to take grooves and cause the ball to spin as it was pushed down the barrel by the rapidly expanding gases produced by the ignited gunpowder. The patch served, too, as a kind of gas check, helping to prevent the gases from escaping around the ball, so that they could propel it properly. The *Jaeger* rifles gained a well-deserved reputation for great accuracy at distances.

It is generally believed that the first Pennsylvania long rifles were developed in or near Lancaster, about 1727–45. The primary source material needed to set the exact dates and makers is missing. There is a fascinating article[27] by Richard Headley, concerning a rifle dated 1705 made by Martin Meillin (AKA Meylin, Meylan, Meily). Headley argues persuasively that the gun, signature, and date are authentic. If this is so, it gives us a much earlier date for the emergence of the distinct form of art we know as the Pennsylvania (or Kentucky) long rifle. There are earlier handcrafted guns that may have been made in Pennsylvania, although none is dated and signed. For a variety of reasons, handcrafted long rifles (and fowling pieces) that are authentically American and were made before the Revolution are exceedingly rare. Dr. George Shumway,[28] has done a great service by identifying, photographing, and describing in detail many extant prerevolutionary guns. At this point it is virtually impossible to distinguish among sporting arms, military contract arms, militia weapons, and hunting weapons. American military arms were made in the same shops and by the same methods as civilian arms.

The highly decorated long rifles of the Golden Age came after the American Revolution, although a few guns of significant artistic merit are to be found in earlier periods. Some early long rifles are ornamented with rococo C-scroll carving in bas-relief and some have the brass patchboxes we associate necessarily with nearly all Golden Age guns. Patchboxes in these early guns are frequently in two pieces, consisting of lid and finial joined by a formed hinge. A few follow the German tradition by using a sliding wooden arrangement. Fowling pieces traditionally did not have patchboxes. Only a handful of military rifles had them; these rifles were usually made for frontier ranger service. Rifled guns, both imported and domestic, were more commonly used on the frontier by experienced riflemen.

What we do find in the colonial period of Pennsylvania history is the beginnings of an arms industry that would, first, arm us during many wars, and, second, train the craftsmen who would produce the guns of artistic merit in the four or five decades after the Revolution. It is hard to imagine how the colonists could have survived the French and Indian wars and the many small engagements with the savages had they not had an arms industry. The existence of an arms industry also served to encourage the development of new types of arms.

The second phase of arms development came about as a result of American participation in the Revolution. Arms were needed to supply colonial militiamen, the Continental army, and people on the frontier who were threatened by the Indians allied to the English. Congress attempted to standardize the basic arms used by all combatants:[29]

> . . . each soldier shall be furnished with a good musket that will carry a one ounce ball, with a bayonet, steel ramrod, worm priming wire and brush, fitted thereunto, a cutting sword or tomahawk, a cartouche box that will contain 23 rounds of cartouches, 12 flints and a knapsack. . . . Each man being provided with one pound of gunpowder and four pounds of ball fitted to his gun. . . . that it be recommended to the makers of arms for militia use that they make good, substantial muskets with barrels three and a half feet in length, that will carry a one ounce ball and fitted with a good bayonet and steel ramrod. July 18, 1775

The gunsmiths of Lancaster were ordered to stop manufacturing rifled arms and to concentrate on the manufacture of muskets for one year.[30] The assumptions were that muskets were easier to manufacture and that each musket required far less shop time than a rifle; that standardization of muskets was expected, but that rifles were made to suit each purchaser; and that uniformity of caliber was expected with muskets, but not necessarily with rifles. Rifles were less adaptable to bayonets and were more easily damaged in hand-to-hand combat. Muskets were less accurate, but could be loaded far more rapidly than rifles and they fouled less easily.

The colonists had a substantial arms-making capacity, as noted by a leading nineteenth-century historian of technology:[31]

> Small arms were made in considerable quantity at Philadelphia, Lancaster and elsewhere. The general insecurity of the frontier settlements, especially during the French and Indian War, the temptations of the chase, and particuliarly [*sic*] the Indian trade, rendered the demand for rifles and other defensive weapons. The manufacture received great impulse during the Revolution. The exportation of firearms, gunpowder and other military stores from Great Britain was prohibited in 1774, and Congress recommended their manufacture in each state.
>
> A letter from Philadelphia to a member of Parliament in December of 1774, soon after the proclamation was received, informed him that the act would be of no avail, as there were gunmakers enough in the Province to make 100,000 stand of arms within a year at 28 shillings each, if needed; and that a manufactory of powder had already been established. Governor Richard Penn, in his examination before the House of Lords in November 1775, stated, in reply to inquiries of the Duke of Richmond on the subject, that the casting of cannon, including brass which were cast in Philadelphia, had already been carried to great perfections; and also that small arms were made in as great perfection as could be imagined. The workmanship and finish of the small arms were universally admired for their excellence.
>
> Some firearms were that year imported from France and the Spanish West India Islands, and pikes were recommended until [fire]arms could be made. Rifles were made in many places in the Provinces at that date, which were thought to be the equal of any imported. . . .

There were several sources of arms available to the Committee of Safety. They could manufacture them on contract with the many gunsmiths in Philadelphia and Lancaster and on the frontier. They could import them from various foreign nations. They could confiscate them from those who refused to sign the oath of loyalty to the new government. Arms that already were held in various stores might be repaired. Various individuals and firms had supplies that could be purchased. And arms confiscated from the enemy might be reissued. The committee ordered that each of these alternatives be explored.

On 12 July 1776, the Committee of Safety at Philadelphia ordered that "whereas the Assembly in one of their late sessions did authorize and direct the Committee of

Inspection and Observation to collect all the arms from non-associators within this Province . . ." this action should be taken.[32] Those who would not deliver their arms were to be arrested and their firearms confiscated. Eventually, many nonassociators were condemned by bill of attainder and their properties sold at vendue. Among such persons were Oswald Eve, a gunpowder maker, and Tench Coxe, later purveyor of public supplies to the new nation. A gunsmith named Henry Mull (Moll) was also on this list.[33]

Pennsylvania created its own gun manufactory in February 1776. The factory was established at French Creek, Chester County and later moved to Allentown.[34] It manufactured gunlocks and restored firearms. A branch operated at Hummelstown, near Harrisburg. The manufactory was operated by Peter DeHaven and his brother, Hugh. It was financed by an advance of £300. It was hoped that the manufactory could turn out muskets with bayonet and steel ramrod at £4/10/0, and barrels at £1/02/0. These prices proved to be unrealistic as inflation raised its ugly head long before the factory began operation. The gunstock makers signed an appeal for high wages. The supplies became more expensive, and were often unavailable. By 1777, the Commonwealth gave up its efforts to make its own guns, liquidated the supplies, and ever after purchased its guns on competitive contract.

Contractors were available to the committee in Lancaster. They agreed to charge "Philadelphia prices" for their work, to confine themselves and their workmen to the manufacture of muskets for the committee, and to "furnish as many as they can possibly complete in the time, and deliver the same to the commissioners of the county or this Committee [of Public Safety at Philadelphia]." Moreover, they agreed to blacklist any and all gunsmiths who refused to accept state contracts, naming nonassociators as "enemies of their country, and published as such, and the tools of the said gunsmiths so refusing shall be taken from them, and moreover the said gunsmiths shall not be permitted to carry on their trades."[35] However, the "Philadelphia prices" did not hold long in Lancaster. Robert Galbraith, lieutenant of the county, complained on 14 August 1777 that "I have consulted the gunsmiths of this county as to making of arms & they in General way hold out for £8/15/0 for Musquetts & byonet; Shocking prices!"[36] The problem seems to have been due more to inflation than to any scarcity of help. There were at least eight gunsmiths in Lancaster at this time; one may add to that number skilled and allied craftsmen such as blacksmiths, nail cutters (or nailors), silversmiths, goldsmiths, pewterers, braiziers, founders, and others. In fact, some six apprentice gunsmiths are listed in Captain Andrew Graff's Company, Lancaster County Militia.[37]

Lancaster County was far better off than the counties on the frontier. The clerk of the Commissioners of Bedford County, David Espy, wrote on 9 February 1776 to the Committee of Safety, "we have but one gunsmith in the County, who has engaged to make 25 Firelocks, and has been employed for these 3 or 4 months past, but he has not got any of them completed. . . . He has been very industrious to procure an Assistant or Journeyman . . . but cannot obtain any."[38] Eventually, William Jones was to join Jacob Saylor, and these men barely kept up with the demands for their services during the Revolution.

As the biographical entries show, such notable long rifle makers as John Philip Beck[39] and Wolfgang Haga (Hagi, Hachen)[40] effected repairs on the northern frontiers of the state. But the cry was heard constantly from the frontiersmen. The inhabitants of Cumberland County reported that "the Provincial Muskets are beforehand with us, and [some militia] having the first choice, took between 60 and 70 of the best of them, leaving a parcel of Rubbish, . . . consisting mostly of Barrels, mostly without locks and stocks, . . . covered with rust . . . unfit for use and scarcely worth repairing. . . ."[41] Colonel Archibald Lochry made much the same report from Westmoreland County on 20 July 1779.[42] Lochry reported continued supply problems on 4 August 1781, [43] and even as late as June 1782, Alexander McClean reported that guns, notably rifles, were scarce throughout the western counties.[44]

In Philadelphia County, the gunsmiths had been ordered, initially, to provide pattern guns to be sent to the committees of the other counties in hopes of standardizing the muskets.[45] Benjamin Franklin was so concerned about the scarcity of firearms that he proposed making pikes of a design he had fashioned, based on extant European designs.[46] Stephen Paschall was directed to make Franklin's pikes, to be sent out to the other counties, although it seems unlikely that many were actually used. It appears that there were eleven gunsmiths in Philadelphia during the war, and we know that several apprentices or indentured servants were pulled out of trades such as silversmithing, goldsmithing, gilding, and cutlery and assigned to gunmaking. Chester County seems to have had only one gunsmith, which, considering its Quaker roots, is not surprising; but it had many cutlers and several iron furnaces, and these were converted to arms production.

The committee contacted gunsmiths, storekeepers, and others and gathered guns and gunpowder for military use.[47] It kept records on those who were buying up guns and/or hoarding arms and ammunition.[48]

There are few reports in the Pennsylvania Archives on importation of foreign arms. One such report, dated 28 June 1780, indicated that "1500 muskets shipped by Mr John Ross in 1776" had just arrived, but on inspection it was found that there were only 350 muskets in the shipment. One thousand of the lot were to have gone to Pennsylvania, and the other 500 to Maryland. One-third of the 350 muskets were taken by the ship as payment of freight charges, leaving only 234. The state received 314 muskets in another shipment, along with 800 locks, and eventually received, and these quite late, only 715 of the 1,500 ordered four years earlier in France.[49] Foreign supplies were not to be counted on.

The archives do not give statistics on the number of arms captured from the enemy and placed in American service. But the prisoners of war taken, notably Hessians, proved to be a resource. There is an accounting of tradesmen among the Hessians held at Lancaster, dated 10 January 1777. It shows one riflemaker, four locksmiths, 2 nailsmiths, a knifemaker, one silversmith, 16 other smiths and 15 carpenters.[50] All of these might be employed in arms making. At the Elizabeth Furnace in Lancaster County, Robert Coleman, the ironmaster, bought 42 German prisoners of war to work at cannon and shot making on 16 November 1782, at £30 each; and 28 more on 14 June 1783.[51]

Pennsylvania had many iron furnaces that were converted to gun-barrel and cannon and shot manufacturing during the Revolution. Cornwall Furnace in Lebanon County, Elizabeth Furnace in Lancaster County, Whitehead Humphreys's Philadelphia Steel Furnace, Martic Forge in Lancaster County, Mary Ann Furnace in Lancaster County, and Reading Forge were among those employed in arms making. The widow Catherine Smith converted her gristmill to the manufacture of gun barrels in White Deer Township, Lycoming County. It was burned by the Indians during the Revolution.

Although the frontier was never quite adequately supplied, the colonial arms makers generally did an excellent job of supplying the needs of the new nation. The eastern counties seem always to have had enough guns, not only for the army, but also for the militia, at least after 1777.

During the third phase of the development of the arms industry in Pennsylvania, there was a burst of artistic expression heretofore unseen in the new world. The best of the guns produced between 1785 and 1835 rival or surpass those made anywhere in the world. Customers were able and willing to pay for high-art guns, and the gunsmiths produced these in quantity. The guns were characterized by much carving, usually in bas-relief, silver inlays or overlays that were artistically conceived and engraved, and elaborately designed and engraved patchboxes. The locks may have been made in the individual shops or imported. Barrels were often made in shops that specialized in this one product, although some were produced in the gunsmiths' shops.[52]

Many of the same craftsmen who produced the fine, artistic long rifles also turned to mass production to help arm the nation as arms contractors to both the central government and the state militias. An excellent article by M. L. Brown[53] shows one arms contract let by Pennsylvania and filled by nearly twenty gunsmiths. These "CP" (for Commonwealth of Pennsylvania) muskets were the state arms during the War of 1812, and many went west with pioneers. Some were used as late as the Civil War. Total production of the "CP" muskets was approximately twenty thousand none of which was turned out by a factory or government arsenal.

Regional variations in guns have allowed the expert to distinguish between gunsmiths' products, even when the guns are not signed, and to locate the guns in area as well as in time. Regional preferences were so strong that, with rare exceptions, most gunsmiths adapted to local styles when they moved. Even when a gunsmith cannot be identified (because the gun is unsigned, the signature is illegible, or the name is unknown), he can be placed within a certain region or county.

As with the decline of any artistic movement, the causes of the decline of artistry in arms making are unclear. Occasional outbursts of artistic merit show that gunsmiths retained the ability to make high-art guns after 1835. The flintlock was being replaced by the percussion lock (or cap lock) by the late 1820s. But this is not necessarily the reason the Golden Age of gunsmithing was coming to an end. There are a few original percussion lock guns of great beauty, exhibiting superior craftsmanship. The responsibility lies with the customers, for the gunsmith would make whatever product could be sold or might be ordered.

With rare exceptions Pennsylvania gunsmiths became little more than gun stockers. Parts were turned out in factories or imported and merely fitted to the wood. Most merchants stocked imported locks and domestically produced cap boxes and barrels. Butt plates and trigger guards were made by brass founders, and the gunsmith had only the final filing to perform in his shop.

Concurrently with the manufacture of guns for the military during the Indian wars, the American Revolution, and the War of 1812, gunsmiths served two purposes. We have already examined the manufacture of sporting arms of artistic merit. The other function, which had deep roots in the eighteenth century, was the repair and occasional manufacture of arms for allied Amerindian tribes. Sir William Johnson recommended the adoption by the English of the French practice of sending gunsmiths among the aborigines to repair arms, sharpen knives, and mend other tools and cooking utensils.[54] This service had been performed early on by Indian traders, such as the Owens family[55] and John Fraser.[56] After the Revolution began, the Americans adopted this same policy and it remained in effect through the nineteenth century.[57] Gunsmiths also provided fancy arms as gifts to the Amerindian leaders.

Trade guns were among the major products of such Pennsylvania firms as Tryon, and Leman and Krider, as well as smaller, independent gunsmithies.

The independent gunsmiths also began experimenting with new methods and technologies to create better arms. The modern percussion cap, a major improvement over the old flintlock rifle ignition system, was invented in Philadelphia by Joshua Shaw between 1812 and 1818.[58] Other inventors began experimenting with repeating and breech-loading arms.

The percussion cap represented the first major improvement to the firearm developed in the United States. The wheel lock had been developed in Europe as had the flintlock. The copper or brass cap developed by Shaw provided a surefire means of igniting the gunpowder in the gun barrel. It led directly to the development of the self-contained cartridge, and with that, the introduction of breech-loading arms.

The revolutionary percussion cap ignition system was not accepted immediately. Few arms were made until after 1830 utilizing Shaw's invention. William Border's daybook showed that he began converting the older flintlocks to percussion in 1832.[59] Flintlocks were made as late as 1840; the system did not really become obsolete until the Civil War.[60] The U.S. Army resisted the new technology, except on an experimental basis, until 1841, and even thereafter manufactured only one series, in 1842, with flintlocks.[61]

By the end of the Civil War all muzzle-loading systems were technologically obsolete, for the breechloader had come into vogue and had been perfected to the point where it was both reliable and feasible.[62] But the cottage industry was not dead. There were probably as many muzzle-loading guns made after the Civil War (ca. 1865–1880) as had been made in comparable periods of time before the war. The opening and settling of the Great Plains and of the areas beyond the Rocky Mountains brought continued demand for muzzle-loading plains rifles. More "Hawken" rifles were made in manufactories in Pittsburgh than in the traditionally ascribed home of these rifles, St. Louis, Missouri.[63]

Muzzle-loading guns were used in beef shoots, target matches, and especially by the poorer class, as hunting arms, long after the Civil War. But some gunsmiths turned their attention to the development of new arms systems and to the improvement of existing systems. The larger manufactories of the cities made most cartridge weapons, although a few breechloaders were made by small shops, such as those operated by the Harder family[64] and John White.[65] Philadelphia became a major center for the manufacture of "Saturday Night Specials," the cheap handguns used for personal protection, in crimes, and by the poor for close range killing of animals. These firms also made or imported shotguns and rifles of modest quality. Gun repairmen were needed to repair the new cartridge guns and to alter for civilian use the many hundreds of thousands of military guns sold at low cost by the government after the Civil War.[66]

We shall now return to the Kentucky long rifle as made in Pennsylvania and examine the major schools or areas of gunsmithing that developed during the last quarter of the eighteenth century and the first half of the nineteenth century.

REGIONAL VARIATIONS

Certain generalizations may be made about the long rifles crafted in Pennsylvania. The more elaborate the gun, the easier it is to place it within one or another school of gunsmithing. We have here a classic case of the picture's telling us more than the proverbial thousand words. A word of caution is needed. The general statements in this book are meant to be a guide only. Although most late guns lack artistic merit, there are exceptions. Occasionally, a gun made within a certain county may not exhibit the normal county characteristics. Schools are not always confined to a single county, nor is the same school necessarily found throughout a county. Even the finest artisans made plain guns, and some mediocre craftsmen made guns of great merit on occasion. Since the entire gun was made by one man (perhaps with an apprentice), some guns show better carving, for example, while others display superior engraving. Few craftsmen excelled at all phases of gunsmithing. The late Joe Kindig, Jr., author of the monumental *Thoughts on the Kentucky Rifle in its Golden Age*, placed much emphasis on varying one's style, and most high-art gunsmiths did indeed offer many variations in their products. The Chambersburg school often used horsehead designs on patchbox finials, but that does not mean that all Chambersburg guns have such finials, nor does it mean that only Chambersburg-school guns have that style of finial. The same is true of carving styles. There are high-art guns that have no patchboxes or that have incised instead of bas-relief carving. Fowling pieces may have neither patchboxes nor carving. But the general concept and styling may still qualify an otherwise plain gun as worthy of consideration and study. Some features are shared among many schools. We may think of a Q-shaped finial as a standard Bedford County feature, but this exists on guns made elsewhere, and not all Bedford County guns have a Q-shaped finial. The carving around the tang varied only slightly among many schools of gunsmithing. Let us now offer certain insights on the schools of gunsmithing located primarily in eastern and southwestern Pennsylvania.

LANCASTER COUNTY SCHOOL

Assuming that the Pennsylvania long rifle emerged sometime about 1740 in Lancaster County, it is wholly appropriate to begin with this school. There were no eyewitnesses to the great event—the creation of the distinctly American long rifle. Many of the early guns shown in Kindig,[67] which were probably made before the Revolution, show the germinal daisy head, even on simple, two-piece patchboxes. They also exhibit simple C-scroll engraving behind the cheekpiece. The earliest carving is often a combination of incised and bas-relief styles. The earlier patchboxes rarely have piercings (internal cutout areas, often on the sideplates or the finial or both).

As the school developed after the Revolution, the carving became more elaborate. Double C-scrolls were used, back to back, with the open face pointing upward toward the comb and downward toward the heel of the stock, and always located behind the cheekpiece. In more elaborately carved guns the carving is extended under and forward of the cheekpiece. Some carving began to emerge around the tang area (ca. 1780–1830). Silver inlays made from coin silver, often representing the star of Bethlehem or a quarter moon, were inlaid on the cheekpiece. The pins or wedges that attached the barrel to the stock were surrounded by silver inlays. Muzzle caps were often decorated.

Before approximately 1800 patchboxes were cast brass. As sheet brass became available, the patchboxes were cut out from these sheets. Coins offered convenient sources of silver, and were frequently rolled out and engraved and inlaid. Inlays were rarely made of metal other than silver, and patchboxes were ordinarily made of brass. A presentation piece might have a silver patchbox, but this must be considered rare.

By the 1830s guns made by the Fordney family emerged with very coarse checkering (or cross-hatching) at the wrist. Better guns sometimes had silver braids or nails inlaid with the checkering in a decorative pattern. A few earlier rifles had these features, but they are found more commonly on the better guns by the Fordneys and their imitators.

A very few Lancaster County guns used rattlesnake motifs. Joe Kindig believed that William Weiss, a Lancaster County gunsmith, carried this design to the Richmond [Virginia] Manufacturing Company, where it was used on a rare type of military rifle, ca. 1805.[68] It is also found on a few later Chambersburg-school rifles. This coiled snake design is a variation of the "Don't Tread on Me" motif found on flags and other materials during the Revolution.

By the 1850s the Lancaster school had become devoted to the manufacture of cheaper hunting rifles and government contract guns, including Indian-trade weapons. Few of the guns made after the Civil War had much artistic merit.

Some Important Lancaster County Gunsmiths

Jacob Dickert	John Graeff
Peter Roesser	Abraham Henry
Melchior Fordney	John Haga
John Gonter	Peter Gonter
John Fondersmith	John Newcomer
Jacob Ferree	William Weiss
Henry Albright	Peter Brong
Jacob Hoak	Jacob Sees
M. Aldendefer	

PHILADELPHIA SCHOOL

The City of Brotherly Love had many gunsmiths (over 500 are listed in this book alone), from the earliest days through the cartridge period. The vast majority who worked independently have left no individual legacy by which we can judge their abilities. Others doubtless worked in one or another of the many gun factories or for government contractors. Many military guns were turned out in Philadelphia for every war from the French and Indian onward.

Among the identifiable features of many Philadelphia guns is the use of pineapples as the primary design element on finials and in other decorations. The pineapples were imitated in late guns made in Huntingdon and nearby counties.

Some Significant Philadelphia Gunsmiths

Deringer family	Tryon family
Henry family	Wurfflein family
Lewis Ghriskey	Christian Sharps
Marine Tyler Wickham	Mathias Hale

READING–BERKS COUNTY SCHOOL

The guns made in and around Reading were influenced by the Bethlehem and Lancaster schools, which is natural enough, considering the geographic location of each of these schools. The guns have a high comb and often have a roman nose. This feature became more pronounced with the passage of time.

The first truly significant gunsmith of this school was Wolfgang Haga (Hagi, Hachen, Houghen), who was active before 1750. Many of the guns attributed to Haga seem to have been made long after Haga's death in 1796. All are unsigned. The dates for Haga have changed significantly since Kindig produced his study of Haga's work in *Thoughts on the Kentucky Rifle in its Golden Age*.[69] Haga may have helped found a school that produced the Pennsylvania rifle somewhat independently of the Lancaster discovery of this style. Haga used both rococo C-scroll carving and other more naturalistic patterns, such as leaf motifs.

The patchboxes were generally more developed and well defined in Berks County than in the Bethlehem school. At its best, the Reading school produced extremely fine relief carving. Wrists tend to be thinner than on Lancaster County guns.

As the school went into decline, gunsmiths turned out austere, workaday guns assembled from commercially made parts. Many of the late guns used stylized animal inlays, notably on the cheekpieces.

Some Important Reading–Berks County Gunsmiths

Wolfgang Haga	John Bonewitz
Andrew Figthorn	Henry Eckler
Peter Belles	Leonard Reedy

BETHLEHEM SCHOOL

The Bethlehem school is notable for having produced the roman-nose stock and a very high, stylized comb that flows out of the wrist area. The sliding wooden patchbox survived for a long time in this area, even into the nineteenth century. Silver-wire inlays were used to highlight other features. Patchboxes remained simple, often made of just two pieces of brass. Embossing was used on the patchbox lid more frequently than in other areas. Early guns had thin, oval-shaped wrists that were flattened on top.

Carving was inferior to that of most other high-art schools. It was emphasized and done less in this area than in other areas, and was usually incised rather than executed in relief. With the notable exception of John Young, one of the finest engravers ever to work on long rifles anywhere, most of the Bethlehem gunsmiths downplayed engraving.

The earlier guns by such craftsmen as Peter Neihardt, A. Herman Rupp, and Jacob Kuntz used stylized Indian (or female) heads near the last ramrod pipe. These curious designs are rarely duplicated anywhere else, in any form.

One gun made by Jacob Kuntz has a coat of arms carved into the area behind the cheekpiece that must rank among the finest and most detailed carving ever executed on an American gun. This is atypical of the school in general.

Kindig observed that the guns made in Bethlehem are most difficult to date because the gunsmiths changed little over the years.[70] This conservatism caused many early design features to be retained long after similar characteristics had become obsolete elsewhere. Side-opening patchboxes, which were rarely used anywhere, are occasionally found on Bethlehem-school guns. Many guns produced by these craftsmen were signed and dated on the patchboxes, another highly unusual feature of the long rifle. This use of data was occasionally carried to the extreme by including the price of the gun. Guns made elsewhere would have had decorative engraving instead.

In his *The Kentucky Rifle*, Merrill Lindsay identified the "Associated Bethlehem" school.[71] He places the better craftsmen of the Bethlehem area in this category. These men used more elaborate carving and engraving than the craftsmen of the other Bethlehem school. Lindsay also places such men as Verner, who frequently listed his name, occupation, and price on the patchboxes, in this grouping.

Related to the Bethlehem school(s) are the gunsmiths who worked in the upper Susquehanna River area. Notable were Peter Smith and John Dreisbach of Mifflin County and the Long family of Snyder County. Guns that are stylistically quite close to the Bethlehem style were produced well into the percusssion lock period in these counties. Many had extensive inlay work, incised or relief carving, or a combination of the two, and pronounced roman-nose designs.

Some Important Bethlehem Gunsmiths

Angstadt family	Moll family
John Rupp	Andrew Verner
George Feder	Pannabecker family

YORK COUNTY SCHOOL

The York County school is closely related to the Lancaster County school, which is reasonable, considering the geographic proximity. The rococo C-scrolls used by Lancaster gunsmiths are found frequently on the better York guns. The splendid patchboxes found on most Golden Age York County long rifles, which show considerable variety, are one hallmark of this school. Piercings are common on the patchboxes, as is fairly elaborate engraving on both the sideplates and the lids. The lockbolt plates are often raised out of the wood and these are often of heavy cast brass and are sculptured and heavily engraved. Thumbpieces and cheekpiece inlays are standard fare. Stars of Bethlehem were used frequently on York County long rifles, but they are usually not so long as on Lancaster County guns. Quarter moons are also found on the cheekpieces.

Kindig noted that one salient feature of York County guns is the tapered lower ramrod pipe.[72] This is a feature of nearly all York County guns, and something not used elsewhere. The carving located around the tangs on better guns resembles that used by the Lancaster and Lebanon schools.

York County gunsmiths frequently lavished much attention on toe plates, piercing and/or engraving them to a significant degree and adding cross-hatching to some. Animal motifs were frequently included on patchbox finials of York County guns.

This school also rapidly declined, and few guns of merit were produced in the percussion period. Later gunsmiths of note often moved to other areas. For example, Andrew Kopp learned his trade in York County, but made most of his guns in Blair County. Frederick Stover moved to nearby Adams County, and his later guns were produced there.

Some Important York County Gunsmiths

Zorger family	Altland family
George Eister (Eyster)	Michael Edwards
George Nunnemaker	Martin Fry
John Leckner	J. Lowmaster
Jacob Ernst	Welshans family
Gobrecht family	Adam Ernst
George Schroyer	Andrew Kleindinst

LEBANON COUNTY SCHOOL

The early, fine flintlocks made in Lebanon County are usually associated with two names: John Philip Beck and Nicholas Beyer [Byer]. There may have been two men named Christian Beck who also made high-quality guns in Lebanon County, although the later man by this name worked in Martinsburg, (West) Virginia and Hamilton County, Ohio, and died in Marion County, Indiana. In any case, the conventional assumption is that Beyer was a late apprentice to J. P. Beck. Beck was a deeply religious man who often placed the initials I.N.R.I (Jesus of Nazareth, King of the Jews) on the bottom of his gun barrels. The same initials are found on guns made by Christian Beck and Nicholas Beyer.[73]

The county characteristics include long wrists, high relief carving that is well executed, great care for detail, a variety of styles that are always well conceived and executed, and a high comb. Carving is based on rococo C-scrolls. Patchboxes are uncarved or only lightly engraved. Piercings are simple and occur usually only in the finial. Sideplates of the patchboxes are plain. A few rifles known to be Beck's have sliding wooden patchboxes, an anachronism by the time J. P. Beck was working. There is great variety to the finials of the patchboxes. The school used finely executed bas-relief carving around the tang area.

The patchboxes are riveted near the upper rear corner of the lids. Rivets are used on the nose caps as well. Ramrod pipes are short and have heavy end moldings. Animal motifs were used occasionally, as a part of the carved area behind the cheekpiece. One magnificant J. P. Beck rifle features a lion in full bas-relief, and finials of Nicholas Byer guns frequently feature birds. Others used househead finials on the patchboxes.

The school rapidly declined, and later guns were primarily half-stock guns assembled from commercially manufactured parts.

Some Important Lebanon County Gunsmiths

Nicholas Byer	John Philip Beck
Christian Beck	

Lebanon County, Annville School

The Annville subschool of the Dauphin and Lebanon County schools has been identified by Dr. George Shumway.[74] Annville is in Lebanon County, about halfway between Harrisburg and Lebanon. The school consists primarily of Peter Berry, Jr. and perhaps a few apprentices who have yet to be definitely identified.

DAUPHIN COUNTY SCHOOL

Two gunsmiths stand out as members of the Dauphin County school: John Brooks and Martin Shell. Their guns are usually, but not always, much slimmer than guns made in nearby areas, and in this resemble some of the Chambersburg school products. The carving is probably among the most elaborate generally executed on long rifles. The intricate patterns may have begun with rococo C-scrolls, but the interplay of curves is outstanding on the best rifles. Combs rise high from the relatively slim wrist area. The daisy or other flower heads found on Lancaster County guns are used frequently. Patchboxes are not necessarily pierced, but are usually engraved, sometimes with flower designs.

Gunsmiths remained in Dauphin County, primarily in Harrisburg, long after the Civil War, but few produced guns of much artistic merit. The number of unsigned and unattributed guns suggests that craftsmen other than Brooks and Shell may have been active in Dauphin County in the Golden Age. Dr. George Shumway and his son Roger have undertaken research in this area, and their forthcoming book will doubtless shed much new light on this school of gunsmithing.

Some Important Dauphin County Gunsmiths

John Brooks	Martin Shell
Michael Poorman	

UPPER SUSQUEHANNA RIVER SCHOOL

There were several significant gunsmiths who made long arms in the area north of Harrisburg along the Susquehanna River, in the present-day counties of Mifflin, Perry,[75] Juniata, Snyder, and Union. The Peter Smith family, later of Huntingdon County, made some beautiful and highly decorated arms. As a group these gunsmiths usually employed excellent engraving and rarely used any form of carving, although there is a very fine Peter Smith rifle that shows superior relief carving. The Shuler family moved to Perry County from Bucks County, bringing a substantial gunsmithing tradition. They were the gunsmiths of that area for several generations. The Dreisbachs were the most significant family of Union County and made guns of considerable artistic merit. The Parks family brought the Bedford County style of gunsmithing to Snyder County.

Some Significant Gunsmiths of the Upper Susquehanna River

Peter Smith, Jr[a]	Dreisbach family
Peter Smith, Sr.[a]	Benfer family
Loudenslayer family	William Sweger
John Parks	Specht family
Valentine Fleck	Joe Long
Samuel Smith[a]	Jacob Bloom
Shuler family	Rothrock family

[a]Spent a considerable part of his working career elsewhere.

LITTLESTOWN SCHOOL

The gunsmithing school of northern Frederick County, Maryland had significant influence on the Littlestown gunsmiths. This is not surprising, considering that Littlestown is less than ten miles from Emmitsburg, Maryland, where John Armstrong and Nathaniel Rowe were active, and about the same distance from Taneytown, Maryland, where Philip Creamer worked.[76] The Sell family, John Shriver, and Frederick Stover are among those prominently associated with the Littlestown school.

Horse or bird heads were commonly used on patchbox finials. Rich details show in the rococo, C-scroll engraving and carving. The Littlestown school exhibits straight, high combs. The toe plates are often elaborately engraved and occasionally pierced. The lock bolt plates are executed in detail and are often mounted above the wood. The tang areas are often carved in details similar to those of the York school. Patchboxes usually show multiple piercings. Overall, the engraving is elaborate and detailed, including that on the patchbox lids.

The Littlestown school consisted of far fewer artisans than the usual school and so its existence depended almost entirely on the passing of the trade down from father to son. As the demand lessened for fine guns, ensuing generations turned to other trades and the school died out. Some later guns were made in the same geographic area, but these were primarily half stocks assembled from store-bought parts.

Littlestown Gunsmiths

Jacob Sell (possibly two men)	J. Shriver
Frederick Sell	Henry Koons
Frederick Stover[a]	

[a]May be classified stylistically as Emmitsburg, Maryland school.

CHAMBERSBURG SCHOOL

Gunsmiths were well established in greater (or original) Cumberland County and doubtless produced some guns of merit as early as the period immediately following the Revolution. With the arrival of Abraham Schweitzer in Chambersburg just before 1800, we find the beginning of a significant school of gunsmithing.[77] The guns tended to be very slim and often displayed superior worksmanship that extended well into the percussion period.

The carving around the tang area was replaced with silver inlays shaped much like the earlier carved designs. Patchboxes on the better guns were highly engraved and had many piercings. Both incised and relief carving were used of the late rococo C-scroll type. Cheekpiece inlays tended toward eagle or human-head designs. Silver thumbpieces were used on most guns.

The Mewhirter family, which supplied three generations of gunsmiths in Chambersburg and which disappeared overnight, offered a considerable variety of styles, including guns with folk art designs. Despite the emergence of Pennsylvania "Dutch" (really German) folk art in many artistic areas, these designs are rarely used on long rifles, except on occasion by gunsmiths of the Chambersburg school.

Joseph Clippinger, who later went to Ohio, offered nearly as much variety in his guns as did the Mewhirters. The Noll family, John and Henry, were both craftsmen of considerable talent. There is little to tie these several quality gunsmiths together, save for the accident of geographic location.

The gunsmiths of Adams, Franklin, and Cumberland counties, the constituent counties of the Chambersburg school, continued work until well after the Civil War, and although relief carving all but disappeared, there were still fine guns made late in the percussion period in this area.

Chambersburg School (broadly defined)

Abraham Schweitzer	Henry Albright
Samuel Shillito	Samuel M. Shillito
Jeremiah Senseny	Jacob Reasor (Resor)[a]
Christian Beck[a]	John Clark[a]
John Noll	Henry Noll
J. H. Johnston	Joshua Yous
Joseph Clippinger[a]	William Mewhirter
John Mewhirter, Sr.	John Mewhirter, Jr
Henry Carlile	George Zollinger
John Settel	John Meals

[a]Spent a significant portion of his career elsewhere.

BEDFORD COUNTY SCHOOL

There were gunsmiths active in Bedford County[78] before the Revolution, notably among the Indian traders, but the history of quality gunsmithing began here with the arrival of Peter White from Emmitsburg, Maryland, about 1803. White's guns often show the lines of the Emmitsburg school, although they slimmed down with the passage of time. White left for Uniontown, Fayette County, in 1818, but Joseph Mills continued the tradition for a decade, and Jacob Stoudenour, the grand master of the Bedford style, influenced the Border and Defibaugh families, who worked in the percussion period.

The romance of the Bedford County rifle is associated with its long, slim lines and consistent use of handmade locks, especially in the percussion period when commercially manufactured locks were almost universally used elsewhere. The carving was modified and simplistic, and the C-scroll was retained even after the Civil War on the better guns. Patchboxes usually had many piercings, and the Q-shaped finials are almost a hallmark of the Bedford style. Silver inlays were used primarily as thumbpieces; there were eagles on the cheekpieces, and as small decorations along the fore-ends. Even late guns were usually full-stock. Occasional guns, notably by William Defibaugh, used a traditional Pennsylvannia Dutch wagon wheel design in the carving. Incised carving is rare among Bedford County long rifles.

Some Significant Bedford County Gunsmiths

Jacob Stoudenour	Peter White[a]
William Defibaugh	William Border
Thomas Oldham[a]	Joseph Mills[a]
Daniel B. Border	John Border
Samuel Border[a]	John Nicholas Matthesiee
Adolphus Drake	Valentine F. Clouse
John Amos	Amos, Border & Co.
Jacob Breigle	(John Amos and D. B. Border)

Fulton County Subschool

James Clark
John Cook
John Whetstone[a]
Jacob Cook
David Whetstone

[a]Spent a significant portion of his career elsewhere.

HUNTINGDON COUNTY SCHOOL

A few high-quality flintlock rifles, such as those made by Levi Fondersmith and Andrew Kopp and his son George, are found within the Huntingdon County school.[79] We think of this school primarily in terms of later guns that are heavily mounted in silver. Few are carved, but the engraving on the many pieces (up to forty pieces on some Douglass rifles) compensates for this omission to some degree. The stocks are very slim, notably at the butt end, often making the commercial percussion locks look disproportionately large.

There is usually a large eagle engraved on an elliptic cheekpiece inlay. There are also many late half-stock rifles that have little to no artistic merit.

Some Significant Huntingdon County Gunsmiths

David Douglass	Joseph Douglass, Sr.
Levi Fondersmith	Joseph Douglass, Jr.

Samuel Gehrett[a]	Andrew Kopp
George Kopp	John Riling
Nicholas Shennefelt	Peter Smith[a]
Samuel Smith	James Stapleton

[a]Spent a significant portion of his career elsewhere.

SOMERSET COUNTY SCHOOL

Two styles predominate in Somerset County: those of the Mier family and the Dunmyer family. The Dunmyer guns have incised carving and resemble Bedford County guns in general conception, although with much shorter barrels (usually thirty-seven to thirty-nine inches in length). Some Dunmyer-style guns have commercial locks and others have handmade locks of a general Bedford County design. The Dunmyer patchboxes usually have long finials that reach up into the wrist area.

The Mier Somerset County rifles do not have remarkable lines, but do have occasional complicated relief carving. Samuel Mier was an unusually fine engraver, and many of his guns show eagle cheekpiece inlays of considerable merit. Several Mier guns have extremely elaborate patchboxes with many streamers rolling out from the sideplates in an acorn design.

The Somerset County designs, notably those of the Dunmyers were continued well into the twentieth century. A few other late long rifle makers assembled guns from manufactured parts.

Some Significant Somerset County Gunsmiths

John Altfather	Samuel Border[a,b]
Elias Crissey[a]	David Dunmyer
Jonathan Dunmyer	Peter Dunmyer[a,b]
Eli Knupp[b]	Charles Monroe Knupp
Jacob Mier	Samuel Mier
Benjamin F. Troutman	

[a]Stylistically part of Bedford school.
[b]Spent a significant portion of his career elsewhere.

WESTMORELAND COUNTY SCHOOL

We are only just beginning to recognize the Westmoreland County school of gunsmithing and to identify its principal artisans. Many of the better, early guns show considerable influence from the Emmitsburg, Maryland school. The carving and engraving on guns made by David Morton, the Kempf (Camp, Kampt) family, and George Kettering are highly decorated with relief carving and first-class engraving. Some of these craftsmen used silver-wire inlay to great advantage. Cheekpiece inlays were commonly engraved with eagles. A few percussion guns of merit exist, but most later guns are workaday guns assembled from commercially available parts.

Some Significant Westmoreland County Gunsmiths

David Morton	John Sherry, Sr.
George Kettering	Jacob Earnest
Solomon Kempf (Camp)	Benjamin Kempf (Camp)
Elisha Pancoast	

PITTSBURGH SCHOOL

The Ferree family of Lancaster County were among those who brought quality gunsmithing to Pittsburgh. Thomas Allison was among the best craftsmen ever to work in western Pennsylvania. He and a few others used rococo C-scrolls, pierced and engraved patchboxes, and smooth, flowing lines to great advantage.

By the time of the Civil War, Pittsburgh had become a major manufacturing center, turning out mass-produced guns of many types, ranging from long rifles to plains rifles. Artistic merit was a forgotten factor as huge shops rivaled those in Lancaster or Philadelphia.

Some Significant Pittsburgh Gunsmiths

Thomas Allison	William and Joseph Craig
Aaron Armstrong	John Fleeger and sons
John White	Samuel McCosh,
Joseph Graham	nephew and uncle
	Charles Flowers

SOUTHWESTERN PENNSYLVANIA

There are many gunsmiths listed early on in Fayette, Greene, and Washington counties,[80] but the products of few have survived. Peter White moved to Uniontown from Bedford County in 1818, and it is all but impossible to tell which of his guns was made where. Among the remaining early gunsmiths, we have almost no surviving pieces to study. Later guns showed traces of the Westmoreland County or Pittsburgh school. Many late guns were made in this area, and the trade continued long after cartridge guns were made.

Some Significant Gunsmiths of Southwestern Pennsylvania

Peter White*	Barnett Engel
John Crossland	Joseph Gibbins
Lewis Donham	Ezra Engel
Michael Edwards	Thomas Hughes

*Also in Bedford County School, and, although beyond the scope of this work, ca. 1800–1805, of the Emmitsburg, Maryland, school.

NORTHWEST PENNSYLVANIA

Shortly after the Revolution, and soon after the Indians in the northwest section were pacified, gunsmiths moved into counties such as Elk, Clarion, Jefferson, Crawford, and Erie. Jacob Resor came to Crawford County shortly before 1800, and was soon followed by Isaac Berlin. However, no known products exist from their workbenches, save for a rifle made by Berlin before he moved to Crawford County. Isaac Mills, after a brief stopover in Westmoreland County, became the pioneer gunsmith in Jefferson County. However, his only known product resembled the work in his native county, Bedford. The Sherry family made guns in Clarion and Jefferson counties, but their best high-art pieces were of the Westmoreland County style.

Many later guns, some with novel decorations and interesting features, exist as proof that gunsmiths did make firearms in northwest Pennsylvania. However, most are half-stock rifles with few distinguishing features.

Erie County had its gunsmiths, but the author has yet to examine a piece that has great artistic merit.

Some Significant Gunsmiths of Northwestern Pennsylvania

John Sherry, Jr.	Isaac Mills
Nicholas Shennefelt	Chauncy Brockway
Horace Warner	David Milliron
Mathias Ringle	Sink family
Showalter family	Alfred Marion Cone
Henry Jarecki	

NOTES

1. Captain J. G. W. Dillin, *The Kentucky Rifle*, National Rifle Association of America (Washington, D.C., 1924). It has never been out of print since, and has gone through four editions. Notable are the photographs of fine, early, high-art guns.
2. Major Ned Roberts, *The Muzzle Loading Cap Lock Rifle* (Washington, D.C., 1944). The current printing is by Dr. George Shumway, York, Pa.
3. Henry J. Kauffman, *Early American Gunsmiths* (Harrisburg, Pa.: Stackpole, 1952). It is unfortunately now out of print, although it has been superseded.
4. Henry J. Kauffman, *The Pennsylvania-Kentucky Rifle* (Harrisburg, Pa.: Stackpole, 1960). The present edition is by Crown books, and the photographs have suffered in the reprinted edition, losing clarity.
5. Joe Kindig, Jr., *Thoughts on the Kentucky Rifle in its Golden Age* (Wilmington, Del., George Hyatt, 1952). A revised, annotated second edition was published by Dr. George Shumway in 1984.
6. S. E. Dyke, *Pennsylvania Rifle* (Lititz, Pa.: Sutton House, 1974).
7. Charles W. Sawyer, *Firearms in American History* (Boston: 1910) and Charles W. Sawyer, *Our Rifles* (Boston: Cornhill, 1920).
8. Stephen VanRensselaer, *American Firearms* (Watkins Glen, N.Y.: Century House, 1940). This work has been in some parts discredited, and it is not documented.
9. Arcadi Gluckman and L. D. Satterless, *American Gun Makers* (Buffalo, N.Y.: Ulrich, 1940). Gluckman also wrote *U.S. Muskets, Rifles and Carbines* (Buffalo N.Y.: Ulrich, 1948), an excellent study of primary and secondary small arms in the U.S. service.
10. Robert Gardner, *Small Arms Makers* (New York: Crown, 1963).
11. Robert Gardner, *American Arms and Arms Makers* (Columbus, Ohio: Heer, 1938).
12. Robert Gardner, *Five Centuries of Gunsmiths, Swordsmiths and Armourers* (Columbus, Ohio: Heer, 1948).
13. Donald Hutslar, *Biographical Information*, vol. 1 of *Gunsmiths of Ohio* (York, Pa.: Shumway, 1973).
14. Daniel Hartzler *Arms Makers of Maryland* (York, Pa.: Shumway, 1977).
15. Russell E. Harriger, *Elk, Clarion and Jefferson Counties*, vol. 1 of *Long Rifles of Pennsylvania* (York, Pa.: Shumway, 1984).
16. H. J. Swinney, *New York State Gunmakers* (Cooperstown, N.Y.: Freeman's Journal Press, 1951) now undergoing major revision.
17. Curtis Johnson, *Illinois Gunsmiths*, 2d ed. (York, Pa.: Shumway, 1983). Much research and full biographical materials are still being prepared for a revised edition.
18. A. W. Lindert, *Gunmakers of Indiana* (Homewood, Ind.: self-published, 1968).
19. John Bivens, Jr., *Longrifles of North Carolina* (York, Pa.: Shumway, 1968).
20. Giles Cromwell, *The Virginia Manufactory of Arms* (Charlottesville: University of Virginia Press, 1975); Harold B. Gill, Jr., *The Gunsmith in Colonial Virginia* (University of Virginia Press, 1974). These men and Wallace Gusler, antique curator at Colonial Williamsburg, and perhaps others, are currently engaged in research on arms making in Virginia. None of them has yet to establish the earliest date of manufacture of Kentucky rifles in Virginia.
21. Lawrence P. Shelton, *California Gunsmiths, 1846–1900* (Fair Oaks, Ca.: Far West Publishers, 1977).
22. William S. Bowers, *Gunsmiths of Pen-Mar-Va* (Mercersburg, Pa.: Irwinton, 1979).
23. Each edition of the Kentucky Rifle Association's directory gives a brief history of the organization and states its purpose.
24. Noah Ludlow, *Dramatic Life as I Found It* (reprint ed., York: Shumway, 1974), 237.
25. George Shumway, "Hunters of Kentucky," *KRA Newsletter* (Spring 1977); and "Hunters of Kentucky," *KRA Newsletter* (Fall 1976).

26. *KRA Directory,* 1986–87, 3. This statement is given in all editions of the directory.

27. Richard Headley, *KRA Newsletter* (Winter 1978).

28. George Alfred Shumway, *The Rifle in Colonial America,* 2 vols. (York, Pa.: Shumway, 1980–81).

29. Col. Rec., 10:292–93 (See Abbreviations, Col. Rec.).

30. 2 Pa. Arch. 13 at 299 (See Abbreviations, Pa. Arch.).

31. J. Leander Bishop, *A History of American Manufactures from 1608 to 1860* (Philadelphia: Young, 1864), 1:572f.

32. 2 Pa. Arch. 1 at 615.

33. A list of persons accused of treason according to the laws of Pennsylvania is given in 6 Pa. Arch. 12. Accusations and rumors were sufficient to cause a name to be posted, accused of treason.

34. See Appendix 1; see also Bishop, *History of American Manufactures* 1:572f.

35. 2 Pa. Arch. 13 at 299.

36. 1 Pa. Arch. 5 at 522.

37. 2 Pa. Arch. 13 at 347.

38. 1 Pa. Arch. 4 at 712–13.

39. 3 Pa. Arch. 6 at 465.

40. 2 Pa. Arch. 1 at 56; 3 Pa. Arch. 6 at 273.

41. 2 Pa. Arch. 13 at 505.

42. 1 Pa. Arch. 6 at 495.

43. 1 Pa. Arch. 9 at 333.

44. 1 Pa. Arch. 9 at 565–67.

45. Col. Rec., 279–82.

46. Col. Rec., 10:284.

47. Col. Rec., 10:586–90; 2 Pa. Arch. 13 at 293.

48. See, for example, the case of Johnston Smith, 1 Pa. Arch. 4 at 731; see also Col. Rec., 10:418.

49. 1 Pa. Arch. 8 at 372.

50. 2 Pa. Arch. 1 at 435.

51. J. M. Swank, *History of the Manufacture of Iron in All Ages.* (Iron and Steel Institute, 1892), 179–80.

52. This was the major point made by Joe Kindig, Jr. in his monumental *Thoughts on the Kentucky Rifle in its Golden Age.*

53. M. L. Brown, *The American Rifleman* (March 1982), 46–47, 81.

54. Sir William Johnson, "A Plan for Defeating the Designs of the French" of July 1754, 2 Pa. Arch. 6 at 216. Johnson recommended that "smiths be sent at proper seasons to all the [allied] tribes to mend and keep in order their arms. . . ."

55. S. K. Bailey, ed., *Ohio Company Papers* (1948), 150–51; see also, Francis Parkman, *Conspiracy of Pontiac* (1912), 214–17.

56. R. C. Brown, *History of Crawford County* (1885), 146–56. W. J. McKnight, *Pioneer Outline History of Northwest Pennsylvania* (1905), 162, 199.

57. 3 Pa. Arch. 7 at 484–85.

58. Gardner, *Small Arms Makers,* 175; see also Scharf-Wescott, *History of Philadelphia* (1884), 2:1054 and 3:2270.

59. Reproduced in part in Calvin Hetrick, *The Bedford County Rifle and its Makers* 2d ed. (York, Pa.: Shumway, 1975), 26–27.

60. Thomas Hall, *Muzzle Blasts* (October 1942), 6–7; McKnight, *Pioneer Outline History of Northwestern Pennsylvania,* 156;

61. Gluckman, *U.S. Muskets, Rifles and Carbines,* chaps. 2 and 3.

62. Ibid., chap. 4.

63. Carl P. Russell, *Guns on the Early Frontiers* (University of California Press, 1958), chap. 4; see also two books by Charles E. Hanson, Jr., *The Plains Rifle* (Highland Park, N.J.: Gun Room Press, 1976), and *The Hawken Rifle* (Chadron, Neb.: Fur Press, 1979).

64. Jacob Harder (1820–98) invented a breech-loading system made by him and his sons under patent 319,482 of June 9, 1885, and made at Lock Haven, Clinton Co. (*Muzzle Blasts* [December 1944], 7).

65. John White, son of well-known gunsmith Peter White, made the transition from muzzle loaders to breechloaders well. He made guns at Rochester, Beaver Co., under David Minesinger's patent, number 6139.

66. Unpublished research by Dr. George Shumway, who has projected a volume in his Long Rifles of Pennsylvania series on Philadelphia. See also Gardner, *Small Arms Makers,* 167, entry on Rupertus company.

67. Joe Kindig, Jr., *Kentucky Rifle in its Golden Age,* 72–154.

68. Ibid., 128–29.

69. See Dr. Shumway's annotation in the revised edition of Kindig, 196–206; see also, Shumway, *Rifle in Colonial America,* 118.

70. Kindig, *Kentucky Rifle in its Golden Age* 155ff.

71. Merrill Lindsay, *The Kentucky Rifle* (York, Pa.: Arma Press, York Historical Society, 1972), part 5.

72. Kindig, *Kentucky Rifle in its Golden Age,* 283ff.

73. S. E. Dyke, "The Beck Family Gunsmiths," *Journal of the Lancaster County Historical Society* (1968), 28–61.

74. George Shumway, *Pennsylvania Long Rifles of Note,* 2d ed. (York: Shumway, 1977), 27.

75. Roy Chandler and Donald Mitchell, *A History of Perry County Guns and Gunsmiths* (Duncannon, Pa.: self-published, 1969). Covers the later gunsmith of this county thoroughly.

76. This area has been studied by S. E. Dyke for Joe Kindig, Jr. and the results published in Kindig, *Kentucky Rifle in its Golden Age* 385–435. See also Bowers, *Gunsmiths of Pen-Mar-Va,* 92–94.

77. Bowers, *Gunsmiths of Pen-Mar-Va,* 1–49.

78. Hetrick, *Bedford County Rifle,* and James B. and Vaughn E. Whisker, *Bedford County Gunsmiths and Gunmakers,* 3rd ed. (Bedford, Pa.: Old Bedford Village Press, 1985).

79. Edith Cooper, *The Kentucky Rifle and Me* (self-published, 1977), showed many fine guns purchased by her late husband, T. J. Cooper, a gun dealer, and it includes photographs of several gunsmiths and their shops and gravestones. Most of the guns shown were of the Huntingdon County style, although they showed many guns made by Joe Long. It is not a scholarly publication and contains a minimum of factual information. See also James B. Whisker, *Gunsmiths of Huntingdon County* (Apollo, Pa.: Closson Press, 1984).

80. See two books by James B. Whisker, *Western Pennsylvania Gunsmiths* (Bedford, Pa.: Old Bedford Village Press, 1983) and *Gunsmiths of Southwestern Pennsylvania* (Apollo, Pa.: Closson Press, 1985).

Abbreviations

AFR. J. S. Africa, *History of Huntington and Blair Counties* (Chicago: Waterman, 1883).

Amer. Arch. American Archives Series IV and V. U.S. Government Printing Office, 1840.

AMM. Daniel Hartzler, *Arms Makers of Maryland* (York, Pa.: Shumway, 1977).

AR. *American Rifleman*, publication of the National Rifle Association, Washington, D.C.

BAR. Barnes Lancaster Directories.

BBHP. F. B. Brandt and H. V. Gummere, *Byways and Boulevards in and about Historic Philadelphia* (Philadelphia: Corn Exchange National Bank, 1925).

BCR. *Berks County in the American Revolution* (Reading, Pa.: Berks County Historical Society, 1894.)

BHAM. J. L. Bishop, *History of American Manufactures, 1608–1860* (Phil. E. Young, 1866).

BLD. Boyd's Lancaster Directories.

BPaD. Boyd's Pennsylvania State Business Directories.

BPD. Boyd's Philadelphia Directory.

BRD. Boyd's Regional Directory, as is appropriate.

BSED. Boyd's Southeast Pennsylvania Directory.

C. U.S. census

CCD. Crawford Co. Directories.

CHAN. Roy E. Chandler et al., *A History of Early Perry County Guns and Gunsmiths* (Duncannon, Pa.: Chandler, 1969).

Col. Rec. Colonial Records of Pennsylvania (Harrisburg: Commonwealth of Pa.)

DAR. Indicates Daughters of the American Revolution registry number.

DB. Deed Book.

DeSD. DeSilver's Philadelphia Directories.

DH. Donald Hutslar, *Biographical Information*, vol. 1 of *Gunsmiths of Ohio* (York, Pa.: Shumway, 1973). Additional volumes have not been published.

DIL. Dillin, J. G. W. *The Kentucky Rifle* (Washington, D.C.: NRA, 1924). There have been four editions. The fourth edition contains Calvin Hetrick's listing of Bedford

County gunsmiths, which has been published separately in a "second edition" with data on Peter White supplied by the publisher, Dr. George Alfred Shumway.

f. family interview.

FER. G. B. Ferree, *Ferree Family Gunsmith* (Indianapolis: Ind.: Ferree, 1967).

FSP. Frank Sellers, *American Gunsmiths* (Highland Park, N.J.: Gunroom, 1983).

GAAAM. Robert Gardner, *American Arms and Arms Makers* (Columbus, Ohio: privately printed, 1938).

GAS-1. George Alfred Shumway, *Rifle in Colonial America*, 2 vols. (York, Pa: Shumway, 1980).

GAS-2. George Alfred Shumway, *Pennsylvania Long Rifles of Note*, 2d ed. (York, Pa.: Shumway, 1977).

GCD. Gopsill's City Directories.

GFC. Robert Gardner, *Five Centuries of Gunsmiths, Swordsmiths and Armourers* (Columbus, Ohio: privately printed, 1948.

GLD. Gopsill's Lancaster, Harrisburg, York, and Lebanon Directories.

GPD. Gopsill's Philadelphia Directory.

GR *Gun Report*, magazine published at Aledo, Ill.

HAN. Charles F. Hannan, *The Wilderness Trail*, 2 vols. (New York: Scribner's, 1912).

HFAML. *History of Lodge 43, F. & A. M., Lancaster, 1785–1935* (Lancaster: F. & A. M., 1935).

HYC. George R. Prowell, *History of York County* (Chicago: Beers, 1907).

IC. Census of Industry/Manufactures.

JKJ. Joe Kindig, Jr., *Thoughts on the Kentucky Rifle in Its Golden Age*, 2d ed. (York, Pa.: Shumway, 1985).

KEA. Henry J. Kauffman, *Early American Gunsmiths*, (Harrisburg: Stackpole, 1952).

KKR. Henry J. Kauffman, *The Pennsylvania-Kentucky Rifle* (Harrisburg: Stackpole, 1960).

KRAN. *Newsletter of the Kentucky Rifle Association*.

KRA-2. *Kentucky Rifles and Pistols, 1750–1850*, ed. J. R. Johnston (Washington, D.C.: Kentucky Rifle Association, 1986). This is the second picture book with a minimum of information published under the sponsorship of the KRA. The first book is *The Kentucky Rifle: A True American Heritage* (1986).

LCD. Lancaster City Directory.

LIN. John Blair Linn, *Annals of the Buffalo Valley* (Harrisburg: Hart, 1877).

LSC. Lawrence Shelton, *California Gunsmiths, 1846–1900* (Sacramento, Calif.: Far West Publishers, 1977).

MB. *Muzzle Blasts*, magazine published by the National Muzzle Loading Rifle Association (Friendship, Ind.: began publication in 1939).

MD. McElroy's Philadelphia Directories.

MPHS. *Memoirs of the Historical Society of Pennsylvania*.

MRB. M. R. Balderson, ed., *Passengers and Ships prior to 1684* (Baltimore: Genealogical Publishers, 1970).

N. newspaper.

NAT. *Now and Then*, publication of the Lycoming Co. Historical Society.

NCAB. *National Cyclopedia of American Biography* (New York: 1916).

NTD. New Trade Directory, Philadelphia

Pa. Arch. Pennsylvania Archives, 9 series. I have cited series first, volume second, and page third. "3 Pa. Arch. 4 at 5" means third series, volume 4, page 5. Harrisburg: Commonwealth of Pa.

PAD. Directory of Pittsburgh and Allegheny Cities.

PaE. Michael Tepper, ed., *Emigrants to Pennsylvania, 1641–1819* (1979).

PaGP. *Pennsylvania German Pioneers*, 2 vols. (Baltimore: Genealogical Publishers, 1983).

PD. Polk's Directory and Gazeteer.

PDP. Polk's Philadelphia Directory.

PIDS. George W. Neible and Historical Society of Pa., indentured servant and apprenticeship records, Philadelphia. These have been printed in Tepper, ed., *Emigrants to Pennsylvania*, and *Proceedings of the Pennsylvania German Society*, vol. 45.

PIR. *Pennsylvania in the Revolution* 2 vols. (Harrisburg: Commonwealth of Pennsylvania, 1880). Service lists from the Pennsylvania Archives, edited by W. Egle et al.

PSSP. M. R. Balderson, ed., *Passengers and Ships* (Baltimore: Genealogical Publishers, 1970).

PXD. Paxton's Philadelphia Directories.

RAG. Robert Gardner, *Small Arms Makers* (New York: Crown, 1963). This is still the best and most accurate listing of gunsmiths nationally.

REH. Russell E. Harriger, *Longrifles of Pennsylvania* (York, Pa.: Shumway, 1984). Vol. 1 covers Jefferson, Elk, and Clarion Counties.

SED. S. E. Dyke, *The Pennsylvania Rifle* (Lititz, Pa.: Lancaster Bicentennial Committee, 1974). Covers only Lancaster County.

SGD. DeSilver's Stranger's Guide Directories.

SHIS. *Swank's History of Iron and Steel* (Philadelphia: American Iron and Steel Association, 1892).

t. tax lists.

WB. Will Book.

WSB. William S. Bowers, *Gunsmiths of Pen-Mar-Va* (Mercersburg, Pa.: Irwinton, 1979).

WWD. Williams Directories.

Biographies of Arms Makers of Pennsylvania

A

Abbott, William. Gunsmith, 618 ½ Market St., Philadelphia (BPD, 1890).

Abel, William. Gunsmith, Williamsport, Lycoming Co. (PD, 1882).

Abendshen, Joseph. 1857–63, 50 Wayne St., Pittsburgh, gunsmith. (PAD; Duncan McConnell, MB 11 [1945]: 7). Mrs. Joseph Abendshen, 1867-68, 168 Main St., Wheeling, W.V. (WWD). She evidently sold out to George Manz, "manufacturer of cutlery, guns, pistols & c.," 1868–69 (WWD).

Acfield, Jacob. 3/26/1776: "upon application of Jno Pollard and Jacob Acfield, for the payment of their accounts for bayonets & repairing of firearms for Capt Harmer's, Jones' & LaMars' Compy's in Colo. . . . by the order of the Board two orders were drawn . . . in favor of Jacob Acfield for £19/9/6." (Col. Rec. 525, 10: 3/26/1776).

Achenbach, Est. J. 1876, Orangeville, Orange Twp., Columbia Co. (Beers, *Atlas of Columbia and Montour Counties, Pennsylvania*) Gunsmith.

Adams, Daniel (b. 1808). On 24 August 1832 William Blain ran an advertisement for his runaway apprentice, Daniel Adams, in the *Westmoreland Republican*. 1837, Somerset Twp., Somerset Co. (t); 1860, Wayne Twp., Crawford Co. Gunsmith (c).

Adams, James. Riceville, Bloomfield Twp., Crawford Co. (CCD, 1871, 1874, 1879, 1890) Gunsmith.

Adams, John. Artillery artificer, 2/8/1777 to 3/24/1781, Capt. Jesse Rowe's Co. at the laboratory at Philadelphia; pvt., Capt. Noah Nicholas's Co. (2 Pa. Arch. 11 at 252; PIR, 2:245).

Adams, Samuel. Artillery artificer, 2/12/1777 to 3/22/1782, Capt. Jesse Rowe's Co. at Philadelphia; pvt., Capt. Noah Nicholas's Co. (PIR, 2:245; 2 Pa. Arch. 11 at 252).

Adams, Tower (1795–1870). Black man, gunsmith, 1825–50, Washington boro. (t); 1850, S. Strabane Twp., Washington Co. (c); 1860, Washington boro. (c); 1870, Canonsburg boro (c).

Addiks, John E. 8/30/1814: "presenting you with a sample of a parcel of muskets, I beg you to observe that the quantity is about 3000, and the price is $14½. The whole of them are made at Philadelphia by Mr Kunkle, whose name is stamped on the locks of them. . . . I have about 60 muskets, second hand, at $11 but in good order & also send a sample of them if desired."

Afflebaugh (Applebaugh), Henry. Germantown Rd. near 4th St., Philadelphia. (PXD, 1816, 1819) Gunsmith.

Afflerbach, John C. 1860, 1136 N. 2d St., Philadelphia. (c; GCD, 1875; PDP, 1892) Gunsmith and locksmith.

Afflerbach, Jost (b. 1774). Smith, arrived in Philadelphia on 7/30/1796 on the brig *Mary*. (2 Pa Arch 17 at 602).

Afflerbach, William. 1141 N. 2d St., Philadelphia (MD, 1846; BPaD, 1861); 1136 N. 2d St., Philadelphia (GCD, 1875; BSED, 1875, 1880; 1890).

Afflerbach, William H. Philadelphia. Gunsmith (BSED, 1880, 1890).

Ahner, Christian. Gunsmith, James St. and College St., Lancaster. (BLD, 1857).

Ailer, Lewis. Germantown Rd. above Front St., Philadelphia. Gunsmith (PXD, 1813, 1819).

Aitkins, John. Goldsmith. Apprenticed on 10/14/1771 to William Taylor, goldsmith, for 1½ years; sold as an indentured servant for £14/14/6 (PIDS).

Aitkins, T. Silversmith, 1660–1710, Delaware Co. (*Proceedings of the Delaware County Historical Society*, 1:113f.).

Ake, Simon. Gunsmith, 1845, McConnellstown, Huntingdon Co. (AFR, 383).

Albert, John. Silversmith, 1814, Huntington Twp., Adams Co. (t).

Albrecht, Andreas (1718–1802). Born on 2 April 1718 in Thurgia, Germany, Andreas Albrecht was a regimental gun-stock maker in Silesia, 1741–46. By 1750 he had become an evangelical Christian and joined a group of emigrants coming to America. On 30 August 1759 he arrived at Christian's Spring and taught music. In 1762 he established a gunshop. In 1766 he married Elisabeth Orthin. In 1771 William Henry, Sr. became his apprentice. From 1772 to 1802 he was in Warwick Twp., Lancaster Co. (t). He died in 1802 at Lititz, and his burial was recorded in the burial book of that church. The Moravian Archives show that he was taxed, 1754–71, Moore Twp., Northampton Co. (GAS-1, 198–294; JKJ, 62, 97, 190).

Albrecht, Charles. 1870–71, 148 12th St., Gunsmith. Pittsburgh (PAD).

Albrecht, Henry, Sr. 1744, ran Durham Iron Works, Lancaster Co. (RAG, 3); 1744–80 (t). Armorer, cannon maker, gun-barrel manufacturer.

Albrecht, Jacob. Naturalized 4/3/1761, Berks Co. (2 Pa. Arch. 2 at 417); gunsmith, 1780–82, Philadelphia (t); 1780 effective supply tax: value £590, tax £15/6/0 (3 Pa. Arch. 15 at 416; 3 Pa. Arch. 16 at 148).

Albrecht, John A. 1750, arrived in Pa. from Germany on ship *Irene* (PaE, 44); gunsmith, Northampton Co., after 1750 (KAE, 1; JKJ, 190–91).

Albright, Henry (1772–1816). Son of Andreas Albrecht, Henry Albright was active before 1796 in Lancaster Co.; 1797–1813, Shippensburg, Cumberland Co. (t); 1814–16, Upper Nazareth Twp., Northampton Co. (t) (WSB, 11–14; JKJ, 97–102) Gunsmith.

Albright, Jacob. 1800, Haines Twp., Centre Co. (c); 1802 (t).

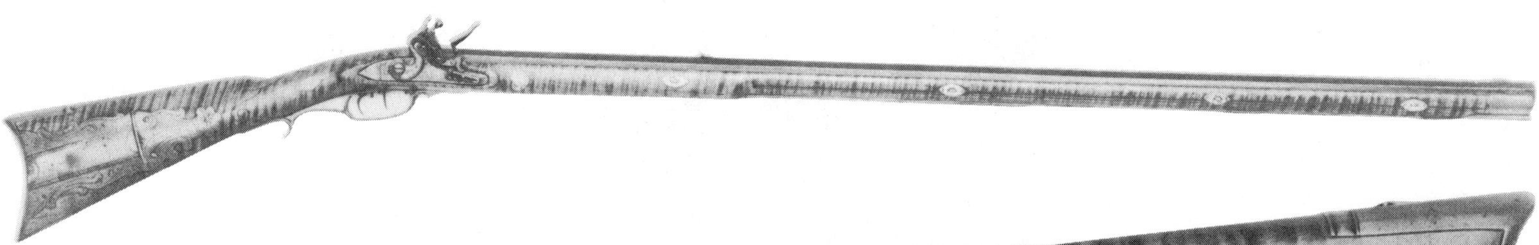

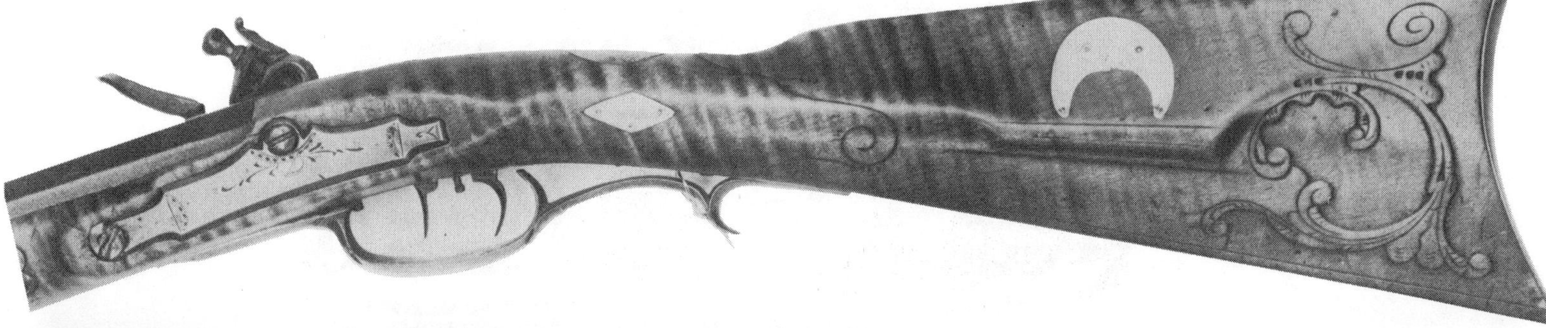

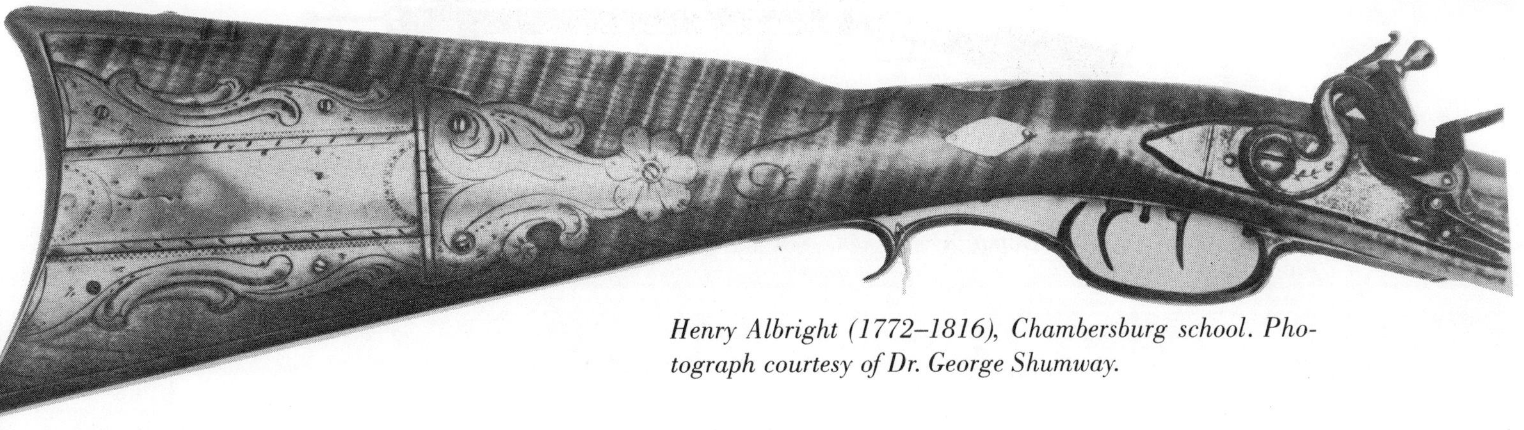

Henry Albright (1772–1816), Chambersburg school. Photograph courtesy of Dr. George Shumway.

Albright, Jacob, Jr. 1803–13, Aaronsburg, Centre Co. (t); 1814–30, Millheim, Centre Co. (t); 1830, Wooster Twp., Wayne Co., Ohio, (DH, 399; c., 1850.)

Albright, John Henry (1772–1835). Gunsmith, before 1800, Lititz, Lancaster Co.; 1800, Clay Twp., Tuscarawas Co., Ohio (DH, 379); returned to Lancaster Co., Pa. by 1816 (JKJ, 190–93; KKR, 169–72).

Albright, Thomas J. Gunsmith, 1835, Stroud Twp., Northampton Co. (t).

Albright, Zachariah, Sr., Gunsmith, Haines Twp., Centre Co. (t); 1840, same (c). His name was missing from tax and census records in the years 1821–39.

Albright, Zachariah, Jr. 1845–47, Ayr Twp., Bedford Co. (now Fulton Co.) (t); 1850, Buckeye, Stephenson Co., Il. (c). Gunsmith.

Aldenderfer, M. 1790, Colebrookedale Twp., Berks Co. (JKJ, 147–48); 1800, Cocalico Twp., Lancaster Co. (t); 1817–20, Colebrookedale Twp., Berks Co. (t). His one known gun, dated 1763, shows a mixture of several schools. Gunsmith.

Alderman. E. M. Inventor, Sharon, Pa. Patent 920,709 of 4 May 1908 for an electromagnetic gun.

Alexander, Donald. 1790–1838, Elizabeth Twp., Allegheny Co. (t) Gunsmith.

Alexander, Thomas. Artillery artificer, enlisted 7/18/1777 in Capt. Nathaniel Irish's Co.; promoted on 4/28/1779 (PIR, 2:247).

Allebaugh, William. 1850, German Twp., Fayette Co.; resided with a widow, Delia Cracraft (c). Gunsmith.

Allegheny Arsenal. Lawrenceville, Allegheny Co. Began

Hiram Berdan's Sharpshooters. Broadside from the training Camp, Weehawken, New Jersey, Summer 1861

operations in 1814; ceased operations, 1829; records began in 1828.

Allegheny Gun Works. 1831–98, 51 W. Diamond St., Allegheny City (Pittsburgh). John and William A. Fleeger, props. (BPaD, PAD).

Alleman, William. Gunsmith, 1825, Bushkill Twp., Northampton Co. (t).

Allen, Richard. 11/21/1745: "Daniel Jones, son of Mary Jones, by consent of his mother, indents himself apprentice to Richard Allen of Philadelphia, brass founder, for six years, from the 20th November 1745, to be taught the trade of brass founder" (PaE, 80; PIDS).

Allen, Samuel. Artillery artificer, enlisted 7/30/1778 in Capt. Nathaniel Irish's Co.; promoted 4/23/1779 (PIR, 2:247).

Allen, Solomon. Gunsmith, born in Pa.; 1870, age 82, Sangamon Co., Ill.; wife Mary (age 72) born in Ky. (c).

Allers, Charles H. 100 Caldwell St., Pittsburgh. (PAD, 1884) Gunsmith.

Allison, John. 12/19/1776: "pay . . . John Allison £40/0/0 for riffles." (Col. Rec. vol. 11)

Allison, Richard (b. 1796). 1860, Sewickley Twp., Allegheny Co. (c) Gunsmith.

Allison, Thomas (1786–1836). 1807–10, Peters Twp., Washington Co. (t); 1810, Pittsburgh (t; c); 1811–18, Peters Twp., Washington Co. (t); 1819–36, New Sewickley, Beaver Co. (t; c). His advertisement in the Pittsburgh *Gazette* on 3/10/1810 says "he will give satisfaction to those who may please to employ him." Gunsmith.

Allison, William (1788–1855). 1811, German Twp.,

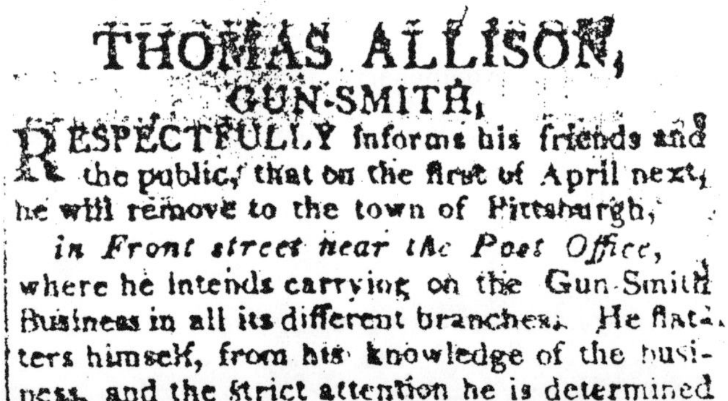

THOMAS ALLISON,
GUN-SMITH,
RESPECTFULLY informs his friends and the public, that on the first of April next, he will remove to the town of Pittsburgh, *in Front street near the Post Office*, where he intends carrying on the Gun-Smith Business in all its different branches. He flatters himself, from his knowledge of the business, and the strict attention he is determined to pay to it, that he will give satisfaction to those who may please to employ him.

He will constantly be supplied with *Ferree's best Powder*, to supply those who may require it. March 22, 1810.

Pittsburgh Gazette, March 22, 1810

Fayette Co. (t); 1813–25 and 1828–55, New Salem, Fayette Co. (t). Gunsmith.

Alloway, Elmer. 23 Green St., Gunsmith. Philadelphia. (DeSD, 1829–30)

Allway, Axariah. 1806–8, Fawn Twp., York Co. (t) Gunsmith. According to tax lists, he died in 1826 in Fawn Twp.

Altfather, John (b. 1834). Son of Daniel and Anna Altfather (Oldfather). 1860, Berlin, Somerset Co. t; c). Gunsmith.

Altland, Andrew (1810–78). Son of Jacob and Catherine (Trimmer) Altland; born 1/11/1810, died 5/6/1878. Dover Twp., York Co. (JKJ, 309–12; f; T; C). Gunsmith.

Altland, Jacob (1785–1830). Born May 3, 1785, died February 15, 1830; married Catherine Trimmer (1788–1861). Shop located at Orts Mill, Dover Twp., York Co. (JKJ, 309–12; f; t; c). Gunsmith.

Altland, John (d. 1853). Brother of Jacob, born and raised in Dover Twp., York Co.; lived at Mummasburg, Butler Twp., Adams Co. His gunsmithing tools were auctioned in 1853 (WSB, 99–100).

Altmaier, Peter. 1855–67, Lewistown, Mifflin Co. (t; BPaD, 1861) Patent 24,774 of 7/12/1859 for a breech-loading firearm. 1868–1904, 2d below Chestnut, Harrisburg. (t; c; BPaD, 1868, 1869, 1880, 1887, 1904) Patents issued to Altmaier at Harrisburg: 261,648 of 7/25/1882 for a breech-loading rifle; 261,802, same date, for a gunlock (RAG, 5; PD, 1882).

Altman, Frederick (d. 6/6/1854). Cowonshannock Twp., Armstrong Co. "the first settler on the Cowanshannock portion was the eccentric Frederick Altman. . . . Besides guns and other things, he made a good pocket knife with 12 blades and invested an auger with a chisel attachment . . . one of his eccentricities was his constant refraining from speaking to any of his children. Their mother was the medium of communication. . . . He bought 160 acres of Wilhelm Willink on 19 December 1833 for $90. He probably settled on it in 1834, built his saw mill on it in 1839. He also manufactured for a while stone pumps . . . made plows with wooden mold boards, as early as 21 September 1825." (R. W. Smith, *History of Armstrong County* [1883], 302, 369; His estate inventory showed: 3 gun barrel augers, 40¢; 1 lot gun barrel augers, 25¢; 3 sets of barrels, 6¢; 1 gun and shot pouch $12.30 1 set smiths tools $11.50 and 5 barrels 6¢ Lydia B. Earhart, *Earhart, Boyd and Allied Families* [1935], 203–16.)

Altman, Gustav (b. 1839). Goldsmith, jeweler, miller. Born at Dessau on Elbe, Prussia, a son of Wilhelm and Ida (Fritsche) Altman. Alexandria, Huntingdon Co. (*History of*

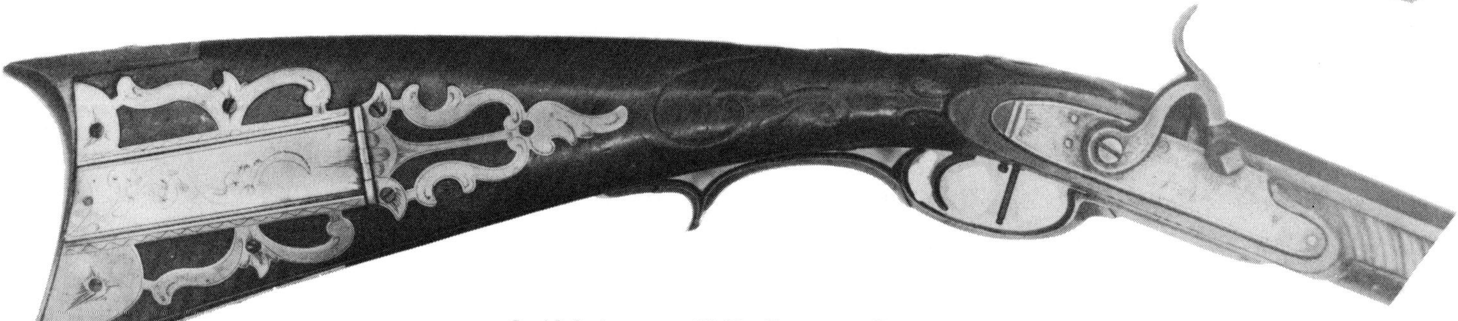

J. Altfather, ca. 1860, Somerset County.

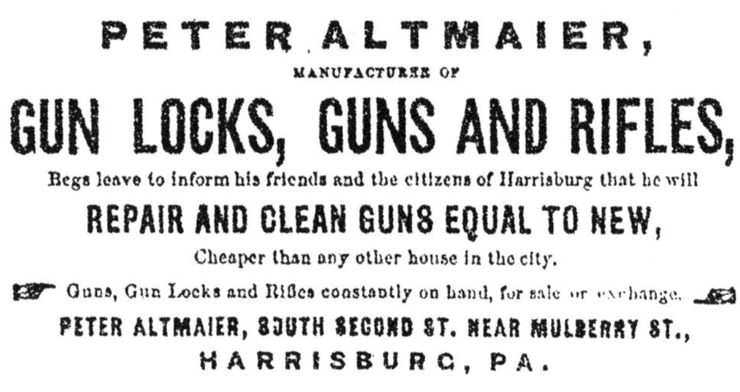

Boyd's Pennsylvania State Directory, 1861

Huntingdon, Mifflin and Perry Counties, Pennsylvania 1:205–6).

Altman, Jonathan (9/2/1802–1871). Firearms inventor, gunsmith. Son of George Philip and Barbara Altman, he married Margaret Shellhammer. (f) Patent 16,634 of 2/17/1857 for a "self-setting hair triggered gunlock." 1853–60, Kittanning Twp., Armstrong Co. (t); 1850, 1860, Kiskimentas Twp., Armstrong Co. (c).

Altman, Peter. Whitesmith and blacksmith, Hempfield Twp., Westmoreland Co. (3 Pa. Arch. 22 at 402).

Alworth, Samuel. 118 S. 4th St., Philadelphia (DeSD, 1793, 1808). Gunsmith.

Ambruster, Matthias. Artillery artificer, transferred from 10th Pa. Regt., revolutionary war (5 Pa. Arch. 5 at 169).

Amos, Border & Co. Probably John Amos and Daniel B.

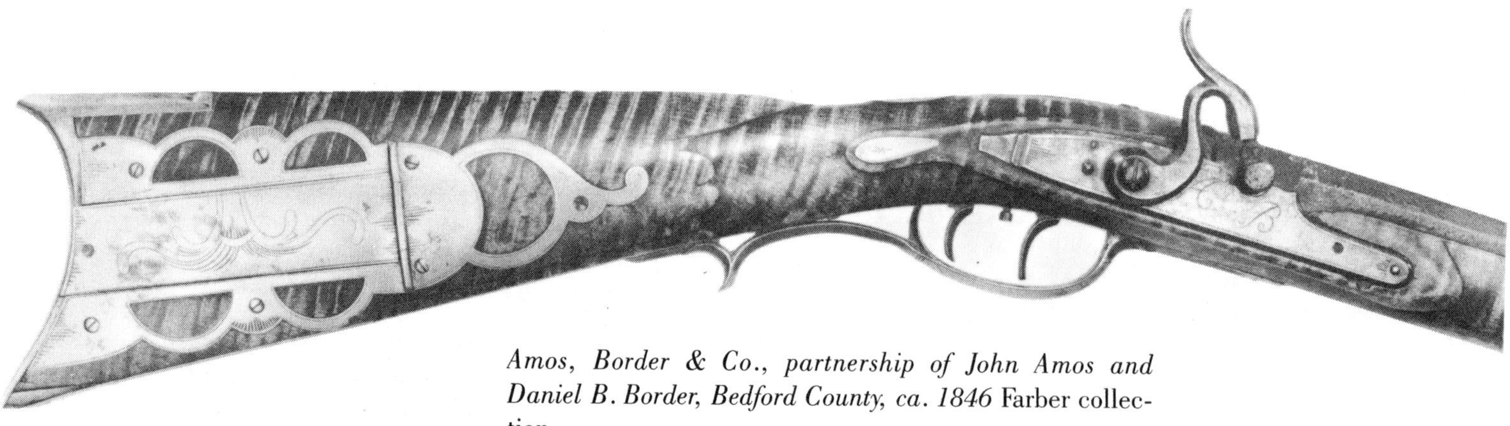

Amos, Border & Co., partnership of John Amos and Daniel B. Border, Bedford County, ca. 1846 Farber collection.

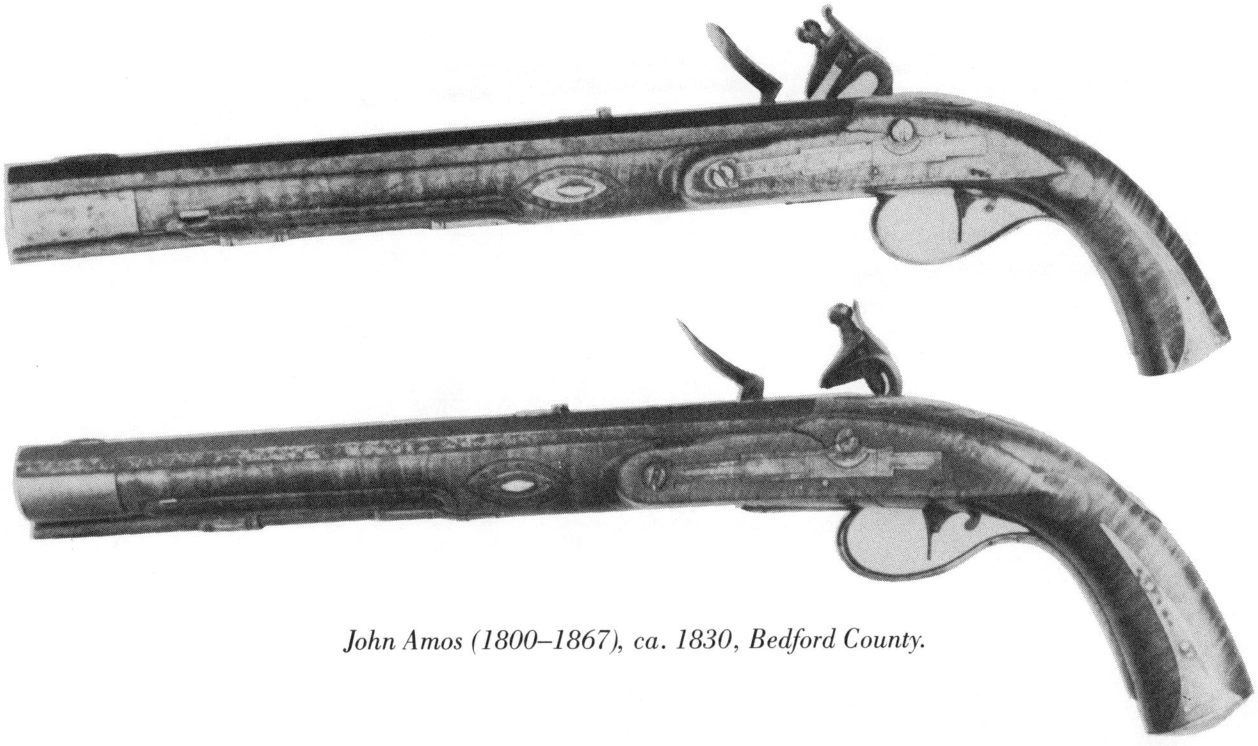

John Amos (1800–1867), ca. 1830, Bedford County.

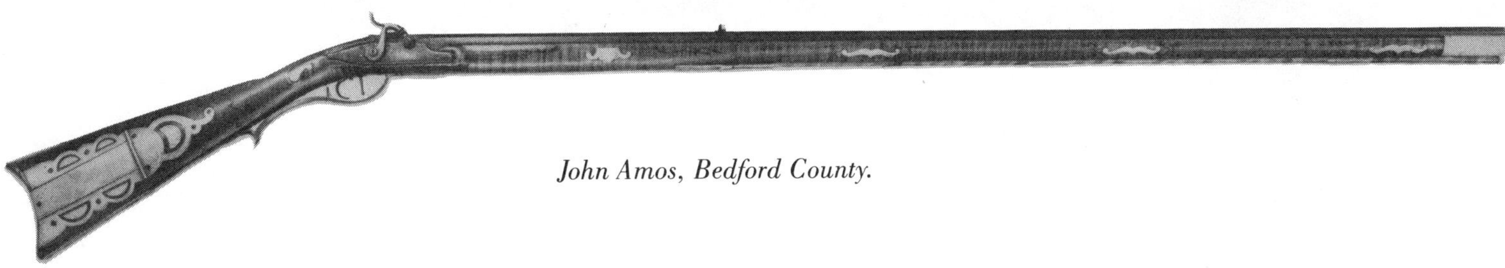

John Amos, Bedford County.

Border, ca. 1846, Bedford Twp., Bedford Co. (t). Gunsmiths.

Amos, John (1800–1867). Gunsmith, Bedford boro, 1829–32; Bedford Twp., Bedford Co., 1835–67 (t; c).

Anderson, Robert. Silversmith, 1796–1800, Washington boro. On 8 August 1798 Robert Anderson advertised in the Washington newspaper that he and William Hutchinson operated a clock-making and watchmaking shop three doors below Purviance's Tavern. In November 1808 Anderson was elected high sheriff of Washington County and George Bertie took over his shop (t; n).

Andreas (Andrew), Abraham. Gunsmith 1786–88, Bethlehem Twp., Northampton Co. (3 Pa. Arch. 19 at 201, 361).

Andrews, Jacob. Gun-barrel maker, 1807, Annville, Dauphin Co. (t).

Andrews, Joel. Patent 33,731 of 11/19/1861 for a gun and bayonet. Norristown, Montgomery Co.

Angel, John. Gunsmith 97 Dillwyn, Philadelphia (PXD, 1816, 1824, 1829, 1830; SGD, 1835–36).

Angstadt, Abraham (b. 1784). Gunsmith before 1850, Berks Co. (t); 1850, age 66, Germantown, Wayne Co., Ind. (c); 1860, East Germantown Ind. (c).

Angstadt, Adam. Blacksmith, 1768, Ruscomb Manor Twp., Berks Co. (3 Pa. Arch. 18 at 148); blacksmith, 1779, Rockland Twp., Berks Co., through 1785 (3 Pa. Arch. 18 at 265, 666, 785); 1800–1810, Kutztown, Maxatawny Twp., Berks Co. (t; c). The Reading *Adler* (Eagle) of August 1810 carried an advertisement tanslated by Joe Kindig, Jr. (JKJ, 162–63) indicating that Adam Angstadt and son offered gunlocks either singly or by the dozen. There was an Adam Angstadt in Germantown, Wayne Co., Ind. in 1850, age 24, gunsmith, who had been born in Pa. (c).

Angstadt, Gideon (1800–1854). Born 3/22/1800 in Berks Co., died 6/17/1854. Active as a gunsmith in Lewisburg, Union Co. (Death certificate no. 148 gives his occupation as gunsmith).

Angstadt, Jacob. 1815–17, Kutztown, Berks Co. (t; GAAAM, 16). Gunsmith.

Angstadt, Joseph. 1790–1817, Kutztown, Berks Co. (t)

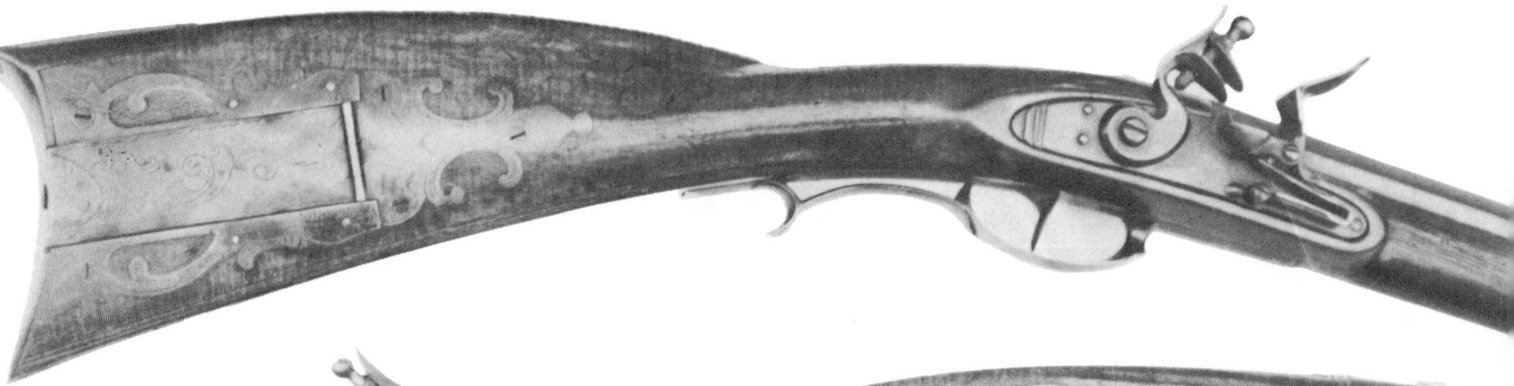

Adam Angstadt, ca. 1800, Berks County.

Dr. Shumway (PaLRN) calls his style "dutchified rococo" (see also KEA, 4). Gunsmith.

Angstadt, Peter. Shumway maintained that there are two early gunsmiths by this name: one worked ca. 1770–77 in Lancaster and the other ca. 1800 in Rockland Twp., Berks Co. Guns are splendid examples of the Lehigh Valley school. (JKJ, 167–68; GAS-2, 13) 1767–85, Rockland Twp., Berks Co. (3 Pa. Arch. 18 at 12, 666, 785). In 1767, single freeman (t). There was a Peter Angstadt, age 41, born in Pa., in 1850 in German Twp., Montgomery Co., Ohio (c); and a Peter Angstadt, gunsmith, in Germantown, Ind. in 1850, also born in Pa. (c).

Angush, James. 1765–75, Gunsmith. Earl Twp., Lancaster Co. (RAG, 7).

Annelly, Thomas. 49 Race St., Philadelphia. (*Stephen's Directory,* 1796) The *Federal Gazette* carried a death notice for his wife, 9/12/1798. Gunsmith. (GAAAM, 4). 1797–98, Trenton, N.J. (KEA, 4).

Anschutz, Edward. 2015 S. 3d St., Philadelphia (DeSD, MD, 1865, 1875, PD, 1882; BSED, 1880). Gunsmith.

Anspach, John. Artillery artificer, 1777–79, Capt. Morgan's militia (3 Pa. Arch. 6 at 291).

Antes, Philip Frederick (7/5/1730–9/20/1801). A son of Henry and Christina (DeWees or DeWesson) Antes, Philip Antes "possessed rare mechanical genius and in 1776 he and Mr [Jonathan] Potts succeeded in casting the first [iron] cannon at Warwick Furnace in America, and their success caused consternation among the British officers. Lord Howe was so incensed that he set a reward of £200 on his head." (William H. Egle, *Notes and Queries* [1898], 224) Impoverished by his efforts for liberty, he moved to Northumberland Co. in 1779. He was a member of the Provincial Council (t, f).

Antes (Antis), William (1735–1810). Prior to 1775, Frederick Twp., Montgomery Co. (t); 1775–76, Mahoning Twp., Northumberland Co. (t); 1778, New Hanover Twp.,

Gideon Angstadt (Anstat) (1800–1854), Berks County.

Montgomery Co. (t); 1781–83, Sunbury, Northumberland Co. (t); 1783–85, Mahoning Twp., Northumberland Co. (t). He was a son of Henry and Christina (DeWees or DeWesson) Antes. In 1809 he was living in Genesee Co., N.Y. (William H. Egle, *Notes and Queries*, 225) On May 1, 1778, Antes reported on arms repairs to the Provincial Council: "I have rec'd your letter of 28th April last past wherein I am requested to give immediate information to Council what number of muskets well fitted are now in my possession and how many more I could completely finish by the 20th May instant. . . . I have 25 in my possession and as touching what number I can finish . . . I have none that will be finished by that time as I am not at present in a proper situation to follow that business . . . excepting in the way of cleaning and repairing arms and that about 500 stand of arms was sent me last fall . . . 361 were sent . . . and 161 muskets were sent to Col. Frederick Antes." (1 Pa. Arch. 5 at 456) On August 1, 1785 he was paid £1 each for repairing 2 rifles damaged "on the way to Sunbury from Philadelphia, being presents from the Commonwealth to two Indian chiefs, according to the Treaty of October 1784." (Edwin McMinn, *On the Frontier with Colonel Antes* [1900], 371–89). On 21 March 1776 he reported to the Provincial Council on the problems of boring cannon barrels (1 Pa. Arch. 4 at 722. See also GAAAM, 16; Shumway, in MB 8 [1978] 16–17).

Antes, William, Jr. In 1810 William Antes, Sr. bequeathed to his son, William, Jr., "all my implements, tools, and utensils of every name, nature or description, belonging to or in any wise appertaining to the mystery or art of a gunsmith" (Ontario, Canadaigua Co., N.Y.; [Surrogate Records, vol. 4]). William, Jr., was living in Northumberland Co., Pa. (t).

Anthony & Ebbecke. Sold guns, ammunition, gunpowder. Allentown (PD, 1882).

Areis, Francis. 60 South St. near 2d St., Philadelphia (*American Advertising Directory*, 1831; DeSD, 1823; KEA, 4). Gunsmith.

Arick, John. Phoenix below George Troad's shop, Germantown, Kensington area (MD, 1846). Gunsmith.

Arkman, Henry. Annville, Lebanon Co., 1860 (c; BRD, 1860). Gunsmith.

Armitage, Joseph. Artillery artificer, enlisted 4/26/1779; in April 1780 stationed at Carlisle, Capt. Thomas Wylie's Co. (PIR, 2:248).

Armstrong, Aaron H. (1819–80). 1841, Bridgewater, Beaver Co. (t); 1850, Mercer, Mercer Co. (c); 1860, Rochester, Beaver Co. (c); 1880, Mercer boro. (c) (*Harris Business Directory*, 1841).

Armstrong, Allen. Philadelphia. (MD, 1840) Pistol and pistol-lock maker and/or importer. Many pistols that contain his locks are known (Shumway, in MB 2 [1969]: 2–3).

Armstrong, Crawford. Gunsmith, Pike St. near Marshall's Row, Pittsburgh (PAD, 1852).

Armstrong, John. 1825–30, Cumberland Twp., Adams Co.; 1832–37, Liberty Twp., Adams Co. (t); 1840, Freedom Twp., Adams Co. (c). The master gunsmith of this name was active simultaneously in Emmitsburg, Frederick Co., Md. The tax lists show that this man was a tanner (1825–26) or had an unspecified trade. There are several guns marked with the John Armstrong name that have been attributed to "the other" John Armstrong of Gettysburg. The evidence is inconclusive. James Serven (AR 6 [1969] 17) conducted an interview with John H. Johnston, a gunsmith whose family is documented herein, who claimed that John Armstrong of Gettysburg made guns. The Momchilov collection (KRA-2, 58–59) included an Armstrong rifle attributed to this man. Other listing books (GAAAM, 16) show this man also.

Arner, Henry. Gunpowder maker, 1832–42, Mahoning Twp., Carbon Co., partner of Alfred Hanline. "between 1839 and 1841 the mill was blown up twice and two lives were lost; the powder mill was abandoned in 1854." (Fred Brenckman, *History of Carbon County* [Harrisburg: Nungesser, 1913] 258; Alfred Mathews, *History of the Counties of Lehigh and Carbon* [Philadelphia: Events, 1884] 746).

Arperman (Apperman), Christian (b. 1787). Gunsmith, described as having a "light complexion," arrived 8 May 1807 on schooner *Betsky* from Bremen. Active 1807–14 in Philadelphia (PaGP 2:195).

Artz, Charles. 1826, Pittsburgh (PAD, 1826). Gunsmith.

Aaron Armstrong (1819–80), Beaver County.

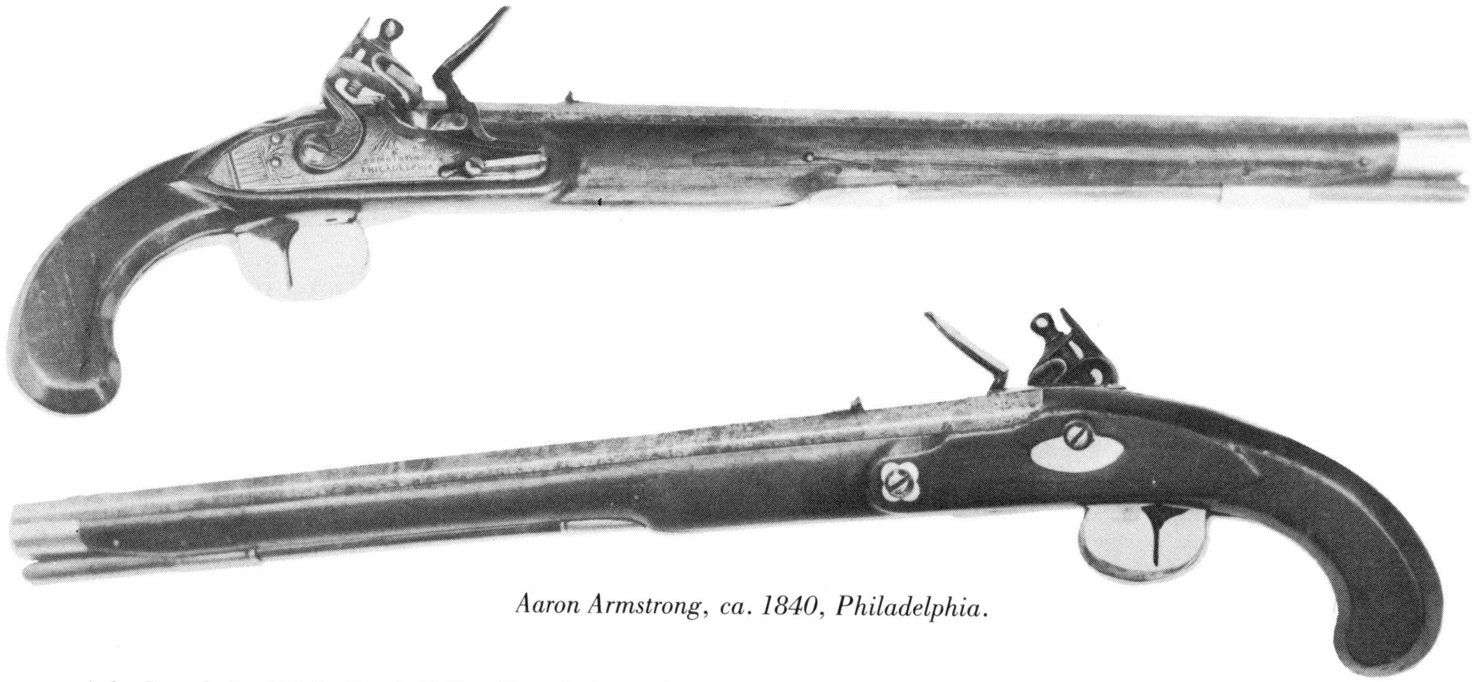

Aaron Armstrong, ca. 1840, Philadelphia.

Ash, Joseph (b. 1808). Brush Valley Twp., Indiana Co. (c, 1850, 1860).

Ashman, Henry (b. 1820). Gunsmith, Richmond, Va. (c, 1850); Annville, Lebanon Co. (BPaD, 1861; MB 10 [1963]).

Ashton, Andrew. 1777: "a very good smith." (2 Pa. Arch. 1 at 507).

Assmus, John. Artillery artificer; 12/1779, Pa. Continental Line at the cannon factory, Philadelphia (2 Pa. Arch. 15 at 384).

Atherton, Cornelius (1736–1809). Born in Cambridge, Mass. In 1763 he ran the Dover Iron Works in Amenia, Dutchess Co., N.Y. In 1769, under Samuel Adams's direction, he opened a gun factory at Boston. In 1772, he was in Afton, N.Y. as a steelmaker. On 20 December 1775 he contracted to provide the N.Y. militia with muskets and bayonets at £3/14/0 each. In 1777, Atherton moved to Plymouth, Luzerne Co., Pa. From 1786 to 1806 he was at Taylorville Depot, Wyoming Co. He returned to Afton, N.Y., where he died on 12/4/1809 (SED; KRAN [1958] 4).

Atkinson, James. 1833–36, Nottingham Twp., Washington Co. (t). Gunsmith.

Atkinson, William. Gunstocker at French Creek. Signed a petition for higher wages dated 10/30/1777 (Arch).

Atkinson, Wilton. Cutler. Son of Matthew Atkinson of Lampeter Twp., Lancaster Co., he was active as a cutler as early as 1763. He married Anna Maria, daughter of Abrum LaRoy, a clock maker, and was active thereafter in that trade. He moved to Baltimore after 1790, and Anna Maria continued his trades until at least 1819 (DB H-204, K-296, K-94).

Atland, John. 1853, Butler Twp., Adams Co. (t). Gunsmith.

Atlee, David. Silversmith, 1792, Middletown Twp., Dauphin Co. (William H. Egle, ed., *History of the Counties of Dauphin and Lebanon*, 2 vols. [Philadelphia: Everts & Peck, 1883] 377).

Atley, Conrad. 1790, Dauphin Co. (c); 1796–1804, Bedford boro (t). Gunsmith.

Auber, Daniel. Gunsmith, 1802–4, Georges Twp., Fayette Co. (t).

Auer, Charles. 906 Poplar St., Philadelphia (BSED, 1880). Gunsmith.

Aulenbach, John G. (b. 1837). 26 N. 8th St., Lebanon boro. Married Mary Arentz (b. 3/10/1845). John was born 2/8/1837 (f; BRD, 1863–64, 1873; BSED, 1880; PD, 1882; GLD, 1863–64). Gunsmith.

Ault, Frederick J. Gunsmith and blacksmith, 1791–1818, Somerset Twp., Washington Co. (t).

Austin, William. Artillery artificer, 2/12/1777 to 3/22/1781, Capt. Jesse Rowe's Co. at Philadelphia Laboratory (2 Pa. Arch. 11 at 252); Capt. Noah Nicholas's Co. (PIR 2:245).

Axer, John. Gunsmith, 1843–44, Lancaster boro (t).

Ayres, Robert, Jr. Gunsmith, 1826–32, Franklin, Venango Co. (t).

Ayres, William G. (1805–43). 1827–43, Franklin, Venango Co. Gunsmith (t).

B

Bach, John. Gunsmith, 1777, 2d St. above Vine St., Philadelphia. On the list of those persons who refused to attend militia camp until compelled by force to do so (2 Pa. Arch. 1 at 726).

Bachman, F. M. Fredericksburg, Lebanon Co. (BRD, 1880). Gunsmith.

Bachmann, Heinrich. Inventor, Lancaster, patent. 7722X of 8/21/1833, for a gunlock.

Backer, Jacob. 1767, Richland Twp., Bucks Co. (3 Pa. Arch. 13 at 703); 1779–1805, Macungie Twp., Northampton Co. (t). Gunsmith.

Backhouse, Richard. 1729–74, prop., Durham Iron Works. Produced musket barrels during the Revolution. (GAAAM, 17) 1/20/1746: "Jacob Wilkens, son of John Wilkens, dec'd, with the consent of his mother, binds him self apprentice to Richard Backhouse of Philadelphia, blacksmith, to be taught the trade of a blacksmith, to be taught to read, write and cypher" (PIDS; PaE, 95).

Badlock, Benjamin. Artillery artificer, enlisted 3/16/1778 in Capt. Thomas Wylie's Co. (PIR 2:248).

Baer, Jacob. 233 Queen St., Lancaster (BLD, 1871–72). Gunsmith.

Baer, John. 1807, Chaceford Twp., York Co. (t); 1815, Jefferson, Codorus Twp., York Co. (t, RAG, 10; JKJ, 354). Gunsmith.

Bailey & Co. Philadelphia. Swordmakers, silversmiths, jewelers. Made the presentation swords of Brig. Gen. King (1861) and Maj. Gen. George McClellan (1864) (RAG, 220).

Bailey, Banks & Biddle, Philadelphia. Jewelers, silversmiths, swordmakers. Successor to Bailey & Co., to 1900 and later (RAG, 220).

Bailey, Elmer E. Inventor, Sinnamokoning, Cameron Co. Air gun patents 487,169 (1892) and 603,549 (1898).

Bailey, George. New Market near Green, Philadelphia (SGD 1830). Gunsmith.

Bailey, Isaac. 1785–90, West Marlborough Twp., Chester Co. (RAG, 10). Gunsmith.

Bailey, Joel (12/16/1732–10/29/1797). Gunsmith, clock maker, surveyor. 1758, Bensalem Twp., Chester Co; 1759–70, West Bradford Twp., Chester Co. (t). Estate inventory of 12/1797 showed: 1 turning lathe $10; 12 chisels and augers $1.50; 1 tenon & compass saw 75¢; 5 large screw-plates $1.50 1 pr bellows $4; 2 stamps, 1 cutting knife 40¢.

Bailey, Major Joel (9/26/1789–10/16/1845). Blacksmith and gunsmith, armorer. Son of Joseph and Lydia Bailey of Penn's Manor, Bucks Co. 1st lt. in War of 1812 in Capt. R. M. Crain's Co., gunsmith and armorer; later, keeper of state arsenal, Harrisburg. Married Elizabeth Seidle (d. 8/14/1875, age 83) on 3/10/1814 (Egle, *History of Dauphin Co.*, 464).

Bailey, John. Gunsmith, Philadelphia (BPD, 1870, 1876); Bristol, Bucks Co. (BRD, 1880; BSED, 1882).

Bailey, Robert. 1777–83, Pottstown (KEA, 5; York DB B).

Bailey, Samuel. 4053 Market St., Philadelphia. (BPD, 1874, 1880) DBA J & S Bailey (Samuel was a son of John) (BSED, 1889).

Bair, Jaffa. 1802, Leacock Twp., Lancaster Co. (t). Gunsmith.

Bair, John. Gunsmith. Prior to 1812, Newberry Twp., York Co. (HYC, 701).

Baker, Andrew (1808–1887). 1831, single freeman, Woodbury Twp., Bedford Co. (t); 1832, married, same (t); 1833–37, Frankstown Twp., Huntingdon Co. (t); 1840–45, Blair Twp., Huntingdon Co. (t); 1845–87, Blair Twp., Blair Co. (t). Gunsmith.

Baker, Caleb. 1719–41, Gunsmith, worked with his father, Robert Baker, on the confluence of Pequa Creek and the Susquehanna River, Lancaster Co. (Lancaster Co. DB B, 577; SED, 14, 53).

Baker, George (d. 1820 or 1821). 1786, Lampeter Twp., Lancaster Co. (t); 1802–9, Georges Twp., Fayette Co. (t); 1810–20, German Twp., Fayette Co. (t; c); 1821, same, noted as dead. Gunsmith. (t).

Baker (Backer), George. 1782, Gunsmith, single freeman, Lancaster boro. (t); 1783–85, with Peter Gonter, Lancaster Co. (t); 1786, "Gonter removed," same, with John and Peter Ferree (t). After 1790–1844, Morgantown, Monongalia Co., (West) Virginia. In March 1789 he married Elizabeth, daughter of William Norris in Monongalia Co. 1811, contract for Va. militia weapons. According to his tombstone in the Nicholson-Baker Cemetery, Cheat Lake, Monongalia Co., he was born 2/6/1762 and died 6/27/1844. He was a private in Maj. Moore's Co., 1st Pa. Regt., in the Revolution (DAR 137775; t, c, f). Gunsmith.

Baker, Godfrey. Artillery artificer, 1/1777, Capt. Will's Co., Berks Co. militia (BCR, 115).

Baker, Jacob. Artillery artificer. Northampton Co., Pa. revolutionary war pension list, 8 May 1786 (3 Pa. Arch. 26 at 39).

Baker, Jacob A. Near Front St., corner of Otter St., Philadelphia. Gunsmith (PXD, 1819; DeSD, 1829, 1836).

Baker, Jacob S. 386 N. Front St., Philadelphia (PXD, 1819); 916 N. Front St. (PXD, 1824; MD, 1859). Gunsmith.

Baker, James. Ca. 1820–30, Mill Creek, Philadelphia (RAG, 11). Gunsmith.

Baker, John. 1728–50, Lancaster boro. (t). He was born in Germany. His shop was on Factory Rd., south of Lancaster, near the Rockford Plantation (SED, 14, 53). Gunsmith.

Baker, John. Artillery artificer, Capt. John Jordan's Co., stationed at Ft. Pitt, 11/1/1782 (2 Pa. Arch. 11 at 262).

Baker, John. 1776–78, Germantown Rd. at North Wales Rd., Philadelphia. 1 July 1776: "John Baker of Norrington has some provincial arms" for repairs. (2 Pa. Arch. 1 at 485) 1779, artillery artificer, through 1782 (2 Pa. Arch. 11 at 236). Active as late as 1830 in Germantown as a gunsmith (RAG, 11).

Baker, John G. Philadelphia. Patent 343,560 of 6/15/1886 for a spring gun (RAG, 11).

Baker, Joseph. 1859–61, Blair Twp., Blair Co. (KKR, 178). Gunsmith.

Baker, Joshua. 1754, Lancaster boro. (t); 1/8/1763, Joshua Baker, gun-barrel maker of Cumberland Co., sold 86 acres in Lampeter Twp., Lancaster Co. (Lancaster DB H-138).

Baker, Melchior (ca. 1760–ca. 1820). 1779–1806, Haydentown, Fayette Co., worked with Albert Gallatin in government arms contracts; 1806, arms contractor at Clarksburg, (West) Virginia (t).

Baker, Nicholas. 1798–1804, Georges Twp., Fayette Co. (t) Son of Melchior Baker. Gunsmith.

Baker, Robert (d. 1728). 1717–28, gunsmith at the confluence of Pequa Creek and the Susquehanna River, Lancaster Co. (SED, 14, 53; t).

Baker, Samuel. 1717–19, worked with his brother, Robert Baker, Lancaster Co. (SED, 14, 53) In 1719 Robert bought out Samuel's interests.

Bakewell, William. Inventor, Pittsburgh. Patent 39,109 of 7/7/1863 for a metallic cartridge.

Baldwin, Anthony, Sr. Silversmith, swordsmith. Chester Co. (t, 1800).

Baldwin, E. A. Tidionte, Warren Co. (BRD, 1873, 1882, 1887; PD, 1882). Gunsmith.

Baldwin, Jacob. Artillery artificer, armorer. 12/1779, Pa. Continental Line (2 Pa. Arch. 15 at 384); Philadelphia gunsmith (Col. Rec., 10:480).

Baldwin, John. Tin manufacturer. 1768, East Caln, Chester Co. (t), 3 Pa. Arch. 11 at 493).

Baldwin, Thomas. Blacksmith and gunsmith, 1697–1730, Chester Twp., Delaware Co. Will dated 3/17/1730; estate inventory showed gunsmithing and blacksmithing tools (WB 1, 334).

Ball, William. Silversmith, edged-weapons maker, ca. 1750–90, Philadelphia. Advertisements in *Pennsylvania Packet* of 7/18/1754 and 10/31/1754 listed him as a goldsmith on Front St., Philadelphia (RAG, 220).

Balsley, Christian. 1781–85, gunsmith, West Pennsboro Twp., Cumberland Co. (t; 3 Pa. Arch. 3 at 501, 655); 1790–93, East Pennsboro Twp., Cumberland Co. (WSB, 57; t).

Bamford, Ralph. Artillery artificer, 1779; superintendent of whitesmiths, Pa. Continental Line (2 Pa. Arch. 15 at 383).

Banick, Jacob. Cutler, associated with John Shugart & Co., Chambersburg. In 1828 Banick won the Franklin Institute Award for Mechanical Arts for his knives (Hazard's *Register,* 2:264).

Banks, John. Gunsmith, 1779–81, Concord Twp., Chester Co. (3 Pa. Arch. 12 at 208, 342, 592) On 6/16/1794 Banks obtained 400 acres in Westmoreland Co. by warrant (3 Pa. Arch. 26 at 401).

Banks, John. Gunsmith, Bushkill Twp., Northampton Co. (t, 1835).

Banks, Uri. Gunlock maker to Pa. Committee of Safety. On 6/17/1775 he contracted to make 50 gunlocks (KEA, 14).

Bannacker, George. Philadelphia (NTD, 1800; KEA, 6). Gunsmith.

Barber, Joseph. Bridesburg, Philadelphia Co. 1859, patent 23,224, with P. C. Reinfried, for a breech-loading rifle.

Barber, Samuel. 1800, Hempfield Twp., Lancaster Co. (t; c); 1807–13 Caernovon Twp., Lancaster Co. (t). Gunsmith.

Barclay, Peter. Armorer on brig *Convention,* Pa. Navy, 4/3/1777 (2 Pa. Arch. 1 at 271).

Bard, George. 1777, Gunsmith, Lampeter Twp., Lancaster Co. (t); 1778, killed in Revolution (SED, 51).

Bardeck, George. On 10/20/1772 George Bardeck was bound as an apprentice to Richard Humphreys, goldsmith, Philadelphia, for 2 years. Bardeck was a Redemptioner who was sold for £16/18/0 to pay for his passage to America. (PIDS).

Bare (Bear), Samuel. Gunsmith, Manheim Twp., Lancaster Co. (Lancaster DB M-426); sold land to Wiston & Miles, merchants of Philadelphia, 8/10/1768.

Barensten, John (d. before 7/19/1661). Armorer at the

Dutch settlement on the Delaware. His wife sold his gunsmithing tools to Hans Block for 980 guilders. When Block reneged, Meyntie Willems, Barensten's wife, asked council to force payment. (2 Pa. Arch. 5 at 429–30) Barensten sold lead bullets to the colony on 3/17/1660 (2 Pa. Arch. 5 at 451). "John Tennison . . . had the use of the [gunsmithing] tools of John Barensten, who was murdered by the savages and whose wife died at Col. Utie's." Their child was delivered to Mrs. Jacob Claeson, a soldier's wife, to raise (Hazard's *Annals*, 1:310).

Barger, Adam. 12/20/1777 to 3/22/1781, Lebanon, Lancaster Co. Artillery artificer in Capt. Jesse Rowe's Co. (2 Pa. Arch. 11 at 252; PIR 2:244).

Barlow, John. Armorer and artillery artificer, 6/9/1779, at the cannon factory at Carlisle (2 Pa. Arch. 11 at 2600).

Barnes, Thomas. 1779–80, Horsham Twp., Philadelphia Co. (3 Pa. Arch. 15 at 53, 54, 422). Gunsmith.

Barnes, T. W. 1779–84, Falls Twp., Bucks Co. (3 Pa. Arch. 13 at 19, 126, 279, 323, 494). Gunsmith.

Barnhardt, Berges. (b. 1823). Gunsmith, 1850, Reading, Southwest Ward (c).

Barnhart, George (1798–1844). Born in Philadelphia; died in Ross Co., Ohio. Son of William Barnhart, Sr.; brother of William, Jr. (RAG, 12; DH, 337) Descendant of Hessian deserter. Gunsmith.

Barnhart, William, Sr. Hessian soldier who deserted to the American side. Gunsmith in Philadelphia. The family moved to Ross Co., Ohio (DH, 337; RAG, 12).

Barnhart, William, Jr. (1802–1867). Born in Philadelphia; died in Ross Co., Ohio (DH, 337; RAG, 12). Gunsmith.

Barnhill, Robert. On 3/18/1789, he was paid "for gun powder, flints, &c., for the use of the militia of Montgomery Co." (3 Pa. Arch. 7 at 300); 1795–1800, 68 N. 2d St., Philadelphia, pistol and rifle maker (NTD, 1795, 1800).

Barns, Luther. 441 3d St., Philadelphia (NTD, 1799); 8 Fayette St. (NTD, 1800; PXD, 1800, 1807, 1808, 1810; KEA, 6). Gunsmith.

Barr, Jacob. Gun-barrel forger, Manheim Twp., Lancaster Co. (t, 1829).

Barr, John. N. Queen St., between Walnut St. and Lemon St., Lancaster (BLD, 1857) Gunsmith.

Barret, Conrad. Gunsmith, Ferguson Twp., Clearfield Co. (t, 1887–90).

Barry, Elijah. 1021 Melon, Philadelphia. In 1860 his firm turned out over 10,000 gunstocks, which it sold to the Union army. Gunstock manufacturer, sold commercially to individuals and large factories (RAG, 13; GPD, 1875).

Barry, Thomas C. 1021 Melon, Philadelphia (GPD, 1886, 1892, 1894; BSED, 1892). Gunsmith.

Bartlett, Edward M. Clock makers, jewelers, silversmiths, edged-weapons makers. West Chester, Chester Co. (t; BSED, 1840); DBA: 1833, H. W. Nutler & Co.; 1834–35, Butler & Bartlett; 1836–40, E. M. Bartlett Co. (t).

Bartling, Conrad. 10/30/1777, gunstocker at French's Creek arsenal; signed a petition asking for higher wages. 12/1779, artillery artificer, Pa. Continental Line (2 Pa. Arch. 15 at 383).

Barton, Isaac. Cutler. 1714, emigrated from Clonmell to Philadelphia, joined Philadelphia Monthly Meeting (*Immigration of Irish Quakers* [Baltimore: Genealogical Publishing Co., 1982] 288).

Barton, James. Smith, 1767–69, Chester, Chester Co. (3 Pa. Arch. 11 at 287, 201, 395, 535).

Bartow, Thomas. Philadelphia, 3/5/1788: "for gunpowder purchased from him, used on announcing the ratification of the Federal Constitution." (3 Pa. Arch. 7 at 206).

Baston, George. Gunsmith, Portersville, Muddy Creek Twp., Butler Co. (PD, 1882).

Batchelor, William R. 1805–12, 85 Water St., Philadelphia; 1813–19, Miller's Ct., Northern Liberties, Philadelphia Co. (PXD, 1805, 1807, 1813, 1816, 1818, 1819). Gunsmith.

Bateman, Thomas. 1847, Butler St., Lawrenceville, Allegheny Co. (KKR, 179). Gunsmith.

Bates, Francis. 1857–58, Gunsmith, Mulberry St., Pittsburgh (PAD, 1857–58); 9th Ward, Pittsburgh (c, 1850, 1860). He may have worked in Ohio for James and Daniel Little (DH, 126).

Battis, Earnest. Buckle filer and finisher, 1779, artillery artificers, Pa. Continental Line (2 Pa. Arch. 15 at 385).

Bauer, George. Apprentice gunsmith, 7/16/1776, in Capt. Andrew Graff's Co., Lancaster Co. Militia (2 Pa. Arch. 13 at 347; 5 Pa. Arch. 7 at 1073); musket maker, repairman for public arms (RAG, 21).

Bauer, Jacob. Gun shown in Shumway (GAS-1, plate 92) marked with this name and dated 9/[?]/1777.

Baugh, John. 1774, Coventry, Chester Co. (3 Pa. Arch. 12 at 67); 1769, Mulberry Ward, Philadelphia (3 Pa. Arch. 14 at 214); 1800, Northern Liberties, Philadelphia Co. Gunsmith (3 Pa. Arch. 15 at 500).

Baum, Adam. Gunsmith. Near Hummellstown, Dauphin Co. Father of Michael (b. 1757), Daniel (b. 1759), and John (b. 1761), all gunsmiths (GAS-1, 380).

Baum & Co. Hummell's Wharf, Monroe Twp., Snyder Co.

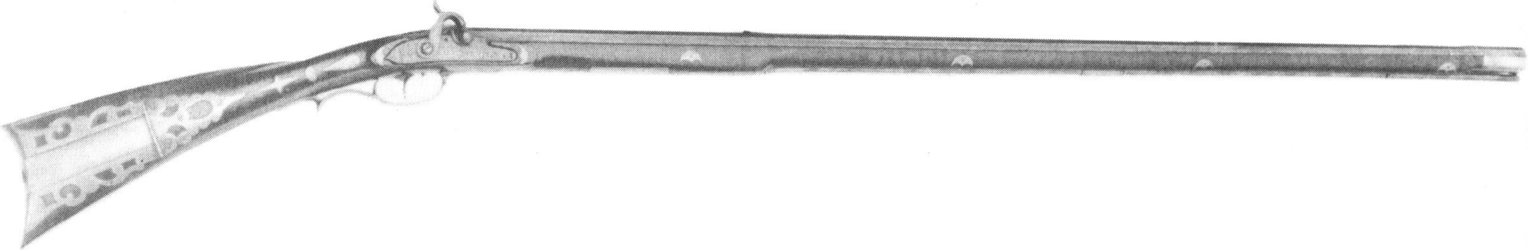

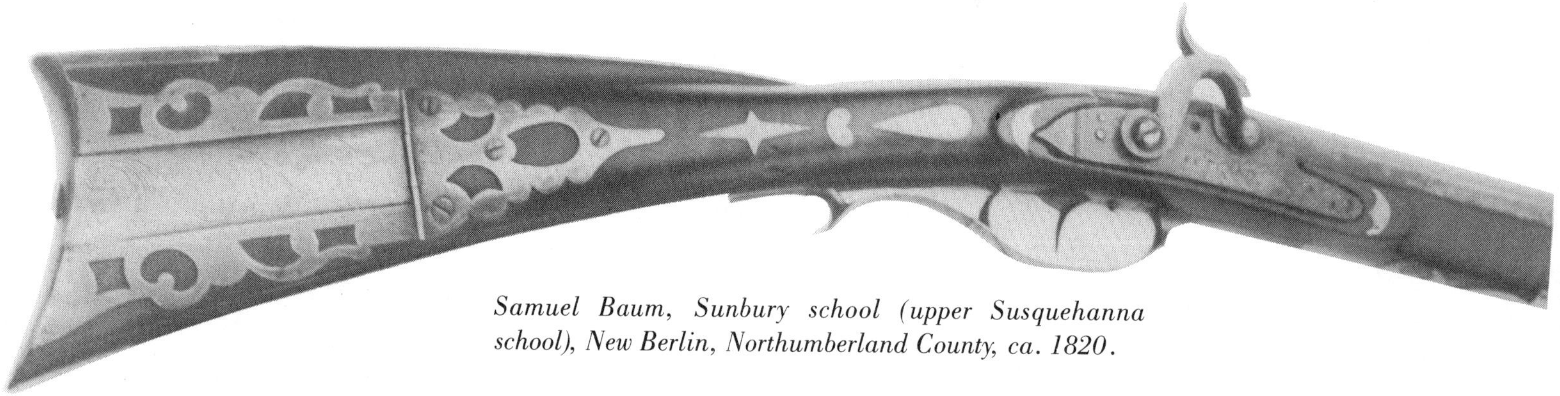

Samuel Baum, Sunbury school (upper Susquehanna school), New Berlin, Northumberland County, ca. 1820.

(BRD, 1887, 1890). Gunsmiths.

Baum, Charles. Gunsmith, Pottsville, Schuylkill Co. In 1850 he had two employees (c, 1850; RAG, 14).

Baum, Daniel (b. 1759). Near Hummellstown, Dauphin Co. (GAS-1, 380). Gunsmith; son of Adam Baum, gunsmith.

Baum, John (b. 1761). Son of, apprenticed to, Adam Baum, gunsmith, near Hummellstown, Dauphin Co. (GAS-1, 380–83, plate 86).

Baum, Samuel. 1789–1840, New Berlin, Northumberland Co. gunsmith. (t; RAG, 14; Ellis and Hungerford, *Susquehanna and Juniata Valleys*, 2:1407).

Baumann, Johannes. Dealer in rifles and guns, Germantown, Philadelphia Co. (*Pennsylvanische Geschichte-Schreiber* [Sower's Newspaper] 10/16/1753).

Baunell, William. Gunsmith, Allegheny Twp., Somerset Co. (t, 1853).

Baur, George. Philadelphia (MD, 1847). Gunsmith. See Bauer.

Bayard & Co. "There is a box of gunlocks in the possession of John Hubley, esq., com'y of stores at Lancaster, which . . . may be procured & delivered . . . they cost 30 shillings apiece purchased at Bayard & Co." (Col. Benj. Flowers to Pres. Wharton, 10/26/1777, written at Allentown [1 Pa. Arch. 5 at 703]).

Bayers, Henry. 1820–45, Indiana Twp., Allegheny Co. (t); 1845–50, Shaler Twp., Allegheny Co. (t; c). Gunsmith.

Baylie, Jonathan. Artillery artificer, Capt. Nathaniel Irish's Co., enlisted 5/1/1778 (PIR 2:247).

Bayly, John. The following advertisement appeared in the *Pennsylvania Packet* for 9/5/1754: "John Bayly of Philadelphia, goldsmith, is about to go to London."

Beakley, Daniel. Gunsmith, Prospect near Butler, Pittsburgh (PAD, 1860–61).

Beaks, John, Sr., and John, Jr. Both artillery artificers, at the laboratory, Philadelphia (2 Pa. Arch. 15 at 385).

Bean, Nathaniel. Swordsmith. From 4/12 through 5/22/1777, one of 16 men in the employ of Lewis Prahl, gunsmith, manufacturing swords for Col. Benj. Flowers's Pa. Horsemen (2 Pa. Arch. 3 at 104).

Bear (Beer); George (1772–after 1850). 1802–6, Allegheny Twp., Armstrong Co. (t); 1807, Kittanning boro. (t); 1808–20, Derry Twp., Westmoreland Co.; 1822–50, Salem Twp., Westmoreland Co. (t; c). Gunsmith, blacksmith.

Bear, Isaac L. Gunsmith, Margaretta Furnace, York Co. (BPaD, 1861).

Bear, Jacob. Gunsmith, 1745–46, Lancaster boro. (Lancaster DB B-418); blacksmith, 1779, Union Twp., Berks Co. (3 Pa. Arch. 18 at 292).

Bear, Jacob. Gun-barrel forger, Manheim Twp., Lancaster Co. (t, 1825–50, c, 1850).

Bear, Jacob. Gunsmith, Lemon Alley between Chestnut and E. Orange, Lancaster boro (GLD, 1863–64; BRD, 1871–72).

Bear, John. Chanceford Twp., York Co. (t, 1807); Codorus Twp., York Co. (t, 1815; JKJ, 354). Gunsmith.

Bear, John. Manheim Twp., Lancaster Co. Gun-barrel forger (t, 1808–41).

Jacob Bare (Bear), Lancaster County.

Bear, Joseph. Gun-barrel forger, Manheim Twp., Lancaster Co. (t, 1825–34).

Bear, Peter. Gunsmith, Warwick Twp., Lancaster Co. (t, 1840–50).

Bear, Samuel. Gun-barrel forger, Lancaster Co. (t, 1809–50).

Bearcock, Thomas W. Inventor, Pittston, Luzerne Co. 1877, patent 195,522 for a revolver, 190,543 for a revolver.

Beatty, William. Cutler. 1813–28, Nether Providence Twp., Delaware Co., along Providence Creek; in 1825 he made 1,600 cast steel axes, 500 broad axes, 500 draw knives, 200 cleavers and choppers, and 500 adzes. In 1828 the shop was converted to a cotton mill. (Charles Palmer, ed., *A History of Delaware County,* 2 vols. [Philadelphia: National Historical Association, 1932] 132) 1866–71, ran Chester Edge Tool Works; after 1871 it was operated by his son, John C. Beatty (ibid., 1:140).

Beaver Falls Cutlery Co. Active 1867–86. Organized by Dr. C. G. Hussey, General Thomas M. Howe, and James W. Brown; located at Beaver Falls and, earlier, at Rochester. At its peak it employed over 400 men, mostly Chinese. Joseph H. Bausman, *History of Beaver County,* 2 vols. [New York: Knickerbocker, 1904] 670).

Beck, Abraham. Son of George Beck, "Abraham Beck ran a blacksmithing shop and also did carriage work and gunsmithing." (E. L. Ross, *A History of Brush Valley Township* [1973], 63) "Abraham had worked with his father, George B. Beck, since boyhood. Their tools were valued at $400 [in 1850 U.S. Census] and [it] was the only shop of its kind in this part of the county" (John A. Caldwell, *History of Indiana County* [Newark, Ohio: Caldwell, 1880] 408).

Beck, Ashbury. Gunsmith. 1827–29, Buffalo Twp., Washington Co. (t).

Beck, Christian Henry (1754–1843). Born in Bethlehem on 7/17/1754 and died on 2/21/1843; buried at the Moravian graveyard at Nazareth. On 9/26/1790 he married Anna Christina Eyerly (1764–1833). From 1787 to 1789 he was in Lebanon Co. (t). He was a son of H. F. Beck, Moravian missionary of note (*History of the Wyoming and Lackawanna Valleys,* 1:127).

Beck, Christian (1787–1845). 1807–11, Jonestown, Dauphin Co. (t); 1813–21, Lurgan Twp., Franklin Co. (t); 1827, Martinsburg, W.Va. (t); 1841–45, Cincinnati, Ohio (JKJ, 250–51; WSB, 25–26) Gunsmith.

Beck, Christian (1813–86). Born in Pa., died in Marion Co., Ind. Capt. in Union army in Civil War. Gunsmith. (Curtis Johnson, *Gunsmiths of Illinois* [York, Pa.: Shumway, 1982] 7).

Beck, Christopher. Locksmith, 1769, Philadelphia (3 Pa. Arch. 15 at 211).

Beck, Daniel (d. 1826). Powder maker. Son of George Beck, Sr. He was killed in an explosion at the powder mill at Kittanning on 6/24/1826 (*Kittanning Gazette,* 7/15/1826). See Beck, George F., Sr.

Beck, George F., Sr. "as early as 1811 . . . [he] commenced the manufacture of powder near the mouth of Pine Run . . . [and continued until] Thursday, June 29, 1826, when an explosion of about 50 pounds of powder in the mortar occurred, caused, it was supposed, by a spark elicted [*sic*] by one of the pounders. John and Daniel Beck were, at the time, employed at the mill. The latter was thrown out of the door and so injured that he afterward died. The former was severely, but not fatally, injured. A part of the roof was also carried away by the explosion, but

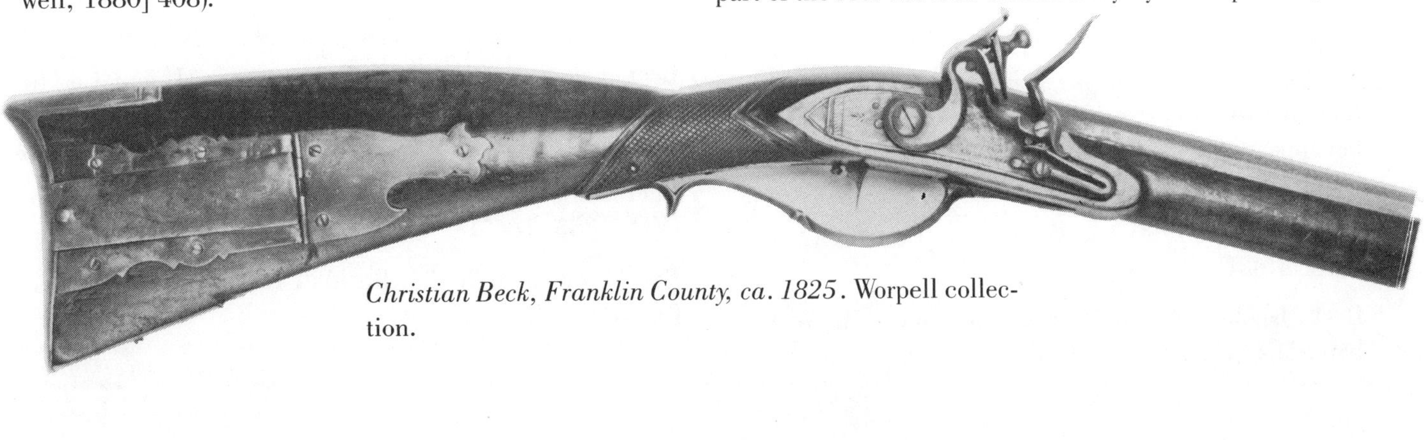

Christian Beck, Franklin County, ca. 1825. Worpell collection.

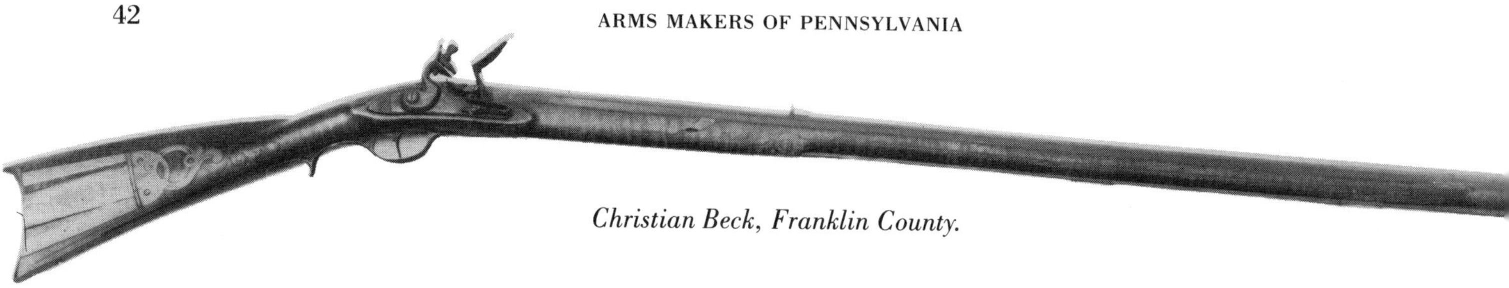

Christian Beck, Franklin County.

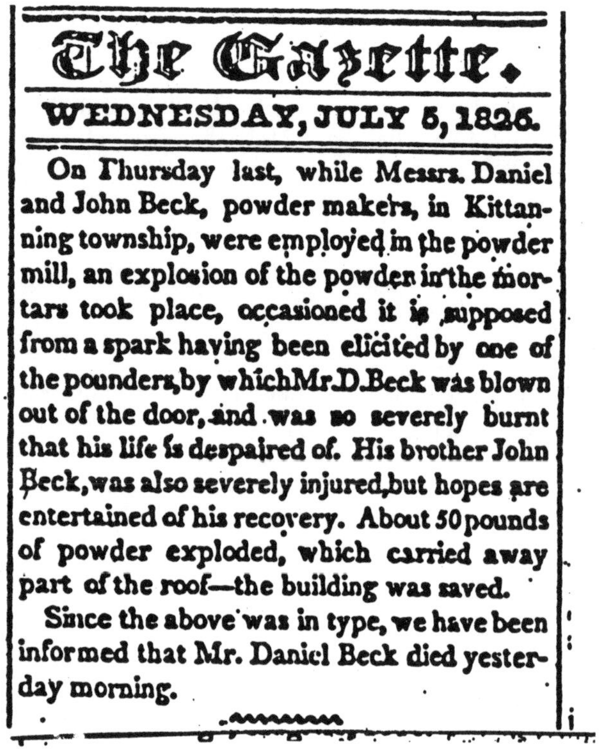

The Gazette.

WEDNESDAY, JULY 5, 1826.

On Thursday last, while Messrs. Daniel and John Beck, powder makers, in Kittanning township, were employed in the powder mill, an explosion of the powder in the mortars took place, occasioned it is supposed from a spark having been elicited by one of the pounders, by which Mr. D. Beck was blown out of the door, and was so severely burnt that his life is despaired of. His brother John Beck, was also severely injured, but hopes are entertained of his recovery. About 50 pounds of powder exploded, which carried away part of the roof—the building was saved.

Since the above was in type, we have been informed that Mr. Daniel Beck died yesterday morning.

Kittanning Gazette, July 5, 1825

the building was saved. Willow charcoal . . . was used in making powder at that mill. The reputation of Beck's powder stood high at home and abroad. Large quantities of it were transported to Pittsburgh in canoes." (Robert W. Smith, *History of Armstrong County* [Chicago: Waterman-Watkins, 1883] 369) He was a farmer and gunsmith and built and operated a gunpowder factory on the Kittanning Road [in Burrell Twp.] not far from the county seat. He married Elizabeth Holsapple (Samuel T. Wiley, ed., *Biographical and Historical Cyclopedia of Indiana and Armstrong Counties* [Philadelphia: Gresham, 1891] 480–81).

Beck, George F., Jr. (d. 1837). Worked with his father George F. Beck, Sr. in the manufacture of gunpowder in Burrell Twp., Armstrong Co. Advertisements for his product appeared in the Greensburgh *Gazette* on 6/13/1817 and 8/22/1817 (f; t; c).

Beck, Gideon. 1780–88, Montgomery Co. (GAAAM, 21; RAG, 15). Gunsmith.

Beck, Jacob (b. 1818). Born in Lebanon Co.; 1870, Hillsboro, Montgomery Co., Ill. (c).

Beck, Jacob. On 12/28/1816 the Greensburg *Gazette* carried an advertisement for Jacob Beck's gunpowder sold by Simon Drum, Jr., Greensburg. "Jacob Beck was born on Crooked Creek and lived there until after his marriage, when he went to Ligonier Valley. He built a powder mill which he operated for sometime" (Wiley, *Biographical Cyclopedia, 481*).

Beck, John Philip (1751 or 1752–1811). Major gunsmith. 1777–1811, Lebanon Co. (t); 1781, was paid for repairing Lancaster Co. Militia weapons (3 Pa. Arch. 6 at 465); 4/29/1789, was paid for repairing 56 stands of arms belonging to Dauphin Co. (3 Pa. Arch. 7 at 299). (See also GAS-1, 429–46; JKJ, 242–45.)

Beck, John Valentine (1731–1791). 1761, arrived in America on ship *Hope*, gunstock maker, single man; arrived in New York on 19 October. (PaE, 52) Activc at Bethlehem, 1761–64. Left on 11/1/1764 for Moravian community in Winston-Salem, N.C.; active there until his death on 3/7/1791 (KEA, 7).

Beck, William. Innkeeper and gunsmith, 1802–12, Hempfield Twp., Westmoreland Co. (t).

Beckel (Bickel), Charles F. (1802–80). Silversmith, watchmakers and clock maker, Bethlehem, Northampton Co. (t, c).

Becker, Joseph. 194 S. Market St., Wilkes-Barre (WBD, 1886, 1890). Gunsmith.

Beddow, Josiah. Gunsmith, 1725 Psyk Ave., Philadelphia (BPD, 1890).

Bedford, Samuel. Artillery artificer and file cutter, Pa. Continental Line, at the cannon factory, Philadelphia (2 Pa. Arch. 15 at 385).

Bee, Benjamin. Gunsmith, Horton, Indiana Co. (BPaD, 1861).

Beerstacher, E. F. 1855–65, Lewisburg, Union Co. (t; BPaD, 1861); held patent 13,592 for a breech-loading rifle.

Beerstecher, E. Gunsmith, 5 Green St., Philadelphia (MD, 1855, 1865).

Beidel, Jacob. 1828–35, Bushkill Twp., Northampton Co. (t). Gunsmith.

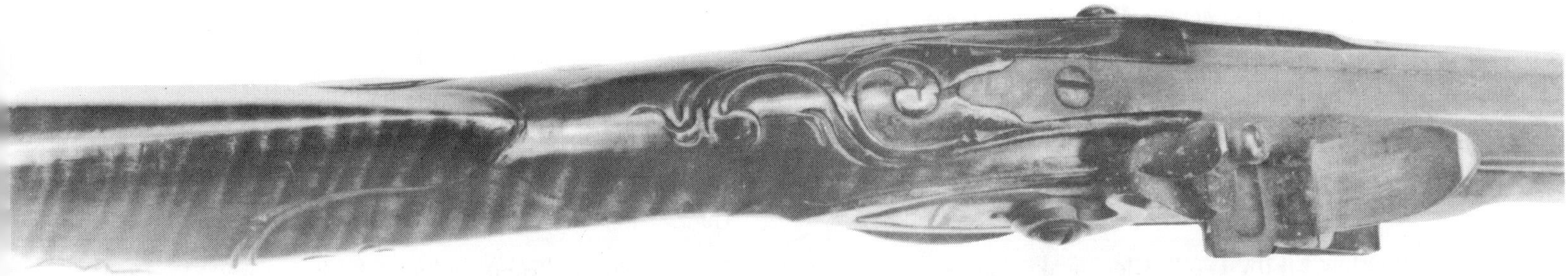

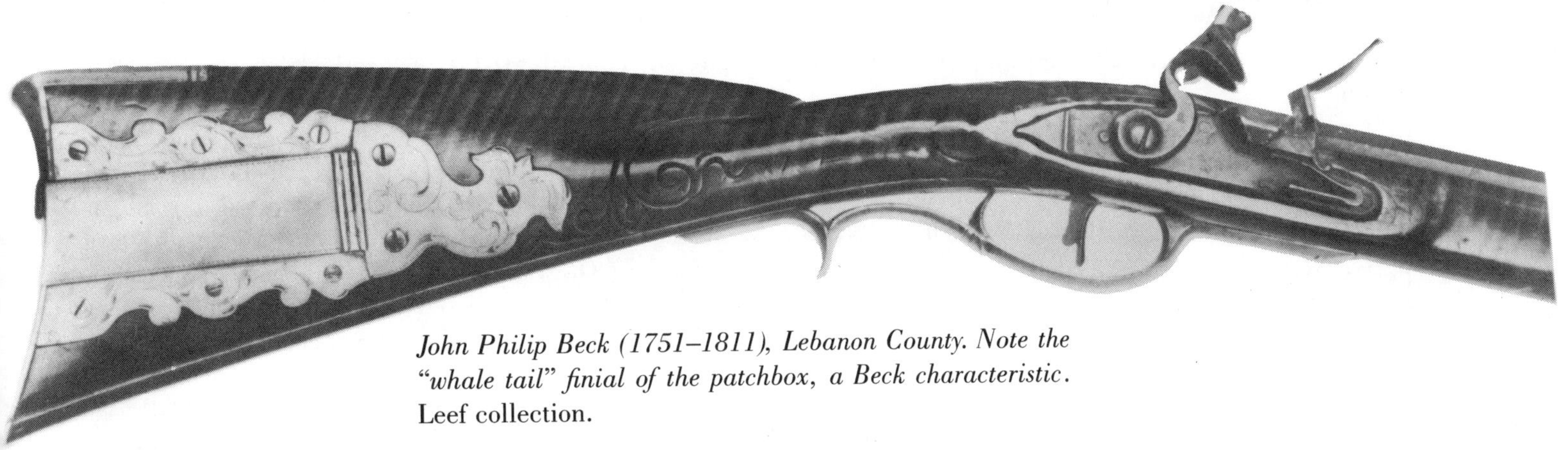

John Philip Beck (1751–1811), Lebanon County. Note the "whale tail" finial of the patchbox, a Beck characteristic. Leef collection.

Beidenhart, Christopher. 1770–73, gunsmith, Lancaster (t; SED, 54).

Beinarsek, Joseph (b. 1777). Gunsmith. Arrived in Philadelphia on 12/3/1807 on the *William Johnson* from Bohemia (2 Pa. Arch. 17 at 664).

Beisel, John. Lilly Alley, Philadelphia (DeSD, 1829–30); 1820, Bushkill Twp., Northampton Co. (t). Gunsmith.

Beisel, Simon. Gunsmith, Lilly Alley, Philadelphia (DeSD, 1829–30).

Bell, John. 1817, Carlisle. (t; KEA, 8; WSB, 57). Gunsmith.

Bell, Robert. 1780, was paid £255 for rifles, Chester Co. (2 Pa. Arch. 6 at 217).

Bell, William W. (1777–1842). Silversmith, Gettysburg, Adams Co. (t).

Bellis, Peter. Gunsmith, Lykens Twp., Dauphin Co. (JKJ, 227; MB 1 [1966] 7).

Belslerling, A. Frankford near Otter, Philadelphia (DeSD, 1816, 1818). Gunsmith.

Belton, Joseph. 1775: "I will now assert . . . I will make a machine by the help of which I will carry a loaded cannon 2 or 3 miles up or down our harbors without any other assistance and all the way there should appear nothing above the surface much larger than a man's hat. . . . My machine would wholly descend under water for some time and by expanding would rise above the surface at pleasure . . . when I had arrived within 150 or 200 yards of a ship I could descend under the surface and go along side of her bottom against which I could discharge the cannon" (1 Pa. Arch. 4 at 650). On 5/3/1777 Belton offered to convert muskets to a multiple charge system of his design (RAG, 16).

Bemper, Abraham. Silversmith, 1772, Bethlehem Twp., Northampton Co. (3 Pa. Arch. 19 at 23).

Bender, Christian. 1814, Lancaster boro (t; SED, 53). Gunsmith.

Bender, Jacob (1/31/1740–12/6/1828). Came to Philadelphia from Hamburg, Germany, landing on 12/19/1798, having served an indenture to pay for the passage of himself, his wife, and 4 children. (f) Moved to Allegheny Twp., Cambria Co., between Bucks Mills and Loretto in 1802, where he was a gunsmith and clock maker until his death. His gunsmithing tools were sold at auction following his death, and the proceeds were used to pay his funeral expenses and to purchase masses for the dead (t; c; f).

Bender, Philip. 1786–93, Manheim Twp., Lancaster Co. (t; SED, 53). Gunsmith.

Benfer, Amos (1841–1916). Gunsmith, Beavertown, Synder Co. (t; f; c).

Benfer, Arnig. Beavertown, Synder Co., gunsmith, 1780–1820 (t; c; f).

Benfer, Moses. 1840–80, Beavertown, Synder Co., gunsmith (c; t; f).

Bennet, Jacob (b. 1713). Born in Bucks Co. Wampum

maker (advertisement in *Pennsylvania Packet*, 1/21/1755).

Bennett, G. A. Gunsmith, Erie boro. (c, 1850).

Bennett, Joshua (b. 1789). Runaway apprentice of Henry Wolf, Pittsburgh. Ran away in February 1808. Wolf offered a $6 reward for his return (Advertisement in Pittsburgh *Gazette*, HJK, 363–64).

Bentley, Eli (b. 1752). Clock maker and gunsmith, 1778–83, Chester Co. (t); after 1783, Taneytown, Md. (t).

Berger (Berges), B. (b. 1823). Gunsmith 1850, West Ward, Reading. (c).

Bergley, William. Gunsmith, 214 W. King St., Lancaster (BAR, 1875–76).

Beringer, John. Gunsmith, Philadelphia (BPD, 1861).

Berlin, Abraham. Gunsmith, 1775–80, Easton, Northampton, Co. (t).

Berlin, Isaac (1755–1830). Blacksmith, swordsmith, and gunsmith. 1775–95, Easton, Northampton Co. (t); 1799, Abbottstown, York Co. (t); 1808, Easton; 1825–29, Mead Twp., Crawford Co. (t); 1830, Woodcock Twp., Crawford Co. (t). Died 6/16/1830. "A Revolutionary soldier, [Berlin] drew tract 1501 in the extreme northwest corner of [Randolph] township. He brought his family to it from across

Berlin, Isaac, Jr. Gunsmith, Northampton Twp., Lehigh Co. (c, 1830).

Bernard, John. Artillery artificer, enlisted on 7/8/1778 in Capt. Thomas Wylie's Co. at Carlisle (PIR 2:248).

Bernds, Philip. Trumbaursville, Bucks Co. (BPaD, 1861). Gunsmith.

Berord, Louis (b. 1843). Axe and edged-weapons maker. After 1871, Beaver Falls, Beaver Co. He previously operated a similar factory in East Douglass, Mass., c. the mountains and commenced a settlement. The solitude or unresponsive character of the soil proved irksome and he soon left, purchasing a farm in Woodcock Township. . . . Of German nativity, he was a pioneer who settled on the east bank of French Creek. He was . . . one of the few who escaped starvation aboard a British prison ship in 1777. . . . He was a gunsmith by trade and died in Woodcock Township on 16 June 1830 in his 77th year" (Robert C. Brown, *History of Crawford County* [Chicago: Wanrer-Beers, 1885] 597, 696).

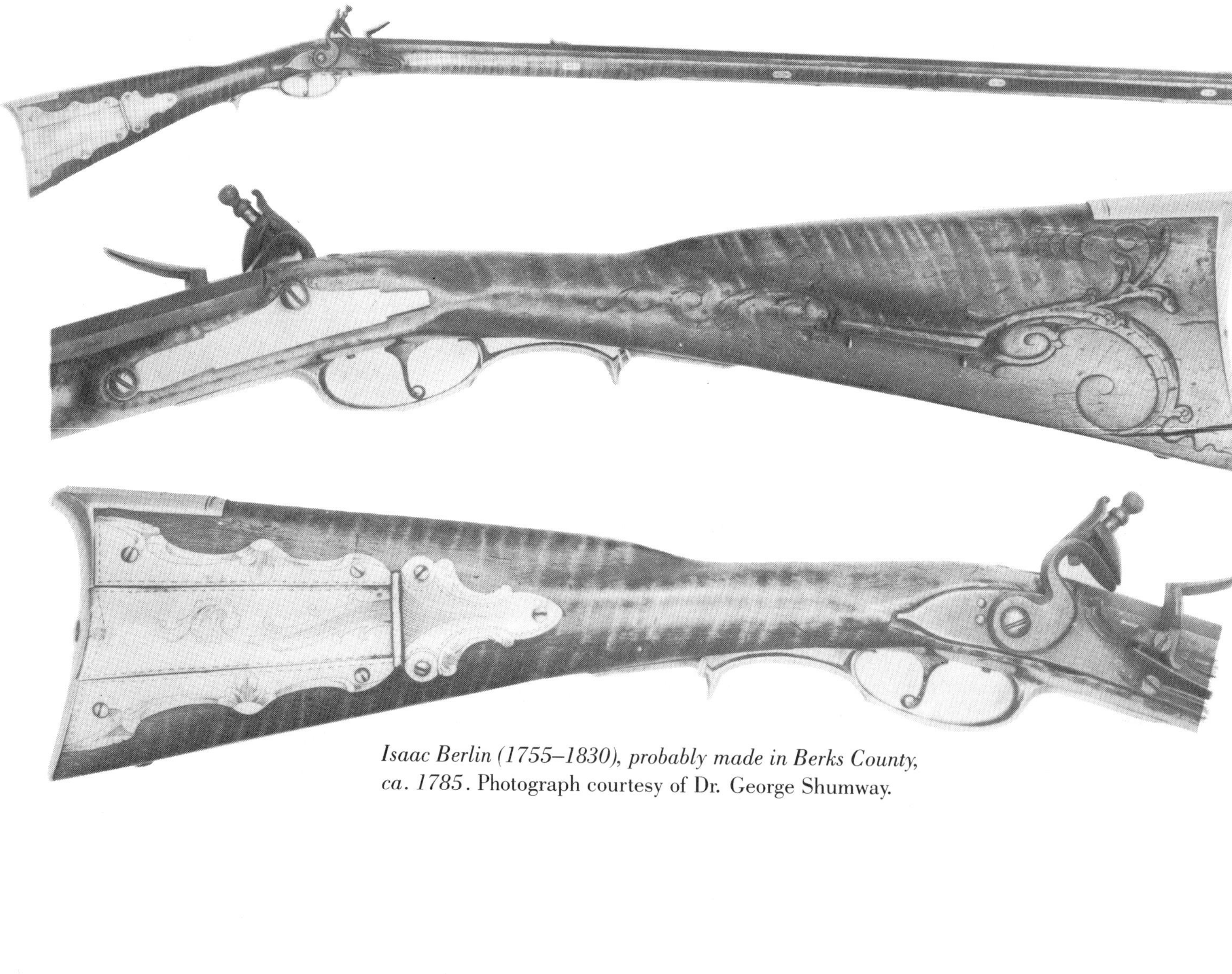

Isaac Berlin (1755–1830), probably made in Berks County, ca. 1785. Photograph courtesy of Dr. George Shumway.

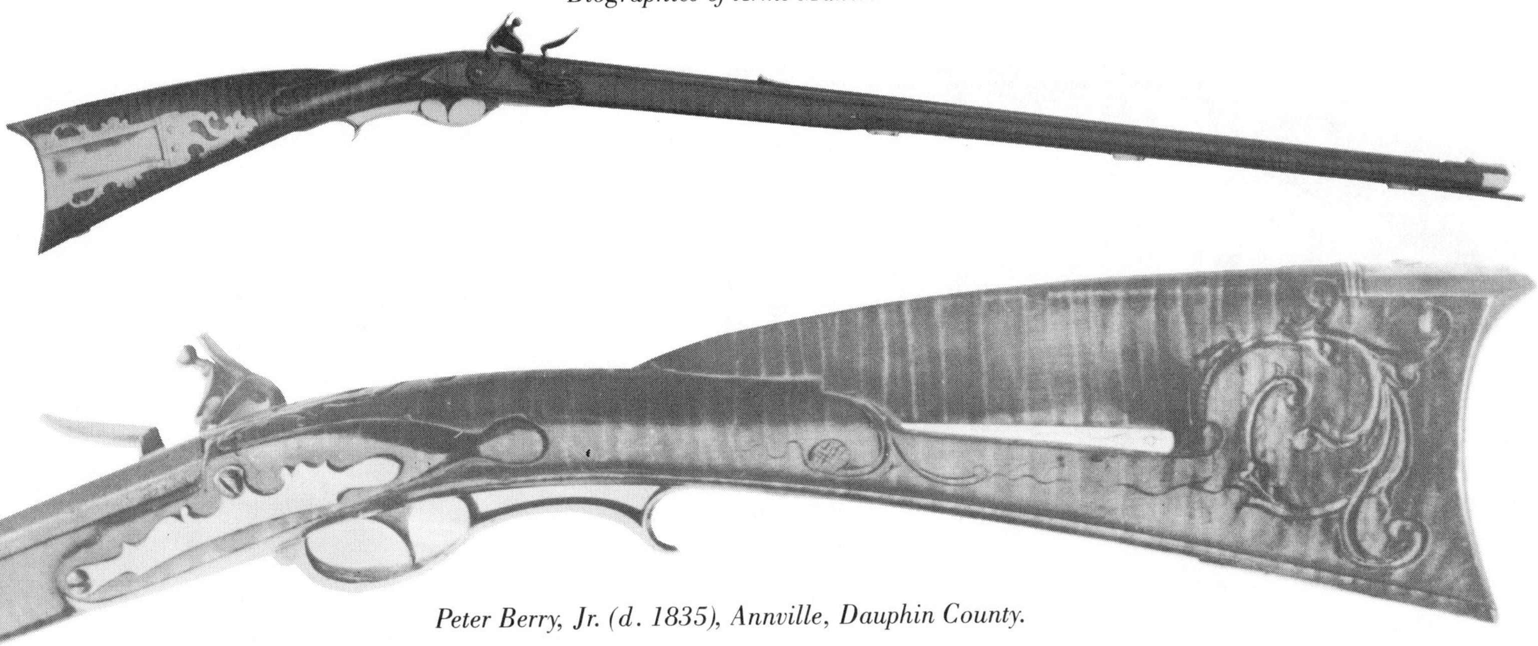

Peter Berry, Jr. (d. 1835), Annville, Dauphin County.

1862; and in Pittsburgh, c. 1865 (Bausman's *History of Beaver County* [1888], 623).

Berrier, Jacob B. (1865–1907). Toboyne Twp., Perry Co. (t; c). Gunsmith.

Berry, Peter, Jr. (d. March 1835). Annville, Dauphin Co. Major gunsmith. His estate inventory showed "all the tools belonging to the gunsmith trade" valued at $25. (t; c; JKJ 266–68) There is a legend that his father, Peter Berry, Sr., was also a gunsmith. There was a Peter Berry who was a smith in Douglas Twp., Philadelphia Co., in 1769 (3 Pa. Arch. 14 at 48), but Peter, Sr.'s estate inventory in Lebanon Co. showed no gunsmithing tools, and the tax lists do not call him a gunsmith.

Bertram, Theodore F. Gunsmith, 1421 Mascher, Philadelphia (GCD, 1875).

Best, John Israel, Gunsmith (1832–1880). 1848–56, Beaver Twp., Clarion Co. (t); 1856–59, Edenburg, Clarion Co. (t).

Bestgen, Christian. Center St. near 14th, Ashland, Schuylkill Co., gunsmith (BRD, 1873, 1880).

Betcher, Gunsmith, N. Black Creek Twp., Luzerne Co. (PD, 1890).

Beyer, D. Gunsmith, Lebanon Twp., Lancaster Co. (t, 1800).

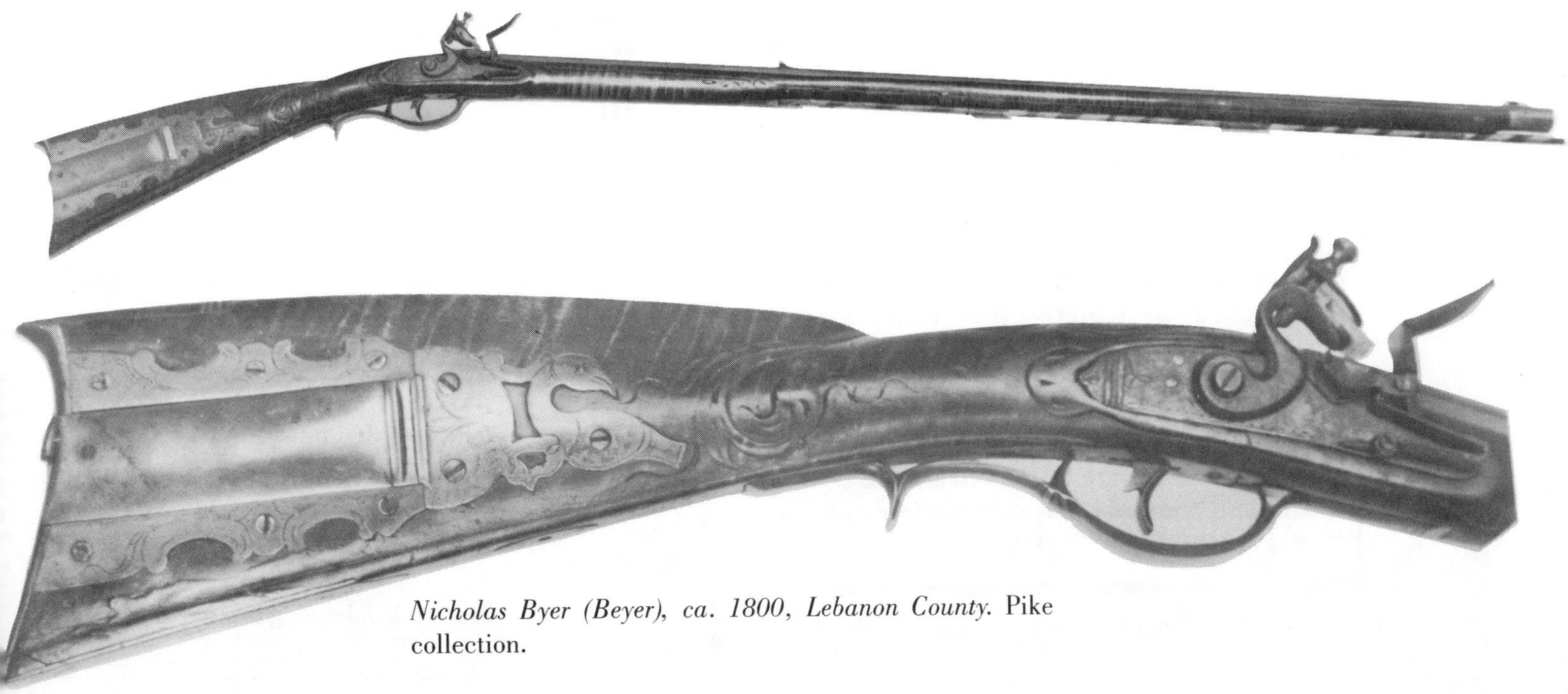

Nicholas Byer (Beyer), ca. 1800, Lebanon County. Pike collection.

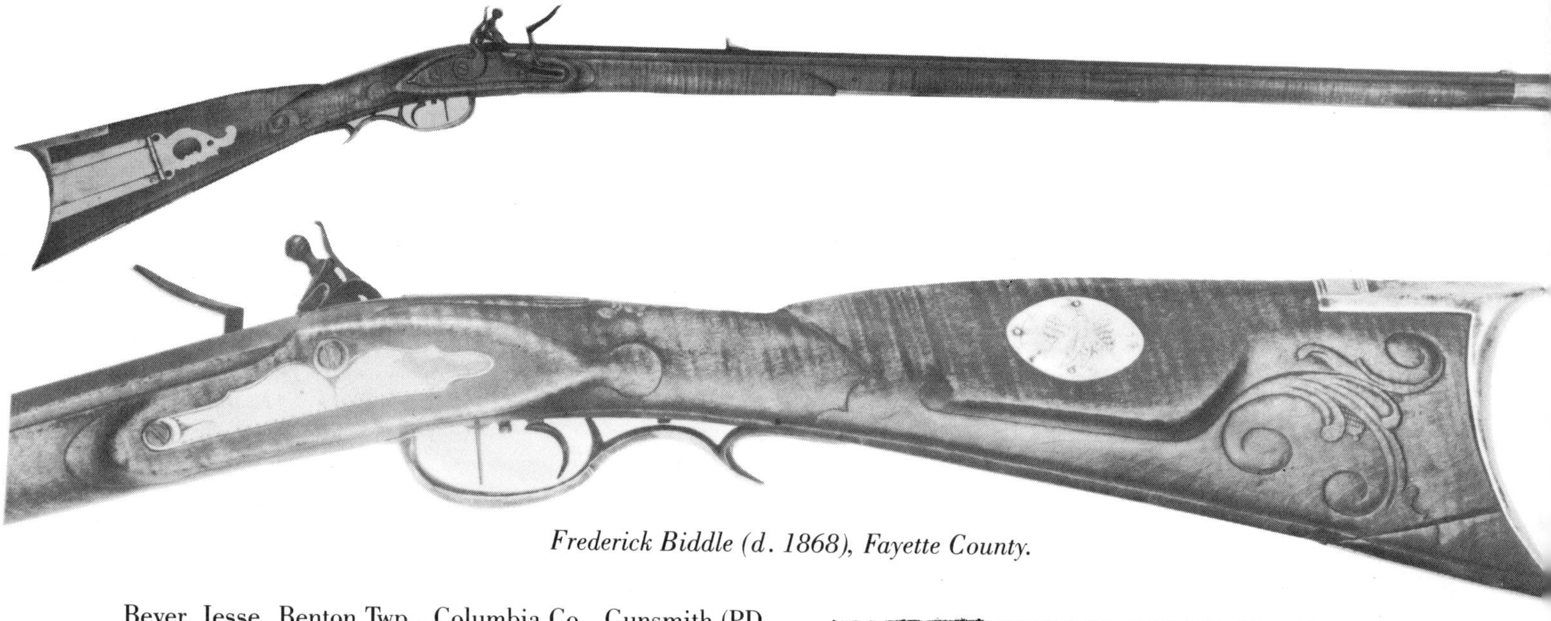

Frederick Biddle (d. 1868), Fayette County.

Beyer, Jesse. Benton Twp., Columbia Co., Gunsmith (PD, 1882, 1890).

Beyer, Nicholas (1780 to after 1850). Probably a late apprentice to John Philip Beck, whose work Beyer's closely resembles. Active 1805–35, Annville, Lebanon Co. In 1850 he was an inmate in the County Home (c). (See GAS-2, 25; JKJ, 232–41.) Major gunsmith.

Bez, Michael. 1758–59, Warwick Twp., Lancaster Co., gunsmith (t; SED, 54).

Bickley, Henry. Artillery artificer, 1/1777, Capt. Will's Co. (BCR, 115).

Biddle, Frederick (d. 1868). Gunsmith 1805–6, Chambersburg, Franklin Co. (t); 1808–15, Bullskin Twp., Fayette Co. (t); after 1816, Sugar Creek Twp., Tuscarawas Co., Ohio (t; c; DH, 380; WSB, 112–13).

Biddle, George (1791–1871). 1808–23, Bullskin Twp., Fayette Co.; 1823–1871, Sugar Creek Twp., Tuscarawas Co, Ohio (DH, 380; WSB, 112–13, c; t). Gunsmith.

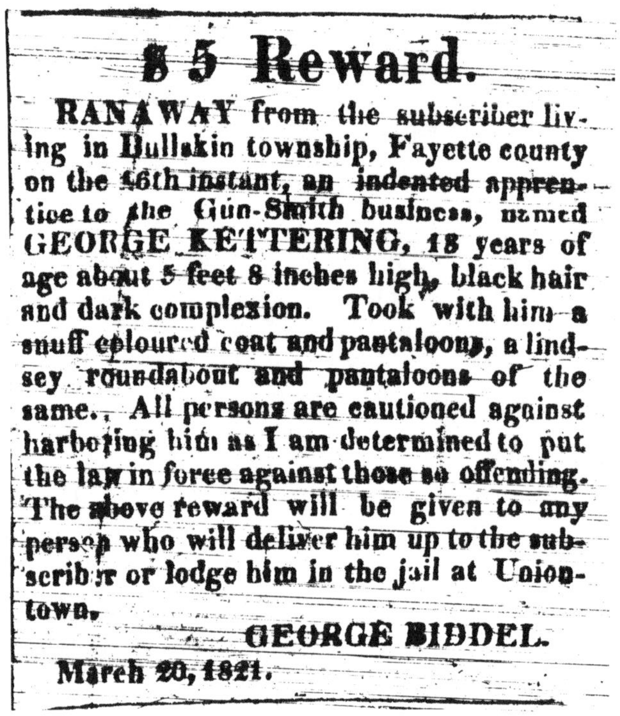

$ 5 Reward.

RANAWAY from the subscriber living in Bullskin township, Fayette county on the 16th instant, an indented apprentice to the Gun-Smith business, named GEORGE KETTERING, 18 years of age about 5 feet 8 inches high, black hair and dark complexion. Took with him a snuff coloured coat and pantaloons, a lindsey roundabout and pantaloons of the same. All persons are cautioned against harboring him as I am determined to put the law in force against those so offending. The above reward will be given to any person who will deliver him up to the subscriber or lodge him in the jail at Uniontown.

GEORGE BIDDEL.

March 20, 1821.

Uniontown Genius of Liberty, March 20, 1821

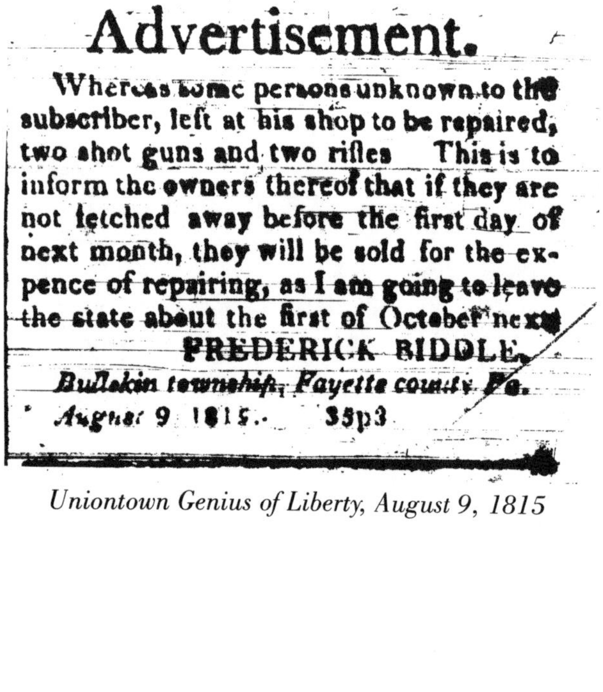

Advertisement.

Whereas some persons unknown to the subscriber, left at his shop to be repaired, two shot guns and two rifles This is to inform the owners thereof that if they are not fetched away before the first day of next month, they will be sold for the expence of repairing, as I am going to leave the state about the first of October next.

FREDERICK BIDDLE.

Bullskin township, Fayette county Pa.

August 9 1815. 35p3

Uniontown Genius of Liberty, August 9, 1815

Biddle, R. & W. C. Lock makers, Philadelphia (MD, 1847, 1855, 1861).

Bideman, Daniel. 79 Wood St., Philadelphia (MD, 1837) Gunsmith.

Bilgharz, Francis W. Gunstock maker, near Apple, below George St., Philadelphia (MD, 1846).

Billhart, Francis. Gunsmith, Kesler's Alley, Philadelphia (MD, 1837).

Binder, G. D. 4th Ward, Philadelphia, gunsmith (c, 1860; RAG, 19).

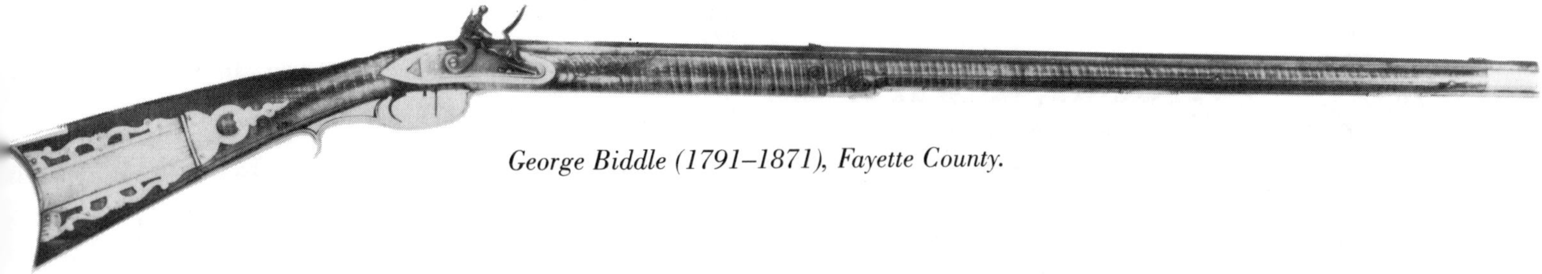

George Biddle (1791–1871), Fayette County.

Binnston (Bennston), William. Gunsmith, 1846–47, Dublin Twp., Bedford Co. (t).

Birchmiller, Robert. Gunsmith 43 N. Queen St., Lancaster (BLD, 1850, 1869). Gunsmith (t; c).

Bird Brothers. Gunlock makers and importers, Philadelphia. DBA: Bird & Ashmore (1800); Charles Bird & Co. (1814); Bird Bros. (1790) (PXD).

Bird, Mark. Arms Contractor to Pa. Committee of Safety, 1776. Birdsboro, Philadelphia Co. His father, William, also was involved (RAG, 19).

Bird, William. 1858–59, Gunsmith. Overton, Bradford Co. (BRD, 1858–59); 1887–90, Forks Twp., Sullivan Co. (PD, 1887, 1890).

Biringer, John. 86 South St., Philadelphia (MD, 1855); 1448 N. 5th St. (BPaD, 1861). Gunsmith.

Birnie, Lawrence. Hummellstown, Dauphin Co. Erected an air furnace, mills, file cutters and related materials for gun making, 1776. (BHAM, 1:572) June 1776 to council: ". . . being a master of the business of file cutting . . . [which is] absolutely necessary for carrying on . . . usefull manufactory of gunlocks . . . [I hope] that your honors will advance the sum of $300 for erecting an air furnace. . . . [In return] your petitioner will engage in 18 months to return the said sum in files manufactured . . . [and I shall work to] oversee the gun lock factory" (1 Pa. Arch. 4 at 777).

Birsel, Samuel. Germantown Rd. near Vager, Philadelphia (MD, 1837). Gunsmith.

Birsel, Simon. Gunsmith, Germantown Rd., Philadelphia (MD, 1837).

Bisbing, Amos S. 1861, Tannersville, Monroe Co. (BPaD, 1861); 1880, White Haven, Luzerne Co. (BRD, 1880; PD, 1870, 1890). Gunsmith.

Bishly, Andrew. 1771, Lebanon Twp., Lancaster Co. (t; SED, 54). Gunsmith.

Bishop, Alexander. Center Twp., Union Co. (c, 1850). Gunsmith.

Bishop, J. Locksmith, Northern Liberties, Philadelphia Co., 1781–82 (t; c, 1790).

Bishop, Moses (1813–90). Linesville, Crawford Co. (t; 1849–90; PD, 1882; CCD, 1874). Gunsmith.

Bisle, Samuel. Gunsmith, 1817–18, Unity Twp., Westmoreland Co. (t).

Bitting, Daniel. 2/23/1777 to 1/5/1780, artillery artificer with Capt. Jesse Rowe's Co. (2 Pa. Arch. 11 at 252; PIR, 2:244).

Bixler, Christian. Silversmith, 1750, Lancaster (t).

Bixler, Christian, Jr. (1732–1811). Silversmith, clock maker, 1760–90, Reading, Berks Co. (t).

Bixler, Christian, III (1763–1840). 1784, Easton. Silversmith (t).

Black, Henry, Jr. Gunsmith, 1807–31, German Twp., Fayette Co. (t).

Black, James. 1/27/1773: "James Black, with the consent of his father, Daniel, indentures himself apprentice to Michael Brother, silversmith, Philadelphia, for 13 years, 8 months and 21 days" (PaE, 213; PIDS).

Black, James (b. 1813). Duncannon, Perry Co., gunsmith (c, 1850; BPaD, 1861).

Black, Michael. Gunsmith, 1799–1800, Union Twp., Huntingdon Co. (t).

Blackman, Anson. Elkland, Tioga Co. (c, 1850); gunsmith, 1870–74, Woodhill, N.Y. (t; RAG, 20).

Blackwood, Marmaduke. 12/5/1775, signed a contract with the City of Philadelphia to supply 200 gunlocks at 22 shillings, 6 pence, each; with Samuel Wigfal (Col. Rec., 10:420).

Bladenberger, Frederick. Regimental Blacksmith, 9/9/1788, to 10th Pa. Regt. (2 Pa. Arch. 15 at 501).

Blain (Blaine), William (d. 1852). Gunsmith 1823, Brothers Valley Twp., Somerset Co. (t); 1827–52, Donegal Twp., Westmoreland Co. (t). On 8/24/1832, Blain advertised 6¢ reward for the return of his runaway apprentice, Daniel Adams, in the Greensburg *Farmers Repository.* Blain's appraisement showed: 1 rifle gun $1.50; 1 shotgun 50¢; 1 wolftrap $1.50; 1 sword & scabbard 50¢; 1 set blacksmith's tools $22; 1 set gunsmith's tools $8.

Blair, Andrew. Gunsmith 1799, East Bethlehem Twp.,

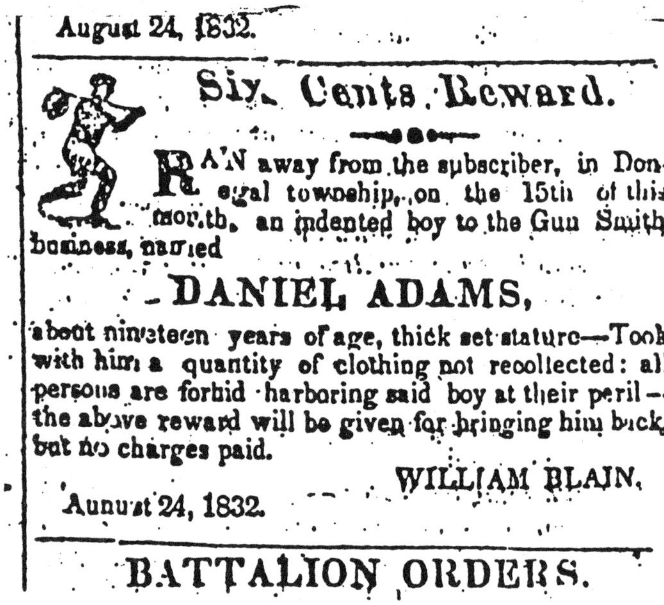
August 24, 1832.

Six Cents Reward.

RAN away from the subscriber, in Donegal township, on the 15th of this month, an indented boy to the Gun Smith business, named

DANIEL ADAMS,

about nineteen years of age, thick set stature—Took with him a quantity of clothing not recollected: all persons are forbid harboring said boy at their peril—the above reward will be given for bringing him back, but no charges paid.

WILLIAM BLAIN.

August 24, 1832.

BATTALION ORDERS.

Westmoreland Farmer's Repository, August 24, 1832

Washington Co. (t); 1802–10, St. Clairsville, Belmont Co., Ohio (t); 1812–17, Mansfield, Richland Co., Ohio (t).

Blair, David. 1785–1800, Fredericktown, East Bethlehem Twp., Washington Co. "David Blair, a gunsmith, had a shop near a mill [in Fredericktown]. To supply his forge, he brought coal in sacks on horseback from Fishpot Run, never imagining that the hills adjacent to the town were full of it." (Boyd Crumrine, *History of Washington County* [Philadelphia: Everts, 1882] 769).

Blakley, Daniel. Artillery artificer, 2/8/1777 to 2/8/1780, Capt. Jesse Rowe's Co. at the laboratory at Philadelphia (PIR, 2:244).

Blanchard, William. Cutler, 1824–36, Fallston, Beaver Co. (Bausman's *History of Beaver County*, 473).

Blautz, John. Gunsmith, Brickersville, Lancaster Co. (BPaD, 1861).

Block, Hans. On 7/19/1661 he bought the tools of John Barensten's estate at the Dutch settlement on the Delaware for 980 guilders. He paid only 300 guilders, and Meyntie Willems, the widow of Block, requested that council order the remainder to be paid. (2 Pa. Arch. 5 at 429–30) On 6/28/1661 council paid Hans Block, "gunner," 837/7/0 guilders for his services (2 Pa. Arch. 5 at 443).

Blum, Ephraim. Lower Smith Twp., Northampton Co. (t, 1776–83). Gunsmith.

Blymire, George. Gun-barrel maker, along Bennett's Run, Newberry Twp., York Co. (KEA, 10; HYC, 624, 632).

Bobb, Anthony. 1779–81, Reading, Berks Co., gunsmith (t; KEA, 10).

Bobb, Conrad. Tinman, 1780, Reading, Berks Co. (t).

Bobb, Jno. Blacksmith, 1781, Alsace Twp., Berks Co. (t).

Bobb, Matthew. Smith, 1779–84, Reading, Berks Co. (t).

Bodin, Joseph. Gunsmith, Philadelphia (MD, 1852).

Bodine, Samuel. Gunsmith, 1844–50, Philadelphia (c, 1850; MD, 1845); 1850, Camden, N.J. (c, 1850).

Boehm, Joseph. Artillery artificer, 1779, at the laboratory, Philadelphia (2 Pa. Arch. 15 at 386).

Boggs, John. Artillery artificer, 4/25/1778, Capt. Nathaniel Irish's Co. (PIR, 2:247).

Boland, Daniel. Armorer, Pa. Navy ca. 1779 (5 Pa. Arch. 4 at 225).

Bollard, Daniel. Armorer, 9/21/1778, Pa. Navy, armed boat *Hancock* (2 Pa. Arch. 1 at 309).

Bollinger, Peter. 1776, Lancaster boro., gunsmith (t; SED, 54).

Bomper, Abraham (1705–1793). Silversmith and clock maker. Born in Herborn, Germany (John G. E. Heckewelder, *Narratives* [Philadelphia: McCarty & Davis, 1820] 399); part of Moravian community at Bethlehem.

Bonewitz, John. Active 1779–1809, Womelsdorf, Berks Co. The occupation of gunsmith is given only in 1792 and 1800 tax lists. One gun is marked Womelsdorftown (GAS-1, 21; JKJ, 207–08). Major gunsmith.

Bonham, John L. (1821–1903). Born in Tioga Co., N.Y. 1862–96, Kersey, Elk Co. (t); 1896–1903, Olean, N.Y. (t). Gunsmith.

Bonner, Barnebas. Artillery artificer, enlisted on 5/3/1778 in Capt. Thomas Wiley's Co. at Carlisle (PIR, 2:248).

Bonnet, Henry A. (1835–1907). Born in Germany, Bonnet came to America in 1852. 1855–57, apprenticed to Nicholas Shennefelt in Clarion boro.; 1859–60, Madison Twp., Clarion Co. (t); 1861–64, Paradise Settlement, Henderson Twp., Jefferson Co. (t); 1864–65, Brookville, Jefferson Co. (t); 1868–78, Clarion boro. (t); 1878–82, Clarion Twp., Clarion Co.; 1882–1907, Brookville boro. Gunsmith, teacher, bicycle repairman (REH, 63–97).

Bonshoff, George W. 344 S. Queen St., Lancaster (BAR, 1875–76). Gunsmith.

Boone, E. Second cousin of Daniel Boone. 1815–18, Oley Valley, Berks Co. (t; RAG, 22). Gunsmith.

Boone, Moses. Artillery artificer, 7/29/1779, at the cannon factory at Carlisle (2 Pa. Arch. 11 at 260); Blacksmith, 11/1/1782, in Capt. John Jordan's Co., assigned to Southern Army (2 Pa. Arch. 11 at 262).

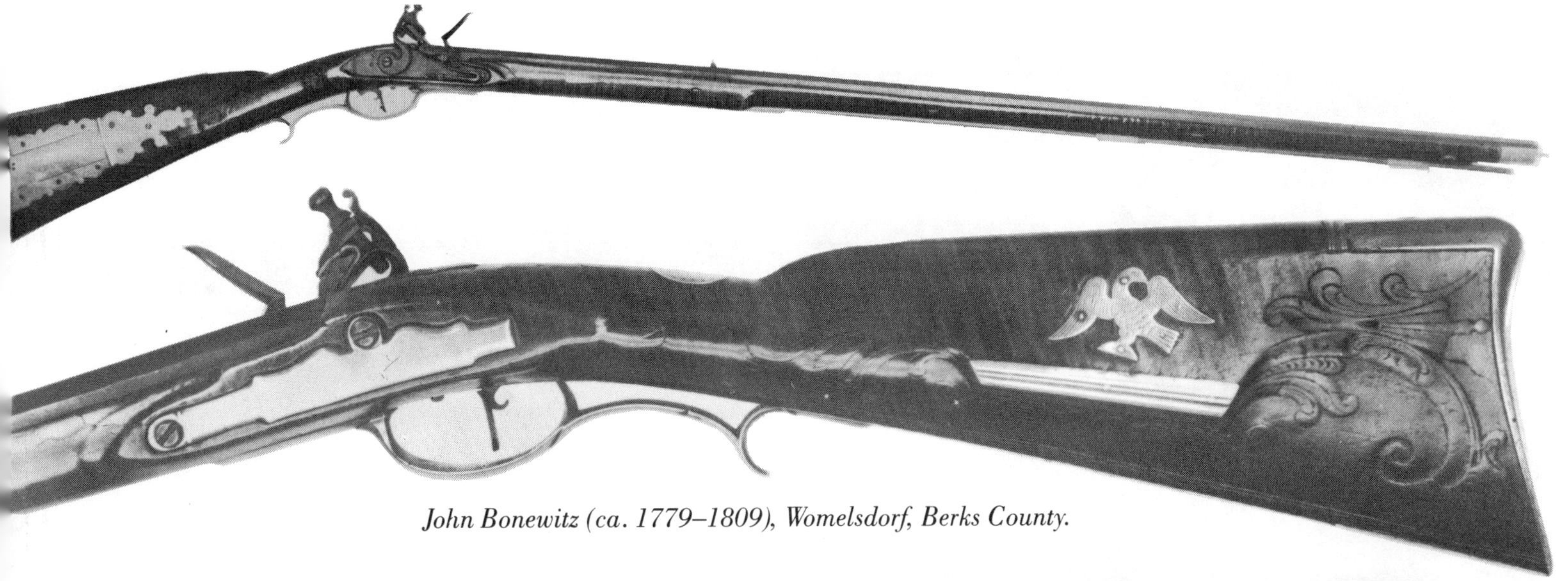

John Bonewitz (ca. 1779–1809), Womelsdorf, Berks County.

John L. Bonham (1821–1903), Elk County.

Henry A. Bonnett (1835–1907) in his gunshop on Pickering St., Brookville, Jefferson County.

Daniel B. Border (1826–91), Bedford County. Note the distinctive hand-forged lock with the Bedford County rat-tail.

William Border (1800–1881), Bedford County.

Boone, Richard, Gunsmith, 1797–1820, 88 S. 2d St., Philadelphia. Made cased pistols (RAG, 22).

Boone, Samuel. 1768–76, Frederickstown, Berks Co.; 1777–78, managed the Maryland state gun repair shop at Frederick (RAG, 22).

Boone, William. Successor to Richard Constable, gunsmith, 88 S. 2d St., Philadelphia (RAG, 22).

Booth & Co. Gunlock makers, 1700–1814, Philadelphia (KEA, 10).

Booth, Richard, Gunsmith, 88 S. 2d St., Philadelphia (PXD, 1813, 1816).

Booth, William. 2d St. and 10th St., Philadelphia, gunsmith (PXD, 1806, 1808, 1810, 1813, 1816).

Border, Daniel B. (1826–91). 1849–54, Bedford Twp., Bedford Co. (t); 1855–91, Bedford boro. Gunsmith and silversmith. Son of William Border. One of the finest Bedford Co. Gunsmiths (t; c; f).

Border, Enos (b. 1820) Gunsmith, 1842–43, Bedford boro. (t); 1849, Napier Twp., Bedford Co. (t).

Border John (1832–64). Son of William Border. Gunsmith, Bedford boro., 1856–64 (t).

Border, Samuel (b. 1815) Gunsmith. Brother of William Border. 1840–42, Bedford boro.; 1843–56, Conemaugh Twp., Somerset Co. (t).

Border, William (1800–1881). 1826–72, Napier Twp.,

Bedford Co. (t); 1873–81, East St., Clair Twp., Bedford Co. (t). Major gunsmith.

Borsser, William. Shrewsburg, York Co. (BRD, 1880). Gunsmith.

Bortner, Philip. Gunsmith, 1814, Ayr Twp., Bedford Co. (t).

Bortree, William. Rear of 433 N. 3d St., Philadelphia (PXD, 1819, 1827, 1833). Gunsmith.

Bosler, & Co. 1817, advertised gun mountings and furnishings in the Pittsburgh *Gazette*.

Bosler (Bolser), Joseph. New 4th St., Philadelphia, gunsmith, 1799–1800 (KEA, 11; RAG, 22).

Bossert, Jacob. Gunsmith, 14th Ward, Philadelphia (c, 1860).

Boston, George. Muddy Creek Twp., Butler Co., gunsmith (PD, 1882, 1887).

Bouchette, John. Gunsmith, Cobb's Ct., Philadelphia (PXD, 1819).

Boughmer, George. Walnut Bottom, Cumberland Co. (BPaD, 1880). Gunsmith.

Bournes (Burns), John. Cutler and cannon maker, 1774–98, Franklin Co. (WSB, 29).

Bovy, Nathaniel, Gunsmith. 1817, Carlisle, Cumberland Co. (t).

Bowen, R. Cutler, 1826, Penn St. and Cecil's Alley, Pittsburgh (t).

Bower, George. Gunsmith. 1 Derringer's Ct., Philadelphia (MD, 1846).

Bowers, John. Gunsmith. Jonestown, Lebanon Co. (BRD, 1860; BPaD, 1861).

Bowers, William. On 3/23/1773 William Bowers was apprenticed to learn the blacksmithing trade from Daniel Evans in Philadelphia. Evans was an established gunmaker (PIDS; PaE, 214). "William H. Bowers, lockmaker of Philadelphia" (FSP, 35).

Bowman, Jeremiah (b. 1848). Gunsmith, 1866–88, Schellsburg, Bedford Co. (t).

Bowman, Peter. Worchester, Montgomery Co. (BPaD, 1861). Gunsmith.

Bown, Albert. 1868, Salem, Ohio (t); 1869–82, Titusville, Crawford Co. (t; PD, 1882). Gunsmith.

Bown & Tetley. Major manufactory, 136 Wood St., Pittsburgh. Established in 1848, it was the "only one in the west[ern part of Pa.] which constructs rifles and guns exclusively for the trade. It has been in operation in the present scale about 4 years, and in that time . . . has attained a reputation throughout the whole western country that is rapidly drawing off trade from eastern houses. . . . Their reputation . . . is conclusive as to their value. . . . This establishment employs 25 hands, whose yearly wages amount to $15,600; and turns out now at the rate of 4000 rifles a year, with a steady increasing demand. . . . This is the first attempt to establish on a large scale the manufacture of this species of arms here, or in the west[ern part of the Commonwealth]" (George H. Thurston, *Pittsburgh As It Is, or, Facts and Figures Exhibiting the Past and Present of Pittsburgh* [Pittsburgh: Haven, 1857]; c, t, PAD, 1857, 1860–61).

JAMES BOWN,
[Successor to Bown & Tetley,]
Manufacturer and Importer of
Guns, Rifles, Pistols,
HARDWARE,
Gunsmiths' Materials, Cutlery, &c.
No. 136 Wood Street, Pittsburgh, Pa.
Agent for Colt's Pistols and Rifles.
May 21-63.1y

Kittanning Mentor, May 4, 1863

Bown, James. Prop., Enterprise Gun Co., 1871–86, 136 Wood St., Pittsburgh. The prices in 1882 ranged from $15 for a "kill buck" model to $32 for the fancy grade, with higher grades made to order (PD, 1882).

Bown, William. Son of James, successor at the Enterprise Gun Works. 1880–90, Salem, Ohio (DH, 77; c).

Bowne, Joseph. Artillery artificer. 1779, at the laboratory at Philadelphia (2 Pa. Arch. 15 at 385).

Bowshinger (Boushinger), Henry. Gunsmith, 1802–23, Berlin, Somerset Co. (t).

Boyd, James and John. 5/20/1758, smiths in the employ of the quartermaster general at Ft. Ligonier (5 Pa. Arch. 1 at 350).

Boyd, Joseph. 1780, Columbia, Lancaster Co. (t); 4/24/1777, enlisted in 11th Pa. Regt. as a blacksmith; 5′ 6″ tall, born in Pa. (2 Pa. Arch. 11 at 60). Gunsmith.

Boyer, Benjamin. Gunsmith, Hamburg, Berks Co. (c, 1850).

Boyer, Catherine. Gunsmith. Daughter of Henry Boyer, gunsmith of Aetna who began operations ca. 1803. She

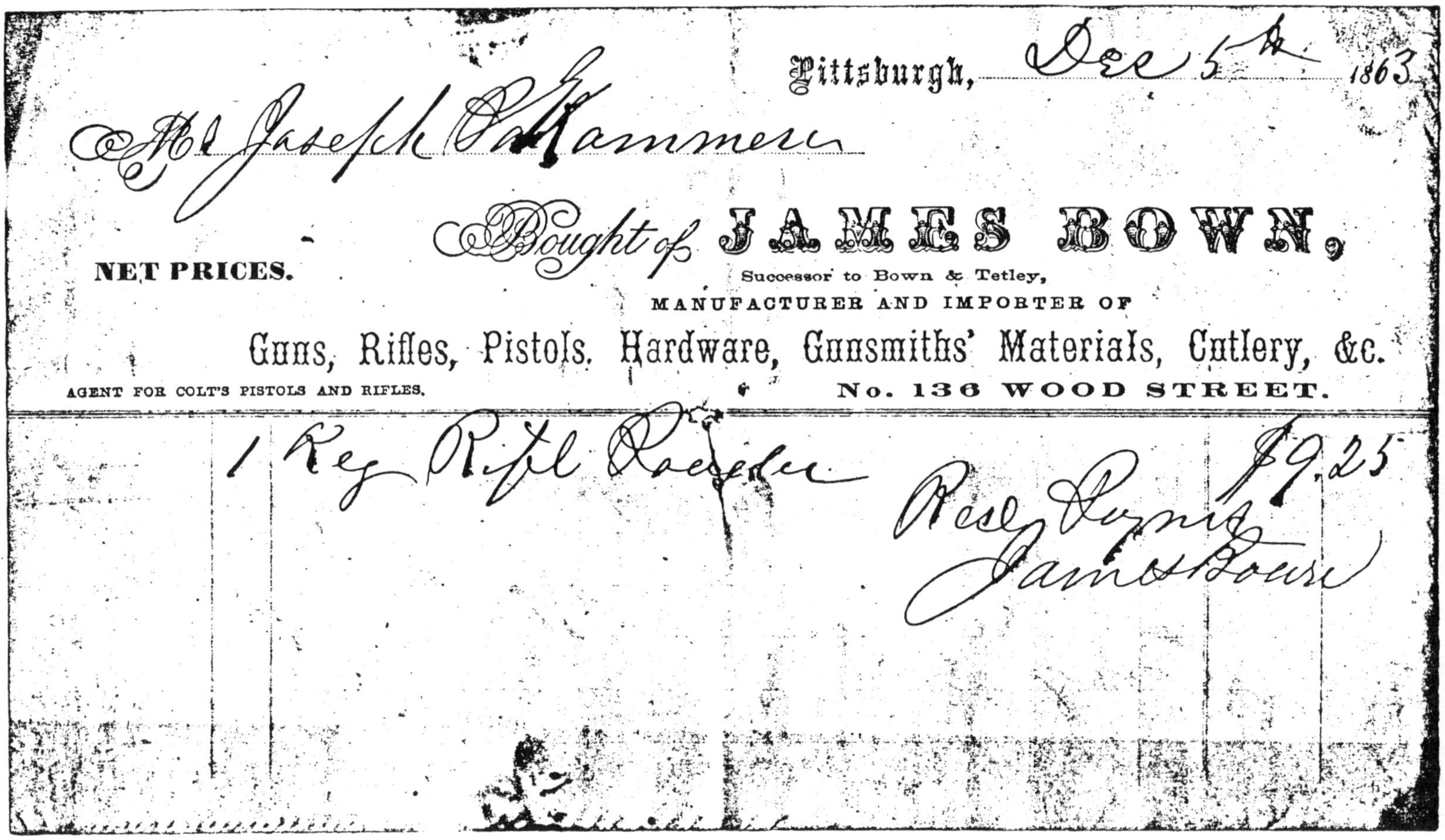

Pittsburgh, Dec 5th 1863.

Bought of JAMES BOWN,

Successor to Bown & Tetley,

MANUFACTURER AND IMPORTER OF

Guns, Rifles, Pistols, Hardware, Gunsmiths' Materials, Cutlery, &c.

NET PRICES.

AGENT FOR COLT'S PISTOLS AND RIFLES.

No. 136 WOOD STREET.

1 Reg Rifl $9.25

Recd Payment James Bown

Receipt for a rifle sold by James Bown

Wanted immediately,

At the ſubſcribers manufactory in Lancaſter,

A number of

Muſket-Barrel & Lock-Makers.

Thoſe that can be well recommended, will meet with generous wages and conſtant employ by

SAMUEL BOYD & THOMAS TURNER.

Sept. 4, 1794.

A few apprentices from 12 to [illegible], will be wanted at the ſaid manufactory.

Harrisburg Oracle of Dauphin, September 8, 1794

was born 8/13/1813, the only surviving child. They operated a gun-barrel and rifle mill, ca. 1840 (f).

Boyer, Daniel. Gunsmith. Before 1817, Orwigsburg, Schuylkill Co.; after 1818, Harrisville, Medina Co., Ohio (DH, 259).

Boyer, David. 1832–35, Mahoning Twp., Carbon Co.; 1836–40, Orwigsburg, Schuylkill Co. (t). Native of Berks Co. His shop in Carbon Co. was on the site of St. John's Catholic Church in Mahoning Twp. Gunsmith. (t; Brenckman, *History of Carbon County;* 747)

Boyer, David, Jr. Gunsmith, Orwigsburg, Schuylkill Co. (BPaD, 1880).

Boyer, George (b. 1840). Orwigsburg, Schuylkill Co. (c, 1860, 1870, 1880) Gunsmith. Son of David Boyer, Sr.

Boyer, Henry. See Catherine Boyer.

Boyer, Henry. 1790–1805, Gunsmith Orwigsburg, Schuylkill Co. (t); 1807–45, Lebanon Co. (t; c).

Boyer, Martine. 1767–68, Reading (t); 1779–80, Long Swamp Twp., Berks Co. (t; 3 Pa. Arch. 18 at 9, 90, 138, 219, 242, 353); 1781–85, Exeter Twp., Berks Co. (t; 3 Pa. Arch. 18 at 478, 605, 741).

Bradford, Cornelius, Pewterer, 2d St., Philadelphia (Advertisements in *Pennsylvania Packet*, 5/3/1753, 10/4/1753).

Bradley, C. H. West Chester, Chester Co. Patent 34,235 of 1/28/1862.

Bradley, Freeman, Jr. Washington boro. Patent 34,126 of 1/14/1862, with John C. Noble, for a magazine gun.

Bradley, John. Artillery artificer from Chester Co. Born in England. 5/7/1778, enlisted in the New Pa. 11th Regt. (2 Pa. Arch. 11 at 62).

Bradley, John (1797–1872). Before 1833, Chester Co., gunsmith. His wife died on 12/1/1833 and he left for Trumbull Co., Ohio. In 1850 he left for Moultrie Co., Ill. (t; c, 1870; *Portrait and Biographical Record of Shelby and Moultrie Counties* [Chicago: Beers & Warner, 1891] 510–11).

Bradley, Simon (b. 1848). Gunsmith, Liverpool, Perry Co., (c, 1870).

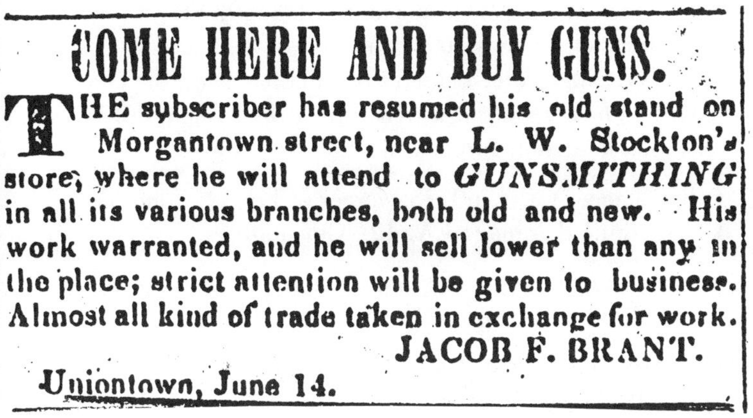
COME HERE AND BUY GUNS.

THE subscriber has resumed his old stand on Morgantown street, near L. W. Stockton's store, where he will attend to *GUNSMITHING* in all its various branches, both old and new. His work warranted, and he will sell lower than any in the place; strict attention will be given to business. Almost all kind of trade taken in exchange for work.

JACOB F. BRANT.

Uniontown, June 14.

Uniontown Pennsylvania Democrat, June 14, 1842

Brand, Peter. Granted a Pa. Militia arms contract by Gov. McKean on 4/17/1801 (8 Pa. Arch. 7 at 301).

Brandt, John. 1834–37, Lancaster boro., gunsmith (t; SED, 54).

Brant, Jacob F. (1811–72). 1835–51, Uniontown, Fayette Co. (t; c); 1854–72, 4 Fifth St., (t; c, 1860; PAD, 1860–61, 1864–65; BPaD, 1861). Gunsmith.

Brashear, Robert Ayres (b. 1811). 1834–60, Franklin, Venango Co. (t) Nephew and apprentice of William G. Ayres (t; c; f).

Breathed, John. Gunsmith, 1823–32, Bethel Twp., Bedford Co. (t).

Brecht, Peter. Accoutrements manufacturer. On 2/22/1776 he filled an order for 150 cartouche boxes for £90; on 8 May 1778, he filled another order for 150 cartouche boxes for £108/15/0 (BCR, 175).

Breidenhart, Christopher (1725–87). Arms maker to Committee of Safety. (4 Amer. Arch. 1 at 198) Apprenticed to Peter Tanner, court gunsmith, Gotha, Germany. Sailed on 9/24/1753 on ship *Neptune* for America, arriving in Lancaster early in 1754. In 1773 he was listed as a tavern keeper. On 5/30/1785 he made his will, and on his death his gunsmithing tools were valued at £63/10/7 (*German Folklore Society Pennslyvania* 12, no. 202 [1947]).

Breigle, Jacob. Gunsmith. 10/15/1828, arrived in New York from Tzarelbergdorf, Bavaria. Uncle of Jacob Breigle, II; brother of Christian Breigle. 1835, single freeman, gunsmith, Union Twp., Bedford Co.; 1849, rifle maker, Middle Woodberry Twp., Bedford Co. (f; t).

Breigle, Jacob, II (1831–95). Born in Bavaria, 9/16/1831, and emigrated to America in 1832 with his father Christian. 1856–62, Queen, Bedford Co. (f; t); 1862–64, served in Army of the Potomac, discharged in 12/1864 suffering from "lung fever"; 1868, near Dubuque, Iowa; 1870–95, Maple Grove, Jasper Co., Missouri, (t; c; f). Apprenticed to Thomas Oldham to learn gunsmithing (f, c. 1880).

Breihen, Christopher. 1759–70, Lancaster boro., gunsmith (t; SED, 54).

Brenberger, Frederick. 1814–16, Martic Twp., Lancaster Co. (t; SED, 54). Gun-barrel maker.

Brenise (Breneissen), George. 1800–1825, York boro., major gunsmith (t, 1800–19; KKR, 19; York *Reporter*, 1/19/1819).

Brenneman, Jacob. Gun-barrel forger. 1828, Earl Twp., Lancaster Co. (t; SED, 54).

Brettell, Francis S. Joint patent holder, with Joseph Frisbie, of patent 16,575 of 2/10/1857 for a gun; Allegheny City, Allegheny Co.

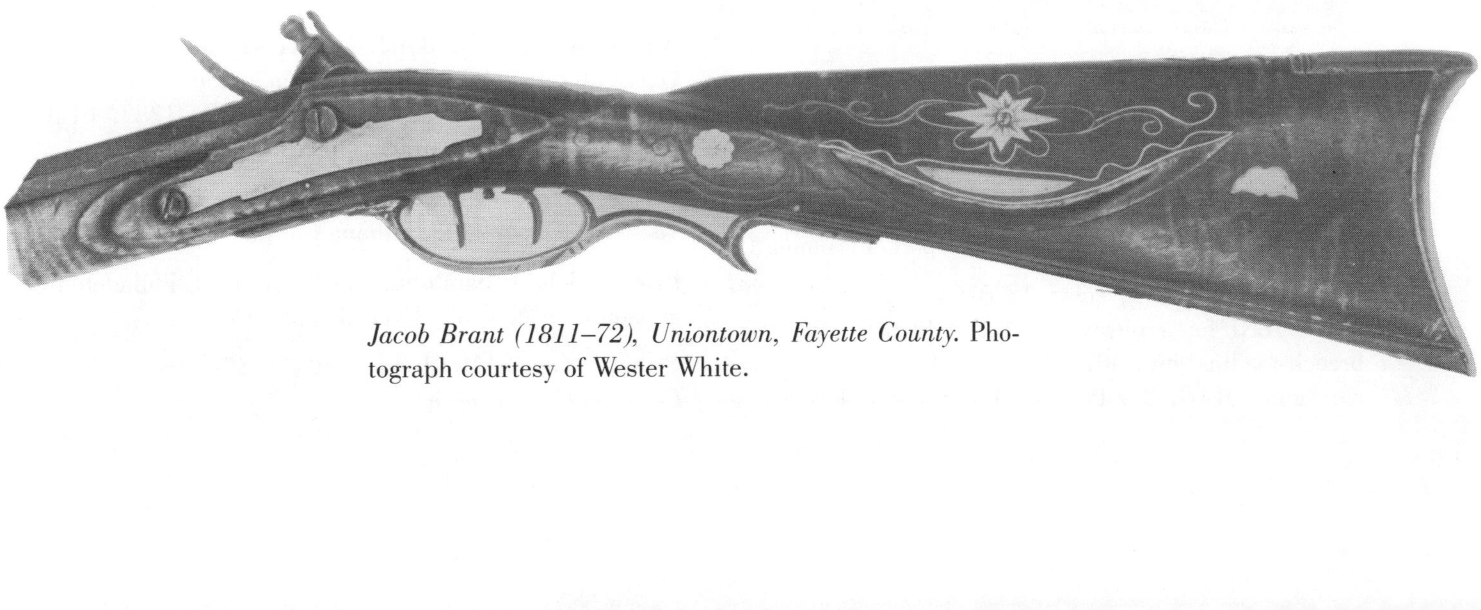

Jacob Brant (1811–72), Uniontown, Fayette County. Photograph courtesy of Wester White.

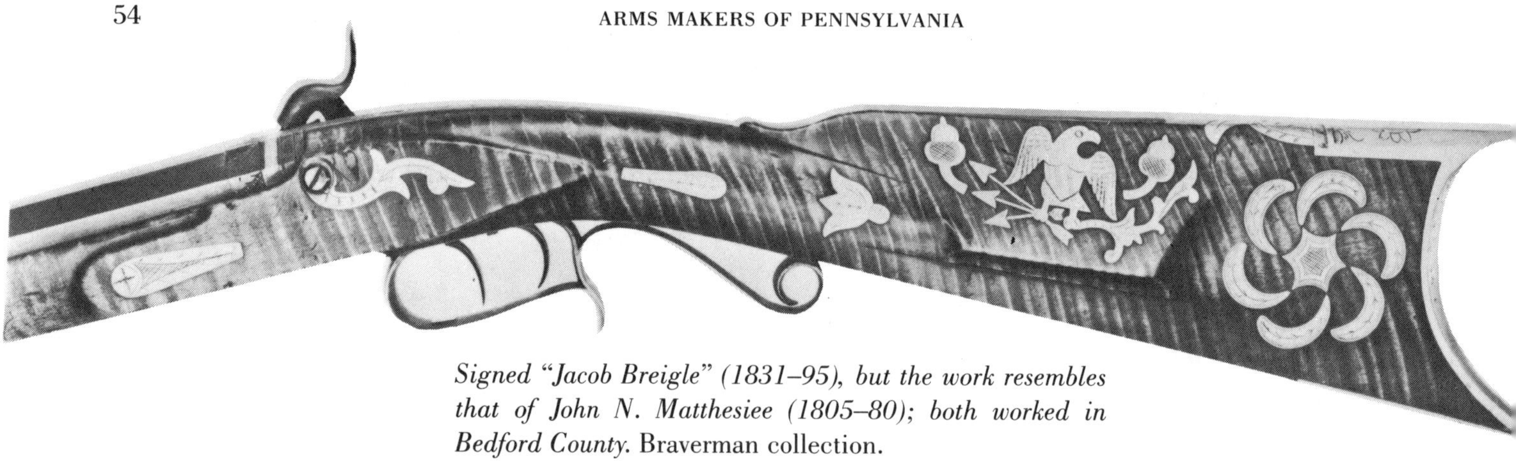

Signed "Jacob Breigle" (1831–95), but the work resembles that of John N. Matthesiee (1805–80); both worked in Bedford County. Braverman collection.

Jacob Breigle (1831–95), Bedford County.

Brewer, Nelson W. Inventor and gunsmith. Held patent 28,646 of 6/12/1860 for a self-loading gun. Active through 1882 (PD, 1882; RAG, 24). Williamsport, Lycoming Co.

Brewer, Roland L. Inventor. 1878–84, patents: 239,914, of 4/5/1881 for a revolver; 279,324 of 6/12/1883 for a breech-loading gun; 307,626 of 10/28/1884 for a magazine gun. (RAG, 24) Pittston, Luzerne Co. His patents were sold to Colt and to James Paris Lee.

Brey, Elias (1817–91). Gunsmith 1838–50, East Greenville, Bucks Co.; 1850–69, Pennsburgh, Montgomery Co.; 1869–91, Krausdale, Lehigh Co. (t; "Elias Brey, the Gunmaker," *The Perkiomen Region* [July 1929]) Patent 20,757 for a double-action gunlock.

Brice, James. Ca. 1760–75, corner of Front St. and Market St., Philadelphia (KEA, 12). Gunsmith.

Bricker, Adam. Low near Lancaster Rd., Lancaster boro (GLD, 1863–64). Gunsmith.

Brickley, D. J. Pine Twp., Indiana Co., gunsmith (PD, 1887, 1890).

Brideman, David. Gunsmith, 79 Wood St., Philadelphia (MD, 1835).

Bridesburg Machine Works. Philadelphia Co. Owned and operated by Alfred Jenks and his son Barton. Principally, the factory made machines to process cotton and wool, but in 1863 it made Springfield rifled muskets on government contract (RAG, 24).

Briggs, Ethan. 19 Green St., Philadelphia, gunsmith (MD, 1837, 1846).

Briggs, Henry. Patent for a breech-loading gun, dated 10/6/1885. Philadelphia.

Briggs, T. L. Gunsmith, Chester, Delaware Co. (PD, 1890).

Briggs, William. 1848–82, Norristown, Montgomery Co. Patent 88,605 of 8/30/1859 for an underhammer gun (J. A. Carderelli in GR [September 1970]; PD, 1882; BPaD, 1861, 1873, 1880). Gunsmith and inventer.

Brilhart, Jacob (1802–67). Gunsmith and carpenter, 1832–67, Georgeville, Indiana Co. (t).

Brindle, John F. Sarah St. near Bedford St., Philadelphia, gunsmith (MD, 1837, 1841, 1846).

Brinkley, George McClellan (1865–1937). Sigel, Jefferson Co. (t; c; f). Gunsmith.

Bristow, John (d. 1694). Cutler. Settled in Chester Co. as early as 1686, coming from Bristol, England. He was a member of council and ran a malt house (George Smith, *History of Delaware County* [1862], 450–51).

Brittain, Lyonel. Blacksmith, 1680, Alny, Bucks Co. (PaE, 24).

Britz, Adam. Gunsmith and locksmith, 1807–10, Heidelberg Twp., York Co. (KKR, 191). He is possibly the same man as Adam Britzius, Lancaster, 1773–82 (3 Pa. Arch. 17 at 3, 455, 605, 754).

Brock, Jacob. Gunsmith, Lafayette St. near Strawberry St., Lancaster boro (BLD, 1857).

Brock, Stephen. 1795–99, Cumberland Twp., Greene Co. (t). Gunsmith.

Brockway, Chauncy, Jr. (1823–1910). Horton Twp., Elk Co. Gunsmith. Civil War veteran, 172d Regt., Pa. Volunteer Infantry. (REH, 219–24)

Broderick, John. Gunsmith, N. Queen St. near W. Lemon St., Lancaster boro (BLD, 1857).

Brong, Peter (d. 1817). Major gunsmith. 1788–1817, N. Queen St., Lancaster boro. About 1800, employed by Jacob Haeffer as a locksmith. On 7/13/1801, contractor on Pa. Militia arms with Henry DeHuff and Abraham Henry (JKJ, 110–11; SED, 54).

Bronoup, James. 1779, Northern Liberties, east part, Philadelphia Co. (t; 3 Pa. Arch. 15 at 101; 220). Gunsmith.

Brooke, I and N. 1808–12, contracted to manufacture 4,000 U.S. M-1808 muskets; delivered 1,257 (GAAAM, 26).

Brooks, Enoch. Inventor. Patent 20,776 of 7/6/1858 for a breech-loading rifle, with G. Walker. Philadelphia.

Brooks, Francis. Gunsmith, Arch St. at 9th St., middle district, Philadelphia. (c, 1790); 87 Front St. (PXD, 1791); 11 Mulberry St. (PXD, 1792; KEA, 14).

Brooks, John. 1799, Philadelphia (PXD, 1799); 1800–1816, Lancaster (SED, 54; KEA, 14); Harrisburg (t, 1795, 1796, 1807, 1814; Rupp, *Lancaster County,* 241; JKJ, 138–39). Major gunsmith. In 1787, John Brooks subscribed 10 shillings to the Harrisburg Reformed Church building fund (Rupp, *Lancaster County,* 278).

Brooks, John. Patent 190,543 of 5/8/1877 for a pistol, with Thomas Bearcock. Pittston, Luzerne Co.

Broschert, George F. Gunsmith and locksmith. Williamsport, Lycoming Co. (BPaD, 1880).

Broswell, Joseph. Gunsmith, West Brownsville, Washington Co. (BPaD, 1861).

Chauncey Brockway (1823–1910), Elk County.

Brother, Michael. On 1/27/1773 James Black was apprenticed to Michael Brother, silversmith of Philadelphia. On 3/6/1773 John Wagg was also apprenticed to Brother to learn silversmithing (PIDS; PaE, 203).

Brown, Albert. New Castle, Lawrence Co., gunsmith (BPaD, 1865, 1873).

Brown & Hirth. (Active 1848–90, 136 Wood St., Pittsburgh. Major manufactory (PAD, 1848, 1875, 1890).

Brown, Anthony Z. 515 N. 3d St., Philadelphia, gunsmith (MD, 1841, 1846).

Brown, David P. 1838–63, gunsmith, 125 Chestnut St., Lancaster boro (t; BLD, 1840, 1857).

Brown, James. Cutler, Pittsburgh (PAD, 1814–15; RAG, 220).

Brown, James. Front St. above Callow Hill, Philadelphia, gunsmith (DeSD, 1829–30).

Brown, James. Cutler, 1785–1816, Newton Twp., Cumberland Co. (t) Revolutionary war pensioner; served in Hays's Light Dragoons (3 Pa. Arch. 23 at 475, 748).

Brown, John. Locksmith, 1769, Philadelphia (3 Pa. Arch. 15 at 210).

Brown, John B. Gunsmith, 1824, Pittsburgh; 1831–32, Smith Twp., Washington Co. (t).

Brown, John F. Gunsmith, 515 N. 3d St., Philadelphia (MD, 1855).

Brown, John H. Reading. "Brown's Patent Gun Cleaner" (PD, 1882); patent 653,208 of 7/10/1900 for an explosive charge.

Brown, Joseph. On 8/19/1772, apprenticed for 13 years to Joseph Lippincott of Gloucester, N.J. to learn the silversmithing trade (PIDS).

Brown, Joseph. Locksmith, 1774, Mulberry Ward, Philadelphia (3 Pa. Arch. 15 at 288; t).

Brown, Matthew. Silversmith, East Pennsboro Twp., Cumberland Co. (t), 1750.

Brown, T. H. Native of Cleveland, Ohio, the only son of Thomas and Sara Brown. Served in Co. E, 31st Ohio Volunteers, in Civil War. 1848–90, Erie boro., gunsmith. (PD, 1882, 1887, 1890; BRD, 1860, 1870) Apprentice of Alfred Marion Cone, gunsmith (FSP, 65).

Brown, Walter (b. 1791). Gunstocker. Rostraver Twp., Westmoreland Co. (c, 1850).

Brown, William. Swordsmith. Employed by Lewis Prahl, gunsmith, in the manufacture of 1,000 swords for Col. Benj. Flowers's Pa. Horsemen, 3/12 to 5/22/1777 (2 Pa. Arch. 3 at 104).

Brown, William H. Pittsburgh, gunsmith (PAD, 1838, 1841).

Brown, W. S. Gunsmith, Pittsburgh (PD, 1890).

Browne, John. Smith, 1782–87, Milford Twp., Chester Co. (t).

Browne, Liberty. "was a silversmith and carried on business on Chestnut St. near Third st. He was an active Democrat and frequently a candidate of that party for a seat in city council. In 1814, he was president of the common council." (MPHS, 8:11).

Brubaker, John. 1802–3, Earl Twp., Lancaster Co., gunsmith (t; SED, 54).

Bruce, James. Artillery artificer, 4/26/1777 to 3/22/1781, Capt. Jesse Rowe's Co. (2 Pa. Arch. 11 at 252; PIR, 2:244).

Bruder, John. Leopard Alley, Lancaster boro (LCD, 1871–72). Gunsmith.

Brundel, John H. Gunsmith, Philadelphia (MD, 1847).

Bruner, George (1720–93). 1739–48, Lancaster boro. (t); 1750–93, Salisbury, N.C. (FSP, 44). Gunsmith.

Bruner, Henry. Before 1750, Lancaster; after 1750, Salisbury, N.C. (FSP, 44). Gunsmith.

Brunner, Caspar. Locksmith, Lancaster (t, 1773–79).

Bryan, James. 1766, West Bradford, Chester Co. (3 Pa. Arch. 11 at 171); 1771, Lancaster boro. (3 Pa. Arch. 17 at 14); 1801, Lampeter Twp., Lancaster Co., gun-barrel smith (t; SED, 44).

Bryan, John. Gunsmith, Derringer's Ct., Philadelphia (MD, 1846).

Bryce, John. Musket-ball manufacturer, Philadelphia (Philadelphia *Evening Post*, 6/10/1775; KKR, 196).

Buchanan, George (1782–1860). Gunsmith. 1840–60, Blair Twp., Blair Co. (t; c).

Buchanan, Samuel. 1814, Halifax, Dauphin Co., gunsmith (KKR, 197).

Buchmiller, D. F. Gunsmith 1875–90, Lancaster (LCD, 1875, 1880, 1890).

Buchmiller, Robert. Gunsmith, 57 N. Queen St., Lancaster (LCD, 1840; BPaD, 1861; LCD, 1869).

Buchmiller, William. 60 N. Queen St., Lancaster (LCD, 1851, 1857, 1860). Gunsmith.

Buck, Peter. 12/28/1776: "pay 55 shillings to Peter Buck for repairing guns." (Col. Rec., vol. 11).

Buckhalder, John. Gunsmith, 1648 N. 4th St., Philadelphia (GPD, 1875).

Buckman, Benjamin. Silversmith, 1782, Wrights Town, Bucks Co. (t).

Buckwalter, Abraham. 1771–80, Lampeter Twp., Lancaster Co. (RAG, 29), Gunsmith.

Buckwalter, David B. (1850–1928). Gunsmith McConnellstown, Huntingdon Co., apprenticed to J. S. Johnston; 1870–84, Jersey Shore, Lycoming Co. (t; c); 1885–95, Houtzdale, Clearfield Co. (t). Died in Bellwood, Blair Co.

Buckwalter, John. Gunsmith 1771–80, Lampeter Twp., Lancaster Co. (RAG, 29).

Bukley, Daniel. 515 Penn, upstairs, Pittsburgh (PAD, 1864–65). Gunsmith.

Bulow, Charles. 1797–1800, Lancaster Co. (GAAAM, 28; RAG, 29). Gunsmith.

Bundle, John F. Sarah near Bedford, Philadelphia, gunsmith (SGD, 1835–36).

Burch, Samuel. Partner in Slotter & Co. with Henry and Frederick Slotterbeck and William Carrigan, ca. 1860–68 (GAAAM, 29). Philadelphia.

Burchard, John. Artillery artificer, 1/1777, Capt. Will's Co. (BCR, 115).

Burden, Calvin. Back of 110 Race St., Philadelphia, gunsmith (PXD, 1819).

Burge, William. 1876–78, Freeport, Armstrong Co., gunsmith (t).

Burger, Adam. Artillery artificer, American Revolution, Capt. Noah Nicholas's Co. (PIR, 2:245).

Burkett, Abraham (1862–1943). R.D. 5, Punxsutawney, Jefferson Co. (t; REH, 131–35). Gunsmith.

Burly, William. Gunsmith, 1854–55, Morris Twp., Washington Co. (t).

Burnett, Henry. Gunsmith, 1817–30, Morgan Twp., Greene Co. (t).

Burney, Thomas. Indian trader and smith to the Amerindians. 1750, Cochake on the Muskingum, Ohio territory; 1752, Pickawillany; July 1754, in Capt. Andrew Lewis's Co., Battle of Great Meadows (HAN, chaps. vol. I 2, 5, 8; Col. Rec. 5:599).

Abraham Burkett (1862–1943) and his wife Lydia (1867–1936), Punxsutawney, Jefferson County.

Burns, James. Gunsmith, 135 Lombard St., Philadelphia (PXD, 1816).

Burt, John. Indian trader and gunsmith. 1727 from Philadelphia; 1745, Donegal Twp., Lancaster Co. (Samuel Dyke and Vincent Nolte, "Lancaster County Gunsmiths," KRAN [Fall 1978]).

> The Petition of Henry Hawkins in the Common Gaol of Chester, humbly sheweth—
>
> THAT WHEREAS your petit'd bound himself an apprentice to John Burtt for five years, to Learn the Trade of Gun Smith, but was not put to the trade, but was forced to go along with the s'd Burtt Indian Trading, at which return was sent to Philadelphia with one Jonas Davenport for goods to go trading again, and was sorely beat and abused by the s'd Jonas Davenport, so that being afraid to go back with him, stayed in Philadelphia. Where the s'd Burtt sold your petition'd to Plantation business & being after that sold to the aforesaid Jonas Davenport, Who sold your petitioner to an Indian, which Indian forced ye petition'd with him back into the woods Several hundred miles, whereafter Some Time was Sent by the orders of s'd Indian with the s'd Jonas Davenport to Justice Mitchels, where on the road, without any Occasion given, the s'd Jonas Davenport bound your petition'ds hands behind him with a rope about the middle fastened to the horses Tale was hald on the Ground a Considerable way through a thick muddy Swamp, alsoe Sorely Beat and bruised on the Body and one Eye almost beat out and like to haven broken both arms, So that your petition'd was in great Danger of Life, and being afger Comited to prison where is now Close Confined and Destitute of friends your petition'd therefore Humbly Prays that he may be brought into Court in order to be further heard, most Earnestly requesting and Humbly Praying this Honourable Bench may be pleased to Take into consideration the poor Condition of your Petition'd and grant him Such relief as the Law affords and your petition'd as in Duty bound Shall Every Pray
>
> Henry Hawkins
>
> To His Majesties Justices of the Court of Quarter Sessions, now setting in Chester this 24th Day of November 1724.

Busch, Frederick L. Gunsmith. 1770–80, Lancaster Co. (RAG, 32).

Bush, Matthias. 6/27/1777: payment "in favor of Matthias Bush for £281/14/10½ . . . for making knapsacks." (2 Pa. Arch. 1 at 61).

Busham, Jacob. Armorer, 1776, Capt. Samuel Miles's Co., Pa. Rifle Regt. (2 Pa. Arch. 10 at 212); armorer, Capt. Philip Albright's Co. (PIR, 1:209).

Bushman, Charles. 38 W. Gay, West Chester, Chester Co., gunsmith (BRD, 1857–58).

Bushman, M. Gunsmith, West Chester, Chester Co. (BRD, 1868).

Bustead, Morgan. 8/4/1775: "Resolved that Morgan Bustead cast and deliver to this Committee 2 howitzers agreeable to the draft offered by him to this Board." 3/1/1776: "Morgan Bustead not having performed any part of his contract for the casting of cannon is to be prosecuted for damages." Council ordered that Capt. Benj. Loxley and Daniel King take charge of Bustead's air furnace and complete the contract. But "Bustead Sister, the Vixin, refuses Capt Loxley admittance to the furnace." On 8/7/1776 Capt. Loxley managed to gain control of the furnace and made cannon throughout the war (see Loxley) (Col. Rec., 10:298; Col. Rec., 11:641; 4 Pa. Arch. 5 at 717).

Butcher, John. Silversmith. On 10/30/1778 the inventory of goods and chattels belonging to John Butcher of Blockley Twp., Philadelphia Co., showed a silver mill and press valued at £3 (6 Pa. Arch. 12 at 619).

Butland, S. B. Gunsmith, Philadelphia (PD, 1882).

Butler, Edward. Regimental Smith, 9/10/1778, 10th Pa. Regt. (2 Pa. Arch. 15 at 498).

Butler, John. Armorer, 5/16/1778, Committee of Safety (2 Pa. Arch. 2 at 699); gunsmith at Lancaster (RAG, 32).

Butler, Richard. Armourer, 1767, Ft. Pitt (WSB, 54–55).

Butler, Thomas (1719 or 1720–1791). 1749–54, Lancaster Co.; 1760, Carlisle; 1765, armorer at Ft. Pitt; 1771–89, West Pennsboro Twp., Cumberland Co. On 4/23/1778 he was appointed a public armorer by William Henry. (W. D. Butler, *Butler Family in America* [1909], 6–7; GAAAM, 30). Armorer with the Bouquet expedition against Ft. DuQuesne (E. D. Williams, *Bouquet's March to the Ohio* [1975], 52).

Butler, William. Armorer at Ft. Pitt, 1766–67. Son of Thomas Butler (WSB, 54–55).

Butterfield, Benjamin. York Rd. near Oxford, Philadelphia (DeSD, 1833, 1835, 1836); Froad and Marlborough, Philadelphia (MD, 1841, 1846). Gunsmith.

Butterfield, Jesse. 1851–65, Philadelphia. Gunsmith and inventor. 1861, DBA Butterfield & Nippes, Kensington, Philadelphia Co. (RAG, 31; BPaD, 1861). Patent 24,372 for a self-priming gunlock.

Butterfield, John. Whitesmith, West Nottingham, Chester Co. (t, 1766).

Butterfield, Sarah. Rd. opposite Marlborough, Philadelphia (MD, 1855). Gunsmith.

Buyers, Robert. Gunsmith, 1801, Salisbury Twp., Lancaster Co. (t).

Buysley, John. Gunsmith, Allentown, Lehigh Co. (c, 1850).

Buzzard, George. Gunsmith, 1802–22, Lower Mt. Bethel Twp., Northampton Co. (t).

Byers, Frederick, Sr. (1731–1801). 1771–1801, Antrim Twp., Franklin Co. (t; WSB, 29). Gunsmith.

Byers, Frederick, Jr. 1789–1809, gunsmith, Antrim Twp., Franklin Co. (t; WSB, 29).

Byers, James. 1775–80, Philadelphia, superintended the brass foundry and cast cannon (2 Pa. Arch. 15 at 385).

Byles, Thomas. Pewterer. 9/5/1754, advertisement in *Pennsylvania Packet:* "Thomas Byles in Market St. buys and sells pewter."

Byor, George Frederick. 1779–80, Upper Delaware Ward, Philadelphia. Manufacturer of cutlery and edged weapon (t).

Byrne, Patrick. Gun-barrel maker, corner of Cherry St. and Sawmill Alley, Pittsburgh (PAD, 1860–61); 15 Booth's Block (PAD, 1864–65).

C

Cable, Jacob (b. 1754). Born in Pa. Blacksmith, 5′ 6″ tall, in the 11th Pa. Regt.; enlisted 9/16/1777 (2 Pa. Arch. 11 at 67).

Cable, William. Lawrenceville, Tioga Co. (BRD, 1847). Gunsmith.

Cahill, John. Artillery Artificer, 1777–82, at the laboratory in Philadelphia (2 Pa. Arch. 11 at 261).

Calderwood, Hugh (1815–83). 1840, Mahoning Twp., Lawrence Co.; 1850–57, Neshannock Twp., Lawrence Co.; 1860–83, Union Twp., Lawrence Co. (t; c). Gunsmith.

Calderwood, William (1780–1860). Born in Ireland; father of Hugh Calderwood. Gunsmith, 1808–11, Bucks Co.; 1813–18, Philadelphia; 1818–20, Liberty Ave. between 7th and Strawberry, Pittsburgh; 1840–50, Mahoning Twp., Lawrence Co.; 1860, Union Twp., Lawrence Co. (MB 11 [1965]: 2; t; c).

Caldwell & Co. Swordsmiths, 1895–1900, Philadelphia (RAG, 220).

Call, George. Gunsmith in Capt. John Wetzel's Co., Westmoreland Co. Militia; enlisted 5/28/1778. (3 Pa. Arch. 23 at 459).

Calladine, Abraham (b. 1727). Locksmith. Born in Germany, age 19 in 1746. In list of provincial soldiers. (5 Pa. Arch. 1 at 6f.).

Calvert, James (1720–90). Born in Cheviot Hills,

Scotland; died in McKean, Crawford Co. gunsmith (t; f; c).

Camey, Thomas. Artificer, 1760–61, Ft. Pitt (Bouquet census).

Cammel, John. Smith, 1763, at Ft. Pitt; killed or taken at Beaver Creek. (Alexander McKee list dated 9/5/1763).

Campbell, Isaac. Gunsmith, Martinsburg, Blair Co. (PD, 1887).

Campbell, James. Goldsmith. Apprenticed for 13 years on 2/12/1772 in Philadelphia to James Lippencott of Gloucester, N.J. to learn the goldsmithing trade (PIDS).

Campbell, James. Gunsmith and silversmith (possibly the same man as above). 1809–14, Uniontown, Fayette Co.; 1817–33, Wheeling, (West) Virginia. (t) The 1814 tax list called him a watchmaker.

Campbell, John. Armorer, provincial ship *Pennsylvania*, Pa. Navy (5 Pa. Arch. 1 at 87).

Campbell, Mungo. Pewterer. "Mungo Campbell, pewterer, corner Chestnut and 2d Sts., Philadelphia, from Dublin." (*Pennsylvania Packet*, 8/29/1751; much the same advertisement appeared in the same paper on 5/7/1752)

Canady, Charles. 20 Budd St., Philadelphia (DeSD, 1820, 1829, 1830, 1833). Gunsmith.

Cane, Christian. Swordsmith in the employ of Lewis Prahl, gunsmith, making 1,000 swords for Col. Benj. Flowers's Pa. Horsemen (2 Pa. Arch. 3 at 104).

Cannon, James. 1865–76, Mt. Morris, Greene Co., wagon maker, gunsmith, blacksmith (t; PD, 1882, 1890).

Capper, John. Blacksmith, 1732, Lancaster (1 Pa. Arch. 1 at 360).

Carlile, Henry (d. 1847). Gunsmith 1818–47, Shippensburg, Cumberland Co. (t; WSB, 52–54).

Carnahan, J. C. Gunsmith, Washington Twp., Indiana Co. (PD, 1882).

Carnman, D. Gunsmith, Aaronsburg, Centre Co. (BPaD, 1861).

Carny, James N. Gunsmith, McAllisterville, Juniata Co. (BPaD, 1861).

Carothers, Thomas. Artillery Artificer, 1777–82, at the laboratory in Philadelphia; from New London, Va. (2 Pa. Arch. 11 at 261).

Carpenter, John. 1771–90, Earl Twp., Lancaster Co. (RAG, 34; GAAAM, 31).

Carpenter, John. Pine Bank, Gilmore Twp., Greene Co. (PD, 1890).

Carr, Charles. 106 South St., Philadelphia (MD, 1855); 250 South St. (MD, 1859). Gunsmith.

Carr, Samuel. Lolars Ct., Philadelphia (PXD, 1819). Gunsmith.

Carr, William. Artillery artificer from 3/3/1777 to 3/22/1781, Capt. Jesse Rowe's Co. at Philadelphia (2 Pa. Arch. 11 at 252; PIR, 2:244).

Carrell, George. Gunsmith, Ross Twp., Northampton Co. (t, 1828–35).

Carrell, John. Offered "Silver mounted swords" in *Pennsylvania Packet*, 9/27/1787.

Carrell, Lawrence. Gunsmith, Southern District, north side, Spruce St., Philadelphia (c, 1790).

Carrigan, William. 1860–68, Philadelphia, gunsmith, partner in Slotter & Co., with Frederick and Henry Slotterbeck and Samuel Burch (FSP, 281).

Carroll, Martin. Brass founder, cannon maker, 12/1779, Philadelphia (2 Pa. Arch. 15 at 385).

Carruth, Adam. 1809–16, South Carolina; 1816–21, Philadelphia. (FSP, 54). Gunsmith.

Carson, John. Armorer. 11/9/1775: "After troops raised by this Province in the late war were disbanded, a number of muskets and military accoutrements were lodged at Mr Carson's in Paxtang, where they remained without any notice being taken of them . . ." (2 Pa. Arch. 13 at 505).

Henry Carlile (d. 1847), Carlisle, Cumberland County.
Varner collection.

JNO. CARSON TO PRESIDENT REED, 1779.

Carlisle, May 24th, 1779.

Sir,

Agreeable to an order from the Executive Council, I have drawn on Mr. Sarjaint, C. M. Stores, at the Publick works, for two Hundred & fifty Rifles, to have them repaired under my direction. I have deposited a number of them in y[e] hands of Gunsmiths for that purpose; one hundred I took to Lancaster. There will be, I expect, a sufficient number of the Guns repaired in two weeks to arm all the Rangers that will be got—from what I can learn there is about sixty of them Enlisted; they will be with Me in a few days for arms. The Arms I drew from Mr. Sarjaint were in very bad order, most of them wants Locks, & a great many Stocks & Locks. I purchased one hundred locks in Lancaster & was obliged to pay Twenty dollars each for them; the Expence attending the repairing the guns will be very great.

I have as yet been able to pay the demands of the workmen, but my Cash is almost run short.

I would be glad you would send Me, by the first safe Conveyance, Eight Thousand dollars; if that proves insufficient, shall draw on you for more; there was not enough Rifles at the works out of Order to compleat my demand, but have engag'd repair'd ones if wanted.

I am, with due respect,
Your Excellency's obed[t]&
Very h'ble Serv[t],
JNO. CARSON.
1 Pa. Arch. 4 at 437.

Carter, James. Armorer, gunsmith. 5/12/1777: payment "in favor of Mr James Carter for £200 for . . . repairing arms for the state" (2 Pa. Arch. 1 at 45); 7/9/1777: payment "in favor of Mr James Carter for £500 for . . . repairing arms of this state" (2 Pa. Arch. 1 at 64); payment "in favor of Mr James Carter for £500 for . . . repairing of arms belonging to this State" (2 Pa. Arch. 1 at 71). Also, Carter received an additional £1,400 for arms repairs in 1777 (3 Pa. Arch. 5 at 53).

Carteret, Daniel. Armorer, U.S. Armory, Philadelphia (2 Pa. Arch. 11 at 258.

Cartwright, Joseph. Artillery Artificer, 12/1779, at Philadelphia (2 Pa. Arch. 15 at 384).

Caster, Matthew. 1800, Plainfield Twp., Northampton Co. (t). Gunsmith.

Cave, Christopher. 1776–80, Dock Ward, south part, Philadelphia, gunsmith (t; c; 3 Pa. Arch 14 at 474, 750); 12/1779, artillery artificer, Philadelphia (2 Pa. Arch. 15 at 383).

Ceheen, Thomas (b. 1731). Earl Twp., Lancaster Co. Gunsmith in capt. Charles McClung's Co., Lancaster Militia; 5′ 9″ tall, age 27 on 5/23/1758 (5 Pa. Arch. 1 at 172).

Center, Jonathan. Artillery artificer, Capt. Thomas Wylie's Co. at Carlisle (PIR, 2:248).

Chabot, Cyprien. Inventor. Patents 47,163 and 49,718 of 1865 for breech-loading gun. Philadelphia (RAG, 35).

Chaloner, John. 10/20/1775: "Capt Robert Towers, Commissary, Reports that . . . he has purchased the following . . . of John Chaloner, 10 cutlasses, at 2 shillings each." (Col. Rec., 10:375).

Chamberlain, Edward. Master of arms, 9/23/1775, Pa. Navy (2 Pa. Arch. 2 at 708).

Chambers, David. Artillery Artificer, enlisted on 1/15/1778 in Capt. Nathaniel Irish's Co. (PIR, 2:247).

Chambers, James (1738–63). 1750–51, Lancaster boro. (SED, 54); 1760–63, Paxton Twp., Lancaster Co. (DB L-26). Gunsmith.

Chambers, James. 8/23/1788: "an order was drawn upon the Treasurer in favor of James Chambers and William Marshall for £50/19/7, payable out of the militia fines of the County of Washington, in full of their accounts for repairing the public arms belonging to said county" (Alfred Creigh, *History of Washington County* [1870], 75).

Chambers, Col. John G. Amwell Twp., Washington Co. 12/14/1812: "invented a gun that can discharge 6 or 8 loads in a succession of space between each sufficient to take another aim, by a single operation of the trigger. The gun has one barrel and does not exceed a common gun in size. On 15 January 1815, the Senate of Pennsylvania appointed a committee to examine the utility of Col. Chambers' gun. . . The gun was considered a very extraordinary piece of machinery and a power instrument in the destruction of the enemy" (Creigh, *Washington County,* 357).

Chambers, Joseph C. West Middletown, Washington Co. On 3/23/1813 he received a patent for "repeating gunnery." The Philadelphia Committee of Defense ordered Tryon & Henry to complete 200 muskets and 20 swivel guns built on this system, at $23 each musket and $18 for the swivel guns. On 4/16/1814 the committee also ordered 100 repeating pistols of Chambers's pattern (GAAAM, 32; 9 Pa. Arch. 6 at 4231).

Chandlee (Chandley), Benjamin. 1747–74 East Nottingham Twp., Chester Co. Watchmakers, silversmith, and goldsmiths, compass and instrument makers. (t)

Chandlee, William. Cutler, 1767–68 Reading, Berks Co. (t).

Chapman (Chapin), Luther. Waman's Ct., Philadelphia (MD, 1844, 1846); gunsmith (MD, 1847, lists "Chapin").

Chapman, James. Revolutionary war arms contractor, Bucks Co. (GAAAM, 32).

Chapman, John. Silversmith and gunsmith, 1784–85, Reading, Berks Co. (t).

Charlton, James. Armorer, 1778, Philadelphia (2 Pa. Arch. 11 at 258).

Chatwin, William. Brass founder and cannon maker, 1777, Philadelphia (2 Pa. Arch. 11 at 260).

Cherrington, T. P., Jr. 1815–58, Cattawissa, Columbia Co. (t). Gunsmith.

Cherrington, T. P., III. 1861, Ashland, Schuylkill Co. (BPaD, 1861); Cattawissa, Columbia Co. (PD, 1890). Gunsmith.

Cherrington, William. Missemer's Mills, Lebanon Co. (BPaD, 1861). Barrel and rifle maker.

Cheyney, Charles. Birmingham, Allegheny Co. (t). Gunsmith.

Chilcote, J. E. 1845–70, Dry Run, Perry Co., gunsmith (t; c).

Chilcote, W. S. (d. 1887). Gunsmith, Houstontown, Fulton Co.

Christ, Caspar. Swordsmith in the employ of Lewis Prahl, gunsmith, Northern Liberties, Philadelphia Co. (2 Pa. Arch. 3 at 104), 1777.

Christ, John. Gun-barrel forger, Cocalico Twp., Lancaster Co. (t, 1805).

Christ, Michael. Smith, 1767–68, Maiden Creek Twp., Berks Co. (t).

Christman, Jacob. Gunsmith, Clark, Mercer Co. (PD, 1880, 1882, 1890).

Christy, William. Inventor, 1866, Philadelphia. Patent 58,064 for a multiple-barrel firearm.

Clark, Andrew. (b. between 1760 and 1769, d. 1827). Artillery artificer, Capt. Noah Nicholas's Co. Buried in Mt. Olivet Cemetery, Harrison Twp., Bedford Co. (PIR, 2:245).

Clark, Charles M. Inventor. Patent 794,852 of 7/18/1905 for a rapid-fire gun. Pittsburgh.

Clark, Daniel. Gunsmith, 212 Green St., Philadelphia (DeSD, 1823).

Clark, I. Jones. 611 Tamarind, Philadelphia (BSED, 1880). Gunsmith.

Clark, James (1746–1841). Born in Lancaster Co. Served as an artillery artificer at the armory at Philadelphia. (2 Pa. Arch. 11 at 258). 1798–1813, gunsmith, Ayr Twp., Bedford Co. (t); 1821, 1828, Hopewell Twp., Bedford Co.; 1833, pensioner in Morris Twp., Huntingdon Co. Buried at Williamsburg, Blair Co., Presbyterian Churchyard; died 7/1/1841).

Clark, John. (1789–1854). Before 1810, Reading, Berks Co. (t); 1811–17, Cumberland Co.; 1818–19, Shippensburg boro.; 1821–36, Tuscarawas St. and Walnut St., Canton, Ohio; 1853, West Brookfield, Ohio (t; c; WSB, 50–52; JKJ, 496). Major Gunsmith.

Clark, N. Gunsmith, 1830–69, Columbia, Lancaster Co. (MB 7 [1978]: 24).

Clark, Peter. 1/9/1777: "pay Peter Clark £5 each for rifles" (Col. Rec., vol. 11).

Clark, Samuel C. Inventor, Philadelphia. Patent 288,548 of 11/13/1883 for a spring gun.

Clark, T. Inventor, Philadelphia. Patent 176,367 of 4/18/1876 for a gun safety.

Clark, William. 1783–91, Dock Ward, south part, Philadelphia (t, c); 1779, armorer, Pa. Continental Line (2 Pa. Arch. 15 at 383).

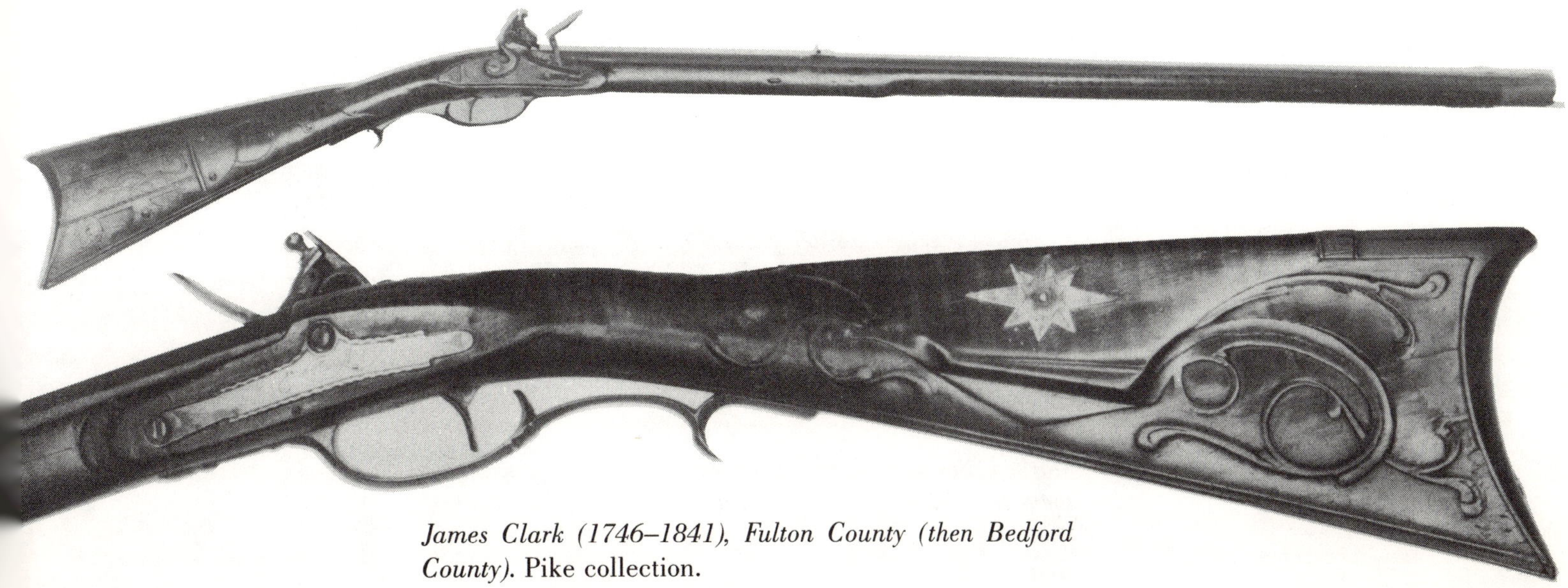

James Clark (1746–1841), Fulton County (then Bedford County). Pike collection.

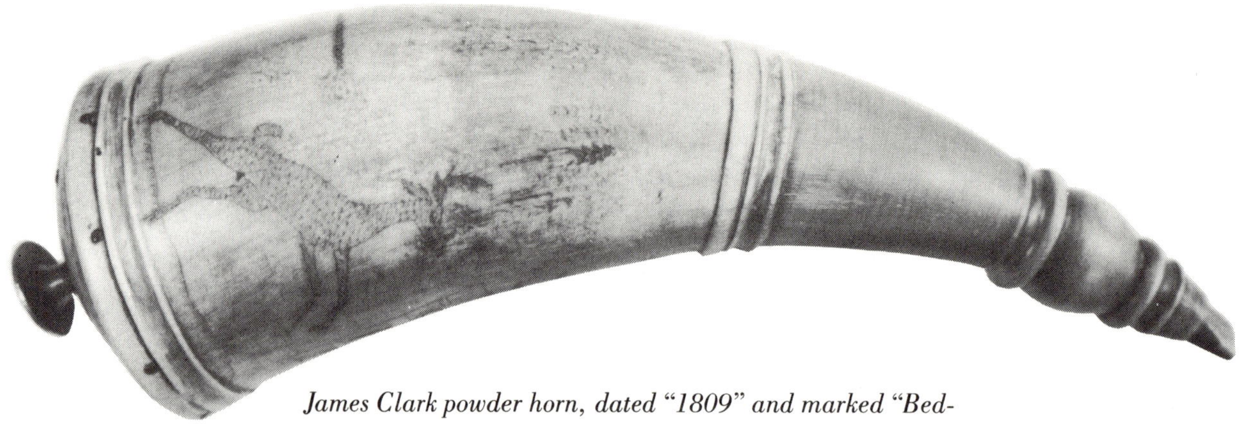

James Clark powder horn, dated "1809" and marked "Bedford Co., Penna." Powder horns that give owner or maker's name, place of manufacture, and date are rare. This horn centers on the "sun in mask," a Pennsylvania Dutch (German) folk-art theme.

Clarke, John. Artillery artificer, 1777–82, at Philadelphia; from New London, Va. (2 Pa. Arch. 11 at 261).

Claus, Daniel. 325 N. Cherry St., Lancaster boro (BLD, 1875–76).

Clause, Henry. 1821, Heidelberg Twp., Lehigh Co. (t, 1821) Gunsmith.

Clauson, Charles, Gunsmith, 6 York Ct., Philadelphia (DeSD, 1826, 1833).

Clauss, Daniel. Gun-barrel maker, Cumru Twp., Berks Co. (c, 1850; BRD, 1855).

Clay, Jonathan. Artillery artificer, 1779, at the cannon factory at Philadelphia (2 Pa. Arch. 15 at 384).

Clay, Slator. 3/4/1755: "Silversmith at the sign of the King's Head, New Castle." (Advertisement in the *Pennsylvania Packet*)

Clayton, William. Smith, 1768, West Bradford, Chester Co. (t).

Cleaveland, Amaziah. Artillery artificer at Carlisle, 1778–80, Capt. Thomas Wylie's Co. (PIR, 2:248).

Clewells, Jacob. 1799–1802, Hanover Twp., Northampton Co. (t). Gunsmith.

Clewells, Jesse. Bustletown, Philadelphia Co. (BPaD, 1861) Gunsmith.

Clewells, Philip. 1820–35, Bushkill Twp., Northampton Co. (t). Gunsmith.

Clewells, William. gunsmith, 1814–35, Bushkill Twp., Northampton Co.

Click, Samuel. Gunsmith Woodward Twp., Clearfield Co. (PD, 1887).

Clifton, Thomas. 6/20/1777: "pay Thomas Clifton £625 for 500 cartouche boxes and 500 belts" (2 Pa. Arch. 1 at 58).

Climer, Jacob, Jr. Smith, 1779–87, Milford Twp., Bucks Co. (t).

Cline, Jacob. 1830–65, Bell Twp., Westmoreland Co. (t). Gunsmith. (BPaD, 1861, listed Saltsburg, Indiana Co., but that was the post office for his locale.)

Clippinger, Joseph. 1821–30, Franklin Co.; 1846–80, New Carlisle, Ohio. (WSB, 39) Major gunsmith.

Closson, Charles. Gunsmith, Philadelphia (DeSD, 1829).

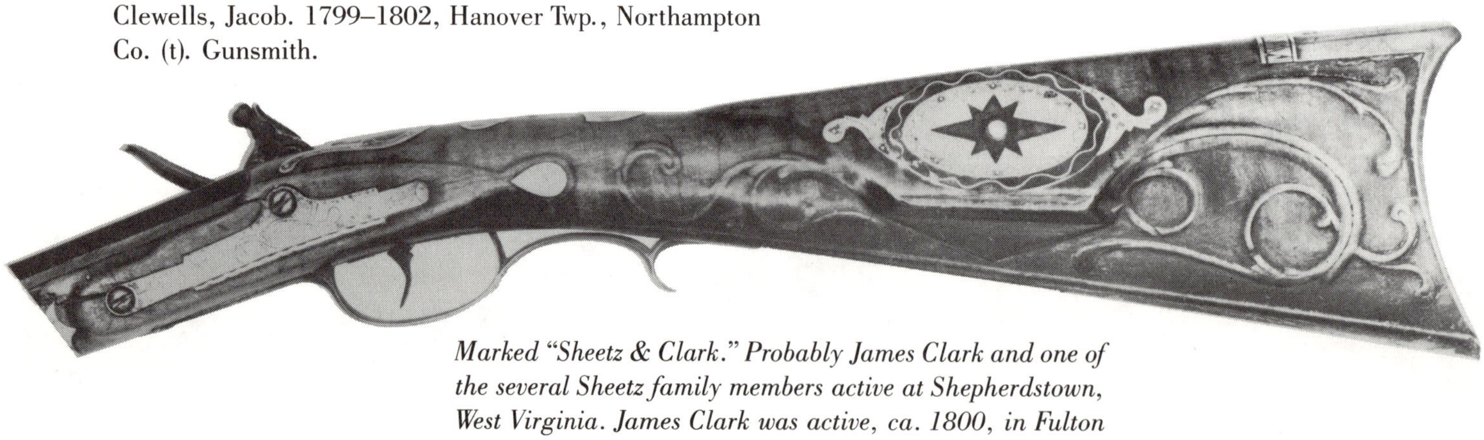

Marked "Sheetz & Clark." Probably James Clark and one of the several Sheetz family members active at Shepherdstown, West Virginia. James Clark was active, ca. 1800, in Fulton County.

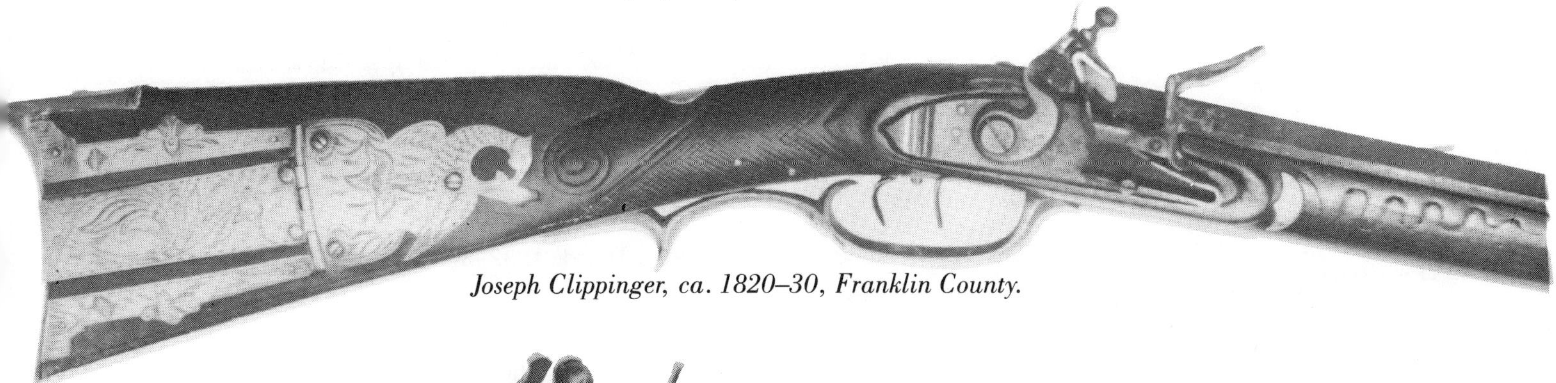

Joseph Clippinger, ca. 1820–30, Franklin County.

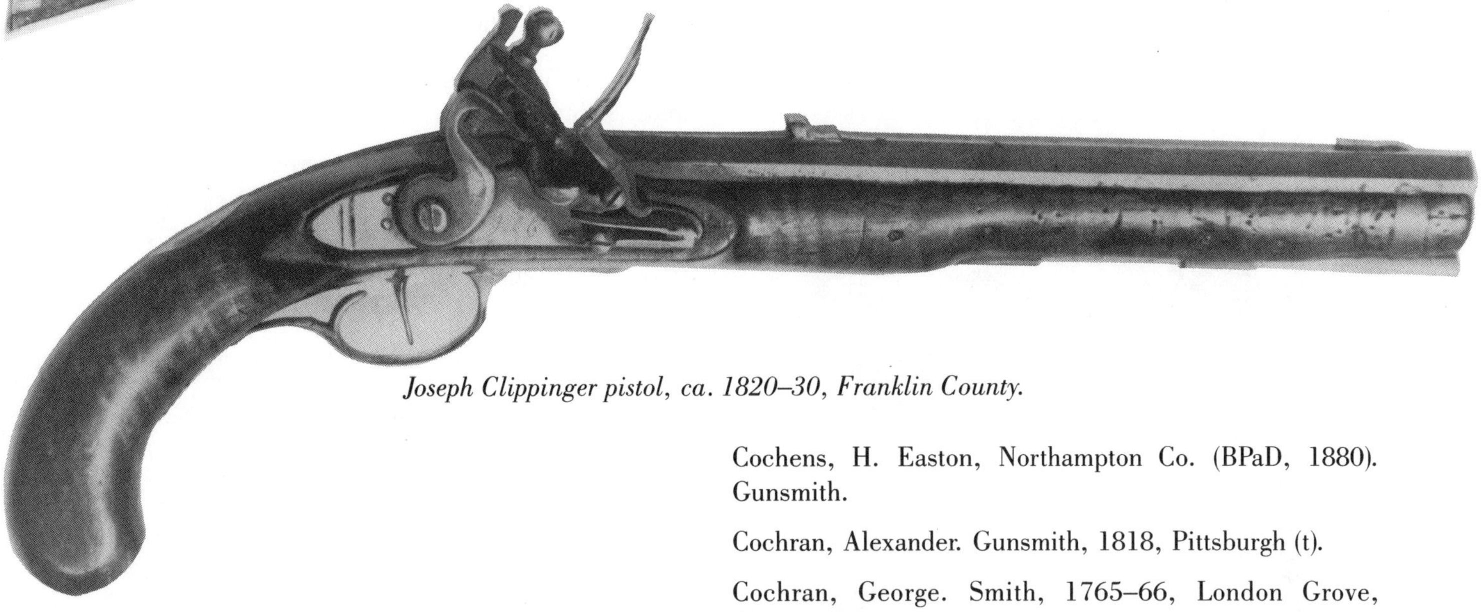

Joseph Clippinger pistol, ca. 1820–30, Franklin County.

Clous, Christian. Gunsmith, 1795–1807, Lynn Twp., Northampton Co. (t).

Clouse, George (1810–56). Born in Germany, emigrated to avoid 7 years of military service; married Christina Friend, whom he met on the trip to America. 1850, South Woodberry Twp., Bedford Co.; 1855, Broad Top Twp., Bedford Co.; 1863, South Woodberry Twp., Bedford Co.; 1867, West Providence Twp., Bedford Co. (t; f; c). Gunsmith, and locksmith.

Clouse, Henry Valentine. 1849–81, South Woodberry Twp., Bedford Co. (t) Gunsmith and whitesmith.

Clouse, Valentine Feltus (1852–1936). Lafayetteville, South Woodberry Twp., Bedford Co. Gunsmith and locksmith.

Coates (Coats), James. Gunsmith, rear of 47 Tammany, Philadelphia (PXD, 1813); worked in 1814 for J. J. Henry as a gunsmith (GAAAM, 33).

Coates, Moses. Belt-axe and edged-tool maker, 1781–90, East Caln, Chester Co. (t; GAAAM, 221).

Coburn, Benjamin. 1816, Little Britain Twp., Lancaster Co., gunsmith (t; SED, 54).

Cochens, H. Easton, Northampton Co. (BPaD, 1880). Gunsmith.

Cochran, Alexander. Gunsmith, 1818, Pittsburgh (t).

Cochran, George. Smith, 1765–66, London Grove, Chester Co. (t).

Cochran, Landon. 1816–36, Brownsville, Fayette Co. (t); 1850, Lost Creek Twp., Vigo Co. (c, 1850). Gunsmith.

Cochran (Cochren), Robert. 1812–18, 458 S. 7th St., Philadelphia (PXD, 1813, 1816, 1818); Rose Alley (DeSD, 1835). Gunsmith.

Cochran, Samuel (d. 1835). 1834–35, Connellsville, Fayette Co. (t). Gunsmith.

Coffee, James. 6/19/1777: "pay James Coffee . . . for iron work for 2 guns, amounting to £44/17/8" (2 Pa. Arch. 1 at 57).

Coffers, August. 1848–57, N. Queen St. near Frederick St., Lancaster boro (BLD, 1857; t; SED, 54). Gunsmith.

Coldren, John (d. 1847). 1822–37, German Twp., Fayette Co.; 1837–47, Brownsville, Fayette Co. (t). Gunsmith.

Coldren, Samuel. 1838–50, Brecknock Twp., Lancaster Co. (t; SED, 54) In 1850 he employed 7–10 men annually (Industrial Census). 1857, gunsmith at N. Queen St., near Lemon St., Lancaster boro (BLD, 1857).

Coleman, Robert (1748–1825). Operated Elizabeth Forge (operational 1750–1856). "After Elizabeth Furnace came into the possession of Robert Coleman, he cast shot and

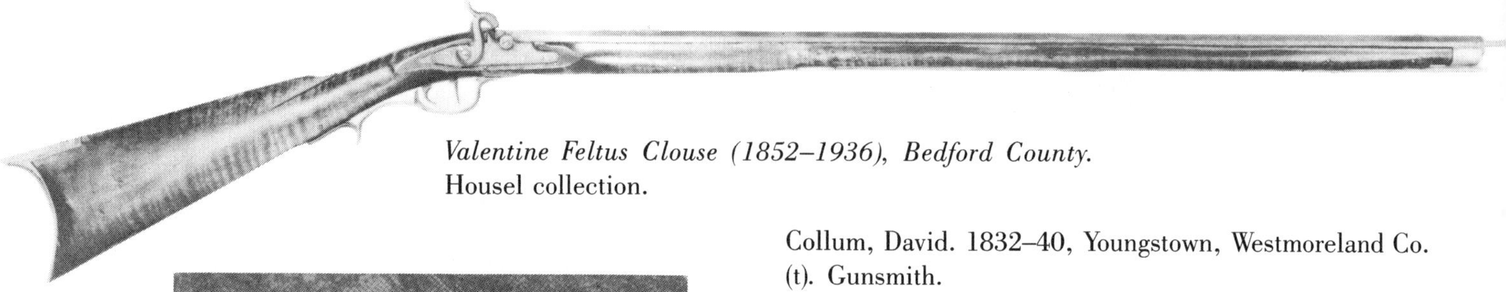

Valentine Feltus Clouse (1852–1936), Bedford County. Housel collection.

Valentine Feltus Clouse (1852–1936), Layfayetteville, Bedford County.

shells and cannon for the Continental Army." On 11/16/1782 he purchased 42 German prisoners of war at £30 each, and added 28 more to the same price on 6/14/1783; a total of £2,100 was paid for these men to work at his forge (SHIS, 179–80). Coleman was the most significant ironmaster of his time.

Collier, Joseph. Was paid £30 for cartouch boxes in February 1776 (3 Pa. Arch. 5 at 9).

Collins, Brazilla. Gunsmith, Philadelphia (c, 1850).

Collins, Edward. Artillery artificer, enlisted 2/11/1777 to 3/21/1781 in Capt Jesse Rowe's Co., making cannon at Philadelphia (PIR, 2:244).

Collum, David. 1832–40, Youngstown, Westmoreland Co. (t). Gunsmith.

Colton, William. Artillery artificer, 1779, at the cannon factory at Philadelphia (2 Pa. Arch. 15 at 385).

Columbian Firearms Co. Philadelphia. Made cheap "Saturday Night Special" revolvers (PD, 1896).

Colvin, Davis. Green Grove, Luzerne Co. (BPaD, 1861). Gunsmith.

Colvin, John. Peckville, Lackawanna Co. (BPaD, 1880; BRD, 1887, 1890). Gunsmith.

Colvin, John R. Olyphant, Lackawanna Co. (BPaD, 1880). Gunsmith.

Colvin, Robert J., Jr. Inventor. Patents 34,740 and 44,784 of 1862 and 1864 for a sword-pistol combination (RAG, 41). Lancaster boro.

Compton, Phineas T. (6/1/1804–7/4/1858). By 1813, Berlin, Somerset Co. Gunsmith in Berlin, 1840–58 (t).

Compton, Samuel (d. 11/27/1902). Berlin, Somerset Co., gunsmith (t; c).

Conch, Andrew (d. 1846). Native of Chester Co., he was a gunsmith in Miller Twp., Huntingdon Co., 1830–46 (t).

Condy, Benjamin. 9/18/1777: "pay Benjamin Condy £9/17/6 for repairing sundry telescopes" (2 Pa. Arch. 1 at 224).

Cone, Alfred Marion (1831–1903). Born on 12/21/1831 in Panama, N.Y.; died 4/1/1903 at Corry, Erie Co. 1860–70, 331 Pa. Ave, Warren boro. (c); 1873–94, N. Center St., Corry, Erie Co. (c; t). Carpenter, oil rigger, cabinetmaker, gunsmith (MB 4 [1942]:7; MB 6 [1945]:5).

Cone & Collins. Warren boro (PD, 1882).

Conestoga Rifle Works, Lancaster Co. Owned and operated by Henry Leman, 1833–87 (RAG, 41).

Conkle, John, Jr. (1791–1864). 1831–46, Rostraver Twp., Westmoreland Co. Born 9/12/1791 at Lancaster; buried at East Salem, Ohio, Cemetery (t; c; f). Gunsmith.

Connelly, William. Philadelphia, gunsmith (DeSD, 1824).

Connor, John. Armorer, 1776, armed boat *Warren*, Pa. Navy. (2 Pa. Arch. 1 at 353).

Conrad, Samuel (1812–53). Gunsmith, 1832–53, Berlin, Somerset Co. (t)

Alfred Marion Cone (1831–1903), Corry, Erie County.

John Conkle, Jr. (1791–1864), Westmoreland County.

Constable, Richard. Gunsmith, worked for, successor to, Booth & Co. (PXD, 1819; DeSD, 1820, 1824, 1830, 1833; MD, 1837, 1841, 1846, 1849) Philadelphia. (RAG, 42) Made revolvers and pistols.

Conway, Arthur. Master armorer, 10/1/1776 to 4/7/1777, schooner *Lydia*, Pa. Navy (2 Pa. Arch. 1 at 321).

Conway, John. Artillery artificer at the laboratory in Philadelphia (2 Pa. Arch. 15 at 385), 1779.

Cooder, Peter (d. 1786). Smith, 1777–85, Millford Twp., (t)

Cook, Alexander. Gunsmith, 1790; between Walnut St. and 3d St., north side, Philadelphia.

Cook, George. Artillery artificer, 1779, at the laboratory in Philadelphia (2 Pa Arch 15 at 386).

Cook, Jacob. 1807–14, Caernovon Twp., Lancaster Co. gunsmith (t; SED, 54).

Cook, Jacob. Gunsmith, 1819–34, Ayr Twp., Bedford Co. (t).

Cook, John. Artillery artificer, 1779, Philadelphia (2 Pa. Arch. 15 at 385).

Cook, John (1755–1839). Gunsmith, 1817–29, Ayr Twp., Bedford Co. (t).

Cook, Michael. Gun-barrel maker; took over Leonard Snider's shop, Warren Twp., Franklin Co. (WSB, 28).

Cook, Thomas. Gunsmith, 36 Coates, Philadelphia (PXD, 1816); Still House Alley (PXD, 1814).

Cookman, W. Gunsmith, Rear of 84 2d St., Philadelphia (MD, 1846).

Cooley, David (1790–1856). 1802–31, Tyrone Twp., Adams Co.; 1831–56, Franklin Twp., Adams Co. Died 7/14/1856 in Huntingdon Twp., Adams Co. (WSB, 95; JKJ, 501–2; t). Gunsmith.

Coons, Joseph. 1810–17, Philadelphia. (RAG, 43, 133) One apprentice was Joseph Mitchell. Gunsmith.

Cooper & Lavely. Pittsburgh. Operated by J. M. Cooper, Sr. and James Lavely, 1861–62 (PAD).

Cooper, B & J. Made wall guns on government contract, Philadelphia (FSP, 1805); after 1805, New York.

Cooper Firearms Manufacturing Co. Operated by James Maslin Cooper, Jr., Pittsburgh, 1850–64. Civil War contractors, making Cooper revolvers and Gatling guns.

Cooper, George. Gunsmith and gun-barrel forger, Cocalico Twp., Lancaster Co., 1834–35 (SED, 54; t).

Cooper, James Maslin, Operated major manufactory, J. M. Cooper & Co., 1839–59, Pittsburgh (GR 6 [1959]).

Cooper, James Maslin, Jr. Operated major manufactory, Cooper Firearms Manufacturing Co., 1850–64, Pittsburgh; 1865–69, Frankford, Philadelphia Co., ran Cooper Firearms Manufacturing Co. (FSP, 68) Patents: pistol, 27,526 of 3/20/1860; revolver, 29,864 of 9/4/1860; revolver, 40,021 of 9/22/1863; revolver, 45,319 of 12/6/1864 (RAG, 43).

Cooper, J. M. & Co. 58 Wood St., Pittsburgh. Manufactory and importer of gunlocks, 1839–59; also dealt in wholesale hardware items. (RAG, 43) Owned and operated by J. M. Cooper, Sr.

Copelen, Thomas. Smith, Indian trader. Reported killed or taken by the Amerindians on 9/5/1763 at Beaver Creek (HAN, 2:379).

Coply, Noah. Smith, Indian trader. 1749, Paxtang, Lancaster Co. (Egle, *History of Dauphin County,* 286).

Corl, Abraham (d. 1842). 1803–26, East Nottingham Twp., Chester Co. Silversmith, watchmaker, clock maker, gunsmith (t; Chester Co. DB F-3). 1831–40, Coventry Twp., (t).

Corley, John. Gunsmith, Philadelphia (MD, 1837, 1841).

Corliss, John. No. 2 Crissman's Ct., Philadelphia, gunsmith (MD, 1837, 1841).

Cornay, Francis. Gunsmith, 89 S. Water St., Philadelphia (PXD, 1811).

Corner, Paul. Gunlock maker, Lancaster (SED, 54; t, 1759).

Cornet, Charles. Gunsmith, 93 S. Water St., Philadelphia (PXD, 1814).

Corns, Abraham. Gunsmith, Chestnut St. near Prince St., Lancaster boro (BLD, 1857).

Cornwall Furnace (established 1742, abandoned 1882). Run by Peter Grubb (d. 1754) as first ironmaster; taken over by Peter Grubb, Jr. and Curtis Grubb, his sons. Supplied cannon balls and shot during the Revolution and the War of 1812 (SHIS, 182–83).

Corolus, [?]. 1801, Donegal Twp., Lancaster Co. (SED, 54; t). Gunsmith.

Correll, George. Gunsmith, Kunkletown, Monroe Co. (BPaD, 1861).

Corwin, W. H. 35 Diamond St., Titusville, Crawford Co. (CCD, 1874). Gunsmith.

Cory, John (d. 1845). 1803–45, Gunsmith Little Beaver Twp., Beaver Co. (t).

Coster, Abraham. Gunsmith; worked for J. Henry in Philadelphia (RAG, 44; MPHS 8 [1867]:149).

Cotton, John. Artillery artificer and blacksmith, 1779, at the cannon factory in Philadelphia (2 Pa. Arch. 15 at 384).

Coulaux, Julien. Gunsmith, 93 S. Water St., Philadelphia (*Stephens Directory,* 1796).

Coutey, Jacob. 7/19/1661: "for muskets delivered according to account and receipt" to the Dutch colony on the Delaware, he was paid 85 florins (2 Pa. Arch. 5 at 455).

Couty, Samuel. 1779–83, gunsmith, Chestnut St., Philadelphia. (t; c) 3/8/1783: "to Perkin & Couty, gunsmiths, the balance due them, £81" (3 Pa. Arch. 5 at 340).

Cover, John. 1859–61, Main St. and Spring St., Connellsville, Fayette Co., gunsmith (t; c; BPaD, 1861).

Cowan, C. Cutler, 1813, Penn St. and Cecil's Alley. (PAD, 1813) He had a slitting mill and tilt hammer, and made iron tomahawks, axes, edged tools, and weapons. He failed financially, and was taken over by Ruggles Whiting of Boston in 1816, who failed in 1819. The business was taken over by R. Bowen in 1826 (SHIS, 228).

Cowan, John. Pittsburgh *Gazette,* 1/14/1792: "Wanted: An Apprentice to the Bow String Manufacturing. John Cowan's business on the southwest corner of the Diamond."

Cowdan, James F. Gunsmith, Chandlerville, Chester Co. (BPaD, 1861).

Cowell, Ebenezer. 1775, Allentown, gunsmith to the Committee of Safety. On 12/1/1777, authorized to receive state arms and refit them for service. (3 Pa. Arch. 15 at 338) 1778, one of the state armorers (2 Pa. Arch. 2 at 699); 1780, gunsmith, Mulberry Ward, east part, Philadelphia (t). As a state armorer, Cowell received £10,788/18/0 for his services, June 1778 through November 1779 (2 Pa. Arch. 3 at 159).

Cowley, John. Cutler. On the roster of Major Joseph Shippen's Co., 2d Battalion, Pa. Regt., dated 11/7/1757. Born in England (5 Pa. Arch. 1 at 195).

Cox, Andrew. Philadelphia Co., 1730: "he is a blacksmith by trade and is recommended by sundry inhabitants" (2 Pa. Arch. 19 at 760).

Cox, Claudius. Pike maker. Employed by Daniel Pettibone, 9/15/1814, "to procure 200 pikes for the defense of this district" (MPHS [1867] 8:143).

Cox, Isaac. 10/20/1775: "Capt Robert Towers, Commissary, Reports that . . . he has purchased the following items . . . of Isaac Cox, 24 cutlasses at 2 shillings, 6 pence each" (Col. Rec., 10:375).

Cox, Martin. Gunsmith, Philadelphia (PXD, 1819).

Coxe, Tench. 1784, on list of persons attainted by treason; surrendered and discharged (1 Pa. Arch. 10 at 250ff.); 1803–12, purveyor of public supplies, Philadelphia. Received and inspected arms (RAG, 31).

Crabbe, Peter. Smith, 1660, at the Dutch colony on the South River, near New Castle, Del. (Hazard's *Annals*, 1:318–19).

Craft, D. S. Gunsmith, Brush Valley, Indiana Co. (BPaD, 1861).

Craft, George W. 1860–75, Morris Twp., Washington Co. Apprentice of Abe Williams (t; BPaD, 1861). Taught gunsmithing at Sam Post.

Craft, Jacob. 116 Budd St., Philadelphia (PXD, 1813). Gunsmith.

Joseph Craig (1822–99), Pittsburgh.

CO-PARTNERSHIP.—The undersigned have formed a co-partnership for the purpose of carrying on the business of Gunsmiths. They will occupy the old stand of William Craig on St. Clair street, opposite the centre of the Exchange Hotel. Guns of all descriptions will be made to order, and every attention given to repairs.

WILLIAM CRAIG,
JOSEPH CRAIG,

mar2-1w

Pittsburgh Gazette, June 4, 1850

Craft, John, Jr. (b. 1714). 1739, Williamsburg, Va.; 1757–63, Fayette Co. (*Horn Papers*, 2: 547–48).

Craig, George (b. 1852). Son of William Craig. 1866–77, Richland Twp., Clarion Co., DBA "William Craig & Son, Foxburg" (REH, 1:201–11).

Craig, Joseph. Artillery artificer, 1779, at the laboratory in Philadelphia (2 Pa. Arch. 15 at 386).

Craig, Joseph Allen (1822–99). Born in Westmoreland Co., England. He came to Pittsburgh in 1831 and was apprenticed, 1838–42, to Christian Wolf. He and his brother, William ("W & J Craig") operated a gunsmithing business in Pittsburgh through 1857. At that time, Joseph went to California and opened a branch. Joseph remained in California thereafter, while William journeyed back and forth (REH, 1:201–11; LSC, 67–68). Gunsmith.

Craig, Robert. 12/5/1775: "The sum of £100 to be advanced to said Wigfal, Blackwood & Robert Craig" to make 200 gunlocks at 22 shillings, 6 pence each, to be delivered within 6 months (Col. Rec., 10:421).

Craig, William (1820–77). 1846, "Craig McDonald," 4th Ward, Pittsburgh (*Harris Directory*, 1846–47); 1848, "W. & J. Craig", Pittsburgh; 1851, "Craig & Musgrave," Pittsburgh; 1857–58, Rochester, Beaver Co.; 1860, Sonora, California; 1861–66, Pittsburgh; 1867–77, Foxburg, Clarion Co. (REH, 1:201–11; LSC, 67–68). Gunsmith.

Cramer, Howard. Inventor. Patent 610,123 of 8/30/1898 for a magazine firearm. Williamsport, Lycoming Co.

Cress, George. 1828, Hamilton Twp., Northampton Co. (t). Gunsmith.

Crider, Felix (1838–1924). 1868, Beaver Creek, Huntingdon Co. (t); 1900–1910, Lebanon boro (c). Gunsmith.

Criner (Kryner, Greiner), William. 1802, Debra Criner bound her son, William, to George Kreps, Hagerstown, Md., for 5 years, 11 months, to learn the art and mystery of gunsmithing. 1812, William Greiner, gunsmith; Montgomery Twp., Franklin Co. (t, 1814, 1829, 1846; WSB, 23).

Crissey, Elias (1835–1915). Apprenticed to Samuel Border, Crissey was a gunsmith all his life in Hooversville, Somerset Co. (t; c; f).

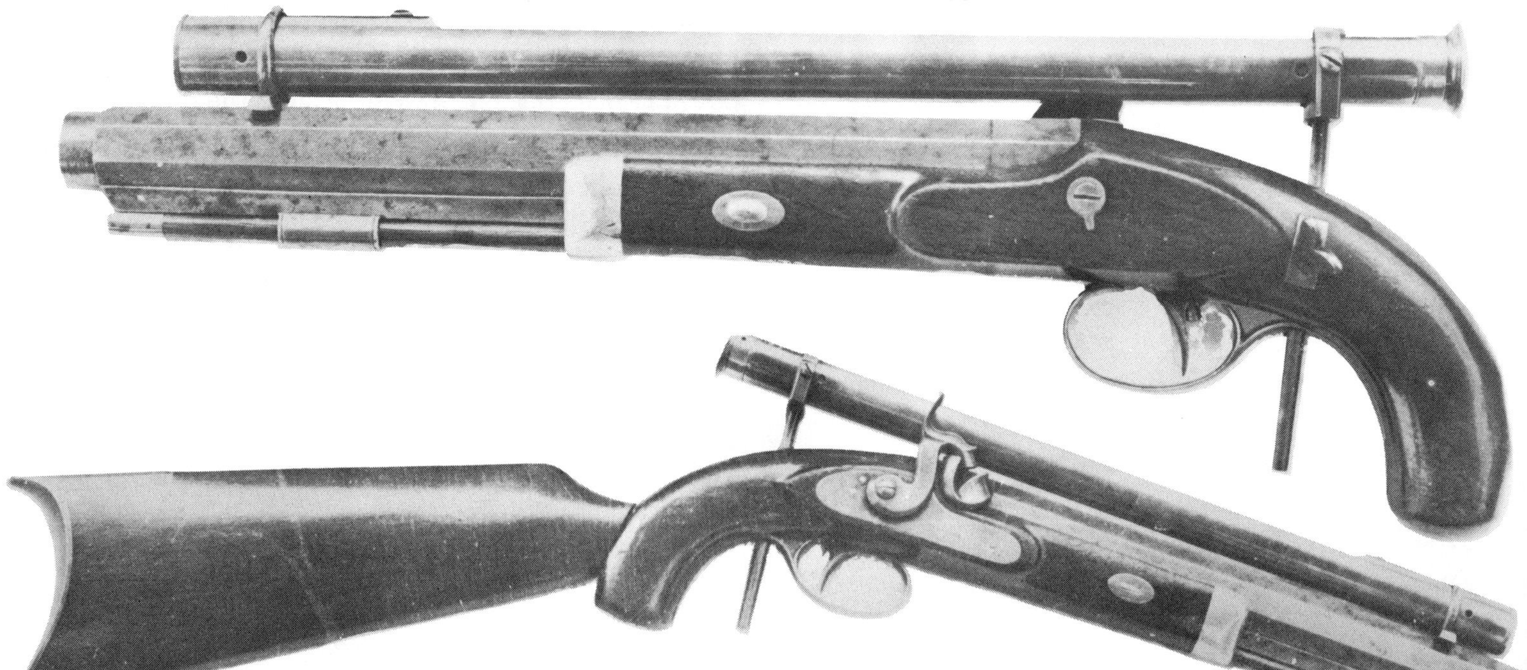

William Craig (1820–77), Pittsburgh, Allegheny County. An unusual pistol carbine with telescopic sight.

Crisswell, Samuel (d. 1810). Carlisle, Cumberland Co. (WSB, 55–57).Gunsmith.

Criswell, Samuel, Jr. (1752–1833). Locksmith, 1776, Capt. Samuel Boyd's Co., Lancaster Co. Militia; 1780–1814, Carlisle, Cumberland Co. (WSB, 55–57; KEA, 21; 2 Pa. Arch. 13 at 334). Gunsmith.

Cristy, Samuel. Cutler, 1779, West Nantmill Twp., Chester Co. (t); 1780, Windsor Twp., York Co. (t).

Croan, John. Gunsmith, 1779, York Co (3 Pa. Arch. 7 at 55).

Crocker, David. Gunsmith, Grove City, Mercer Co. (PD, 1890).

Croll, Daniel. 1790–1800, Providence Twp., Montgomery Co. (KEA, 21) Gunstock maker.

Croner, Abraham (1791–1861). 1837–43, Carrol Twp., Washington Co. (t); 1845–61, Cross Creek Twp., Washington Co. (t; BPaD, 1861). Gunsmith.

Crookham, James. Artillery artificer, 1778, Capt. Thomas Wylie's Co. at Carlisle (PIR, 2:248).

Crosbee, Edward. 4082 Market St., Philadelphia (BSED, 1880). Gunsmith.

Crosbee, Walter L. 4079 Filbert St., Philadelphia (BSED, 1880). Gunsmith.

Croslet, Richard. Goldsmith, 1681–82, Penn's Grant (MRB, 198); 1701, "from Bristol, England," goldsmith (3 Pa. Arch. 3 at 330).

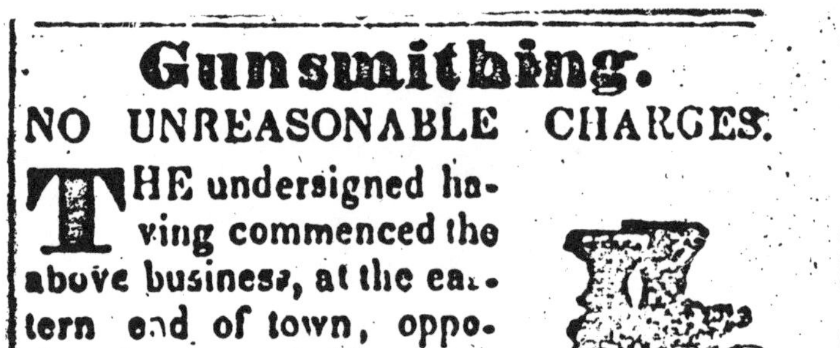

Gunsmithing.

NO UNREASONABLE CHARGES.

THE undersigned having commenced the above business, at the eastern end of town, opposite the stone bridge, tenders his services to his friends and the community at large. He intends keeping on hand an assortment of *RIFLES, SMOOTHBORES, SHOT GUNS, PISTOLS &c.* which he will sell LOW for cash or approved produce.

ALSO—All kinds of repairing done in the neatest manner and on the shortest notice, such as Pistols, gun locks, door-locks, coffee-mills, or any other kind of repairing, whatever.

☞ His charge shall be low and work well done.

JOHN CROSSLAND.

June 2, 1840.

Uniontown Weekly Conservative, June 9, 1844

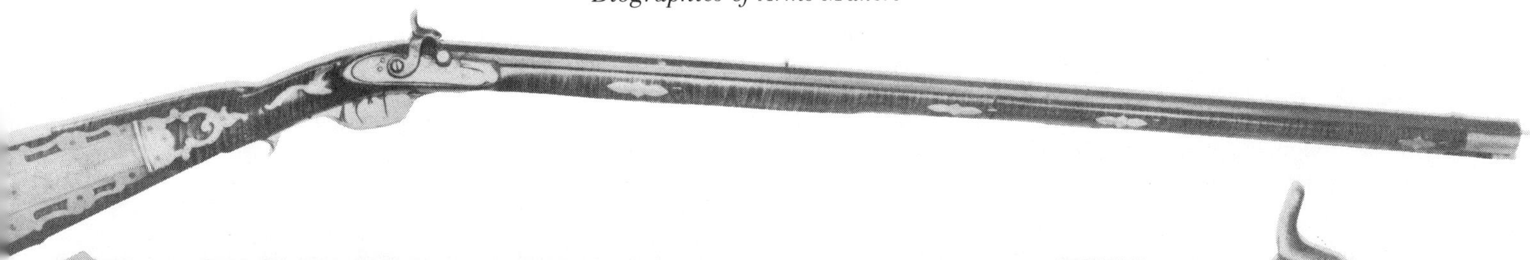

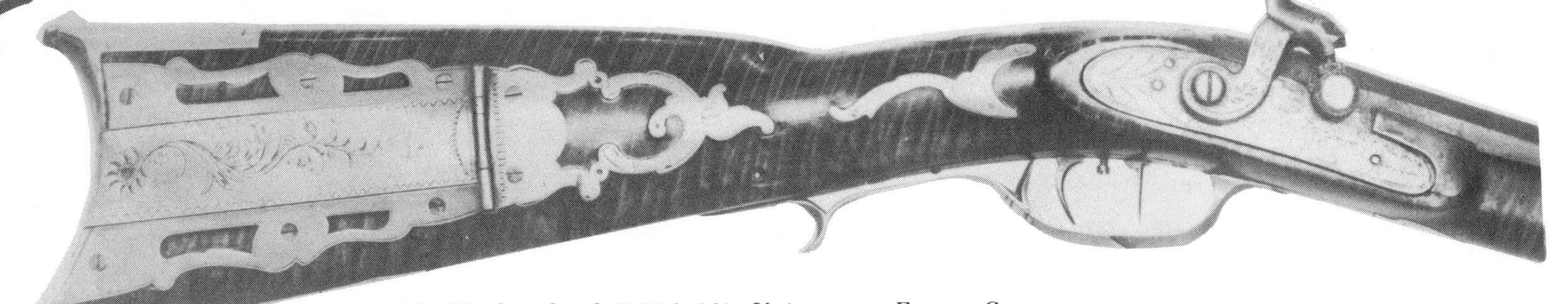

J. M. Crossland (1816–90), Uniontown, Fayette County. Noble collection.

Crossland, John M. (1816–90). 1840–90, Uniontown, Fayette Co. (c; t) Gunsmith. Probably late apprentice of Peter White.

Crum, Daniel (1841–1926). Near Newport, Perry Co., gunsmith (t; c).

Crum, Samuel (1840–1924). Sugar Run, Perry Co. (c; t). Gunsmith.

Crum, William. Cass Twp., Huntingdon Co. gunsmith (PD, 1890).

Crunkleton, John. Gunsmith, 1772–98, Antrim Twp., Franklin Co. (WSB, 29).

Cryscher, Daniel. Gun-barrel forger. Cocalico Twp., Lancaster Co., 1830 (t; SED, 54).

Culbertson, Alexander. Gunsmith, 1805–8, Scrub Grass Twp., Venango Co. (t).

Culver, Edward. Gunsmith, 1835, Bushkill Twp., Northampton Co. (t).

Cummings, William. Gunsmith, 45 Green St., Philadelphia (DeSD, 1829, 1830, 1833, MD, 1835–36, 1837, 1841, 1846); 107 Charlotte St., Philadelphia (MD, 1855).

Cunkle, George. North St. and 3d St., Harrisburg (BPaD, 1861). Gunsmith.

Cunningham, Alexander. Gunsmith, 1802–5, German Twp., Fayette Co. (t).

Curry, Archibald. Gunstocker, 1777, at French's Creek Arsenal.

Curry, Charles (d. 1863). Born in Ireland. 1837–52, Philadelphia (t); 1852–63, San Francisco, Calif. (LSC, 73). Gunsmith.

Curry, William (1739–1820). 1778–83, Carlisle, Cumberland Co. Gunsmith. 1783–1820, Washington Co. (t; WSB, 55) "William Curry was a native of Dauphin County, Pa., and was born in 1739. He learned the trade of gunsmith and cooper. He was a soldier in the Revolutionary War, after which he married, and in 1783, emigrated from Dauphin County, with his wife to Washington County and purchased 21 acres. . . . [T]he land at the time was not patented and later Mr Curry rode on horseback to

Daniel Crum (1841–1926), Perry County.

Lancaster and secured the patent. On this tract he lived until his death in 1820, aged 81 years" (Crumrine, *History of Washington County*, 657).

Custer, Nicholas. Gunsmith, Providence Twp., Montgomery Co., 1800 (t; c).

D

Dahlgren, John A. (1809–70). Invented the Dahlgren cannon and Dahlgren naval knife. In 1863 and 1869 he was chief of naval ordnance (GAAAM, 31).

Dalby, Alexander (1827–80). 1850–80, gunsmith, Millsboro, Washington Co. (t; c).

Dalby, Enoch. 1824–78, Millsboro, Washington Co., gunsmith (t; c).

Dalby, Henry. 1838–39, East Bethlehem Twp., Washington Co., gunsmith (t).

Dalby, James M. (1837–1906). 1860–1906, Millsboro, Washington Co., gunsmith (t; PD, 1882).

Dalby, William (d. 1828). Gunsmith, 1815–28, Fredericktown, Washington Co. (t).

Daley, Peter (d. 1804). 1782–96, Franklin Twp., Greene Co. (t); 1797–1804, Ohio Co., (West) Virginia. (t).

Dalton, John. Artillery Artificer, 2/14/1777 to 3/21/1781, in Capt. Jesse Rowe's Co. at the cannon factory, Philadelphia. (PIR, 2:244).

Damer, Henry. Gunsmith, N. Queen St. below Frederick St., Lancaster boro (BLD, 1857).

Daniel, Adam. Gunsmith, Wayne Twp., Crawford Co. (CCD, 1874).

Daniel, Phineas. Owned a shot tower in Philadelphia, purchased later by Thomas Sparks. Patent of 3/31/1804 for a method of making shot (RAG, 48).

Daniels, Daniel. "Daniel built a small powder mill prior to 1810 and supplied his neighbors, fond of hunting, with gunpowder." Randolph Twp., Crawford Co. (Brown, *History of Crawford County*, 597).

Danseth, Andrew. 1799, Pittsburgh (t); 1800–1820, Cincinnati (t). Gunsmith (FSP, 75).

Darlington, Benjamin. In Pittsburgh *Gazette* of 11/29/1816 he advertised parts and accoutrements for sale in Pittsburgh.

Darroch, Thomas. 12/9/1776: "pay Thomas Darroch £6 each for rifles" (Col. Rec., vol. 11 *of the*, 39:34).

Dash, Jacob (d. 1790). About 1780, Montgomery Co., he operated a gunpowder mill *Proceedings* German Society Pa. 34).

Daub, Jacob. Gunsmith, Upper Salford Twp., Philadelphia Co. 1779–83 (t).

Daugherty, Absalom. 1842, Annville, Lebanon Co. (KKR, 204). Gunsmith.

Davidson, John. 1776: "pay John Davidson £75/04/6 for cartouch boxes" (Col. Rec., 10:685); 1777–78: "paid John Davidson's bill for scabbards, slings, cartouche boxes, belts & c. £163/12/6" (3 Pa. Arch. 5 at 381).

Davis, Casper Ealy (1848–1928). New Paris, Bedford Co. (t; f; c) Gunsmith and handyman. Son-in-law of Benjamin Vore.

Davis, Elisha. Artillery artificer, 1778, at the cannon factory at Carlisle (PIR, 2:248).

Davis, James. Artillery artificer, 1778, Philadelphia (2 Pa. Arch. 11 at 258).

Davis, Joshua. Inventor. Patents: 112,127 of 2/28/1871 for a magazine rifle; 132,357 of 10/22/1872 for a revolver; 182,646 of 9/26/1876 for a revolver. Limestone Twp., Montour Co.

Davis, Thomas. 11/30/1752: "cutlery and pewter for sale, where Hugh Roberts was formerly" (*Pennsylvania Packet*, Philadelphia).

Davison & Jones. Gunsmiths, Allegheny City (Pittsburgh) (PAD, 1887).

Dawson, John. Armorer, armed boat *Experiment*, Pa. Navy (2 Pa. Arch. 1 at 289) 1777.

Dawson, Paul. Swordsmith in the employ of Lewis Prahl, gunsmith, making swords for Col. Benj. Flowers's Pa. Horsemen, 1777 (2 Pa. Arch. 3 at 104).

Dawson, William. Cutler, 1769, North Ward, Philadelphia (3 Pa. Arch. 15 at 195).

Day, Richard. 68 Marker St., Philadelphia. Gunsmith (SGD, 1835–36; MD, 1837, 1841, 1846).

Dean, John. Gunsmith, 1795–1805, Morris Twp., Huntingdon Co. (t).

Dean, S. O. Rockdale Twp., Crawford Co. Gunsmith (PD, 1882; CCD, 1890).

Dearth, Lacy (b. 1812). 1840–60, Brownsville, Fayette Co., gunsmith (t; c).

DeBarrier (DeBearer), Henry. Gunsmith, 1774–80, Mulberry Ward, Philadelphia (t).

DeBinder, George. Gunsmith, 120 South St., Philadelphia (MD, 1855, 1859); 337 South St. (MD, 1861; BPaD, 1861; MD, 1865).

DeBolt, Henry. Gunsmith 1869, Monongahela Twp., Greene Co. (t); 1876, Mapletown, Greene Co. (t).

DeCoursey, John G. Patent 53,582 of 4/3/1866 for a new type of bullet. Philadelphia.

Deeds, Henry W. (1811–88). Rifle-barrel maker. 1860, Leacock Twp., Lancaster Co. (c, 1860) "About 1838 Cornelius Reichwein opened a gun shop where is now Ruth's grist mill, in which Henry Deeds worked for many years as one of the most skillful gunmakers in the county" (Montgomery, *History of Berks County*). Cumru Twp., Berks Co. 1847–51. Principally a gun-barrel maker.

Defibaugh, David (1841–1914). 1860–64, 1870–1914, Everett, Bedford Co. (t; f; c). Gunsmith.

Defibaugh, James E. Silversmith, 1829, Bedford boro. (t); gunsmith, 1852, Bedford Twp., Bedford Co. (t).

Defibaugh, James Lawrence (b. 1842). 1860–70, Monroe Twp., Bedford Co. (t; c); 1870–80, Bedford Twp., Bedford Co. Gunsmith.

Defibaugh, Milton, 1870–80, Monroe Twp., Bedford Co. (t; c); 1882–83, Bedford boro. (t); 1886, Everett, Bedford Co. (t). Gunsmith.

Defibaugh, William (1814–91). 1836–50, West Providence Twp., Bedford Co. at the widow Defibaugh's. 1851–91, Robinsonville, Monroe Twp., Bedford Co. Master gunsmith.

DeGuff, Heinrich (d. 1799). Gunsmith, Lancaster (Burial Book of the Moravian Church, 1744–1821, 22).

DeHaven, Hugh. 3/14/1777, 4/8/1777, assistant to his brother, Peter DeHaven, at the Pennsylvania Arsenal (2 Pa. Arch. 3).

DeHaven Peter. 1768, gunsmith, Lancaster boro. (t; SED, 54); 1777–79, principal gunsmith at the Pennsylvania gun repair facility at French's Creek and later at Hummellstown, Dauphin Co. (BHAM, 1:527; RAG, 50–51). On 3/10/1777 the French Creek gunpowder factory exploded (1 Pa. Arch. 5 at 255). In early 1777, DeHaven had 19 workmen at French's Creek (1 Pa. Arch. 5 at 155). The factory was moved in 1778 to Hummellstown due to British pressures in the Philadelphia area. It was abandoned in early 1779, and DeHaven applied for a position as a wagonmaster (1 Pa. Arch. 7 at 174). His final accounting of the disposal of equipment was given in April 1779, and amounted to £4,000 (3 Pa. Arch. 5 at 35). In 1790 he was a gunsmith in Philadelphia (c, 1790).

DeHuff, Abraham. 1771–82, Lancaster boro. (He is listed as a gunsmith by RAG, 51; but not as such in tax lists.) Abraham DeHuff was married on 5/26/1756 at Philadelphia to Mary Finch (1734–90). He was born on 2/13/1735 in eastern Md. and died in Lancaster on 3/11/1821 (S. M. Sener, in Egle, *Notes and Queries* [1898]).

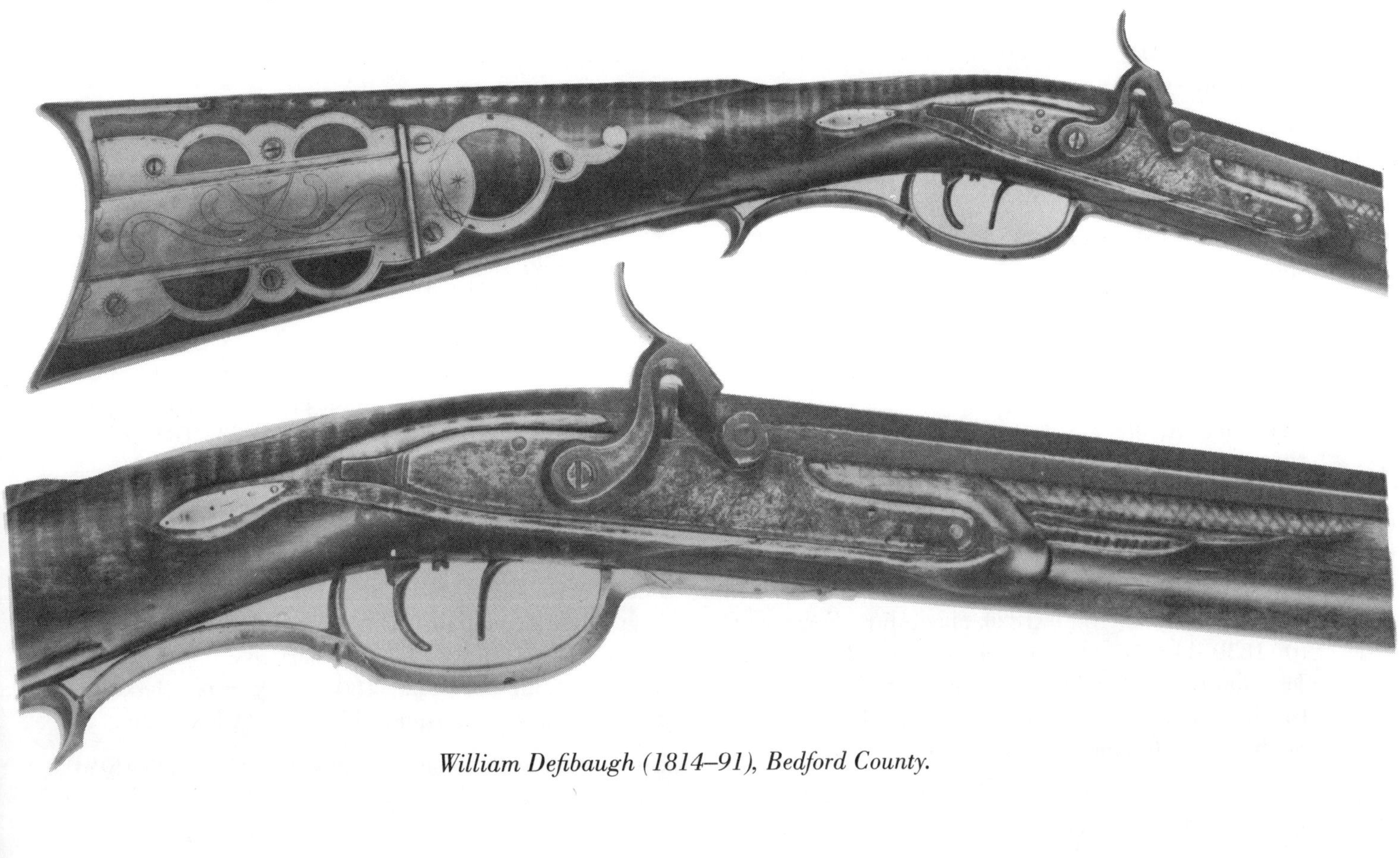

William Defibaugh (1814–91), Bedford County.

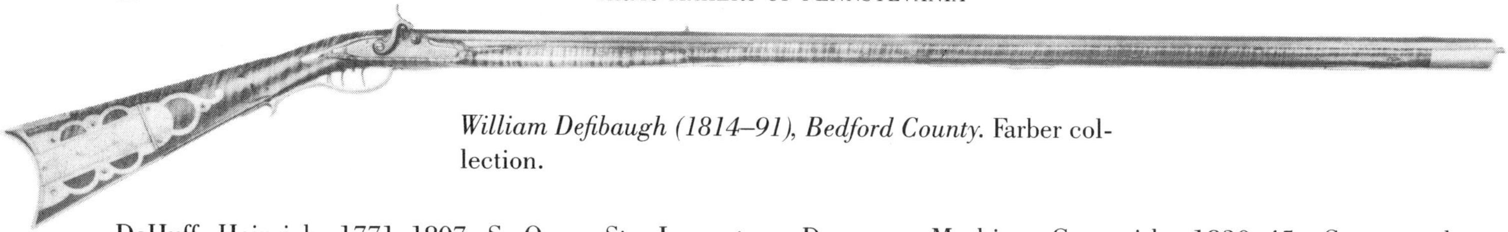

William Defibaugh (1814–91), Bedford County. Farber collection.

DeHuff, Heinrich. 1771–1807, S. Queen St., Lancaster boro. He married Philippina Eberman (d. 3/25/1795) on 3/14/1768. On 4/17/1801 he received a contract from Gov. McKean for M-1795 "liberty cap" muskets. He worked with Abraham Henry and Peter Brong. In 1807 he was a contractor with Peter Gonter and Jacob Dickert (t; SED, 54; KEA, 22; RAG, 51; JKJ, 103).

DeHuff, John. 1793–1803, Washington boro. (t); 1803–19, Lancaster boro (t). Gunsmith.

Deille, Edward. Artillery artificer, 1779, at the cannon factory in Philadelphia (2 Pa. Arch. 15 at 384).

Deisinger (DeSinger), Wendell. N. 2d St. at Dauphin St., Philadelphia (BPaD, 1861); 7 Green St. (MD, 1846); 1500 N. 2d St. (MD, 1855).

Deisler, John. On 2/22/1776, given an order for 150 cartouche boxes; on 5/8/1776, given an order for an additional 150 cartouche boxes at £108/15/0 (BCR, 175). Berks County.

Deitz, Adam. Gunsmith, 1822–33, Warren boro (t).

Delaney, I. F. Reading boro (PD, 1890). Gunsmith.

Delaney, Nelson. 1840–60, Church Alley, Reading (t); 8 N. 5th St. (BPaD, 1861); 650 Walnut St. (BRD, 1880; PD, 1882). Gunsmith.

Delavan, George W. (b. 1832). Brother of Eliza, wife of Joseph Craig, his apprentice to the gunsmithing business, 4th Ward, Pittsburgh (c, 1850).

Delavan, Joseph. Gunstocker, 1777, at French's Creek Arsenal; 1779, artillery artificer, Philadelphia (2 Pa. Arch. 15 at 384).

Dellet (Dellot), Peter. Prince St., corner of Lemon St., Lancaster boro (c, 1850; BLD, 1857). Gunsmith.

Dely, John. Artillery artificer, 1778, Capt. Thomas Wylie's Co. at Carlisle (PIR, 2:248).

Demler, Henry. Silversmith. On 5/22/1772, apprenticed for 5 years to George Dowig, Philadelphia, to learn silversmithing (PIDS).

Dempsey, John (b. 1800). 1828, Shade Twp., Somerset Co. (t); 1830–44, Conemaugh Twp., Cambria Co. (t); 1845, Johnstown (t); 1846–50, Conemaugh Twp., Cambria Co. (t); 1850–52, Conemaugh boro. (c; t); 1853–54, Summerhill Twp., Cambria Co. Gunsmith.

Dempsey, Mathias. Gunsmith, 1830–45, Conemaugh Twp., Cambria Co. (t).

DeMuth, Christopher. Gunsmith, 1790–1804, Lancaster boro (t; SED, 54).

DeMuth, Johannes (d. 1794). 1790, gunsmith, Lancaster boro (SED, 54; t). Burial Book of the Moravian Church lists his date of death.

DeMuth, Jonathan. Gunsmith, 1828, Bushkill Twp., Northampton Co. (t).

DeMuth, Joseph. 1800–1813, Lancaster, gunsmith (SED, 54; t).

DeMuth, William. Gunsmith, 1820, Bushkill Twp., Northampton Co. (t).

Dene, Thomas. 5/4/1749, *Pennsylvania Packet:* "Thomas Dene, hired man, born in Ireland, silversmith, age about 30, ran away from George Fling of Northern Liberties, reward by Fling or John Clace at the Golden Fleece in 2d St."

Dening, William (1736–1830). 1777–81, made cannon for the Continental army, Mifflin Twp., Cumberland Co.; also made gun barrels. Died on 12/19/1830, age 94, in Cumberland Co. (WSB, 50; PIR, 2:246).

Dennis, William. 1815–18, Wilkes-Barre, gunsmith (t; KEA, 22); 1818–32, Montrose, Susquehanna Co. (t).

Derby, David B. Erie Co. (c, 1850). Gunsmith.

Deringer, Henry, Sr., 1794–98, Easton, Northampton Co. (t); gunsmith, 30 N. Front St. and 29 Green St., Philadelphia (PXD, 1810, 1813, 1816).

Deringer, Henry, Jr. (1786–1869). 370 N. Front St. and 612 N. Front St., Philadelphia. Frequent government contractor, ranging from Indian trade guns to military muskets (PXD, 1810, 1813, 1816; DeSD, 1823, 1833, 1837; RAG, 51–52).

Deringer, John H. N. Front St. near Coates, Philadelphia (DeSD, 1835–36). Gunsmith.

Derr, Christian, Sr. (1748–after 1820). 1777–81, 11th Pa. Regt.; transferred from Thompson's Riflemen. Wounded at Germantown; 3 broken ribs. He carried 3 musket balls in his body until his death. 1805, Oley Twp., Berks Co.; 1820, Union Co. Gunsmith (PIR, 1:749; KKR, 207).

Derr, Daniel. Bellefonte, Centre Co. Gunsmith (BPaD,

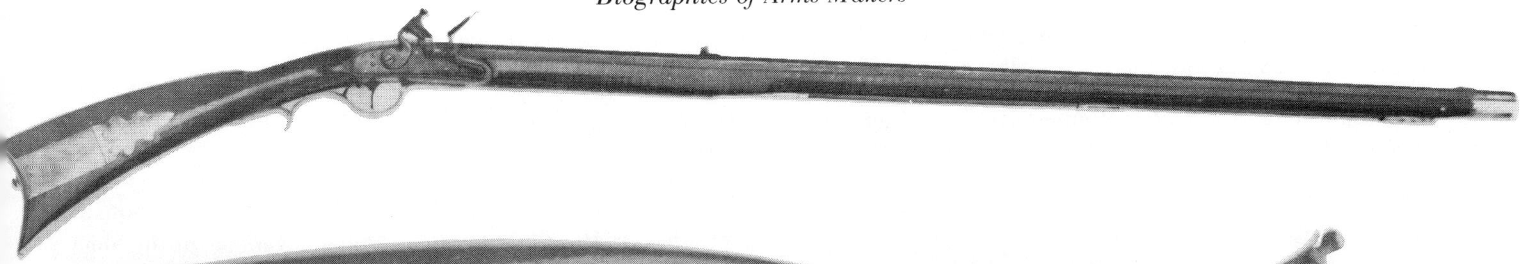

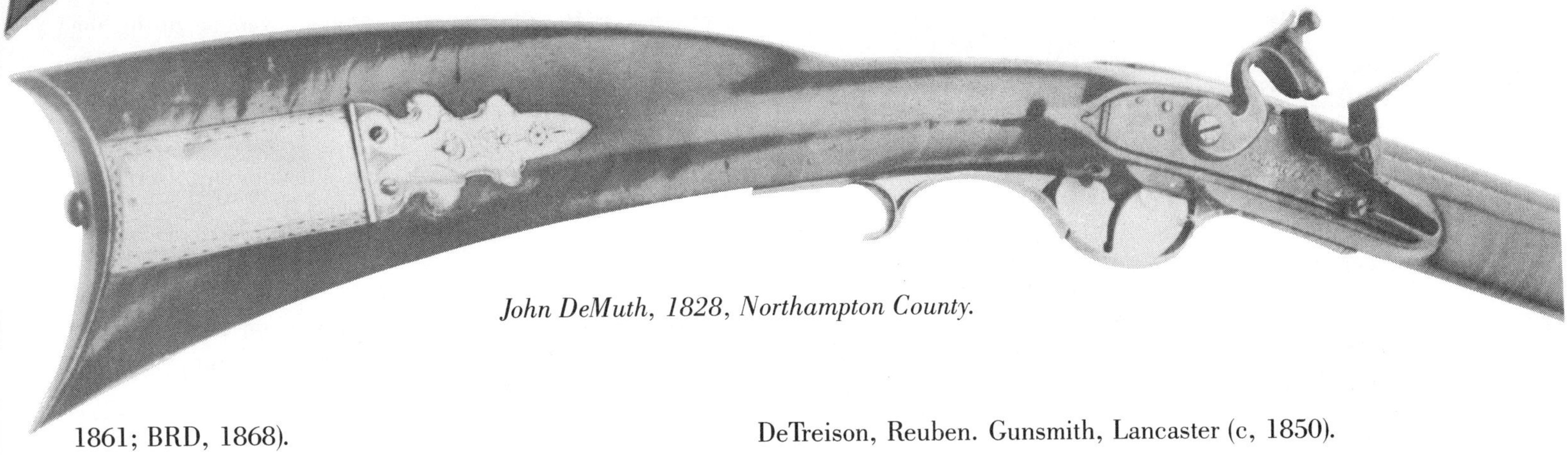

John DeMuth, 1828, Northampton County.

1861; BRD, 1868).

Derr, Johannes. Gunsmith, 1800–1807, Weisenberg Twp., Northampton Co.; 1818, Berks Co. (t; GAAAM, 42).

Derr, John. Eldest son of Michael Derr, a Palatine Protestant. 1834, Durham Twp., Bucks Co.; 1839, Noxamixon Twp., Bucks Co.; 1849, Shamokin, Northumberland Co.; 1853, Upper Augusta Twp., Northumberland Co. (t; *History of Wyoming and Lackawanna Valleys* [Chicago: Beers, 1883] 1:313–14). Gunsmith.

Derr, Peter. Tulpehocken, Berks Co. Gunsmith (BRD, 1860; BPaD, 1861).

Derringer, Jacob. Gun factory, 19th Ward, Philadelphia. (c, 1860).

Derringer, John. A tailor whose name was used by Slotter & Co. (q.v.) in order to make "Derringer" pistols. Philadelphia, ca. 1860–68 (RAG, 52).

Derringer, Philip. Gunsmith, Bushkill Twp., Northampton Co. (t, 1828); N. 4th above Washington St., Reading (BRD, 1860; BPaD, 1861).

Derringer, T. T. Gunsmith, 8 Coates, Philadelphia (MD, 1841, 1846).

DeSaney (DeLaney?), Nelson. 1846–49, Upper Saucon Twp., Lehigh Co. Gunsmith (t).

Deschner, Theodore. Gunsmith, 5 Bush Block, Bellefonte, Centre Co. (BRD, 1871, 1873, 1875; PD, 1882, 1887).

DeShera, William. Gunsmith, Jersey Shore, Lycoming Co. (PD, 1882).

DeSinger, Valentine. Gunsmith, 362 N. Front St., Philadelphia. (MD, 1837).

Deterer, Adam. Committee of Safety contractor, 1775–76 (RAG, 52).

DeTreison, Reuben. Gunsmith, Lancaster (c, 1850).

Detrick, Jacob. ca. 1800, Queen and Front St., Chambersburg. Sold parts, accoutrements, and supplies (WSB, 50).

Detwiler, Andrew. Springfield Furnace, Blair Co. (BPaD, 1861). Gunsmith.

Detwiler, Anthony (b. 1808). 1850–61, Huston Twp., Blair Co. (t); 1866–76, Middle Woodberry Twp., Bedford Co. (t); 1879–92, Bloomfield Twp., Bedford Co. (t). Gunsmith.

Detwiler, Christian. Bellville, Mifflin Co. (c, 1860, 1870). Gunsmith.

DeVeese (Device), John. 1816–20, Springhill Twp., Fayette Co. (t). Gunsmith.

DeVries, David Pieterszen. 1630–33, New Netherlands: "a bold and skillful seaman" and a "master of artillery in the service of the United Provinces" (Hazard's *Annals*, 1:24).

Dickens, Fowler. Lilly Alley, Philadelphia, gunsmith (DeSD, 1829, 1830; MD, 1833, 1835–36, 1837, 1841).

Dickens, John. 31 New Market, Philadelphia (DeSD, 1829, 1830). Gunsmith.

Dickens, Richard. 11 Derringer's Ct., Philadelphia (DeSD, 1835–36; MD, 1837, 1841). Gunsmith.

Dickert & Gill. Benjamin Gill and Jacob Dickert, ca. 1800 (KEA, 24). Lancaster, gunsmiths.

Dickert, Jacob (1740–1822). Born in Maintz, Germany; apprenticed at age 15 and a journeyman by age 21. Arrived in America by 1769. 1769–1822, Lancaster boro., master gunsmith. Government contractor with Peter Gonter and Henry DeHuff for 557 muskets, order dated 12/9/1807. (t; GAS-1, 204–19; JKJ, 81–84) 1/14/1777:

"£8/09/10 to Jacob Dickert for repairing arms, &c." and 9/23/1777, £24/2/4 for the same (3 Pa. Arch. 6 at 376).

Dickey, David. 1779–91, Middletown Twp., Cumberland Co. (t; WSB, 55). Gunsmith.

Dickey, Moses. (d. 1814). 1801–14, Salem Twp., Westmoreland Co. (t). Gunsmith.

Dickey, William. Silversmith, 1787, Louisburg, Dauphin Co. (Egle, *History of Dauphin and Lebanon Counties*, 297).

Dickinson, A. Gunsmith, Gray's Ct., Philadelphia (MD, 1841).

Dickson, W. H. (b. 1844). 1864–68, Cory, Erie Co.; 1869–83; 1109 State St., Erie (t; c; PD, 1882). Gunsmith and locksmith.

Diddy, John. Artillery artificer, 3/21/1777 to 3/22/1781, Capt. Jesse Rowe's Co. at the cannon factory at Lebanon (PIR, 2:244).

Diebenberger (Dippenberger), Heinrich. 8/14/1770: "Henrich Diebenberger, gunsmith, has removed from Second St. to market St., Philadelphia, the second house from the Black Bear Inn." An advertisment on 9/4/1772 "Heinrich Dippenberger" as a gunsmith (*Wochenlicher Pennsylvanischer Staatsbote* [Heinrich Miller's newspaper]).

Diefenderfer, Philip. Lower Macungie Twp., Lehigh Co. (t, 1847); Lowhill, Lehigh Co. (BSED, 1880). Gunsmith.

Diemer, Jacob. 1828–35, Moore Twp., Northampton Co. (t). Gunsmith.

Diemer, Peter. Gunsmith, 1807–35, Moore Twp., Northampton Co. (t).

Dietz, John. 1800, Somerset Co. (t; c). Gunsmith.

Dietz, Theodore. W. King near Mary St., Lancaster boro. Powder-horn maker (BLD, 1857).

Diggin, Edward. Artillery artificer, 1778, Capt. Thomas Wylie's Co. Made cannon at Carlisle (PIR, 2:248).

Dillon, James. Apprenticed in Chambersburg, ca. 1819; 1826–44, Bedford boro. Gunsmith, clockmaker and watchmaker, silversmith (t; f).

Dimond, Daniel. Pershing, Cambria Co. (BPaD, 1861). Gunsmith.

Dimond, Levi. South Fork, Cambria Co., gunsmith (c, 1860, 1870).

Dingler, John. Gunsmith, 1807–14, Easton (t).

Dingler, Samuel. Gunsmith, 1823, Easton (t).

Dingler, William. Gunsmith, Green St. at Front St., Philadelphia (MD, 1846); rear of 337 N. 3rd St., Philadelphia (MD, 1855).

Dinsman, John. 404 N. 3d St., Philadelphia (PXD, 1813, 1816, 1819). Gunsmith.

Diston, Henry. Operated firm of H. Diston & Sons, founded 1840, Philadelphia. Made bayonets during the Civil War. Otherwise, tool and edged-weapons makers (RAG, 221).

Doblenar, F. F. G. (b. 1783). Goldsmith. Arrived 8/26/1805 on ship *Margaret* from Württemberg, Germany, age 22, "dark" (2 Pa. Arch. 17 at 637).

Doddridge, John (1745–91). Born in Maryland, 3/30/1745. He lived in Bedford Co., 1771–73. (t) He moved to Independence Twp., Washington Co., where he received a Va. land grant. His son wrote of him:

> Not possessing sufficient health for service on the scouts and campaigns, his duty was that of repairing the rifles of his neighbors when they needed it. In this business he manifested a high degree of ingenuity. A small depression on the surface of a stump or log and a wooden mallet were his instruments for straightening the gun barrels when crooked. With the aid of a bow string he could discover the smallest bend in a barrel. With a bit of steel he could make a saw for deepening the furrows when requisite. A few shots determined whether the gun might be trusted. (Joseph Doddridge, *Notes and the Settlement and Indian Wars* [Pittsburgh: McClain, 1912] 115, 272–73).

Doll (Dull), Casper (1724–93). Arrived on 8/27/1739 on the ship *Samuel* from Bern canton, Switzerland. Active ca. 1845–93, Plainfield Twp., Northampton Co. *(Proceedings of the Pennsylvania German Society)* Gunsmith.

Doll, Daniel. ca. 1790–1800, York (KEA, 26). Gunsmith.

Doll (Dull), Henry. 1776, pvt., 1st York Regt.; 1818, gunsmith in York Co. (2 Pa. Arch. 10 at 221; 3 Pa. Arch. 23 at 519; t).

Doll (Dull), Jacob (d. 1847). On 4/17/1801 he was a contractor with Conrad Welshans and Henry Pickel for 1,000 "liberty cap" muskets. 1838–47, West Manchester Twp., York Co. (KEA, 26; t). In 1827 he was paid $41.37 for cleaning and repairing York Co. Militia arms (Hazard's *Annals*, 2:59).

Doll (Dull), Joseph (b. 1768). 1797–1811, Harrisburg, Dauphin Co. (t). Silversmith.

Donald, Alexander. 1800, Gunsmith. Elizabeth Twp., Allegheny Co. (t); 1810–13, Rostraver Twp., Allegheny Co. (t).

Donham, Lewis N. (1833–1902). Schoolteacher who made guns on holidays. Nephew of Barney Engle of Greensboro, Greene Co. Twice married, he fathered 14 children. Greene Co. (t).

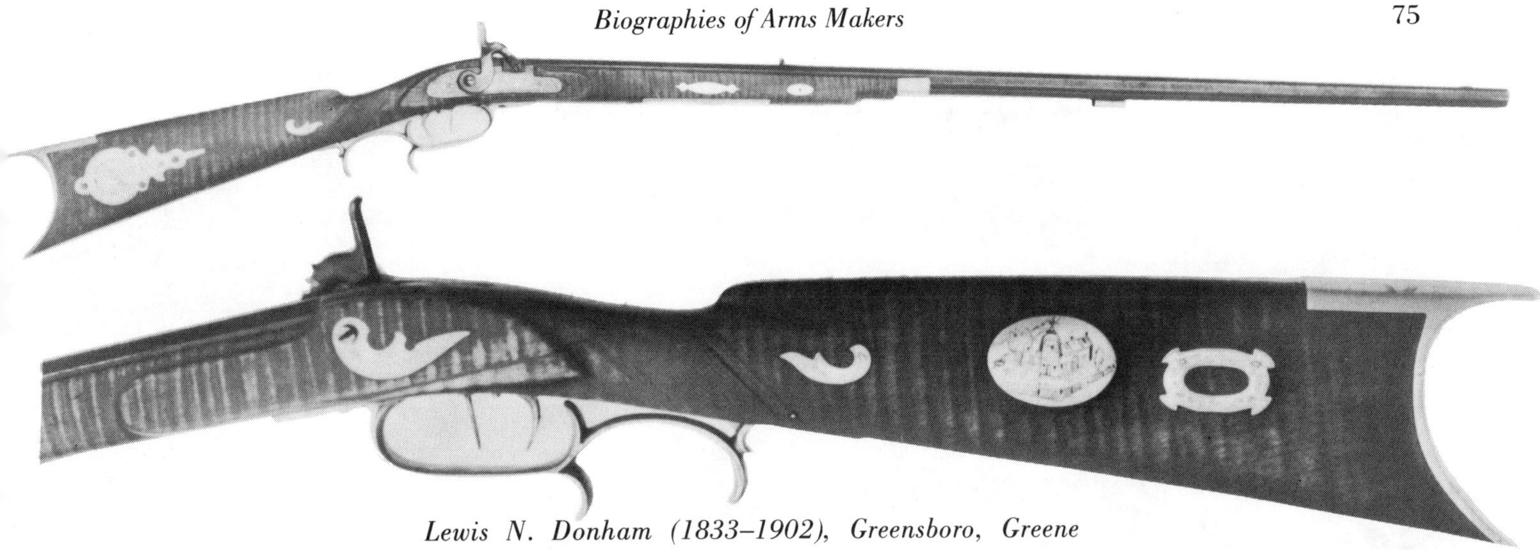

Lewis N. Donham (1833–1902), Greensboro, Greene County. Horvath collection.

Doning, Wilhelm (1736–1830). Born in Westphalia, Germany, he was ironmaster at Mt. Holly Forge, Middlesex, Cumberland Co., 1774–1818, after which he received a revolutionary war pension. He produced the first wrought iron cannon made in America. He was recruited by the English, but refused to desert. One piece captured at Brandywine is in the Tower of London museum. The barrels were made of 4 layers of wrought iron staves, hooped like a barrel with iron bands, and firmly boxed and breeched like other cannon (GFC, 105).

Donner, Joseph. Gunsmith, 1818, Haines Twp., Centre Co. (t).

Dorman, Harman (b. 1797). Gunsmith, Covington Twp., Clearfield Co. (c, 1850). Born in Germany.

Dornbach, Anthony. 1803, gunsmith, Cocalico Twp., Lancaster Co. (t; SED, 54).

Dorner, John (b. 1796). Born in Md., he was active as a gunsmith at Sabbath Rest, Antis Twp., Blair Co. (BPAD, 1861; t).

Dotterer, Conrad. Gunsmith, 1789, Skippack Twp., Montgomery Co. (t).

Dotterer, Henry. Gunsmith, 1784, Skippack Twp., Montgomery Co. (t).

Dougherty, Absalom. ca. 1840–45, Annville, Lebanon Co. (KEA, 26). Gunsmith.

Douglass, David (b. 1823). 1850–52, Frankstown Twp., Blair Co. (t; c); 1860–64, Warrior's Mark Twp., Huntingdon Co. (t); 1866–67, Frankstown Twp., Blair Co. (t). Gunsmith.

Douglass, James. Gunsmith, 1840, Tyrone, Huntingdon Co. (t).

Douglass, John. Gunsmith, 1830–33, Huntingdon boro. (t).

Douglass, Joseph, Sr. (d. 1841). 1813–41, Juniata Twp., Huntingdon Co. (t). Gunsmith.

Douglass, Joseph, Jr. (b. 1819). 1840–49, McConnellstown, Huntingdon Co.; 1850–80, Huntingdon boro (t; c). Gunsmith.

Douglass, Thomas. 1833–40, McConnellstown, Huntingdon Co., gunsmith. Apprenticed to Nicholas Shennefelt (t; AFR, 384). Africa says he went to Ohio.

Douglass, William. Inventor, Corry, Erie Co. Patent 43,903 of 8/23/1864 for a cannon.

Douthert, John. Gunsmith, 1826–28, Conemaugh Twp., Indiana Co. (t).

Dow, Alexander. Capt.-lt., artillery artificers, 1777–82 (2 Pa. Arch. 11 at 250).

Dowdel, James. Artillery artificer, 8/12/1777 to 3/22/1781, Capt. Jesse Rowe's Co. at the Philadelphia cannon factory (PIR, 2:244).

Dowig, George. Silversmith, Philadelphia. (PIDS).

Dowler (Doular), John. 365 N. 4th st., Philadelphia, gunsmith (PXD, 1816, 1819); 174 Coates (DeSD, 1823, 1829, 1833).

Downard, John. Gunsmith, 1820–34, Georges Twp., Fayette Co. (t).

Downward, Thomas (d. 1825). Gunsmith, 1785–1825, George Twp., Fayette Co. (t).

Doyle, John (1750–78). Born 1/23/1750, a son of Thomas and Mary (Young) Doyle, he was trained as a gunsmith. On 9/5/1776 he formed his own independent company as part of the 1st Pa. Battalion (he was reassigned in 1778 to the 6th Pa. Regt.). In 1781 at Chickahoming Church, Va., he was wounded and never fully recovered (HFAML, 329).

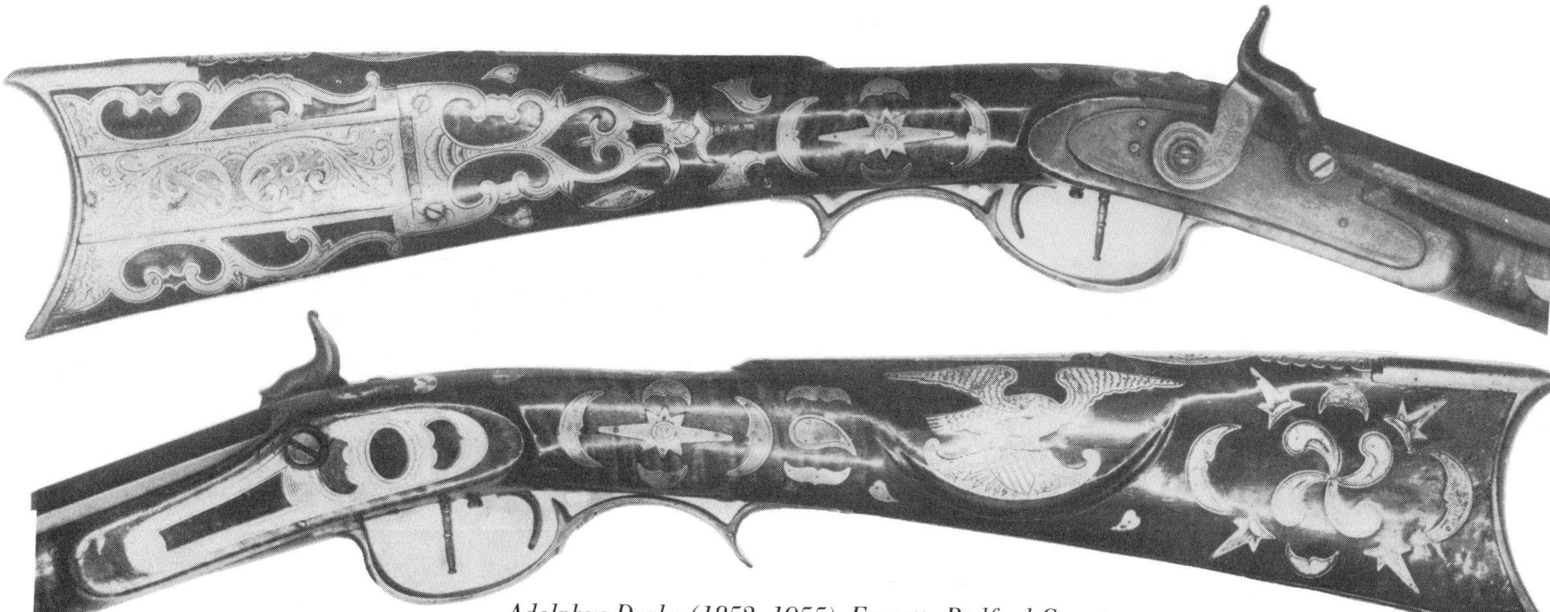

Adolphus Drake (1852–1955), Everett, Bedford County.

Huntingdon County school, marked "Kough." Possibly the work of Joseph Douglass or one of his sons or apprentices.

Adolphus Drake, Bedford County.

Drake, Adolphus (1852–1955). Born near Flintstone, Md., a son of Isaac Drake. A reputed gunsmith, he was active, ca. 1880–1930, in Everett, Bedford Co. He was 103 years old and blind at his death (t; f; c).

Dreisbach, Henry. Blacksmith, Alsace Twp., Berks Co. (t).

Dreisbach, John (d. 1823). Mifflinburg, Union Co. Active 1793–1802. Gunsmith, Buffalo Twp., Union Co. (LIN, 2:279, 332; t).

Dreisbach, John, II. Gunsmith, nephew of John Dreisbach the elder and probably his apprentice. 1812–61, East Buffalo Twp., Union Co. (t; c; LIN, 2:279, 332).

Dreisbach, Martin. 1760–80, Alsace Twp., Berks Co., smith (2 Pa. Arch. 2 at 403; 3 Pa. Arch. 18 at 306).

Dreisbach, Samuel. (b. 1790). Mifflinburg, Union Co. Took over John's shop in 1823; active 1812–38, gunsmith. (LIN, 2:279, 332) There was a Samuel Dreisbach in Circleville, Pickaway Co., Ohio, 1839–74 (DH, 315).

Dreppard, Andrew. W. King St., corner of Mulberry St., Lancaster, 1843–57 (t; SED, 54; BLD, 1847). Gunsmith.

Dreppard, George (b. 1787). 1849–50, Lancaster (t; c; SED, 54). Gunsmith.

Dreppard, John (b. 1803). W. King St., Lancaster, 1837–75. (SED, 54; BLD, 1861, 1857, 1873–74, 1855) Home: 15 Water St., Lancaster. Gunsmith.

Dreppard, William (b. 1816). Lancaster, 1843–50. (t; c; SED, 54) Gunsmith.

Drinkard, C. M. Bloomsburg, Columbia Co., gunsmith (PD, 1887, 1890).

Drinker, John. In 1751, in Philadelphia, he built the first gunshop on the location of the later Krider gunshop. Drinker was supposedly the first white child born in Philadelphia. In 1785, Robert Fulton occupied the same location, 2d and Walnut St. (BBHP, 36, 147).

Driver, Samuel. Patent 18,424 of 10/13/1857 for a cannon shell. Philadelphia.

Dubbs, A. Pine Grove, Schuylkill Co. (BSED, 1880) Gunsmith.

Dubbs, Aaron (d. 6/22/1874). Lower Milford Twp., Lehigh Co. Ran a *Scheifmuehle* (machine shop) on the Hosensack, a branch of Perkiomen Creek. Gun repairman (t; c).

Dubbs, Daniel. Gunsmith, 1812, Lower Milford Twp., Northampton Co. (t).

Dubbs, Jacob (1710–75). After 1732, Lower Milford Twp., Lehigh Co. (*Proceedings of the Lehigh County Historical Society,* 18:76).

Dubbs, John. Gunsmith (b. 1788). Lower Milford Twp., Lehigh Co. (t).

The Dubbs Family Gunsmiths

The first of the Dubbs line was an early convert to Protestantism. The family had its origins in Bohemia, one of the provinces of present-day Czechoslovakia. The name was then Dubsky, derived from the Czech word for oak. The new Protestant convert went to Switzerland, leaving behind a family who had remained Catholic. On October 11, 1531, the first of the Dubs line was killed in battle in the service of the Calvinist reformer Zwingli. Throughout the ensuing generations we find military leaders and armourers, the Swiss listing for gunsmiths as well as arms suppliers.

Jacob Dubs (I) married Anna Glatti, and their one surviving son was *Jacob II*. He was born on 31 March 1710 in Aesch, parish of Birmensdorf, Zurich Canton, Switzerland. On 30 September 1732, along with 185 other "Palatines" he landed in America, arriving on the *Dragon*. Jacob II was ill on his arrival, and did not personally appear before the customs officials. Thus, the British had a field day, making several variations of his name, including Tups, Dupps, and Dubbs.

He emigrated to the Great Swamp area of Bucks County, in what is today Lower Milford Township, Lehigh County. He acquired 150 acres on the Hosensack run, a division of Perkiomen Creek. This swift stream provided water power for his gunsmithing and farming activities. Known as a *Tau-*

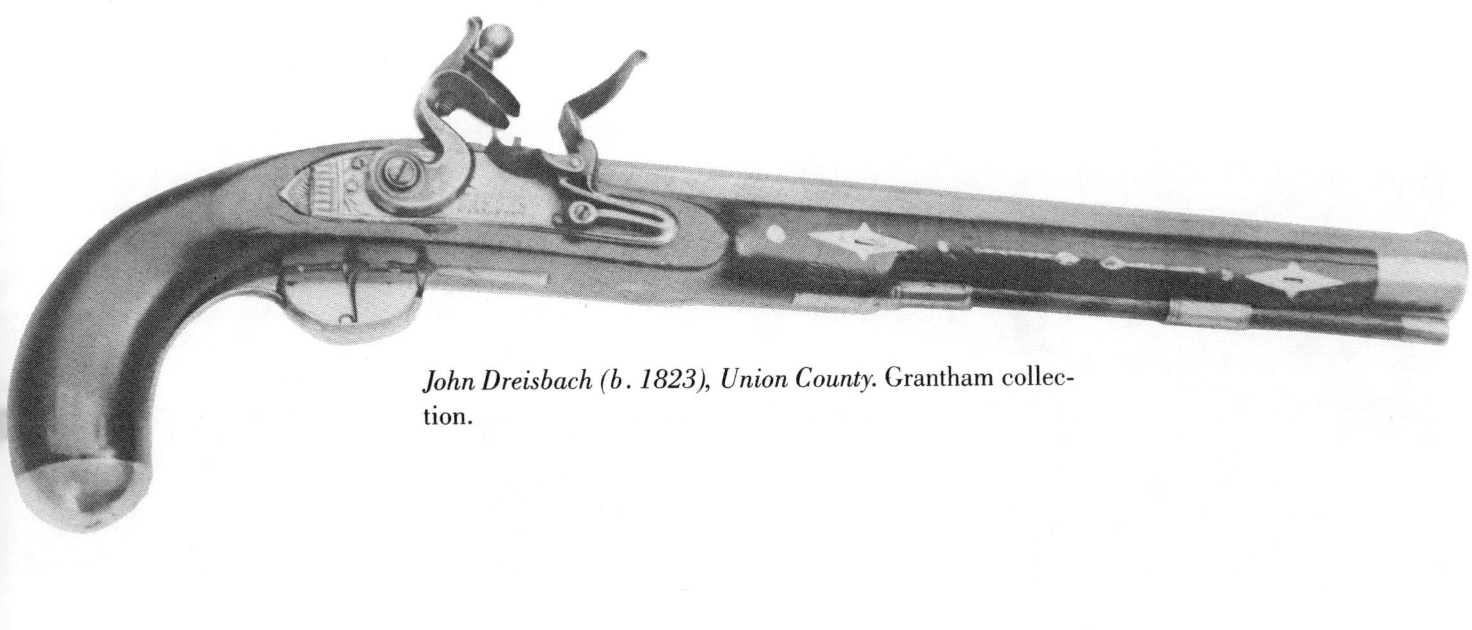

John Dreisbach (b. 1823), Union County. Grantham collection.

sendkunstler (roughly, "jack of all trades"), making everything from darning needles to ploughs and firearms. He made and played a harpsicord, so was evidently an educated and literate man.

In 1734 Jacob Dubbs II was naturalized. He married soon after that to Veronica Welker, another Palatinate. They had five children: Felix and Daniel and three daughters. In 1759 Felix drowned, falling in a neighbor's well before daybreak. Daniel remained with his father, who, in turn, sold him the family business in 1772 for 350 pounds. Jacob died about 1775 and his wife sometime thereafter; both are buried at the Great Swamp Church.

Daniel Dubbs married Elizabeth Schwenk, whose family owned the well in which Felix had drowned. Daniel expanded the family mill, creating a *Schleifmuehle* (machine shop), which made sickles, screw-augers and, during the War of 1812, muskets on federal contract.

Daniel had seven children, including *John*, born 5 September 1788. *John Dubbs* worked with his father on the War of 1812 contract, and in 1824 he took over the shop. John had a son *Aaron Dubbs*, who worked with his father, running the shop until Aaron died on June 22, 1874. It then passed out of family ownership. Note that the *Boyd Directory of 1880* for southeastern Pennsylvania listed an "A. Dubbs" in Schuylkill County, possibly a collatoral relative. ["Jacob Dubbs of Milford," Rev. Dr. J. H. Dubbs, in *Genealogies of Pennsylvania Families* (1981), 272–83]

Duden, Daniel. Gunsmith, 1801–7, Gettysburg, Adams Co. (t); 1808–32, McConnellstown, Huntingdon Co. (t); 1834–37, Henderson Twp., Huntingdon Co.; 1841, Martinsburg, Bedford Co. (t).

Duff, George. Partner in Whitmore, Wolff, Duff & Co., Pittsburgh, importers and distributors of hardware and gunlocks, after 1860.

Duffa, Joran. "gunner," 1649, New Sweden (PaE, 4).

Duffield, William. Armorer, 1780–81, Capt. John Orbison's Co., Cumberland Co. Militia (3 Pa. Arch. 23 at 767).

Dufford, Joseph (b. 1804). Gunsmith, 1850, N. Butler Twp., Butler Co. (c).

Duffy, Frank. Gunsmith, Everett, Bedford Co. (*Farquar Directory*, 1886).

Dulibar, Christian. Gun-barrel forger, 1847–50, Lancaster boro. (t).

Dunbar, Alexander. Providence Twp., Montgomery Co. (c, 1800; KKR, 214). Gunsmith.

Dunham, Job. Gunsmith, New Geneva, Fayette Co. (BPaD, 1861).

Dunkin, James (b. 1751). Armorer, 1777–80, Pa. German Regt.; 1818, Huntingdon Co., pensioner (2 Pa. Arch. 11 at 83).

Dunkle, George (b. 1805). 1825–28, Shippensburg, Cumberland Co; 1829–53, Metal Twp., Franklin Co. (WSB, 26; c; t); Oxford, Ill. (c, 1860). Gunsmith.

Dunlap & Maderia. 1820–30, Chambersburg, Franklin Co. Cutlers (t; c). Owned and operated by John Shugart & Co.

Dunlap, G. & R. Owners and operators, 1837–58, Pennsylvania Rifle Works, Pittsburgh. Gun and lock makers and importers. Robert, Sr., Robert, Jr., and George Dunlap, props (FSP, 88).

Dunmyer, David (1825–1912). 1856–76, Sipesville, Somerset Co. (t; c). Gunsmith. Member, G.A.R. Buried in Casebeer Lutheran Cemetery.

Dunmyer, Jonathan (1828–1885). 1850–85, Somerset Twp., Somerset Co. (t; c; f). Master gunsmith.

Dunmyer, Peter (1795–1859). 1821–32, Conemaugh Twp., Cambria Co. (t); 1834–35, Union Twp., Bedford Co. (t); 1839–59, Somerset Twp., Somerset Co. Master gunsmith.

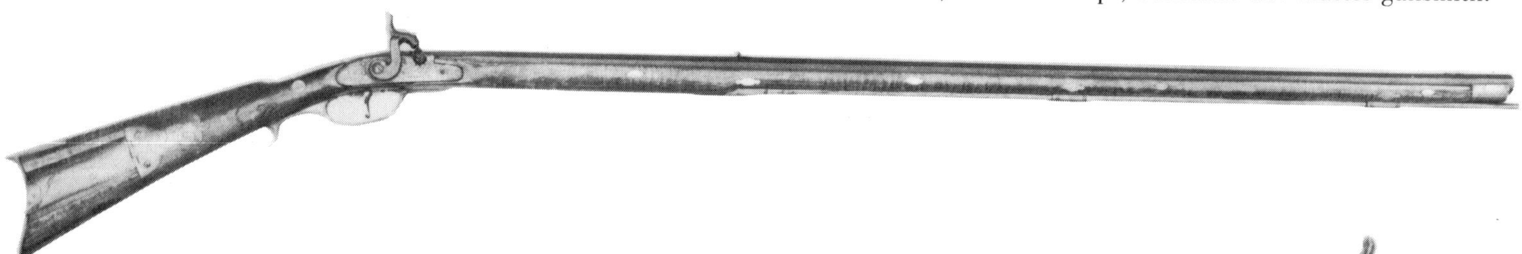

George Dunkle, ca. 1840, Franklin County. Bowers collection.

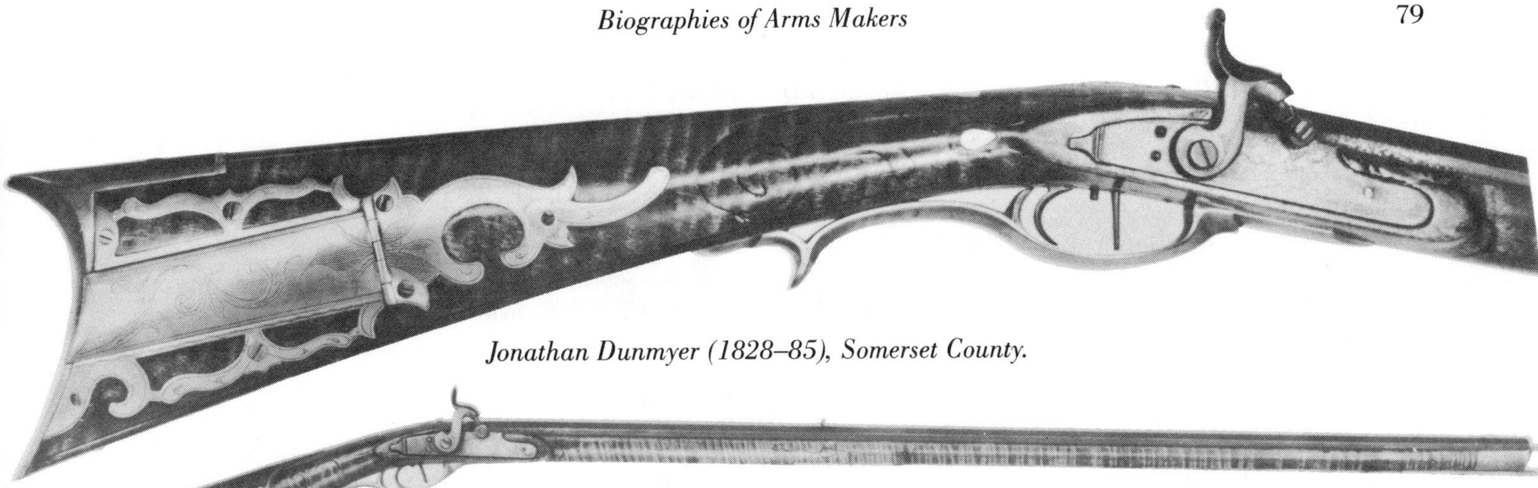

Jonathan Dunmyer (1828–85), Somerset County.

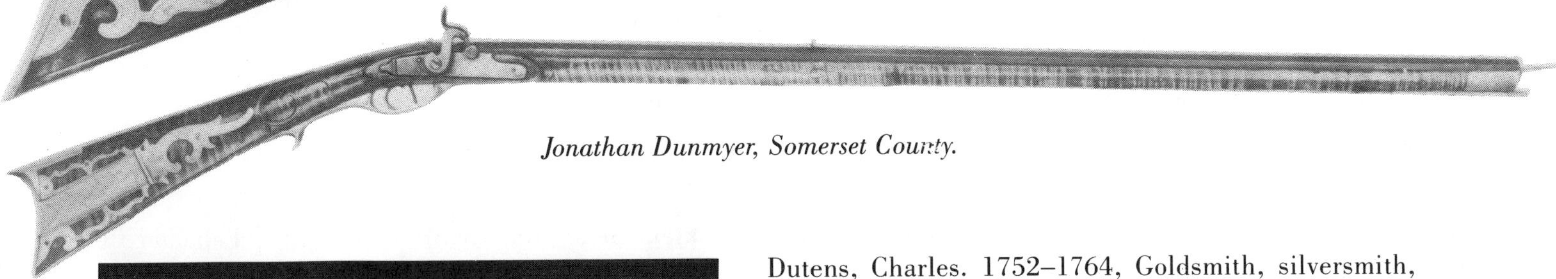

Jonathan Dunmyer, Somerset County.

Inlay from the cheekpiece of an otherwise plain gun. This is the only signed inlay the author has ever seen. Peter Dunmyer (b. 1795) was active in Bedford and Somerset counties.

Dunwick, William. 10/6/1775: "Upon the application of Mr Dunwick, gunsmith now employed in the making of Provincial muskets for Chester County, an order was given on the Commissary for 2 lbs of gunpowder to prove some of them ready now. Ordered, That the Commissary attend Mr Dunwick in proving the muskets for Chester County and furnish the powder necessary for that purpose." By 1/3/1776 he had completed 86 muskets and had 600 more under way. On 10/6/1776 he proved these (3 Pa. Arch. 14 at 211, 765).

DuPuy, Daniel & Son. Silversmiths. In 1772 they had apprentices. (PIDS) On 6/22/1749, advertisement in the *Pennsylvania Packet:* "Daniel DuPuy, goldsmith, Second St., Philadelphia."

Durham Iron Works (active 1727–1791). Durham Twp., Bucks Co. 1774–81, operated by George Taylor, making musket balls and cannon for the Continental army (GFC, 112).

Dutens, Charles. 1752–1764, Goldsmith, silversmith, Philadelphia. Advertisement in the *Pennsylvania Packet*, 7/23/1752: "Charles Dutens, jeweller, has set up business at the sign of the Ring and the Dove in Market St," Advertisements ran also on 3/27/1753, 2/4/1755, 5/9/1754, 1/27/1757, 4/16/1761, 8/6/1764. (t).

E

Eachus, F. H. Gunsmith, West Chester, Chester Co. (PD, 1884).

Eadelman, Charles. Gunsmith, 1847, Salisbury Twp., Lehigh Co. (t).

Eadelman, William. Gunsmith, 1847, Salisbury Twp., Lehigh Co. (t).

Eagle, Conrad. Gunsmith Russel's Alley above May St., Lancaster boro (LCD, 1871–72).

Eagle, Malachi. Gunsmith, Centre Twp., Butler Co. (PD, 1890).

Eagle Rifle Works (active 1833–61). Operated by James Golcher, Philadelphia. In 1850 there was an investment of $2,000. The company employed 17 workmen with a monthly payroll of $350. The yearly output was 500 guns valued at $3,500 and 600 dozen gunlocks valued at $3,600 (IC).

Ealer, Franklin A. (b. 1832). N. Queen St., near Lemon St., Lancaster Gunsmith, (BLD, 1857).

Ealer, Lewis W. (b. 1791). Born in Philadelphia, 1791. By 1822 he was active in Baltimore, Md. where he was active through 1833. 1837, gunsmith, Philadelphia, through 1856; 1857, gunsmith at Lancaster; 1865, Philadelphia (MD, 1856, 1865; BLD, 1857; DeSD, 1837, 1845, 1856; AMM, 150).

Earley, Amos (1840–1905). West Hanover, Dauphin Co. (BRD, 1869; BPaD, 1861, 1880; t; c) Gunsmith (PD, 1882, 1890).

Earley, Jacob. Gunsmith 1790, West Londonderry Twp., Lebanon Co. (t; c).

Earnest, George. 126 Brown, Philadelphia (PXD, 1810). Gunstocker.

Earnest, Jacob (1805–84). 1826–30, Hempfield Twp., Westmoreland Co. (t); 1830–84, Salem (Delmont), Westmoreland Co. Gunsmith (t).

Earnest, John (1827–1904). Son of Jacob. Gunsmith, 1850–1904, Delmont, Westmoreland Co. (t).

Earnest, Simon Peter (1834–80). Son of Jacob. Gunsmith, 1870, 1880, Delmont, Westmoreland Co. (c).

Earnhard, John. Gunsmith, 1814, Moore Twp., Northampton Co. (t).

Earnhardt, Charles. Gunsmith, Philadelphia (MD, 1847).

Ernst, Jacob. 1820, Paradise Twp., York Co., Gunsmith (JKJ, 356–57; t).

Eastburn, Joseph. Gunstocker, 1777, French Creek Arsenal (2 Pa. Arch. 15, 384).

Easterly, George. 1777: "George Easterly, blacksmith, applies to do iron work on cannon" (2 Pa. Arch. 1 at 507).

Eberle, Henry (d. 1822). Sword and bayonet maker, 1776–83, Elizabethtown, Lancaster Co.; 1785, Barree Twp., Huntingdon Co. (GFC, 98; t).

Eberle, Jacob. 1810, gunsmith, Warwick Twp., Lancaster Co. (t; SED, 55).

Eberle, Jacob (d. 1861). 1840–61, Petersburg, Huntingdon Co. Cutler. Son of Henry and Barbara (Kauffman) Eberle (t; f; c).

Eberman, Henry. 1810–50, Lancaster boro., gunsmith (SED, 55; t).

Eberman, John. 1759, gunsmith, Lancaster (t; SED, 55).

Ebersole, W. H. (1846–1920). 1870–1920, Duncannon, Perry Co. (t; c). Gunsmith.

Eberts, George John. 1800–1807, Plainfield Twp., Northampton Co. (t). Gunsmith.

Eby, Abraham. 1781, Leacock Twp., Lancaster Co., gunsmith (SED, 55; t).

Eby, Christian. 1768, Warwick Twp., Lancaster Co., gunsmith (t; SED, 55).

Eby, John. 1801, gunsmith, Leacock Twp., Lancaster Co. (SED, 55; t).

Eckler, Henry (d. 1862). Pine Grove, Lebanon Co., Gunsmith (BRD, 1860; BPaD, 1861).

Eddie (Eddy), James. Noble St. and Third Sts., Philadelphia, gunstocker (PXD, 1813, 1814). Worked for J. J. Henry.

Eddy, Ira B. 1854–57, Philadelphia. Associate of Christian Sharps (t; c). Manufacturer, gunsmith.

Edgington, Daniel. Gunsmith, 1829, Brownsville, Fayette Co. (t).

Edmond, Jacob. Gunsmith, Aaronsburg, Centre Co. (BPaD, 1861).

Edmond, John. 1850, Bushkill Center, Northampton Co. (c); 1861, Aaronsburg, Centre Co. (BPaD, 1861).

Edmonds, William. Gunsith, Monroe Co. (c, 1850).

Edwards, Michael (1767–1826). 1800, apprenticed to Conrad Welshans, York. 1807–26, Washington boro. (t). In the years 1814–16 he was listed as a butcher. His sons became gunsmiths in West Virginia (WSB, 112).

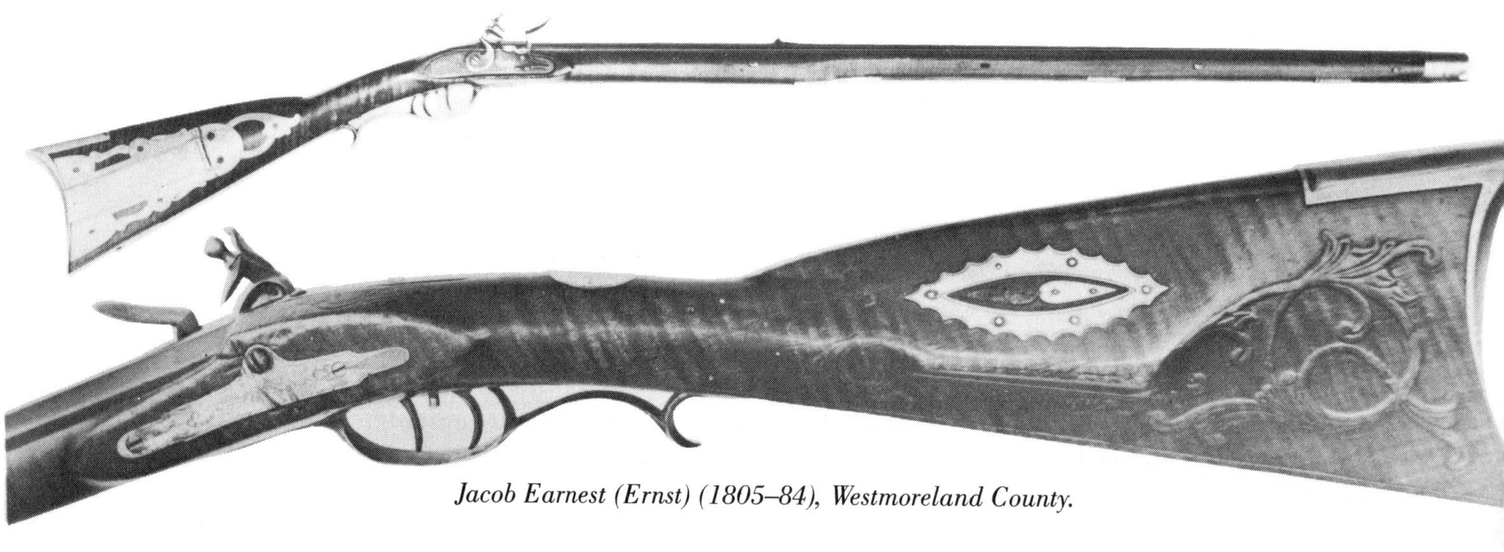

Jacob Earnest (Ernst) (1805–84), Westmoreland County.

MICHAEL EDWARDS,
GUN SMITH.

RESPECTFULLY informs the Gentlemen of Washington, and its vicinity, that he has commenced the above Business, next door to *Christian Keefer, Sign of General Washington*, Front street, in the Town of Washington, where he intends carrying on this Business in all its various branches, from his serving a regular Apprenticeship, in the town of *Little York*, and has wrought in some of the principal towns of the United States, that he hopes to give general satisfaction, to all those that will favor him with their work, shall have it done on shortest NOTICE, and in Newest Fashion, and on the most Reasonable Terms.

Washington, February 20, 1807 3t

Western Telegraph [Washington], February 20, 1807

Edwards, Michael, Jr. (b. 1777). 1807–36, Washington boro. (t); 1838–60, Moundsville, W. Va. (t; c). Gunsmith.

Edwards, William (b. 1814). 1835–36, Washington boro. (t); 1838–60, Moundsville, W. Va. (t; c). Gunsmith.

Ege, George (1748–1829). 1776–1804, owned 4,000 acres in Heidelberg and Tulpehocken Twps., Berks Co. Made cannon shot during the Revolution (BCR, 176–77). Joseph Olds, ironmaster, married Ege's daughter, Rebecca.

Egler, Jacob D. Patent 403,096 for a spring gun. Pittsburgh, 1889.

Eichel Brothers. Powder-horn makers, W. King between Mary St. and Charlotte St., Lancaster. (BLD, 1857) Eicholtz & Brothers, 59 N. Queen St. (BPaD, 1861). H. C. and Robert L. Eicholtz.

Eicholtz, H. C. 58 N. Queen St., Lancaster (BLD, 1874–75; *Ferris Directory*, 1884; BAR, 1875–76).

Eicholtz, H. C. Gunsmith, Kutztown, Lebanon Co. (PD, 1882).

Eicholtz, John. Gunsmith, 1843–44, 59 N. Queen St., Lancaster (t).

Eicholtz, Robert L. 46 S. Lime St., Lancaster, gunsmith (BLD, 1873–74; BAR, 1875–76; t, 1848–76).

Eisenhard, Henry. Gunsmith, Lehigh Co. (c, 1850).

Eisenhower, Israel. Gunsmith Union Deposit, Dauphin Co. (BSED, 1880).

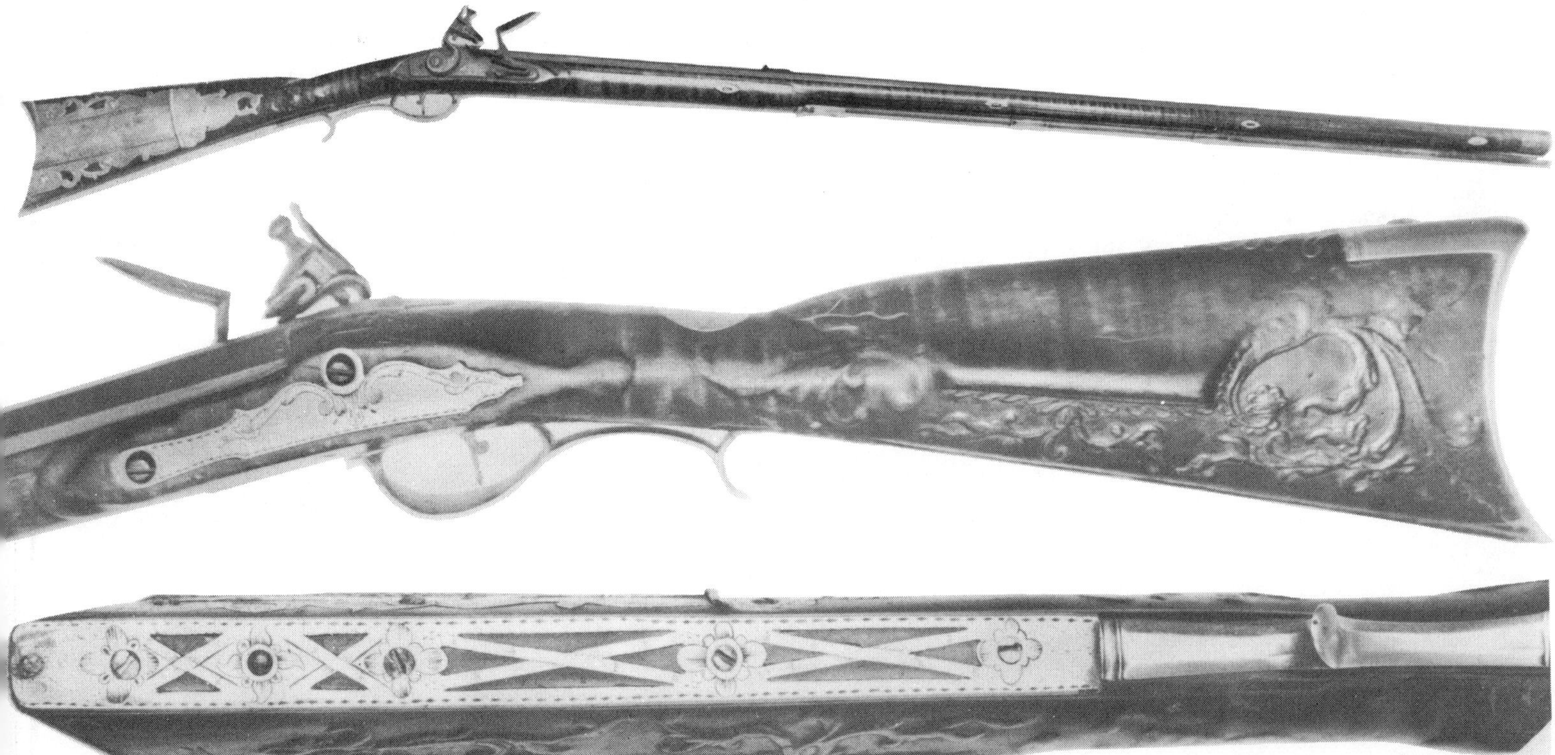

Michael Edwards (1777–after 1860), Washington County. This is one of the finest guns ever made and shows the best of the arms maker's art. Edwards was trained in York County, and the styling reflects that school of gunmaking. Pike collection.

Eister (Eyster), George (1762–1831). Manchester Twp., York Co. (GAS-2, 43; JKJ, 296–99). Major gunsmith.

Elder, John. Gunsmith, 1823–28, Conemaugh Twp., Indiana Co. (t).

Elizabeth Furnace (active, ca. 1750–1856). Near Brickerville on Middle Creek, Lancaster Co. Robert Coleman, ironmaster, made cannon and shot during the Revolution (GFC, 106; SHIS, 180).

Elliott, James. Locksmith, 1787, Louisburg, Dauphin Co. (t).

Elliott, Michael. Gunsmith, 1854–56, Morris Twp., Washington Co. (t).

Ellis, Richard. Artillery artificer, 3/3/1777 to 3/22/1781, Capt. Jesse Rowe's Co. at the cannon factory at Philadelphia (PIR, 2:244).

Ells, Joseph. Gunsmith and inventor, 1851–57, Pittsburgh. Patents: 10,812 (1854); 11,419 (1854); 17032 (1857). Made repeating rifles.

Elton, A. Next to 43 Zane St., Philadelphia, gunsmith (PXD, 1810, 1813, 1816).

Elton, Thomas. 1780, Mulberry Ward, west part, Philadelphia (3 Pa. Arch. 15 at 302). Gunsmith. Partner of Isaac Johns.

Embrick, C. Reighart's Landing, Lancaster. Gun polisher (BLD, 1873–74, 1875–76).

Emerson, James E. Cutler, ca. 1830–40, Beaver Falls, Beaver Co. (SHIS, 387) Made steel knives from blister steel. DBA Emerson, Smith & Co.

Emery, Andrew & Co. Freedom, Beaver Co. (t, 1841). Gunsmith.

Engle, Barnett (1808–87). Front between Miner and Walnut, Greensboro, Greene Co. (t; c; f). Gunsmith.

Engle, Christian. 1824–33, Monongahela Twp., Greene Co. (t); 1845, Morris Twp., Greene Co. (t); 1834–42, Monongalia Co, W. Va. (t). Gunsmith.

Engle, Ezra (b. 1791). 1817–21, Greene Twp., Greene Co.; 1824–37, Monongahela Twp., Greene Co. (t); 1838, Greene Twp., Greene Co. (t); 1850, 1860, Tyler Co., W. Va. (c). Gunsmith.

Engle, Michael. Master smith, 3/13/1777 to 3/24/1781, at the cannon factory in Philadelphia. Engle was from Reading (PIR, 2:244).

Engle, Peter (1767–1833). Gunsmith From Allegany Co., Md. 1801–3, Springhill Twp., Fayette Co. (t); 1802–23, Greene Twp., Greene Co. (t); 1824–33, Monongahela Twp., Greene Co. (t).

Engle, Sylvester (1826–1900). 1848–49, Wilkes-Barre (t); 1850–96, Hazleton, Luzerne Co. (t; c). Silversmith, gunsmith, jeweler.

Engle, Thomas. Artillery artificer, 3/3/1777 to 3/22/1781, Capt. Jesse Rowe's Co. at the cannon factory in Philadelphia. (PIR, 2:244).

Enterprise Gun Co. 1848–79 or later, Pittsburgh. Made many different kinds of muzzle-loading rifles, from plains to boys'. (t)

Enters, Lewis. Buttonwood, Philadelphia Co. Gunstock manufacturer. (DeSD, 181, 1830, 1833)

Erb, John. 1842–50, Mahoning Twp., Carbon Co. Owned powder mill. (t; c)

Ernfighter, David. Artillery artificer, 1778, Capt. Nathaniel Irish's Co. (PIR, 2:247)

Ezra Engel (Engle) (b. 1791), Greene County. Jenion collection.

Ernfighter, Jacob. Artillery artificer, 1778, Capt. Nathaniel Irish's Co. (PIR, 2:246)

Ernhart, Charles. Gunsmith, Charlotte St. at Beaver St., Philadelphia. (MD, 1841)

Ernst, Adam (1781–1857). 1803–11, Berwick Twp., Adams Co. (t); 1811–20, worked with George Eister at York (t); 1820–57, Franklin Twp., York Co. (WSB, 100; JKJ, 358–60). Major gunsmith.

Eshelman, John. 1843–50, Lancaster, gunsmith (t; SED, 55); 1856–57, 37 4th St., Pittsburgh (PAD, 1856–57); Harrisburg Pike near Mulberry, (GCD, 1863–64).

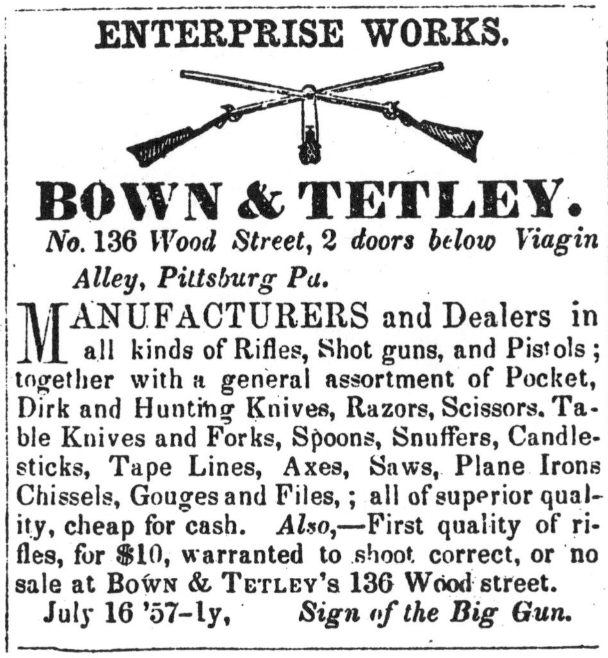

ENTERPRISE WORKS.

BOWN & TETLEY.

No. 136 Wood Street, 2 doors below Viagin Alley, Pittsburg Pa.

MANUFACTURERS and Dealers in all kinds of Rifles, Shot guns, and Pistols; together with a general assortment of Pocket, Dirk and Hunting Knives, Razors, Scissors, Table Knives and Forks, Spoons, Snuffers, Candlesticks, Tape Lines, Axes, Saws, Plane Irons Chissels, Gouges and Files, ; all of superior quality, cheap for cash. *Also,*—First quality of rifles, for $10, warranted to shoot correct, or no sale at Bown & Tetley's 136 Wood street.

July 16 '57-ly, *Sign of the Big Gun.*

Pittsburgh Gazette, July 16, 1857

Espy, William. Artillery artificer, 1778, Capt. Thomas Wylie's Co. (2 PIR 2:248)

Estep, Cornelius. 1849–50, Donegal Twp., Washington Co. (t) Gunsmith.

Estle, Elias. 1777–78, made rifles at Martin Hough's shop, Augusta Town, Yohogania Co., Va. (Greene Co., Pa.). *Horn Papers*, 2:584)

Evans, Brooke, Frequent government contractor, musket maker, 1800–1828, Evansburg, Montgomery Co. (GAAAM, 48)

Evans, Daniel. 5th near Market, Philadelphia, 1769–78 or later. Did iron work on cannon during the Revolution. (t; 2 Pa. Arch. 1 at 42) In 1773 Evans had an apprentice named William Bowers, studying black-smithing. (PIDS)

Evans, Daniel. Gunsmith, 1817–24, East Bethlehem Twp., Washington Co. (t)

Evans, David. 4/18/1777: "pay David Evans £13/7/0 for ammunition boxes, &c." (2 Pa. Arch. 1 at 35). Iron monger.

Evans, Edward. Government contractor, 1790–1815, Evansburg, Montgomery Co. (GAAAM, 48). Made cannons, swords, and gun barrels.

Evans, George & Co. 132 N. 5th St., Philadelphia. Swordsmith; made edged weapons, cutlery (BSED, 1887, 1891).

Evans, Israel. Smith, Upper Dublin, Philadelphia Co. On list of those attainted by treason (1 Pa. Arch. 10 at 250ff.).

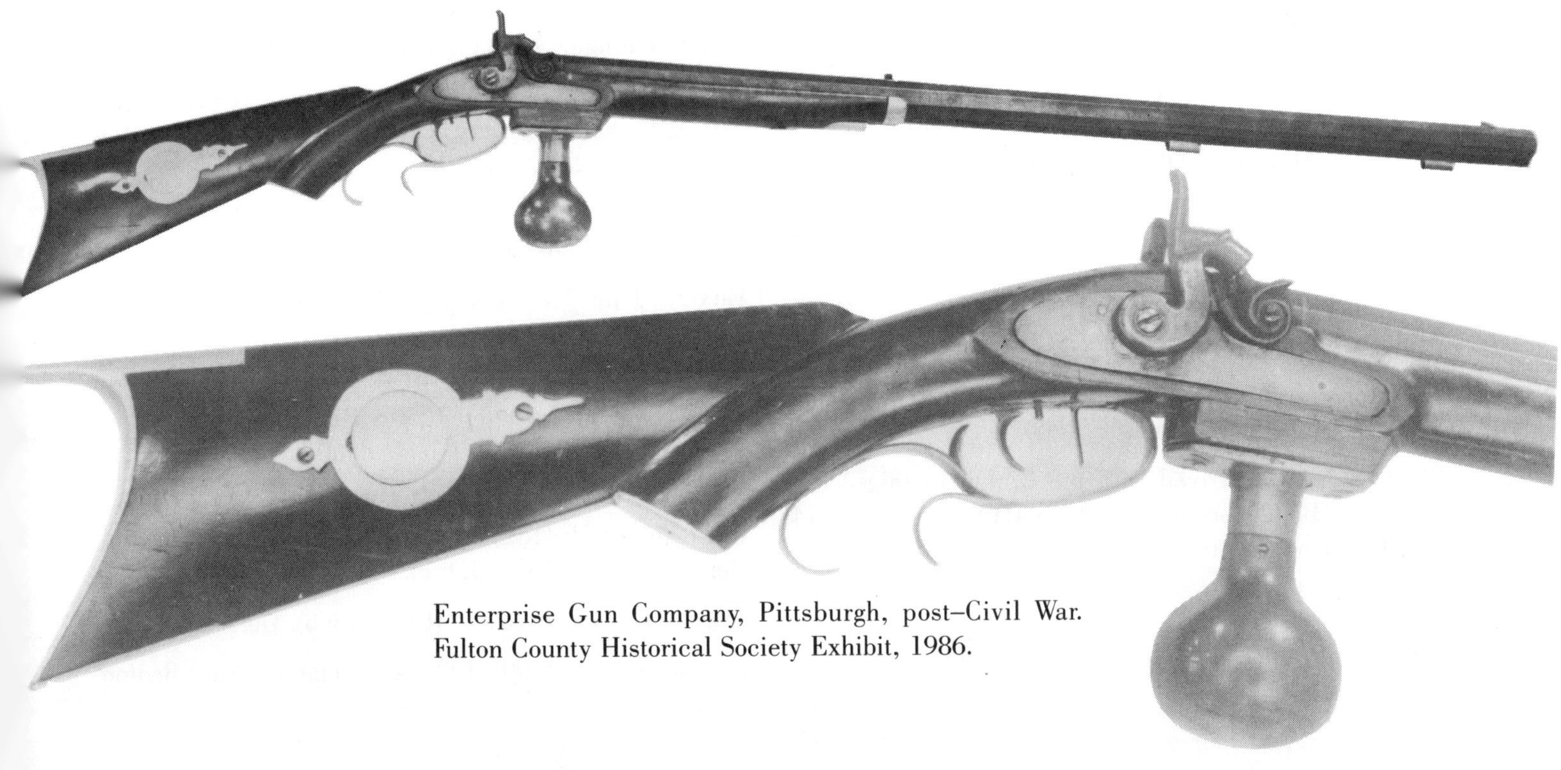

Enterprise Gun Company, Pittsburgh, post–Civil War. Fulton County Historical Society Exhibit, 1986.

Evans, James. Government arms contractor, ca. 1800–1825 Evansburg, Montgomery Co. (GAAAM, 48).

Evans, James E. (1825–70). 230 South St. and 25 N. 2d St., Philadelphia (MD, 1855, 1859, 1861; BPaD, 1861). Government contractor, gunsmith.

Evans, Joseph. 10 Harmony St., Philadelphia, gunsmith (MD, 1846, 1847).

Evans, Joseph. Silversmith. 4/27/1772, apprenticed to Daniel DuPoy & Sons, Philadelphia, to learn silversmithing, for 7 years, 7 months, and 22 days (PIDS).

Evans, O. & E. Government arms contractors. Evansburg, Montgomery Co. On 10/7/1812 they contracted to produce 1,960 U.S. M-1808 muskets for use in the War of 1812. (GAAAM, 48) Owen and Edwards Evans, props. (GAAM, 48).

Evans, Owen (1758–1812). Government arms contractor. Evansburg, Montgomery Co. (GAAAM, 48).

Evans, Peter. Cutler, 1777–80, North Ward, Philadelphia (t).

Evans, Stephen. Established and ran Mt. Joy Armory (later, Valley Forge Armory). Government arms contractor. Order from Pa. of 3/8/1797 for 20,000 muskets of M-1795 pattern (GAAAM, 48).

Evans, William. Artificer in the detachment of Royal Artillery under Capt. David Hays, 1758–60 (*Minutes of Yohogania County Court*, 3/24/1780, in W. A. Crozier, *Virginia Colonial Militia*, 10).

Evans, William L. (1797–1861). Associated with John Rogers at Valley Forge. Made M-1827 naval pistols dated 1830 or 1831 (GAAAM, 48).

Eve, Oswald. Gunpowder supplier to the Committee of Safety, 1775–77, Northern Liberties, Philadelphia Co. (Col. Rec. 10:398; 1 Pa. Arch. 4 at 696) He was on the list of persons attainted by high treason according to the laws of Pa. (1 Pa. Arch. 10 at 250ff.), and his property was seized and sold at auction, with the proceeds going to the Commonwealth (6 Pa. Arch. 13 at 267).

Everitt, Thomas. Gunsmith, 1848, Salisbury Twp., Northampton Co. (t).

Everson, Washington. Gunsmith, 1834–36, Belle Vernon, Fayette Co. (t); Logan, Ind. (c, 1860).

Eyerly, Jacob. Arrived with the Gotlob Koenigsdoefer group at Bethlehem on 9/14/1753 (PaE, 46); 1754–74, Northampton Co. (t).

F

Faber (Farber), Edward. Gun-barrel factory, 1834–35, Pittsburgh. He and his brother met with financial difficulties, as shown in their advertisement in the *Allegheny Democrat* on 4/14/1835. They went bankrupt (FSP, 98).

Fagaler, George M. (b. 1769). Goldsmith, arrived 8/26/1805 on the ship *Margaret* in Philadelphia. Born in Würtemberg, Germany (2 Pa. Arch. 17 at 635).

Fahringer, Herman. Gunsmith, 1799, Macungie Twp., Northampton Co. (t).

Fainot, Franklin. Gunsmith, 1779–82, Lancaster. (t).

Fainot, George Frederick. 1779–1803, major gunsmith, N. Queen St., Lancaster (t; SED, 55; GAS-1, 344).

Fairlamb, Nicholas. Armorer, Chester Co., 1775–78 (1 Pa. Arch. 4 at 730, 755; G. Smith, *History of Chester* [1862], 289).

Faivre, Alexander C. Meadville, Crawford Co. Patent 19,553 of 3/19/1858, and 172008 of 11/1/1876, for gun parts.

Fanckleberry, Joseph. Swordsmith in the employ of Lewis Prahl, gunsmith, 1777 (2 Pa. Arch. 3 at 104).

Fane, John C. Gunsmith, Chester, Delaware Co. (PD, 1879).

Faries, William. Gunstocker, 1777, French Creek Arsenal (1 Pa. Arch. 15 at 381).

Farleman, Henry. Born in Lancaster Co., he received marriage license 14 in 1852. Gunsmith, living in Walker Twp., Juniata Co.

Farley, William. Armorer, 1776, armed boat *Ranger*, Pa. Navy (2 Pa. Arch. 1 at 334).

Farlow, William. Artillery artificer, 1782, stationed at New London, Va. (2 Pa. Arch. 11 at 261).

Farrell, Michael. 7 Shingess, Pittsburgh (PAD, 1864–65); 1878, West Hoboken, N.J. (FSP, 98). Gunsmith.

Farver, John. 1814–45, Liverpool, Perry Co. (t). Gunsmith.

Farver, William (1818–63). 1820, Liverpool, Perry Co. (t); 1850, Benton, Iowa (c, 1850), Ripley Ohio (t); 1856, Perry Co. (t). He was knocked off a canal boat and drowned, 9/16/1863 (GAAM, 49).

Faust, Joseph H. (1817–88), gunstock maker, Alsace Twp., Berks Co. (GAAAM, 49; t; c).

Fay, A. Altoona, Blair Co. (BRD, 1868). Gunsmith.

Fay, George (1811–90). 1835–44, Hopewell Twp., Bedford

William Farver (1818–63), Perry County.

Co. (t); 1848–58, Allegheny Twp., Blair Co. (t); 1859, Gaysport, Blair Co. (t); 1860, Hollidaysburg Blair Co. (c); 1861, Duncansville, Blair Co. (t); 1862–90 (c; t).

Fay, Richard. Philadelphia. Patent, 209,613 of 11/5/1878 for a breech-loading gun.

Fay, William. 1860–64, Altoona, Blair Co. (t) Killed in Civil War. Gunsmith (t, f).

Feather, George. Gunsmith, 1820–22, Elizabeth Twp., Lancaster Co. (t).

Fehr, John. 1830–50, gunsmith, Nazareth, Northampton Co. (t, c).

Fehr, R. H. Gunsmith, Bethlehem, Northampton Co. (PD, 1882; BRD, 1890).

Feinebecker, Daniel. Gunsmith, 1796, Cocalico Twp., Lancaster Co. (t).

Fell, Thomas. Smith, 1780, Capt. Thomas's Co., Bucks Co. Militia (3 Pa. Arch. 6 at 73).

Felton, Robert. Silversmith, 1683, Philadelphia (Col. Rec. 1:85, 86).

Fenner, Thomas. N. Front St. below Master, Philadelphia, gunsmith. (DeSD, 1829, 1833); 6 Courtlin Place (MD, 1837; DeSD, 1835–36).

Ferguson, Anthony. Gunsmith, 1816–17, South Huntingdon Twp., Westmoreland Co. (t).

Ferguson, William (b. 1755). Artillery artificer, 1778, Capt. Nathaniel Irish's Co.; 1835, age 80, Cumberland Co. (PIR, 247).

Ferree, George Spencer (1818–96). 1838–40, Clinton, Allegheny Co. (t); 1850, Pittsburgh (c, 1850); 1869–96, Laconia, Ind. Apprenticed to John Fleeger. (FER).

Ferree, Isaac (1786–1821). Gunsmith 1808–10, Mifflin Twp., Allegheny Co. (t); 1810–12, Pittsburgh (t); 1814–15, Harrison Co., Indiana Territory (FER); 1816, Bath Co., Ky (FER); 1818, Cincinnati, Ohio (FER); 1821, died in Baton Rouge, La. (FER).

Ferree, Jacob. 1782, Strasburg, Lancaster Co., gunsmith (t; SED, 55).

Ferree, Jacob (1750–1807). 1782, Strasburg Twp., Lancaster Co. (t); 1784, Jefferson Twp., Allegheny Co. (t); 1786–88, Dickinson Twp., Washington Co. (t); 1788–1800, Mifflin Twp., Allegheny Co. (t); 1801–7, Moon Twp., Allegheny Co. (t). Owned gunshop and powder mill.

Ferree, Joel Sr. (b. 1731). 1745, apprenticed to Philip Lefevre (FER); 1752–75, Lancaster boro., gunsmith and revolutionary war contractor (t; SED, 55).

Joel Ferree to Benj. Poultney

Leacock Township, August 17th, 1775
Friend Poultney:—I take this opportunity of informing you that sometime ago as I had a Letter from George Ross, Esqr., by order of the Committee, making enquiry what Qt. or Number of Barrels I could supply 'em weekly, to which I return'd answer in the Particulars he demanded. I was of opinion then to have been able to provide 15 or perhaps 20 pr. week, but as I am determined to use my endeavors to promote the Business and serve my Country in the common Cause, I am about to enlarge my works in so extensive a Manner as to turn out between 30 & 40 weekly. My Diligence in the Affair shall be as quick as possible. It will not require much time to complete my works in order for such Dispatch of Business; I hope a few days will do.

The Patterns were sent to Stewart's store the same day you was here. I intend you shou'd let the Committee know what I am about doing to supply 'em as punctual as in my Power. You may let them see this, by which they judge whether or not it will tend greatly to further the Dispatch of the finishing off the Arms so much needed.

This from your Friend,
Joel Ferree

[2 Pa. Arc. 1 at 1]

Ferree, Joel, Jr. (1774–1815). 1786–1812, Mifflin Twp., Allegheny Co. Gunsmith (t; FER).

Ferree, Joel Thornton (1815–82). 1835–38, Clinton, Alle-

ISAAC FERREE,

GUN-SMITH,

RESPECTFULLY informs his friends and the public, that on the first of April next, he will remove to the town of Pittsburgh. (in Fifth Street, between Market & Wood Streets) where he intends carrying on the Gun-Smith Business in all its different branches. He flatters himself, from his knowledge of the business, and the strict attention he is determined to pay to it, that he will give satisfaction to those who may please to employ him.

He will constantly be supplied with *Ferree's best Powder*, to supply those who may require it. February 28, 1810.

Pittsburgh Gazette, February 28, 1810

LAND FOR SALE.

THE subscriber proposes to sell the following tract of Land, on Peters creek, two and a half miles from the mouth, in Mifflin township, Allegheny county, two miles from Elizabeth town, and twelve from Pittsburgh.—The tract contains three hundred acres, and the usual allowance; the land is of the first quality; upwards of one hundred acres cleared, and under good fence, of which about twenty acres are meadow. On the tract are an orchard of upwards of two hundred bearing apple trees, a grist mill, saw mill, and powder mill and other necessary buildings to accommodate three or four families.—The land will divide well into two farms. From eight hundred to one thousand weight of sugar, is made yearly.

For terms apply to the subscriber living on the premises.

Joel Ferree.

Feb. 28, 1809. tf.

Pittsburgh Gazette, May 31, 1809

GUN POWDER.

THE subscribers have on hand, and intend constantly to keep, at their Powder Mill, on Peters creek, two miles from Elizabeth town, a supply of their best manufactured RIFLE POWDER, which they will sell low for cash or on short credit.

From the long experience of this Powder, in most parts of this western country, and its high repute among our greatest gunners, we flatter ourselves with the hope of receiving a part of the public patronage.

Joel & Isaac Ferree.

Feb. 28, 1809. tf.

Pittsburgh Gazette, May 31, 1809

gheny Co. (t) Gunsmith (FER).

Ferree, Philip (ca. 1692–1755). Before 1712, Esopus, N.Y.; after 1712, Lancaster Co. (Lancaster DB D-489, D-209; FER).

Ferree, Squire Joseph. 1776–77, NE corner Main St. and School House Lane, Germantown, Philadelphia Co. He was an armorer, in charge of lead and gunpowder (*Guidebook to Historic Germantown*, 71–72).

Zanesville, April 28, 1813.

DIED, on Monday morning last, at the house of his friend Mr. Alexr. Culbertson in this place, col. JOEL FERREE, of general Crooks' brigade, Penn. militia, after an illness of something more than two weeks, aged about 44 years

Col Ferree had performed an ardous tour of military duty in the service of his country; and with fond expectations had arrived thus far on his way home, where affluence, and a tender and a numerous family would have afforded a solace for past privations and sufferings. But the hand of death arrested his progress, and he was forbid to see them more! His health was blasted, and his life forfeited on the altar of his country's good! His eyes were closed in the presence of a daughter and a young son, who had journeyed hither from their former residence with great rapidity, on hearing of their father's illness. An affectionate wife, and twelve children (two of them born during his absence) are left to mourn the prop of ther future prosperity and happiness.

Pennsylvania Republican [Harrisburg], May 11, 1813

Fesig, Conrad. Gunsmith and silversmith, 1780–84, Reading (t).

Fesig, Peter. Silversmith, 1781–84, Reading (t).

Fetter, George. Gunsmith, 1831–48, Elizabeth Twp., Lancaster Co. (t).

Fetter, George, Jr. 1844–50, Elizabeth Twp., Lancaster Co., gunsmith (t; SED, 55).

Fetter, Henry. 1847–48, gunsmith, Elizabeth Twp., Lancaster Co. (SED, 55; t).

Fetter, Jacob. 1777, drafted as a smith into the artillery artificers of the 10th Pa. Regt.; "injured in the service" (PIR, 1:730).

Fetter, John. Between Ferry St. and Redoubt Alley, Pittsburgh (PAD, 1810). Gunsmith.

Fetter, Perry. 1860–1908, Kratzerville, Snyder Co. (t; c). Gunsmith.

Fetter, Peter. 1837–50, Elizabeth Twp., Lancaster Co., gunsmith (t; SED, 55).

Fetters, William. Discharged 6/12/1776 from the 5th Pa. Battalion to be employed as a gunsmith (PIR, 1:158).

Few, Richard. Smith, 1771–77, Bethel, Chester Co. (t).

Fiedler, Josef. Patent 181,830 of 9/5/1776, for a breech-loader, Pittsburgh.

Field & Langstreth. Gun and lock makers, importers, 1846–70, Philadelphia. After 1855, Field, Longstreth & Co.

Fiersch, Joseph. Gunsmith, 1828–29, Bushkill Twp., Northampton Co. (t).

Figthorn (Fichthorn), Andrew, Sr. gunsmith Naturalized 4/3/1763, Berks Co. (2 Pa. Arch. 2 at 442).

Figthorn (Fickthorn), Andrew, Jr. (d. 1822). Artificer during the Revolution. 1779–92, Reading, Berks Co. (JKJ, 213–14; GAAAM, 50; t); 1777, artificer, Wills's Artillery Co. (BCR, 207).

Fighthorn, Andrew, III (d. 1827) gunsmith 1800–27, Reading (JKJ, 213–14).

Filby, James. 1810, Morgan Twp., Greene Co. Gunsmith (*Horn Papers*, 1:387) in the village of Clarksville.

Fillman, William A. 1820–67, Milton, Northumberland Co. (t; BPaD, 1861). Gunsmith (t).

Finch, Joseph. Franklin, Venango Co. (c, 1850) Gunsmith.

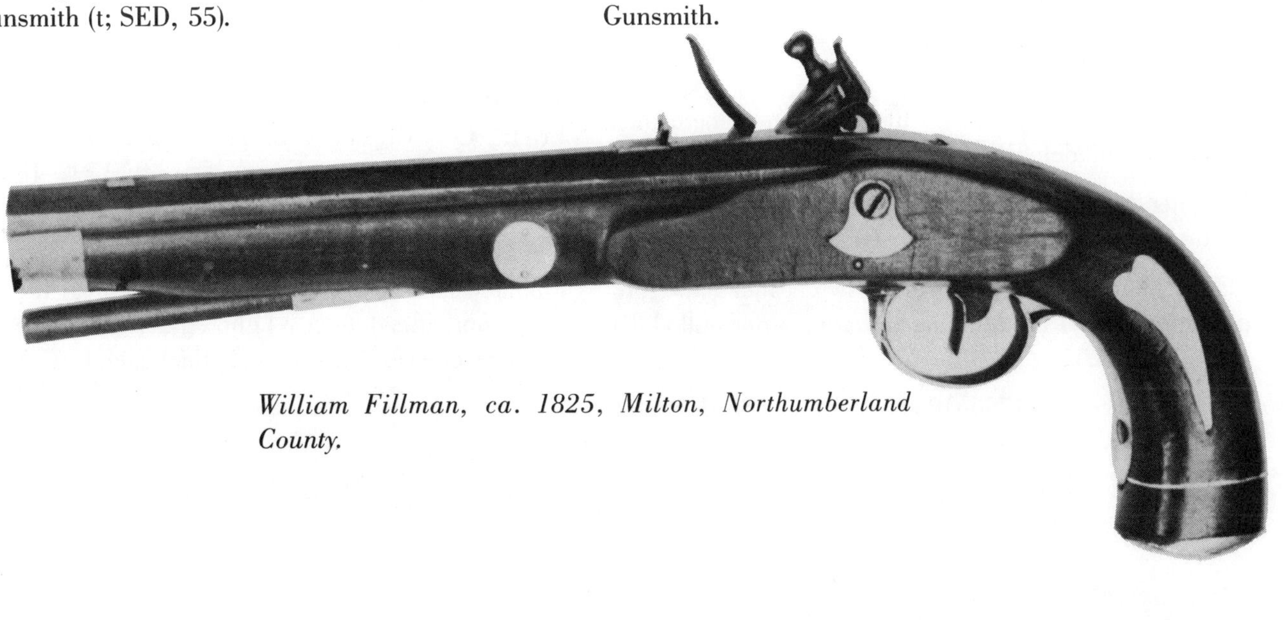

William Fillman, ca. 1825, Milton, Northumberland County.

Fink, Michael (b. 1759). Smith, 1778–81, Bankson Co.; 1835, age 76, living in Huntingdon Co. (2 Pa. Arch. 10 at 430).

Finley, David. Smith, 1786, Rostraver Twp., Westmoreland Co. (t). 11/19/1776: "Pay Capt Finley £9/19/0 for repairing arms" (Col. Rec., vol. 11).

Firth Sterling Steel Co. McKeesport, Charles Yandes Wheeler (1843–99), president. Developed the Wheeler-Sterling armor piercing projectile (NCAB, 2:11).

Fishburn, Philip (1722–95). Arms Contractor to Committee of Safety, Dauphin Co. (GAAAM, 51).

Fishel, Jacob (1813–61). 1841–61, Hopewell Twp., Bedford Co. (t). Gunsmith.

Fisher, David. Gunsmith, Macungie Twp., Northampton Co. (PD, 1887).

Fisher, Francis. 1859–60, corner Main St. and Chestnut St., Pittsburgh. (PAD, 1859–60) Gunsmith.

Fisher, George (d. 1821). 1800–1816, Washington boro. (t); 1819–21, Cadiz, Ohio (t). "George Fisher of Harrisburg requests that those who borrowed muskets from him to return them" (*Oracle of Dauphin*, 8/14/1798).

Fisher, Harrison. Gun-barrel maker, 528 N. Duke St., Lancaster. (BLD, 1871–72). There was a gunsmith named Harrison Fisher in Lower Heidelberg Twp., Berks Co. (c, 1850).

Fisher, Jacob. 1810, Bethel Twp., Bedford Co. (t); 1826–36, Belfast Twp., Bedford Co (t); 1841–56, St. Clair Twp., Bedford Co. (t); 1857–58, Hopewell Twp., Bedford Co. (t). Gunsmith and blacksmith.

Fisher, Paul. White Deer, Lycoming Co. (BPaD, 1861) Gunsmith.

Fisher, William (1746–1822). Cumberland Twp., Greene Co. Employed at McCullough Iron Works, making guns during the Revolution (*Horn Papers*, 2:717).

Fitch, John (1743–98). Gunsmith and silversmith, 1777–88, Warminster Twp., Bucks Co. (t) 5/15/1777: "pay John Fitch £20 . . . for repairing arms last summer" (2 Pa. Arch. 1 at 46).

Fitzpatrick, Felix. Artillery artificer, 1779, at the cannon factory at Philadelphia (2 Pa. Arch. 15 at 386).

Fitzwater, L. Gunsmith, 109 Shippen, Philadelphia; bench hand to George W. Palmer (MD, 1846).

Flack, John. 1776, Buckingham Twp., Bucks Co. Proposed to Council the building of a gunpowder mill (1 Pa. Arch. 4 at 709).

Fleck, George K. (1796–1880). 1840–77, Shirley Twp., Huntingdon Co. (t; c); 1880, Mt. Union, Huntingdon Co. (c). Gunsmith.

Fleck, Valentine (b. 1805). 1850, Shirley Twp., Huntingdon Co. (c); 1860, Menno Twp., Mifflin Co. (c; BPaD, 1861); 1870, 1880, Union Twp., Mifflin Co. (c). Gunsmith.

Fleeger, Christian. McCandless, Butler Co. (BPaD, 1861) Gunsmith.

Fleeger, John (1807–82). 1831–82 Owner, Allegheny Gun-Works, Pittsburgh. Gun manufactory (c; t; PAD, 1860–61, 1873).

Fleeger, Peter. Gunsmith, between Godrich and Hope, Pittsburgh (PAD, 1856–7).

Fleeger, William A. Gunsmith, Pittsburgh (PAD, 1860–61, 1864–65, 1884).

Fleeger, William, Jr. 174 Shanapin, Pittsburgh (PAD, 1864–64; PD, 1882). Gunsmith.

Flegle, George. Master armorer, U.S. Arsenal, Philadelphia, during the War of 1812 (t).

Flegel, J. G. 221 St. John's, Philadelphia. Gunsmith (DeSD, 1829, 1830, 1835–36).

Fleming, Robert. Blacksmith, 1776, in Capt. Cookson Long's Co., Northampton Militia (2 Pa. Arch. 14 at 330).

Flinn, John. Artillery artificer, 1779. (2 Pa. Arch. 15 at 384).

Florus, Peter. Gunsmith, 1845, Lower Milford Twp., Lehigh Co. (t).

Flower, Col. Benjamin (1748–81). Artillery artificer, 1777, commanding officer. Oversaw casting of cannon and preparation of accoutrements. Flower was killed in action and buried at Christ Church, Philadelphia (2 Pa. Arch. 11 at 249; 2 Pa. Arch. 15 at 382).

Flowers, Charles (1821–97). 1841–97, gunsmith, Harmony, Butler Co. (t; c).

Floyd, James. Gunsmith, 1759, Col. Burd's Co. (5 Pa. Arch. 1 at 352).

Fluharty, James, Sr. (d. 1822). 1773, Springhill Twp., Fayette Co. (t); 1774–75, gunsmith in Dunmore's War; 1790, Bullskin Twp., Fayette Co. (c); 1792–1822, West Liberty, Ohio Co., (West) Virginia. His estate in Ohio Co. court showed gunsmithing tools (t, c, f).

Fluharty, James, Jr. Gunsmith 1794, German Twp., Fayette Co., apprenticed to his father (t); 1800–1803; German Twp., Fayette Co. (t); 1806–14, Rockdale, Crawford Co. (t, f).

Fluharty, Stephen (1783–1816). Gunsmith and blacksmith, apprenticed to his father, James Fluharty, Sr. Drowned in Short Creek, Ohio Co., W. Va., after a drunken brawl (t; f).

Foehl, Charles. 1873–1912, gunsmith, 2927 Thompson St., Philadelphia (GPD, 1892; PD, 1882; BRD, 1888; BSED, 1900).

Fogle, Heinrich. Gunsmith, Low St., Lancaster (BLD, 1857).

Foher, Ludwig. 1773–80, Philadelphia. Musket maker for Committee of Safety (Col. Rec. 10:550; KEA, 31).

Follett, Thomas. Artillery artificer, 1779. (2 Pa. Arch. 15 at 383). Drum maker and preparer of skins.

Fondersmith, George. 1803–26, Strasburg Twp., Lancaster Co., Gunsmith (t; SED, 55).

Fondersmith, George, II (b. 1814). Born and trained in Lancaster Co. 1860, McKean Twp., Licking Co., Ohio (c, 1860); Urbana, Champaign Co., Ill. (c, 1880). Gunsmith.

Fondersmith, Isaac (b. 1835). Burgettstown, Washington Co., gunsmith (c, 1860, 1870).

Fondersmith, Jacob. 1811, Martic Twp., Lancaster Co. (t); 1819–21, Germany Twp., Adams Co. (t); 1833–43, East Liberty, Allegheny Co. (t); 1847, Lawrenceville, Allegheny Co. (t; WSB, 99). Gunsmith.

Fondersmith, John, Sr. (b. 1746). Gun contractor during the Revolution (I.D. Rupp, *History of Lancaster and York Counties* [Lancaster: Hill] 307). 1799–1803, Strasburg Twp., Lancaster Co. (t; SED, 55; JKJ, 113; GAS-1, 312–15) Master gunsmith. Contract for 500 "liberty cap" muskets dated 1/14/1799, and one dated 4/16/1801 for 500 additional muskets.

Fondersmith (Fendersmith), Levi. "Near McConnellstown, Levi Fendersmith had a gunshop shortly after 1800, which was discontinued after his removal" (AFR, 384).

Fondersmith, Lewis (b. 1778). 1803, Strasburg Twp., Lancaster Co. (t); 1820, Woodberry Twp., Huntingdon Co. (t; c); 1829–31, Walker Twp., Huntingdon Co. (t; c); 1839–42, Eldersville, Washington Co. (t; c); 1845–52, Burgettstown, Washington Co. (t; c). Major gunsmith.

Fondersmith, Valentine (1746–1820). Major gunsmith, 1779–1824, Strasburg Twp., Lancaster Co. (t; SED, 55; GAS-1, 312–15).

Fondersmith, William (b. 1828). 1849–64, Burgettstown, Washington Co. Gunsmith and silversmith (t; c).

Ford, John (1780–1862). Gunsmith and silversmith, 1801–62, Harrisburg. (t; c) 1817, appointed state armorer and superintendent of the state arsenal. "Pay John Ford $704.12 ½ for cleaning and repairing public arms." (Hazard's *Annals*, 2:59) Married Mary Schnevely (1781–1830) of Cumberland Co. (Egle, *Dauphin and Lebanon Counties*, 492–93).

Ford, Matthew. armorer, 1777, on the armed boat *Dickinson*, Pa. Navy (2 Pa. Arch. 1 at 278).

Ford, Peter. Silversmith, York Co. (Egle, *Dauphin and Lebanon Counties*, 492–93); father of John Ford.

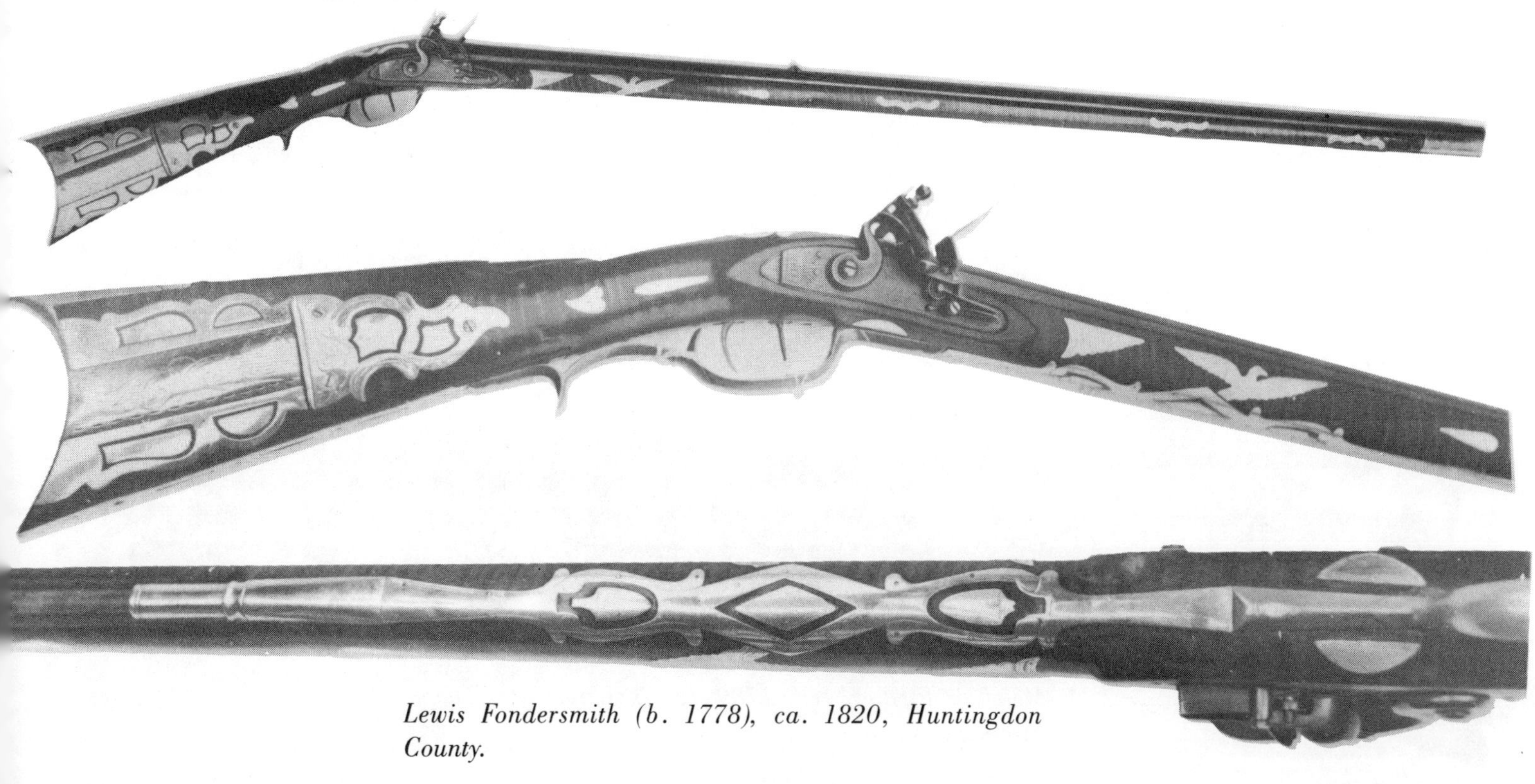

Lewis Fondersmith (b. 1778), ca. 1820, Huntingdon County.

Fording, John J. 38 Ferry St., Pittsburgh (PAD, 1860–61). Gunsmith.

Fordney, Casper (b. 1807). 1828–30, Lancaster boro., gunsmith (t; SED, 55); Knox Co., Ohio (c, 1850).

Fordney, Charles. Gunsmith, Leopard Alley, Lancaster (LCD, 1871–72); 323 N. Cherry St. (LCD, 1873–74; BLD, 1875–76).

Fordney, Henry. Gunsmith, Lancaster boro., 1833–38 (t; SED, 55); 1840–53, Philadelphia (t; c).

Fordney, Isaac. Gunsmith, 1840, Lancaster (t).

Fordney, Jacob, Sr. (d. 1819). Gunsmith, Lancaster, 1800–19. (WSB, 1; t).

Fordney, Jacob, Jr. (1808–78). 1834–78, Lancaster, gunsmith (t; c; BLD, 1857, 1873–74, 1875–76; BPaD, 1861).

Fordney, Jacob, III. Gunsmith, 1843, Lancaster (t).

Fordney, John J. 528 N. Duke St., Lancaster (BLD, 1875–76); 22 W. King St., (PD, 1899–1900). Gunsmith.

Fordney, Melchior (ca. 1789–1846). 1807–46, Lancaster. (t; SED, 55) He had moved in with a woman to whom he was not married, and a man named Haggerty struck him with an axe, killing him as punishment for his adultery (JKJ, 130–32). Major gunsmith.

Fordney, Philip. 1836, gunsmith, Lancaster (t; SED, 55).

Forest, Casper. 1855–77, powder-horn manufacturer, corner W. King St., Lancaster (t; BLD, 1873, 1875–76; BPaD, 1861). In 1860 he made 8,000 powder horns (IC).

Forker, Israel (b. 1807). 1825–28, Meadville, Crawford Co. (t); 1838–66, Portage Co., Ohio (DH, 322). Gunsmith.

Forker, John (1794–1864). Gunsmith, 1816–64, Mercer, Mercer Co. (t; c; BPaD, 1861).

Forker, John. Philadelphia (BRD, 1875). Gunsmith.

Forker, Joseph (b. 1829). 1850–57, Mercer, Mercer Co., gunsmith (t; c).

Forker, Samuel (1798–60). Meadville, Crawford Co. (t; c; BPaD, 1861—the year after he died). Gunsmith.

Forker, William H. 1859–1900, 28 Chestnut St., Meadville, Crawford Co. (t; c; BRD, 1873; CCD, 1874). Gunsmith and inventor.

Forman, John. 562 Frankford Ave., Philadelphia (BPaD, 1861). Gunsmith.

Forrester, John (b. 1801). 1824–38, Monongahela Twp., Greene Co. (t); Tyler Co., W.Va. (c, 1850, 1860). Gunsmith.

Forsyth, Joseph. Gunsmith, 1802–3, Somerset Twp., Somerset Co. (t).

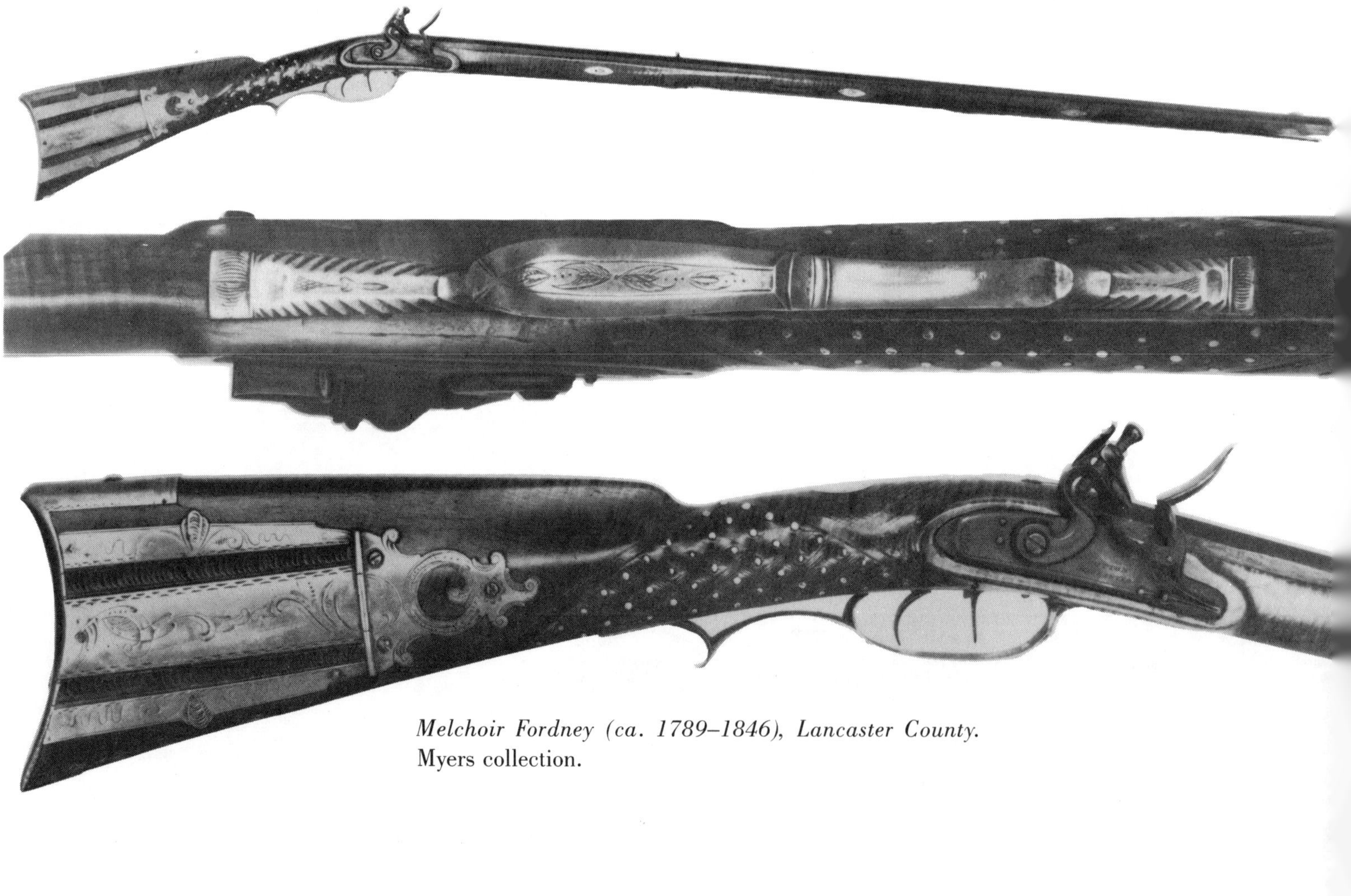

Melchoir Fordney (ca. 1789–1846), Lancaster County. Myers collection.

Fort Pitt Cannon Co. (operated 1803–74). In 1813 it cast the cannon used by Oliver Hazard Perry on Lake Erie. In 1814, it built 20 guns of 42-pound size, 40 of 32-pound size, 20 of 24-pound size, and 28 of 18-pound size. Founded and operated by Joseph McClurg and family. The company name was changed to Ft. Pitt Cannon Co. in 1816. In 1849 Lt. Thomas Jefferson Rodman devised a method for casting large cannon at this factory. During the Civil War it cast 2,000 pieces of 20-inch cannon. In 1874 the last weapons were made and it was taken over by Mackintosh-Hemphitt (Killikelly, *History of Pittsburgh*, 62; V. S. Clark, *History of Manufactures in the U.S.*, 2nd ed. [New York: Scribner's, 1934] 16–17).

Foster, John. Gun-barrel factory, 1783, Newberry Twp., York Co. (t).

Foster, Thomas. Gunsmith, 1817–27, German Twp., Fayette Co. (t).

Foster, Thomas. 267 S. 4th St., Philadelphia. Gunsmith (MD, 1865, 1841, 1856).

Foster, William (b. 1809). 1846–50, Wolf Creek Twp., Mercer Co. (t); 1860–80, North Liberty, Liberty Twp., Mercer Co. (t; c; BPaD, 1861) Gunsmith.

Fottrell, Patrick. Armorer, 1777, Capt. John Marshall's Co., Pa. Foot Regt. (2 Pa. Arch. 10 at 271); stationed at U.S. Armory at Philadelphia, 1778 (2 Pa. Arch. 11 at 258).

Foulke, Adam. Gunsmith, 1770–76, Salisbury Twp., Northampton Co. (t)

Foulks (Foulkes), William. 1759, Lancaster Co. (t; SED, 55); 1775 (KEA, 33).

Fowler, John. Inventor, Lancaster. Patent of 1/4/1810 for a plating technique. (RAG, 67).

Fowler, Stacy. Poplar Lane at St. John's, Philadelphia. Gunsmith (DeSD, 1829, 1830, 1833).

Fox, Benjamin. Gunsmith, 1843–76, 448 N. Queen St., Lancaster (t; SED, 55; BLD, 1871–72, 1873–74, 1875–76).

Fox, George. Gunsmith, 1847, Lawrenceville, Tioga Co. (t).

Fox, Horace (1815–87). 1835, Frewsburg, N.Y.; 1850, Fredensburg, Jefferson Co. (now Forest Co.); 1855, Chautauqua Co., N.Y.; 1860 through 1882, Frewsburg, N.Y. (REH, 142–43, 159).

Fox, John. Contractor to Committee of Safety, 1775–77 (Col. Rec., 10:514; KEA, 33).

Fox, John. Cutler, Philadelphia; bought land in Lancaster Co. (DB M-364-b) 7/11/1776: "John Fox and Wigfall, cutlers, attend the meetings [of the Torys] and are busy [at sedition]" (2 Pa. Arch. 1 at 614). On the list of persons attainted by high treason, Philadelphia (1 Pa. Arch. 10 at 250ff.).

Fox, Patrick. Artillery artificer and blacksmith, 1778, U.S. Arsenal at Philadelphia (2 Pa. Arch. 11 at 258).

Fox, Philip. Gunsmith, Lancaster, 1843 (SED, 55; t).

Fox, Reuben (b. 1794). 1816–36, Metal Twp., Franklin Co. (t; WSB, 23); 1837, Corry, Erie Co. (t); 1850, Stark Co., Ohio (c); 1855, Tuscarawas Co., Ohio (t); 1860, Stark Co., Ohio (c). Gunsmith.

Frailey, Andrew J. 334 N. Price St., Lancaster (SED, 55; c; t; BLD, 1857, 1871–72, 1873–74, 1875–76). Gunsmith.

Frailey, Frederick. 1772–82, 5203–5 Main St., Germantown. Made gun carriages during the Revolution; later made drums (*Guidebook to Historic Germantown*, 45).

Fraley, H. J. Gunsmith, North Branch Twp., Wyoming Co. (PD, 1882).

France, J. A. 1850, gunsmith, Lancaster (c, 1850; SED, 55).

Francis, William. 1779–82, artillery artificer, cannon maker, Philadelphia (1779) and New London, Va. (1882). Entitled to 200 acres of donation land (2 Pa. Arch. 11 at 258, 262; 3 Pa. Arch. 3 at 650).

Frank, Abraham. Newberry, Lycoming Co. (BPaD, 1861). Gunsmith.

Frank, Charles Lewis (b. 1778). Arrived at Philadelphia, 12/3/1807, on ship *William Johnson*. Born in Königsburg, Prussia. Locksmith; age 25, black hair, middle sized (2 Pa. Arch. 17 at 665).

Frankford Arsenal (founded 1816). Tacony Rd. at Bridge St., Bridesburg, Philadelphia Co. In World War I it employed 6,100 people on a 62-acre site. Manufactured ordnance and ammunition (FSP, 109).

Fraser, John (1721–73). Born in Scotland, Fraser was an Indian trader by 1747 (1 Pa. Arch. 2 at 14) from Lancaster Co. Shortly thereafter he established a trading post at Venango and was expelled in December 1753 by the French. He then went to land he owned on the Braddock battlefield. He warned the settlers near Redstone Fort of an impending Indian massacre. He was an armorer in Dunmore's War and later established a gunshop near Cumberland, Md. By 1768 he was in Friend's Cove, Bedford Co., and when Bedford Co. was formed in 1771, he was appointed a justice of the peace. His tavern (used also as a court room) is on the corner of Pitt St. and Richard St., Bedford boro. (McKnight, *History of Northwest Pennsyl-*

vania, 199; t; Brown, *History of Crawford County,* 146, 155–56, 162) Indian trader, gunsmith and blacksmith.

Frayley, Henry. Arms contractor, 1776–77, Philadelphia, to Committee of Safety. (2 Pa. Arch. 1 at 478); gunsmith, 175 Coats St., Philadelphia (DeSD, 1816).

Frazier, Francis. Gunsmith, Dunnings, Luzerne Co. (BPaD, 1861).

Frederick, John. 1859–79, Cumru Twp., Berks Co., gunsmith (t; c).

Frederick, Samuel. Gunsmith, 1821, Bloom Twp., Columbia Co. (t).

Freed, David. Gunsmith, 10 Green St., Philadelphia (MD, 1846).

Freed, George (b. 1825). Gunstocker, Rostraver Twp., Westmoreland Co. (c,. 1850).

Freed, Jacob (b. 1819). Musket-stock maker, Rostraver Twp., Westmoreland Co. (c, 1850).

Freively, Joseph. Gunsmith, 1835, Bushkill Twp., Northampton Co. (t).

Fresh, Joseph (b. 1803). Born in Switzerland. Gunsmith, 1845–49, Cambria Twp., Cambria Co. (t); 1850–58, Summerhill Twp., Cambria Co. (t; c); 1860–70, Croyle Twp., Cambria Co. (c).

Frey, Jacob. Swordsmith, in the employ of Lewis Prahl, gunsmith, Philadelphia Co., 1777. (2 Pa. Arch. 3 at 104); Artillery artificer, 1779, at the cannon factory at Philadelphia (2 Pa. Arch. 15 at 384).

Frey, Martin (b. 1769). Gun-barrel boring mill, ca. 1800, York Co. (KEA, 33).

Friebele, Christian. 1835–50, gunsmith, Upper Nazareth Twp., Northampton Co. (t; c).

Fries, William. 1850, gunsmith, Lancaster. (c; t; SD, 55)

Frisbie, Joseph D. Patent 16,16,575 of 2/10/1857 for a revolver. Pittsburgh.

Froggatt, William. Gunsmith, 392 N. Front St., Philadelphia (MD, 1846).

Fronefield, John (1770–1853). 1800, Providence Twp., Montgomery Co. (KEA, 34) Died in Schuylkill Twp., Chester Co. (Registrations of Death, 2/25/1853, no. 183). Gunsmith.

Fry, Edward (1869–1955). Ligonier, Westmoreland Co. Active, 1890–1930. Gunsmith (t; c; f).

Fry, George. Gunsmith, 1841–44, Hopewell Twp., Bedford Co. (t).

Fry, John (1821–89). 1843–52, Washington Furnace, Westmoreland Co. (t); 1853–55, Ligonier, Westmoreland Co. (t); 1860, Ligonier Twp., Westmoreland Co. (c). Gunsmith.

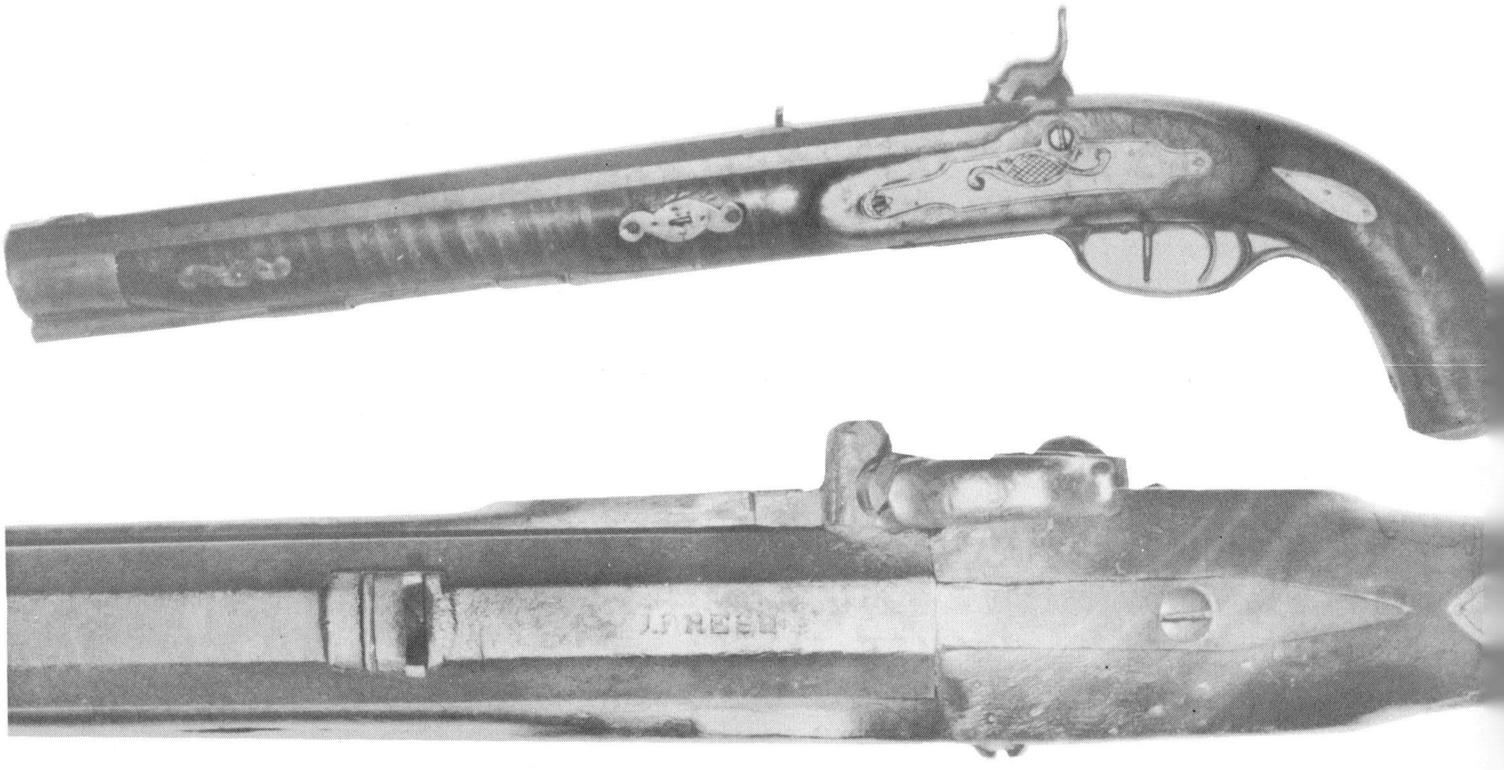

Joseph Fresh (b. 1803), Cambria County.

Joseph Fresh, Cambria County.

Fry, Joseph (1826–91). 1850, Ligonier Twp., Westmoreland Co. (t); 1855–62, Ligonier Boro., Westmoreland Co. (t; c); 1866–91, Ligonier Twp., Westmoreland Co. (t; c). Gunsmith.

Fry, Martin (b. 1739). Locksmith and gunsmith, York. Apprenticed to Jacob Leather, 1756 (JKJ, 325–27).

Fry, Martin (b. 1769). 1799–1801, York. Gunsmith, and ran a gun-boring mill. 1821, Carlisle, Cumberland Co. (t; WSB, 59).

Frymer, John. 1850, gunsmith, Cocalico Twp., Lancaster Co. (c, t; SED, 55).

Fullerton, Robert. Artillery artificer, 1779, at the lead furnace (2 Pa. Arch. 15 at 385).

Fulmer, Jonathan. Gunsmith, 519 Northampton St., Easton (BRD, 1889).

Fulton, Robert. Silversmith and gunsmith, 1785, Gray's Alley, Philadelphia. Located in old Kirder gunshop (t).

Fulton, Seibert. Gunsmith, Perry Co. (c, 1850).

Fultz, Valentine. Artillery artificer, 1777, Capt. Will's Co. (BCR, 115).

Fulwiler, John. Artillery artificer, 1777, Capt. Will's Co. (BCR, 115).

Funk, George. Gunsmith, 1844, Bethel Twp., Bedford Co. (t).

Funkhouser, Abraham (1816–54). North Sewickley Twp., Beaver Co. (t; c). Gunsmith.

Funkhouser, Jonathan (1808–75). Gunsmith, North Sewickley Twp., Beaver Co., 1831–75 (t; c).

Furney, John. 1826, Belfast Twp., Bedford Co. (t); 1829–48, Woodberry Twp., Bedford Co. (t); 1852, Thompson Twp., Fulton Co. (t). Gunsmith.

Furney, William. Gunsmith, Pittsburgh. 1816–18, (t).

G

Gable & Allen, West St. and 3d St., Williamsport, Lycoming Co. (PD, 1882). Gunsmiths.

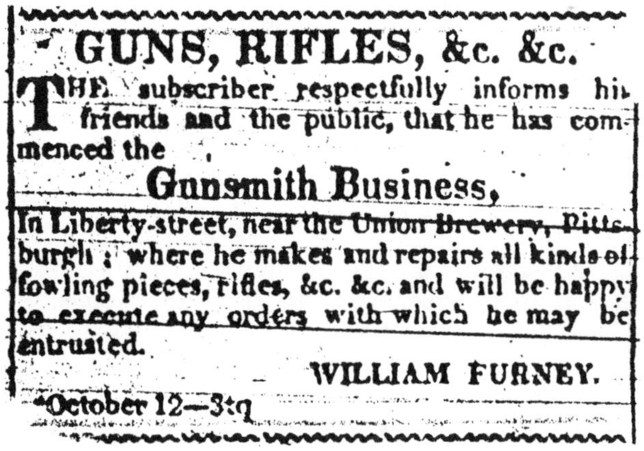

GUNS, RIFLES, &c. &c.

THE subscriber respectfully informs his friends and the public, that he has commenced the

Gunsmith Business,

In Liberty-street, near the Union Brewery, Pittsburgh; where he makes and repairs all kinds of fowling pieces, rifles, &c. &c. and will be happy to execute any orders with which he may be entrusted.

WILLIAM FURNEY.

October 12—3tq

Pittsburgh Mercury, October 12, 1816

Gable, Henry, S. (d. 1854). 1805–54, Williamsport, Lycoming Co., gunsmith. His shop was located on the south side of E. 3d St., between Mulberry and State St., on the present-day site of the Grit Publishing Co. (t; NAT, 214).

Gable, Henry, Jr. 1850–80 West St., corner of 3d St., Williamsport, Lycoming Co. Gunsmith (BPaD, 1861; BRD, 1875; BSED, 1880).

Gable, Henry, III. Gunsmith, West St. and 3d St., Williamsport, Lycoming Co. After 1860, "Henry Gable [Jr.] & Son" (BPaD, 1861; BSED, 1880; NAT, 214).

Gable, Thomas. Rear of 230 W. Walnut St., Lancaster (LCD, 1871–72). Gunsmith.

Gahm, Gustavus. Gunsmith, Titusville, Crawford Co. (PD, 1882).

Galbraith, Robert. In the will of John Galbraith, dated 12/28/1757, he ordered that his son, Robert, was to be apprenticed to Henry Willis or some other gunsmith to learn that trade (Cumberland Co., WB A-40).

Galbreath, J. H. Gunsmith and supplies, 1851–58, Lancaster (t).

Galbreath, John. 1757, Lancaster (M. Dyke, KRAN [1979], 7). Gunsmith.

Galbreath, Samuel. Gunsmith, 1800, Newton Twp., Cumberland Co. (t); 1804–9, Letterkenny Twp., Franklin Co. (t; WSB, 26–27); 1810–13; Pittsburgh (t).

Gale, William. Chillisquaque Twp., Northumberland Co. (KKR, 235), 1826. Gunsmith.

Gall, John. Gunsmith, 1857, Lancaster Co. (t); 1860–61, Pine Alley, Washington Co. (KKR, 235; BPaD, 1861).

Gallagher & Co. 1528 Frankstown Rd., Philadelphia. (MD, 1861) Gunsmiths. Mahlon J. Gallagher, prop. Manufactory (FSP, 113).

Gallatin, Albert (1761–1849). Born in Geneva, Switzerland. American politician, diplomat, gun manufacturer. 1799–1803, contract for U.S. M-1795 muskets, Springhill Twp., Fayette Co. (Gallatin papers, New Geneva, Fayette Co., Pa.).

Galloway, Henry. 1800–1802, Morgan Twp., Greene Co. (t); 1803–6, Amwell Twp., Washington Co. (t); 1806–11, Jefferson Twp., Greene Co. (t); 1813–18, Morgan Twp., Greene Co. (t); 1818–19, Jefferson Twp., Greene Co. (t). Gunsmith.

Gannon, Thomas. Gunsmith, Huntingdon Co. (c, 1850).

Gappen, Michael. Gunsmith, 1841–69, Dunkard Twp., Greene Co. (t).

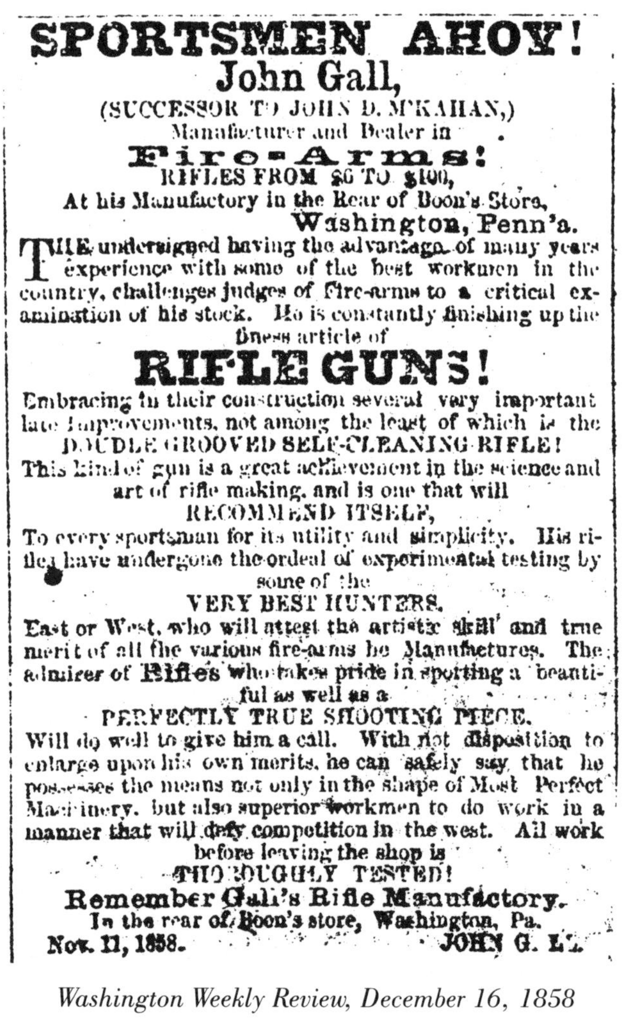

SPORTSMEN AHOY!
John Gall,
(SUCCESSOR TO JOHN D. M'KAHAN,)
Manufacturer and Dealer in
Fire-Arms!
RIFLES FROM $6 TO $100,
At his Manufactory in the Rear of Boon's Store,
Washington, Penn'a.

THE undersigned having the advantage of many years experience with some of the best workmen in the country, challenges judges of Fire-arms to a critical examination of his stock. He is constantly finishing up the finest article of

RIFLE GUNS!

Embracing in their construction several very important late improvements, not among the least of which is the

DOUBLE GROOVED SELF-CLEANING RIFLE!

This kind of gun is a great achievement in the science and art of rifle making, and is one that will

RECOMMEND ITSELF,

To every sportsman for its utility and simplicity. His rifles have undergone the ordeal of experimental testing by some of the

VERY BEST HUNTERS.

East or West, who will attest the artistic skill and true merit of all the various fire-arms he Manufactures. The admirer of Rifles who takes pride in sporting a beautiful as well as a

PERFECTLY TRUE SHOOTING PIECE.

Will do well to give him a call. With not disposition to enlarge upon his own merits, he can safely say that he possesses the means not only in the shape of Most Perfect Machinery, but also superior workmen to do work in a manner that will defy competition in the west. All work before leaving the shop is

THOROUGHLY TESTED!

Remember Gall's Rifle Manufactory.
In the rear of Boon's store, Washington, Pa.
Nov. 11, 1858. JOHN G. L.

Washington Weekly Review, December 16, 1858

Gardiner, William. Artillery artificer, 1779 (2 Pa. Arch. 15 at 383).

Gardner, John N. Gunsmith, 1865–94, 610 Lack Ave., Scranton (t; c; BRD, 1865, 1894).

Gardner, J. W. Gunsmith, Hyde Park, Scranton, Lackawanna Co. (PD, 1882).

Garland, John. Armorer, 1776–77, armed ship *Montgomery*, Pa. Navy. He ran away (2 Pa. Arch. 1 at 322).

Garlick, Absalom. (d. 1877). Silversmith, Brush Creek area, West Providence Twp., Bedford Co. (t).

Gardner, Lewis (b. 1791). 1829–30, Menallen Twp., Fayette Co. (t); 1850, Parker Twp., Butler Co. (c). Gunsmith.

Garrett, William. 1765–79, Cutler and smith, Upper Darby, Chester Co. (RAG, 221; t).

Garrick, Adam. Artillery artificer, 1777, Capt. Jesse Rowe's Co. at the cannon factory at Philadelphia (2 Pa. Arch. 11 at 252; PIR 2:244).

Garver, John. Gunsmith, Norristown, Montgomery Co. (BPaD, 1816).

Gaskill, Budd (b. 1770). Gunsmith and blacksmith. Born in N.J. 1793–1800, Georges Twp., Fayette Co. (t); 1800–1829, Belle Vernon, Fayette Co. (t); 1840–43, Carroll Twp., Washington Co. (t); 1850–60, Greenwood Twp., Crawford Co. (c).

Gaskill, Henry (b. 1810). 1831–32, Belle Vernon, Fayette Co. (t); 1834–51, Carroll Twp., Washington Co. (t; c); 1860, Greenwood Twp., Crawford Co. (c). Gunsmith and boat builder.

Gass, Valentine. Gunsmith, Dry Run, Franklin Co. (BPaD, 1861).

Gatling, Richard Jordan (1818–1903). Inventor of the Gatling gun. His patent of 11/4/1862 was first made by Greenwood & Co., Cincinnati, Ohio. Only 12 guns were made there before fire destroyed the building. His improvement of 5/9/1865 was made in Philadelphia by the Cooper Firearms Manufacturing Co. (GAAAM, 72).

Gatterneyer, J. L. 1749, arrived in Bethlehem from Bavaria. Moravian, blacksmith (PaE, 42).

Gaughler, Nicholas. Gunsmith, 1799, Penn Twp., Union Co. (t) There was also a man by this name in Selinsgrove, a gunsmith, in 1855 (t).

Gaunt & Monet. 1813, Philadelphia (t; PXD, 1813) Made Chambers system multishot guns under contract.

Gears, Joseph. Green St. at 3d St., Philadelphia, gunsmith (DeSD, 1835–36).

Gearson, Joseph. Gunsmith, 85 Dillwyn, Philadelphia (DeSD, 1829, 1830, 1833).

Geary, William. Washington St. above 2d St., Philadelphia, 1833 (KEA, 34; RAG, 72).

Gehbe, Ernest. 1799–1800, Bethlehem boro (t). Gunsmith.

Gehrett, James Wilson (1832–1914). 1850, Todd Twp., Huntingdon Co. (c); 1860–83, Hopewell Twp., Huntingdon Co. (c; t). Gunsmith.

Gehrett, John (b. 1806). 1840–50, Todd Twp., Huntingdon Co. (t; c).

Gehrett, Samuel (1789–1854). 1817–43, Union Twp., Huntingdon Co. (t; c); 1850–54, Plumcreek Twp., Armstrong Co. (t; c). "payment of the claim of Samuel Gehrett of Armstrong Co., the sum of $127 in full for repairing state arms in 1843" (*Laws of Pennsylvania* [1853], Act 345, Sec. 41).

Geiger, Joseph V. 1861–90, Towanda, Bradford Co. (BPaD, 1861; PD, 1882, 1890). Gunsmith.

Geirsach, Christian. Gunsmith, 1804–14, Bushkill Twp., Northampton Co. (t).

Geirsch, Charles. Gunsmith, 1835, Bushkill Twp., Northampton Co. (t).

Geirsh, William. Bushkill Center, Northampton Co. (BPaD, 1861). Gunsmith.

Geise, George. Gunsmith, 1800, Weisenberg Twp., Northampton Co. (t).

Geisinger, Henry. 1856–60, 19th Ward, Philadelphia, ran gun manufactory. (MD, 1857; c, 1860) In 1860 he employed 15 workers (IC).

Gelfinger, George. Locksmith, 1779, Heidelberg Twp., Berks Co. (t).

George, Charles. 1847–61, Maxatawny (Rothrockville), Berks Co., gunsmith (t; c; BPaD, 1861; BSED, 1880).

George, Henry, 1850–61, Nora, near Molltown, Greenwich Twp., Berks Co., gunsmith (t; c; BPaD, 1861).

George, Jacob. 1793–1839, Greenwich Twp., Berks Co., gunsmith (t; KEA, 35, c).

George, James (d. 1819). 1812–16, ran gun-barrel factory on Kishacoquilla Creek, near Jack's Mountain narrows, Brown Twp., Lycoming Co. (Ellis and Hungerford, *Susquehanna and Juniata Valleys*, 639).

George, John. 1813–15, Brown Twp., Mifflin Co. (Ellis and Hungerford, *Susquehanna and Juniata Valleys*, 499). Gunsmith and gun-barrel maker.

Gerhart Brothers (active 1873–98). 430 Penn, Reading, Berks Co. (BRD, 1873; BSED, 1880; PD, 1898; t; c).

Gerhart, Daniel (b. 1844). 1st Ward, Reading. Apprenticed with Nelson Delaney; worked with Delaney until 1873. 1874–82, had his own shop in Reading (RAG, 72; GAAM, 57; t; c). 1882–98, Tamaqua, Schuylkill Co. (t; c; PD, 1882).

Gehrett, Jacob F. (b. 1836). 1850–60, Todd Twp., Huntingdon Co. (c); 1861–93, Orbisonia, Huntingdon Co. (c; t). Gunsmith.

Getz, Frederick (d. 1813). Gunsmith, 158 St. John's, Philadelphia (PXD, 1813).

Getz, Jacob F. 1860–90, 2415 Washington Ave., Philadelphia, gunsmith (BPaD, 1861; PD, 1882; BSED, 1880, 1888).

Getz, John, 1779–1833, Lancaster, gunsmith (t; SED, 56).

Getz, Peter (d. 1810). 1799–1810, Lancaster (SED, 56; t).

Geultner, Frederick. Gunstock maker, Budd below Coates, Philadelphia (DeSD, 1835–36).

Ghiselin, Caeser. Before 1715, Philadelphia; apprenticed and first practiced his trade there, before moving to Baltimore in 1715 (Carl Bridenbaugh, *The Colonial Craftsman* [New York: Reynal & Hitchcock, 1952] 142). Silversmith.

Ghiselin, William. Goldsmith. Advertisement in the *Pennsylania Packet*, 11/14/1751. "William Ghiselin, goldsmith, is removed from Second St., to the house where the Widow Bright lately lived in Second St."

Ghriskey, Lewis. Gunsmith, Kunchel St. and Tammany St., Philadelphia (PXD, 1813); 351 N. Front St. (PXD, 1814, 1816, 1819).

Gibbins, Henry C. (b. 1834). 1860–80, Brownsville, Fayette Co., gunsmith (BRD, 1873; t; c).

Gibbins, Joseph (1808–97). 1830–97, Brownsville, Fayette Co., gunsmith (PD, 1882; c; t). Buried at Redstone Cemetery, North Bend.

Gibbs, Adam. Gunsmith 1847–57, 22 N. Prince St., Lancaster (t; c; BLD, 1855, 1857).

Gibbs, Mrs. Ann. On 11/8/1775, the Committee of Safety purchased "imported firearms from Mrs Ann Gibbs." price: £47. (Col. Rec. 10:398).

Gibbs, Henry (d. 1843). 1820–43, Lancaster, gunsmith (KEA, 35; SED, 56; GAAAM, 57; t; c).

Gibbs, Henry, II. ca. 1820–74, 22 N. Prince St., Lancaster (t; c; GCD, 1863–64; BLD, 1873–74, 1875–76). Gunsmith.

Gibbs, John (b. 1797). Gunsmith, 1821, Lancaster boro (t; SED, 56); 1822–24; Honesdale, Wayne Co. (t; c); 1826–53, Lancaster, Ohio (t; c).

Gibbs, William. 1834–40, gunsmith, Lancaster (SED, 56; t).

Gibson, Francis. 1805–30, Tyrone Twp., Cumberland Co. Gunsmith, distiller, politician (t; WSB, 57).

Gibson, Henry. Gunsmith, Washingtonville, Montour Co. (c, 1870).

Gibson, James. Artillery artificer, 1777–82 (2 Pa. Arch. 11 at 250).

Gibson, James. "a man of color," gunsmith, Wharton near 2d St., Philadelphia. (PXD, 1816; t).

Gibson, Nathan. 11/29/1776: "pay Nelson Gibson £6/05/0 for powder horn" (Col. Rec. vol. 11).

Gibson, Thomas. 139 Cherry St., Philadelphia (MD, 1846). Gunsmith.

Giersach, Christian. Gunsmith, 1804–14, Bushkill Twp., Northampton Co. (t).

Giersch, Charles. Gunsmith, 1835, Bushkill Twp., Northampton Co. (t).

Gilder, D. F. Williamsport, Lycoming Co. (PD, 1890). Gunsmith.

Patent Springs for Windows.

I HAVE purchaſed the patent right (for Cumberland county and town of Chambersburgh) to make and ſell ſprings for faſtening and ſupporting window ſaſhes, and will ſupply any number of them that may be wanted. This ſpring will anſwer the purpoſe of a lock when the window ſaſh is down, and effectually prevents it from being raiſed from the outſide of the houſe. It ſupports the ſaſh at any place it is raiſed to. It is not ſubject to wear ſo as to injure it, & the expence of the ſpring & putting it in will not coſt one fourth part of the ſum that the weight and pully will. Thoſe who wiſh to put them in their windows, will pleaſe to call on me at my ſhop, oppoſite the ſign of the ſhip in Carliſle.

FRANCIS GIBSON.

Carliſle, 14th May, 1804.

N. B. I alſo carry on the *Gunſmith Buſineſs*—Likewiſe caſt *Gun* and *Rifle Mounting* at 3ſ. 6d. per ſett.

Carlisle Gazette, May 14, 1804

Gilder, Frederick, Gunstocker, 20 Callow Hill, Philadelphia (PXD, 1816).

Gill, Benjamin. 1843–50, gunsmith, Lancaster (t; SED, 56).

Gill, Henry. 1843, gunsmith, Lancaster (t; SED, 56).

Gill, J. D. 1783–1819, gunsmith, Lancaster (SED, 56; t).

Gillman, Daniel. 1779–1812, gunsmith, Donegal Twp, Lancaster Co. (t; SED, 56).

Gillman, Nicholas. 1801–8, gunsmith, Donegal Twp., Lancaster Co. (SED, 45; t).

Gilmore, Jonathan. Artillery artificer, 1778, Capt. Thomas Wylie's Co. at Carlisle (PIR, 2:248).

Gilmour, James. 1872–75, Tamaqua, Schuylkill Co. (BRD, 1872; PD, 1875). Gunsmith.

Ginereth, Daniel. Gunsmith, 1819–32, Dublin Twp., Bedford Co. (t).

Ginerich, Daniel. 1785–1801, gunsmith, Warwick Twp., Lancaster Co. (SED, 56; t).

Gingerich, George. Gunsmith, Pleasant Gap, Centre Co. (BPaD, 1861).

Gingerich, Henry. 1175–78, arms contractor, gunsmith, Lancaster (GAAAM, no. 57; S. E. Dyke, KRAN 11, no. 3 [1985] 7).

Ginter, C. Gunsmith, Smith's Mills, Clearfield Co. (BPaD, 1861).

Girsch, Joseph. Rear of 137 Green St., Philadelphia, gunsmith (DeSD, 1829, 1830, 1833).

Glass, Daniel. Gun-barrel manufactory and gunsmith. 1848–59, Wyomissing Creek, Berks Co. Sold out to Gouglar & Haberling in 1858 or 1859, which continued operation through 1862 (RAG, 74).

Glass, Hugh Cook (b. 1775) 1795, runaway apprentice from Henry Wolf at Pittsburgh. (Pittsburgh *Gazette*, 4/23/1795). An interesting article speculated whether this might be the same man who was an explorer in the West and battled a she-bear near Ft. Kiowa in 1823 (MB 6 [1963]).

Glass, John. 1797–99, Gunsmith Pittsburgh (t); 1812–14, and 1818–20, Springfield, Ohio; 1814–17, Putnam, Ohio (t).

Glass, John. Barrel maker, Goughlersville, Berks Co. (PD, 1887).

Glass, Michael. Gunsmith, 1772–80, Allen Twp., Northampton Co. (t).

Glass, Samuel. 1807–9, Gunsmith Washington boro (t);

1810–14, Springfield, Ohio (t); 1814–15, Putnam, Ohio (t).

Glass, Valentine. Gunsmith, Huntington Twp., Adams Co. (c, 1850).

Glassbrenner, David (1798–1872). Gunsmith, 1825–72, West Hanover, Dauphin Co. (Shumway, MB 2 [1971]: 2–3).

Gleist, Daniel. Gunsmith, 1770–83, Bethlehem boro (t).

Glenn, J. D. Gunsmith, Oley Twp., Berks Co. (PD, 1882, 1890).

Glenn, Thomas. 1788, East Earl Twp., Lancaster Co. (t; SED, 56). Gunsmith.

Glockler, Bernard. 40 Duncan, Pittsburgh, gunsmith (PAD, 1864–65).

Glover, Benjamin. Gunsmith, 1823, Brownsville, Fayette Co. (t).

Glover, James (1753–1844). Served in N.J. and Pa. Continental lines during the Revolution, engaged at Germantown and Princeton. He was with Washington when he made the famous crossing of the Delaware River, and at Valley Forge. Later settled at Allegheny City and finally in Adams Twp., Butler Co. "A skillful blacksmith and [he] was engaged much of the time as an armourer" (McKnight, *Pioneer Outline History of Northwestern Pennsylvania*, 453–54).

Glover, Thomas, Artillery artificer, 1779, Pa. Continental Line. Buckle filer and finisher (2 Pa. Arch. 15 at 385).

Glutton, Richard. 1681–82, "Gunton" at Penn's Grant (MRB, 205).

Gobrecht, Samuel (1750–1836). Gunsmith, 1807, Adams Twp., York Co. (t); 1810–36, Germany Twp., Adams Co. (t; WSB, 99).

Godschall, Nicholas. 1762–70, locksmith, Reading. Naturalized 9/12/1762 (2 Pa. Arch. 2 at 439; t; GAAAM, 58).

Goetz & Tryon. Gunsmiths, 163 N. 2d St., Philadelphia. (PXD, 1811, 1813)

Goetz & Westphall. Gunsmiths, Philadelphia. By 10/7/1812, delivered 1,019 U.S. M-1808 muskets for Pa. Militia use. Frederick Goetz and Charles Westphall (PXD, 1808, 1811; RAG, 75).

Goetz, Frederick, Sr. (d. 1811). Juniper St. near High St., Philadelphia, gunsmith (PXD, 1806, 1808, 1810).

Goetz, widow of Frederick. Juniper St. near High St., Philadelphia, gunsmith (PXD, 1813).

Goetz, Frederick, Jr. Gunsmith, 32 Sassafrass St., Philadelphia. (PXD, 1816) 163 N. 2d St. (PXD, 1818).

Goff, William. Gunsmith, 1826–30, Pittsburgh (t).

Gohrmann, Henry (d. 1854). Born in France, died in Pittsburgh. A gunsmith by trade; death certificate no. 627 of 1854.

Golcher, George W. 2032 Frankford Ave., Philadelphia (PD, 1882; PCD, 1892); 1934 Frankford Ave. (BSED, 1880). Gun parts manufacturer.

Golcher, James (d. 1805). 2d St. and Poplar St., Philadelphia. Ran gun manufactory; revolutionary war contractor (RAG, 75).

Golcher, James (d. 1880). Prop., Eagle Gun Works, 138 N. 2d St., Philadelphia. In 1850 he employed 17 workers who made 600 dozen gunlocks. (IC) Gun and gunlock maker (MD, 1841, 1846, 1855, 1849).

Golcher, Joseph. Gunlock manufacturer, ca. 1850, Philadelphia (RAG, 75, 167).

Golcher, Manuel. Gunsmith, 1823–24, 8th St. near Green St., Philadelphia (DeSD, 1823); 2 Will's Row, Philadelphia (DeSD, 1824, 1833, 1835).

Golcher, Thomas L. 1868–82, gunsmith, 116 W. Girard, Philadelphia (PD, 1882; BSED, 1880; MD, 1868).

Golcher, William (d. 1915). 1840–90, Girard near 4th St., Philadelphia; after 1890, superintended Naval Yard, Philadelphia (RAG, 75).

Golden, James (b. 1800). Born in Va. Gunsmith, 1823–55, Waynesburg, Greene Co. (t; c).

Gonter, John. 1810, Columbia, Lancaster Co. (t); 1824–25, Penn St., Reading (KEA,36); 1776, apprentice gunsmith in Capt. Andrew Graff's Lancaster Militia (2 Pa. Arch. 13 at 347; 5 Pa. Arch. 7 at 1073); 1790–1801, Hagerstown, Md. (WSB, 129–30). Major gunsmith.

Gonter, Paul. 1800, gunsmith, Columbia, Lancaster Co. (t; SED, 56).

Gonter, Peter, Sr. (1711–1768). 1750–68, Lancaster.

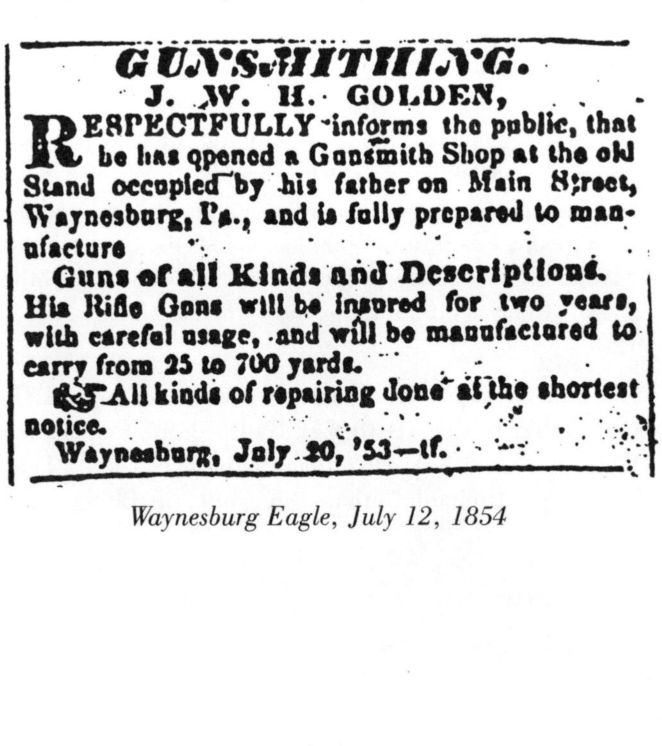
GUNSMITHING.
J. W. H. GOLDEN,
RESPECTFULLY informs the public, that he has opened a Gunsmith Shop at the old Stand occupied by his father on Main Street, Waynesburg, Pa., and is fully prepared to manufacture
Guns of all Kinds and Descriptions.
His Rifle Guns will be insured for two years, with careful usage, and will be manufactured to carry from 25 to 700 yards.
All kinds of repairing done at the shortest notice.
Waynesburg, July 20, '53—tf.

Waynesburg Eagle, July 12, 1854

Whitesmith, blacksmith, and gunsmith (JKJ, 121–22; t).

Gonter, Peter, Jr. (1751–1819). 1773–1819, Lancaster. (SED, 56, t) Major gunsmith. Gunsmith, Capt. Andrew Graff's Lancaster Militia, 1776 (2 Pa. Arch. 13 at 347; 5 Pa. Arch. 13 at 1073). His shop was located on N. Queen St., corner of Orange St., Lancaster. Treasurer of Lancaster boro., 1803–7. Contractor for M-1795 muskets; delivered 557 (JKJ, 121–22).

Good, Peter. Gunsmith, Alsace Twp., Berks Co. (c, 1850).

Goodling, Peter (1761–1815). Son of Peter and Catharina Goodling of Manchester Twp., York Co. 1781–1815, Shrewsberry Twp., York Co. Major gunsmith. In 1792 General Hand refused to accept guns made on government contract by Goodling (KRAN [Spring 1977]; Shumway, 6 [1979]: 25).

Goodman, John. Artillery Artificer, 1779, Pa. Continental Line (2 Pa. Arch. 15 at 384).

Goodman, Michael. 1777, Artillery artificer, Capt. Jesse Rowe's Co., Philadelphia; discharged by Judge McKean (2 Pa. Arch. 11 at 252; PIR, 2:246).

Goodman, Nicholas. Gunsmith, 1826–34, North Huntingdon Twp., Westmoreland Co.; 1834–38, Sewickley Twp., Westmoreland Co. (t).

Goodsell, Albert G. (b. 1823). 1844–70, Coudersport, Potter Co. (t; c; BPaD, 1861). Gunsmith.

Gordon, James. Goldsmith, 1770–74, Philadelphia. Gordon left his wife in Pa. and eloped with Nancy Thompson to Hartford, Conn. where he went to work with James Tiley. His bigamy uncovered, Gordon and his second bride fled to R.I. (Bridenbaugh, *The Colonial Craftsman*, 137–8).

Gore, Daniel. 1768–79, Wyoming Co. (*Wyoming and Lackawanna Valleys*, 1:417). Blacksmith, gunsmith, and whitesmith.

Gore, Obediah (1714–1779). 1768–79, Wyoming Co. Blacksmith, gunsmith, and whitesmith. Brother of Daniel. The brothers were Conn. partisans in the border dispute, and often resorted to arms. They designed a cannon made from "a large pepperage log bored and then hooped from breech to muzzle with stout iron bands. It was shot once, and then given a heavier powder charge. It burst and one set of hoops was thrown 1000 feet across the Susquehanna River and afterwards was found in the willows on the other shore" (Charles Miner, *History of Wyoming County* [1845], 128).

Gorham, Richard H. Shamokin, Northumberland Co. (PD, 1882; BSED, 1880); 522 Railroad St., Conemaugh, Cambria Co. (Clark, *Johnstown Directory*, 1889). There was also a Richard Gorham, gunsmith, in Phillipsburg, Clearfield Co. active in 1882 (PD, 1882).

Gorner, John. Locksmith, 1773 (t); lock maker, 1759 (SED, 56).

Gortner, Joseph. Locksmith, 1763–64, Lancaster Co. (DB K-94).

Goss, Ebenezer. Artillery artificer, 1778, Capt. Thomas Wylie's Co. at Carlisle (PIR, 2:248).

Gouger (Goucher), Thomas. 1777, Market St., Philadelphia. On 6/25/1776, he was paid £30 for disclosing his art "of boreing and grinding gun barrels." (Col. Rec., 10:615). In 1777 council "agreed to allow Mr Goucher 28 sh each piece for 300 barrels he now has finished and that he be allowed 30 sh each piece hereafter, while iron continued at £60 a ton" (2 Pa. Arch. 1 at 496).

Goughnour, Daniel. Gunsmith, Pershing, Cambria Co. (BPaD, 1861).

Gougler & Haberling. Gun-barrel makers and gunsmiths, Berks Co. on Wyomissing Creek, 1859–62 (RAG, 74, 176; MB 2 [1963]).

Gould, John. Gunsmith, Clinton, Wayne Co. (PD, 1882, 1890).

Gouny, Nicholas. Gunsmith, 174 S. 2d St., Philadelphia (PXD, 1813, 1816, 1819).

Grabe, Joseph (b. 1818). Gunsmith and locksmith, middle sized, arrived at Philadelphia on ship *William P. Johnson* on 12/3/1807, black hair and grey eyes (2 Pa. Arch. 17 at 664; Strassburger, *Pennsylvania German Pioneers*, 2:200); 1808, in the employ of Owen and E. Evans (RAG, 77).

Grace, Robert. 1135 Pine St., Philadelphia, gunsmith (BRD, 1867, 1886; BSED, 1880).

Graef, John. Major gunsmith, 1772–1802, Lancaster boro. (SED, 56; t) On 4/11/1798, contractor for M-1795 muskets, with Abraham Henry. In 1803 he was among those who lobbied Congress to retain a tax on foreign-made firearms. His shop was on E. King St. (JKJ, 87–88; GAS-1, 304–07) Naturalized on 4/21/1762 at Lancaster (2 Pa. Arch. 2 at 432). On 5/4/1775: "Mr Graff agrees that the country shall have his powder . . . at the rate of £15/cwt" (2 Pa. Arch. 13 at 293).

Graeff, George. Gunsmith, 1785–86, Lancaster (SED 56; t).

Graeff, William. Gunsmith, 1751–79, Reading (t). Naturalized at Berks Co., 9/18/1761 (2 Pa. Arch. 2 at 423).

Graeff, William. 1807–9, Lancaster, Gunsmith (t; SED, 56).

Graff, Jacob. 1743, naturalized in Lancaster Co. (2 Pa.

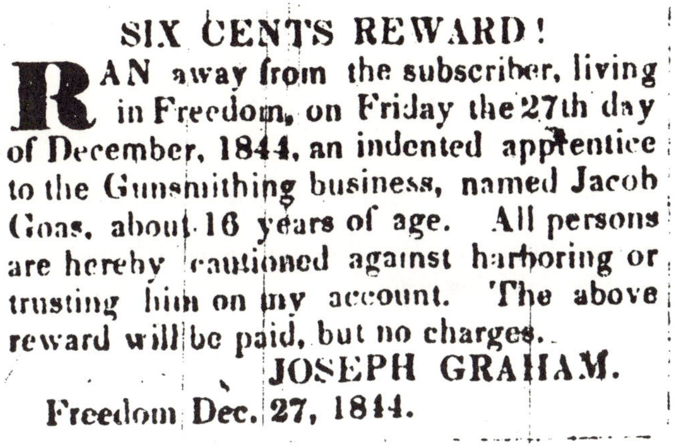
SIX CENTS REWARD!

RAN away from the subscriber, living in Freedom, on Friday the 27th day of December, 1844, an indented apprentice to the Gunsmithing business, named Jacob Goas, about 16 years of age. All persons are hereby cautioned against harboring or trusting him on my account. The above reward will be paid, but no charges.

JOSEPH GRAHAM.

Freedom Dec. 27, 1844.

Western Argus, January 1, 1845

Arch. 2 at 359) Quaker. 6/1/1743, Strasburg Twp., Lancaster Co. (DB H-14); 1776, home guard under Capt.-Lt. Crawford (2 Pa. Arch. 13 at 349). Gunsmith, Lancaster Co., 9/13/1761 (2 Pa. Arch. 13 at 422). Blacksmith, Warwick Twp., Lancaster Co., 10/28/1751 (DB L-127).

Graffelder, Peter Napoleon. Gunsmith, SW corner of Penn St. and 5th St., Reading (c, 1850).

Graham, Joseph (b. 1818). Gunsmith, 1840–54, Freedom, Beaver Co. (t).

Granger, John. Gunsmith Guthrieville, Chester Co. (BPaD, 1861).

Grant, John. Artillery artificer, 1779, at the laboratory at Philadelphia (2 Pa. Arch. 15 at 385).

Grant, Lewis. Government contractor. 5/7/1777: "pay Lewis Grant £132/8/0 for his account of gun ladles and copper ferrels for rammer heads." (2 Pa. Arch. 1 at 43); was paid for similar materials on 7/21/1777 (2 Pa. Arch. 1 at 67, 147; Col. Rec., 10:38).

Graves, John. 1769–73, Lancaster, gunsmith (t; SED, 56).

Gray, George. 8/24/1775, provided 1,500 brushes and priming wires for provincial firelocks (Col. Rec., 10:314).

Gray, George (b. 1829). Gunstock maker, Rostraver Twp., Westmoreland Co. (c, 1850).

Great Western Gun Works. Major manufactory, ca. 1865–90, J. H. Johnston, prop. 179 Smithfield St., Pittsburgh (office); 621 Smithfield St. (plant). Operated by John's son James perhaps as late as 1921, by which time the guns were cartridge arms.

Greaves, William & Son. 1849–63, cutlers, 25 N. 6th St., Philadelphia (t; c; MD, 1849, 1855, 1863).

Green, Lott. Gunsmith, 1812–15, Luzerne Twp., Fayette Co.; 1816, Redstone Twp., Fayette Co.; 1817–44, Merrittstown, Fayette Co. (t; c).

Green, Riley (b. 1808). Gunsmith, Meade Twp., Crawford Co. (c, 1850).

Green, William. Gunsmith, Meade Twp., Crawford Co. (c, 1850).

Greenawalt, Abram. Artillery artificer, 1778, Capt.

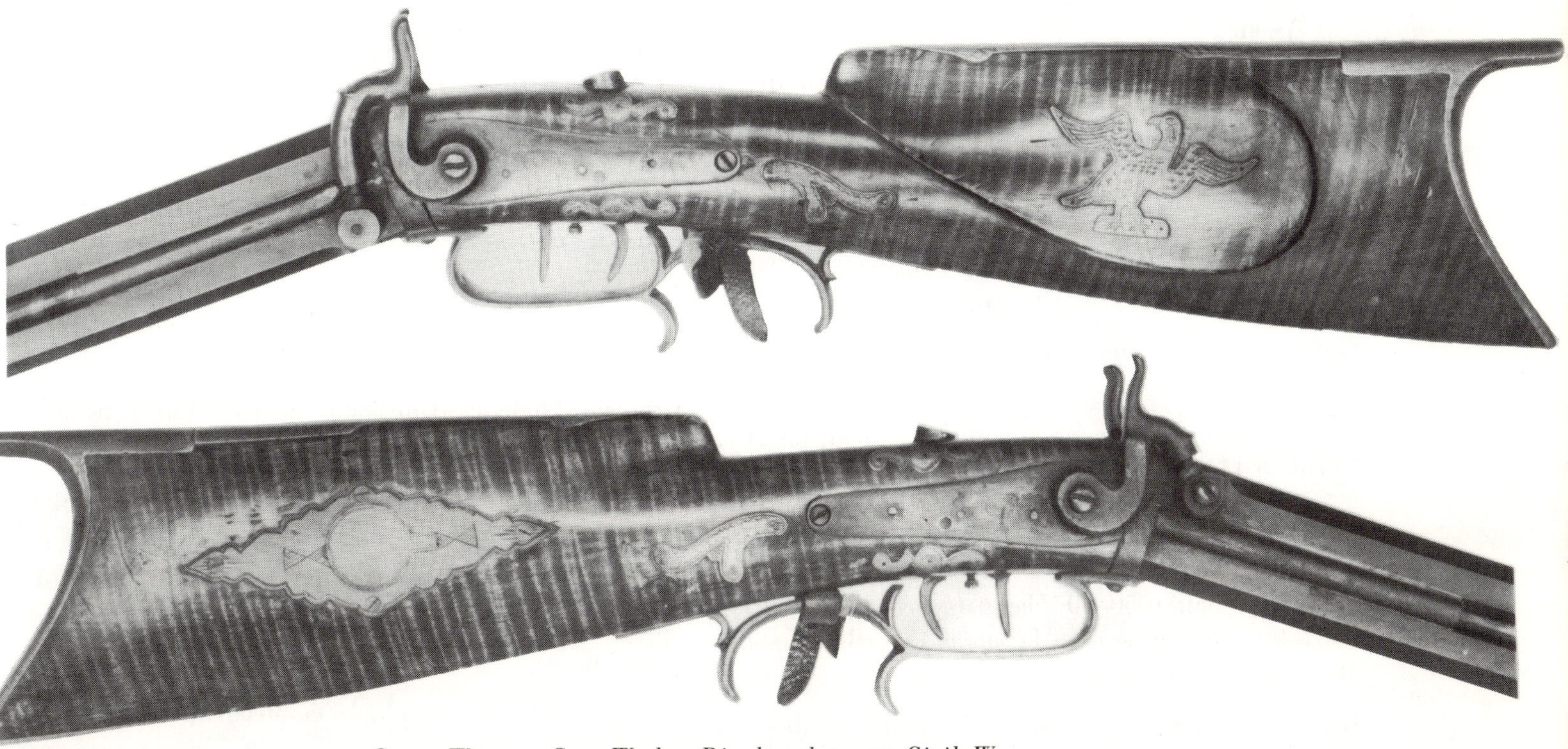

Great Western Gun Works, Pittsburgh, post–Civil War.
Fulton County Historical Society Exhibit, 1986.

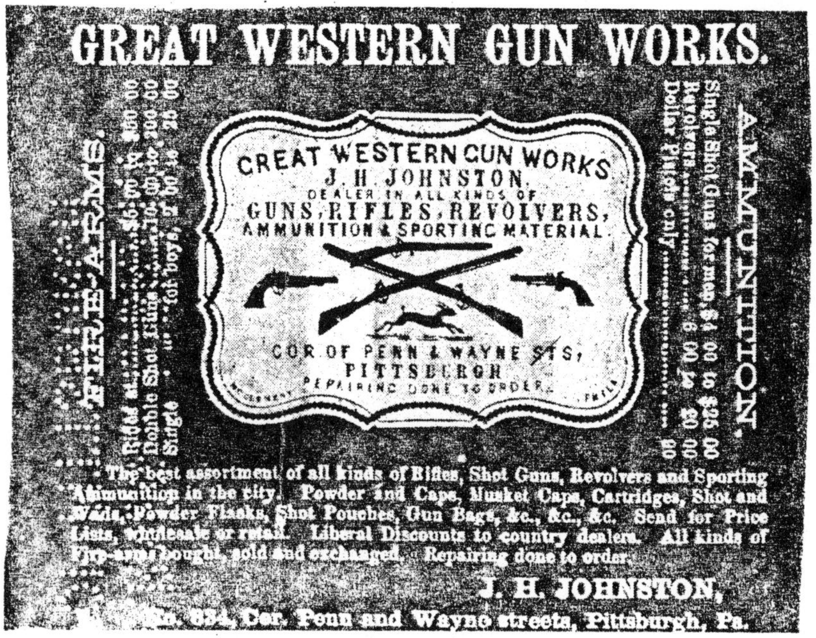

Directory of Pittsburgh and Allegheny City, 1868–69

Thomas Wylie's Co. at Carlisle (PIR, 2:248).

Greentree, Alexander. 6/12/1776, discharged from the 5th Pa. Regt. to be employed as a gunmaker (PIR, 1:158); employed by Lewis Prahl (RAG, 78).

Greibel, Henry. 1834–36, Stoneycreek Twp., Somerset Co. (t). Gunsmith.

Greninger, Thomas. Gunsmith, Logan Twp., Clinton Co. (PD, 1890).

Gress, George. Gunsmith, 1820, Northampton Co. (t).

Gress, John. Gunsmith, 1820, Easton boro (t).

Griffith, John. Gunsmith, 14 Poplar Lane, Philadelphia. (PXD, 1823); Griffith & Turner, gunsmiths (DeSD, 1824).

Griffith, Joseph. 14 Poplar Lane, Philadelphia (DeSD, 1823; PXD, 1816, 1819). Gunsmith and locksmith.

Griffiths, William. Gunsmith, 11th Ward, Philadelphia (c, 1860).

Grillet, Alexander. Inventor, Bridesburg, Philadelphia Co. Patent 45,152 of 11/22/1864 for a breech-loading firearm.

Grim, Frederick. Gunsmith, Lancaster (BLD, 1855, 1857).

Grimes, Daniel. 1850–80, Sheridan, Lebanon Co. (t; c); 1887–90, Newmanstown, Lebanon Co. (BRD, 1887, 1890). Gunsmith.

Grimes, Matthew. Swordsmith, 1777, Northern Liberties, Philadelphia Co., in the employ of Lewis Prahl, gunsmith (2 Pa. Arch. 3 at 104).

Grimm, Jacob. Gunsmith, 1829, Adamsburg (Beaver Springs), Snyder Co. (t).

Groff, Henry. 1847, West Earl Twp., Lancaster Co., gunsmith (t; SED, 56).

Groh, Tobias. Gunsmith, 1814, Allentown boro (t).

Groomes, Benjamin. 1/1/1856, Carmichaels, Greene Co., patent for repeating arm.

Gross, Samuel (1810–87). Born in Union Co.; 1831–87, Seneca Co., Ohio (RAG, 80).

Grotz, Sigismund. Gunsmith, corner of Main St. and Pine Sts., DuQuesne (PAD, 1864–65); Rochester, Beaver Co. (PD, 1887).

Grove, John. Gunsmith, 1807–8, Cumberland Twp., Greene Co. (t).

Grove, Samuel (d. 1822). 1779–1822, Bottstown, Newberry Twp., York Co. Gun-barrel factory along Bennett's Run (JKJ, 348–51; HYC, 521). Gunsmith.

Grove, Samuel, Jr. (1782–1834). Lewisberry, York Co. (JKJ, 348–51). Gunsmith.

Grover, James. Set up arms, 1655–57, Dutch Colony on Delaware (2 Pa. Arch. 5 at 416–17).

Grubb & Co. 712 Market St., Philadelphia. Made "Saturday Night Specials." Armorers and importers of cheap guns (t; c; BSED, 1880, BPaD, 1861).

Grubb, Daniel. Gunsmith, 1824, Meadville, Crawford Co. (t).

Grubb, George. Gunsmith and cutler, 316 S. 2d. St., Philadelphia (PXD, 1806).

Grubb, Joseph C. 1847–55, 76 Market St., Philadelphia; 1861–86, 236 Market St. Importer of guns, pistols, parts, and accoutrements (PD, 1882; BPaD, 1861; MD, 1847, 1855; t; c).

Grubb, Tobias (1792–1872). 1820, Hamilton Twp., Lehigh Co. (t); 1825, Northampton Twp., Lehigh Co. (t). Died 6/15/1872 in Crawford Co. (KEA, 38; RAG, 80; t; c).

Grunigal, Frederick. Gunsmith, 1858–59, East Lane near 3d St., Pittsburgh (PAD, 1858–59).

Grup, Philip. Gunsmith, 1792, Upper Mt. Bethel Twp., Northampton Co. (t).

Guest, John. 1802–11, gunsmith, Lancaster (t; SED, 56).

Guion, James P. 1856–61, Penn Twp., Lycoming Co., gunsmith (t; c).

Gumay, David. Silversmith, 1805, Meadville (t).

Gumpf, Andrew. 1829–49, gunsmith, Lancaster (SED, 56; t).

Gumpf, Andrew J. Gunsmith, 36 N. Price St., Lancaster (BRD, 1873, 1874–75; BSED, 1880; BLD, 1875–76; PD, 1899–1900).

Gumpf, Christian. 1800–34, gunsmith, Lancaster (SED, 56; t). Apprenticed at age 18 to Jacob Messersmith in 1779. Lobbied Congress in 1803 to retain import duty on foreign firearms (HJK, 246–48).

Gumpf, George. 1825–49, gunsmith, Lancaster (SED, 56; t).

Gumpf, Henry. 1814–43, gunsmith, Lancaster (SED, 56; t).

Gumpf, Jacob. 1789–93, gunsmith, Lancaster (SED, 56; t).

Gumpf, James (d. 1887). Gunsmith, Lancaster, 1850–87 (t; c; BLD, 1861).

Gumpf, Mathias. 1789–93, gunsmith, Lancaster (SED, 56; t).

Gumpf, Mathias. Gunsmith, 42 N. Water St., Lancaster (FSP, 1863–64).

Gumpf, Michael. 1843, gunsmith, Lancaster (t; SED, 56).

Gumpf, Stophel. 1789–97, gunsmith, Lancaster (SED, 56; t).

Gunter, Peter. 1813, Lurgan Twp., Franklin Co. (t); 1817–23, Letterkenny Twp., Franklin Co. (t); 1835–37, Orrstown, Franklin Co. (t; WSB, 26). (Not the same as Peter Gonter of Lancaster.)

Gurnfay, David. Silversmith and watchmaker (*Crawford Weekly Messenger*, Meadville, Crawford Co., 6/25/1807).

Guterman, C. (b. 1763). Born in Germany, came to America on 12/3/1807 on the *William Johnson*. Gunsmith and locksmith (2 Pa. Arch. 17 at 664).

Guyer, Clinton. Inventor, Muncy, Lycoming Co. Patent 226,744 for a breech-loading rifle. 1875–90, "J. P. Guyer & Son."

Guyer, James P. 1861–82, Muncy, Lycoming Co. (BRD, 1873; BPaD, 1861; NAT, 214–15) Gunsmith. Father of Clinton Guyer.

Gwinn, Alexander. Gunsmith, McCoysville, Juniata Co. (c, 1850).

H

Haas, George ca. 1790, Robeson Twp., Berks Co. (RAG, 81). Gunsmith.

Haas, Nicholas. Gunsmith, 152 Wood St., Philadelphia (MD, 1837, 1841, 1846).

Hachen (Haga, Hagi), Nicholas (1718–58). Born in Bern canton, Switzerland. 1751–58, Hanover, York Co. (GAS-1, 393).

Hachen (Hagi, Haga, Houghen), Wolfgang (1721–96). Born in Kirchdorf, Bern canton, Switzerland. By 1751 he was in Reading, where he remained the rest of his life. Notice in the *Pennsylvania Packet*, 7/2/1752, said that Wolfgang Haggan, gunsmith, had accidentally ignited some gunpowder and had blown up his house. On 6/18/1777: "pay Wolfkong Hoghen £7/0/4 for repairing arms for Northumberland Co. militia" (2 Pa. Arch. 1 at 56), and on 8/12/1779: "paid Wolfgang Hacha his bill for repairing arms, £649/3/7," Berks Co. Militia (3 Pa. Arch. 6 at 273). It is likely that many of the guns attributed to him were actually made by his apprentices long after his death (Shumway, MB 1 [1979] and 9 [1978]; GAS-1, 83; JKJ, 197–99).

Hadden, James. Gunsmith and armorer, 1769–77, Lower Delaware Ward, Philadelphia (t); 1778, armorer at Carlisle at the cannon factory (2 Pa. Arch. 11 at 260); 1780: "pay James Hatton for repairing arms £135/12/0" (3 Pa. Arch. 6 at 703).

Hadley, John. Smith, 1769, Southwark Twp., Philadelphia Co. (t); artificer, 1760–61, Ft. Pitt (Bouquet census).

Haeffer, Jacob. 1803–22, Lancaster, gunsmith. (SED, 56; t) On 4/17/1801 Gov. McKean signed a contract authorizing Haeffer to make M-1795 muskets (JKJ, 103–05; RAG, 81).

Haeffer, John. Gunsmith, 1803, Lancaster (SED, 56; t); 1819–20, Liberty St., Pittsburgh (t).

Haeny, Samuel. Gunsmith, 1823, Bushkill Twp., Northampton Co. (t).

Hagerman, James. Gunsmith, 229 5th St., Philadelphia (DeSD, 1823–24).

Haggerty, James. Artillery artificer, 1778, at the cannon factory at Philadelphia, transferred to the Invalid Corps (2 Pa. Arch. 11 at 259).

Hagi, John. 1796–1850, Cocalico Twp., Lancaster Co., gunsmith (t).

Hahn, Jacob. Gunsmith, 1800–1811, Berwick Twp., Adams Co. Associated with Adam Ernst. (WSB, 100; t).

Hahn, John. Star, Warren Co. (BPaD, 1861); Warren boro., 1860–82 (c; t; PD, 1882). Gunsmith.

Hahs, J. H. (b. 1784). Locksmith, age 20, 5′ 8″ tall, arrived at Philadelphia on the brig *Newton* (2 Pa. Arch. 17 at 609).

Haines, Isaac (1750 or 1751–1831). Gunsmith, 1772–1831, Lampeter Twp., Lancaster Co. (t; GAS-1, 326, 343).

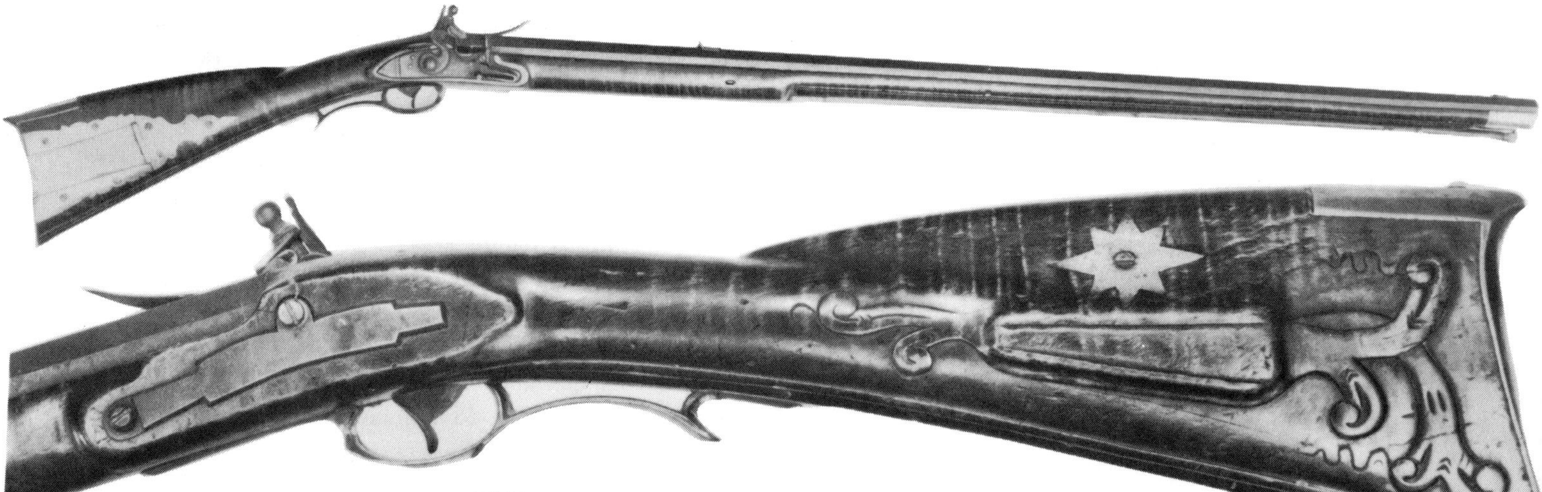

Wolfgang Haga (Hachen) (1721–96), Reading, Berks County. Many examples of this school of arms making were made long after Haga's death.

Isaac Haines (1750–1831), Lancaster County.

Haldeman, Daniel. Gunsmith, 1835–40, Union Twp., Fayette Co. (t).

Haldeman, F. 1860–88, Heidelberg Twp., Berks Co. (t; c; BRD, 1871, 1880) Pistol and rifle maker.

Haldeman, Joseph (b. 1818). 1839–41, Union Twp., Fayette Co.; 1850, 1860 Reedsville, Preston Co., W. Va. (t; c). Gunsmith.

Haldeman, Peter S. (b. 1832). 1850–72, gunsmith and blacksmith, Georges Twp., Fayette Co. (t; c).

Hale, Mathias. Gunsmith, Juniper St. above Race St., Philadelphia (DeSD, 1819).

Hall, Charles. Silversmith, 1785–95, N. Queen St., Lancaster (t).

Hall, Charles (d. 1887). 1873–87, Lancaster boro (t; c; GAAAM, 63). Gunsmith.

Hall, David. Silversmith, 1795–1800, Lancaster (t).

Hall, John. Gunsmith, 1748–81, Philadelphia (t).

Hall, William B. Inventor, Lancaster. Patents: 264,827 (1882), 276,808 (1883), 414,213 (1889), for breech-loading arms.

Hallowell & Meyer. Gunsmiths, Philadelphia (BPD, 1865).

Hallowell, Harrison. 1846–1915, gunsmith and cabinetmaker, Langville, Beaver Twp., Jefferson Co. (REH, 130, 145).

Hamelin, Charles (b. 1817). Born in Mass. Gunsmith, Rye Twp., Perry Co. (c, 1850).

Hamilton, Charles. Armorer, supplier. 5/4/1775: Lancaster Co. "shall have his powder, being 26 casks, at the rate of £14/cwt . . . and that the county shall have his lead at 45 pence/cwt" (2 Pa. Arch. 13 at 293).

Hamilton, F. M. Gunsmith, Richmond Twp., Crawford Co. (CCD, 1874).

Haminge, John (b. 1757). 1777–79, blacksmith, 11th Pa. Regt., 5′ 6″ tall (2 Pa. Arch. 11 at 60).

Hammer, David (b. 1818). Gunstock maker, Rostraver Twp., Westmoreland Co. (c, 1850).

Hammer, John H. Inventor, Pittsburgh. Patent 584,222 of 6/8/1897 for a cane gun.

Hammerlae, William. Coopertown, Venango Co. (BPaD, 1861). Gunsmith.

Hampton, Filson J. (d. 1910). 1840–1906, Blair Mills, Tell Twp., Huntingdon Co. (MB 12 [1963]).

Hampton, John N. 1837–60, gunsmith, Dauphin Co. (MB 4 [1963]; KEA, 42).

Hampton, Phylse, 1852–96, Tell Twp., Huntingdon Co., gunsmith. (t; c; MB 4 [1964] 8) Apprentice to David Wagner, Perry Co.

Hanatter, Jacob (d. 1847). Allen Twp., Cumberland Co. (KEA, 42; RAG, 83).

Hancock, Robert. 150 N. 5th St., Philadelphia (PXD, 1810, 1813).

Hancock, Thomas. Artillery artificer, enlisted on 7/30/1777 for the duration in Capt. Nathaniel Irish's Co. (PIR, 2:247).

Handlyn (Handlin), John. Apprenticed on 1/8/1773 to John Nevell of Philadelphia. (PIDS) 1775–1803, Philadelphia, gunsmith. Among those who, in 1803, petitioned Congress to retain the import duty on foreign-made arms. Committee of Safety contractor (KEA, 42; Col. Rec., 10:471; t; c).

Hanible, Godfrey. Cutler. On 11/18/1771 apprenticed to Samuel Wheeler in Philadelphia to learn the cutler's trade for 1 year, 7 months, and 24 days (PDIS).

Hankins, William C. Inventor, Philadelphia Co., west bank of the Schuylkill River, Fairmont section. Made rifles under his patent 22,753 ca. 1863–72 (RAG, 174).

Hanline, Alfred. Powder mill operator, 1832–54, Mahoning Twp., Carbon Co. Partner of Henry Arner. An explosion in 1839 killed Daniel Arner and John Synder. The mill was rebuilt and operated until 1854. In 1843 the name of the firm was changed to Hanline & Rothrock. John Erb was the shop foreman. Another explosion killed one man (Brenckman, *History of Carbon County,* 746–74).

Hanna, Isaac. 1782, came to Union Twp., Northumberland Co. from Lancaster Co. He made guns while in Lancaster Co., but is said to have only repaired them in Northumberland Co. (LINN; 62, 214).

Hannah, James. Artillery artificer, 1779, at the laboratory at Philadelphia (2 Pa. Arch. 15 at 386).

Hannah, John. 1769–70, Hopewell Twp., Cumberland Co., gunsmith. Assessed as "poor" (t; WSB, 55).

Hanni, Herronimus. Gunsmith, Reading (Reading *Adler,* 1/4/1824).

Hanning, Conrad. Artillery artificer, 1778, Capt. Thomas Wylie's Co. at Carlisle (PIR, 2:248).

Hannis, Joseph. 193 St. John's, Philadelphia, gunsmith (DeSD, 1830, 1829, 1835–36, 1831).

Hannis, William. Advertisement by James O'Hara in Pittsburgh *Gazette* for a runaway apprentice, 5/20/1797: "William Hannis, by trade a blacksmith, but professes the locksmith branch particuliarly, a robust, well made man, not exceeding 25 years of age, much inclined to idleness, smoaking, marked with smallpox, of a brown complexion, short hair, a forbidding look, he usually wore at work a short coat and overalls."

Hansen, Lourens. 1657, killed and scalped by the Amerindians. Captain de armes, Christina, New Sweden (Altena) (HAN, 346).

Hansson, Mats. Gunner, 1641–48, Ft. Christina, New Sweden (PaE, 1).

Hantz, Jacob. 1800–1810, Berwick Twp., Adams Co. (t). Gunsmith.

Harder, Frank E. (b. 1855). Gunsmith. Second son of Jacob Harder, active at Lock Haven, Clinton Co. (t; c; f).

Harder, George W. (1852–1930). 1876–84, Tyrone, Blair Co. (t; c); after 1885, Williamsport, Lycoming Co. Made rifles, shotguns, and combination guns. Later: Harder Sporting Goods store (PD, 1882, Tyrone).

Harder, Jacob (1820–98). Born in Claverick, N.Y. Served a 6-year apprenticeship with Bartlett Brothers in Binghampton, N.Y. 1844–58, Athens, Bradford Co. (t; c); 1860–74, Lock Haven, Clinton Co. (t; c). Patent 319,482 of 6/9/1885 for a breech loading gun. Died 9/1/1898 in Clearfield. In 1850, he employed 15 men in his shop. (IC) He exhibited in 1876 at the Philadelphia Fair (f).

Harder, John E. (b. 1856). Third son of Jacob Harder. 1875–87, Clearfield boro. Made guns under his father's patent (t; c; f).

Harder, William H. Son of Jacob (t; c; f). Active at Clearfield.

Harding, Josiah. Covington, Tioga Co. (PD, 1882, 1890). Gunsmith.

Hardinger, Peter Paul. 1779–81, Windsor Twp., Berks Co. (t). Gunsmith.

Harning, L. Gunsmith, Schuylkill Haven, Schuylkill Co. (BPaD, 1861).

Harper, David. Advertisement in *Pennsylvania Packet* on 2/4/1755: "David Harper, gold and silver smith, next door to the Indian Head on Market St., Philadelphia."

Harper (Harpur), William E. Silversmith and clock maker, Philadelphia (MD, 1845, 1849).

Harring, Christian. Gunsmith, 1821, Bloom Twp., Columbia Co. (t).

Harrington, Thomas. Maker of locks and gun mountings,

1850–66, Philadelphia (GAAAM, 64; RAG, 85).

Harriot, Andrew. Artillery artificer, 1778, Capt. Thomas Wylie's Co. at Carlisle (PIR, 2:248).

Harriot, Azahel. Artillery artificer, 1778, Capt. Thomas Wylie's Co. at Carlisle (PIR, 2:248).

Harriot, Reuben. 1777–79; artillery artificer, cast cannon (PIR, 2:248; 2 Pa. Arch. 11 at 256).

Harris, Charles W. 1852–55, Hartford, Conn.; 1863, patent 39,771 for Cooper Revolvers, Pittsburgh.

Harris, George. NE corner of Liberty St and Springfield St., Pittsburgh. Swordsmith and silversmith (PAD, 1815; t).

Harris, Henry. Gunsmith, 1777–81, Paxton Twp., Lancaster Co. (SED, 56).

Harris, John (1727–91). 1778, Capt. Nathaniel Irish's Co., artillery artificers; with Maj. Craig at Ft. Pitt. (PIR, 2:246) After the war, a gunsmith at York (t).

Harris, Dr. Robert. Gunpowder supplier and manufacturer, Springfield Twp., Chester Co., 1776–80 (1 Pa. Arch. 4 at 709; Col. Rec., 20:644).

Harris, Samuel. 1/13/1777: "to Samuel Harris, £16/7/0 for repairs to the arms of the Northumberland Militia." Augusta Twp., Northumberland Co. (KKR, 249) 1766–67, Newberry Twp., York Co. (Lancaster Co. DB H-335, L-261).

Harris, William. (1760–1834). Revolutionary war soldier. 1780–1820, Harrisburg (t; c); 1820–25, Seneca Co., Ohio (DH, 352). After 1825, he received a pension for his services.

Harshman, George (b. 1810). 1835–47, Gunsmith Connellsville, Fayette Co. (t); 1850, Bullskin Twp., Fayette Co. (c).

Hart, Aaron (1780–1843). Trained as a gunsmith in Hesse Cassel, Germany. 1810–43, Pittsburgh. Ferry boat operator, iron dealer, gunsmith. Advertisement in Pittsburgh *Gazette*, 12/18/1812, offering various types of guns for sale (t; f; c; KKR, 250).

Hart, John. Cutler, 1762, Lancaster Co. (DB H-78).

Hart, Levi N. Dorwart near King St., Lancaster (GCD, 1863–64; LCD, 1870–71, 1872–73). Gunsmith.

Hartman, Conrad. Artillery artificer, 1777–82, Ft. Pitt (2 Pa. Arch. 11 at 261).

Hartman, John. Artillery artificer, 1779, at the laboratory at Philadelphia (2 Pa. Arch. 15 at 385).

Hartman, John. Gunsmith, 1800, Weisenberg Twp., Northampton Co. (t).

Hartman, J. P. Gunsmith, West Pikeland Twp., Chester Co. (PD, 1890).

Hartman, Peter. Gunsmith, 1856–87, East Ward, State St. and 7th St., Erie; gun and scales manufacturer (PD, 1882; BPaD, 1861; BRD, 1856; c; t).

Hartsock, George. Gunsmith, 1802–24, Springhill Twp., Fayette Co. (t).

Harvey, James (b. 1819). 1840–48, Little Beaver Twp., Beaver Co. (t); 1848–76, Darlington Twp., Beaver Co (t; c); 1880, New Galilee, Beaver Co. (c). Gunsmith.

Harvey, Samuel. Gunsmith, Bushkill Twp., Northampton Co. (c, 1850).

Hasaman, William. Gunsmith, Mechanic St. near George St., Philadelphia (MD, 1841, 1846).

Haskell, George Richards (1825–97). Born in Geneva, N.Y. In 1854 he developed the first steel, breech-loading cannon, the first production of which was purchased by Mexico in 1855. In 1855 he began to experiment with multilayered loads in cannon. In April 1891 he invested $55,000 in the Haskell multicharge gun, believing it to be perfected. It was made at Reading, Berks Co., of hydraulically forged steel. By 1862 he had worked on a practical machine gun. Congress appropriated $100,000 to cover part of the $300,000 of his personal money spent on weapons development (NCAB, 231).

Haslett, James (1773–1833). Born in Ireland. In 1798–1801 he was the foreman at Globe Mills, Philadelphia Co., for Robert McCormick. When McCormick defaulted on a contract for Virginia Militia weapons, Haslett took over the operation and successfully completed it. After 1803 he was in Baltimore, Md. (AMM, 162–68; RAG, 122).

Hass, George. Blacksmith and artillery artificer, 1779, at the cannon factory at Philadelphia (2 Pa. Arch. 15 at 384).

Hasson, Samuel (d. 1833). 1811–33, Rostraver Twp., Westmoreland Co. (t). Gunsmith.

Hasson, William (b. 1802). 1824–36, Rostraver Twp., Westmoreland Co. (t); 1837–60, Belle Vernon boro (t; c). Gunsmith.

Hattin, Henry. Gunner and artillery artificer, 1779, Capt. Thomas Wylie's Co. at Carlisle cannon factory (PIR, 2:248).

Hauck, Bernard. Gunsmith, 1800, York Co. (KRAN, 1976).

Havener, George. Gunpowder maker for the Committee of Safety (Col. Rec., 10:381, 382) 10/25–26/1775.

Haver, George M. (b. 1898). Rice's Landing, Fayette Co. (MB 7 [1943]: Contemporary maker of muzzle-loading long rifles.

B. Hauck, York County.

Haverstick, John M. 725 E. Fulton St., Lancaster. (LCD, 1925–26) Twentieth-century long rifle maker.

Havick, Henry. Artillery artificer, 1779, at the laboratory in Philadelphia (2 Pa. Arch. 15 at 386).

Hawk, Charles. Gunsmith, 1835, Bushkill Twp., Northampton Co. (t).

Hawk, Nicholas (1782–1844). 1809–35, Chestnut Hill Twp., Northampton Co. (t); 1840–44, Gilbert, Monroe Co. (t). Major gunsmith. Many unmarked guns similar to his work are labeled "Nicholas Hawk school."

Hawk, Peter S. Gunsmith, 1835, Bushkill Twp., Northampton Co. (t); 1861, Long Valley, Monroe Co. (BPaD, 1861; *Harris' Directories*, 1861).

Hawken, Henry. 1796–1815, Hempfield Twp., Westmoreland Co. (t); 1815–28, Columbus, Ohio; 1835, Springfield, Ohio (t; c; f). Gunsmith.

Hawken, John. 1823–27 Gunsmith Mifflin Twp., Franklin Co. (t); 1837–47, Clark Co., Ohio (DH, 67–68).

Hawkins, George D. Gunsmith and locksmith, 1872–76, Bedford boro (t).

Hawks, Thomas. Gunsmith, Green St. and Budd St., Philadelphia (MD, 1837).

Hay, Peter. Gunsmith, 1820, Bushkill Twp., Northampton Co. (t).

Hayden, Creed H. Gunsmith, Dunbar, Fayette Co. (PD, 1882).

Haye, John. Armorer, 1777, Pa. Navy, on armed boat *Chatham;* ran away, 3/12/1777 (2 Pa. Arch. 1 at 262).

Hazard Powder Co. 111 Chestnut St., Philadelphia (BPD, 1890).

Hazleton, Hamilton. Armorer, 1776, Pa. Navy, on armed ship *Montgomery;* discharged 9/25/1776 (2 Pa. Arch. 1 at 322).

Headman, John. 18 Cherry St., Philadelphia, gunstock maker (DeSD, 1830, 1831).

Heasley, Joseph (10/4/1789–2/21/1823). 1804–19, Hempfield Twp., Westmoreland Co. (t); 1820–23, Mercer Co. (t; f). His estate was not inventoried until 9/23/1840, but included: vise, bellows, smith's tools, and an anvil. His wife, Magdalena, later married Michael Uber.

Hecker, Andrew. Gunlock maker and gunsmith, 162 Hanover St., Carlisle (BRD, 1871; RAG, 88) 1865–72.

Heckert, Philip (d. 1779). 1760–79, Gunsmith York boro (KEA, 48; RAG, 88).

Heckert, Philip, Jr. (d. 1812). 1781–99, 1812, York. Gunsmith (3 Pa. Arch. 21 at 6, 330; RAG, 88; KEA, 48; JKJ, 326–72).

Heckman, George M. Gunstock maker, 18 Cherry St., Philadelphia (DeSD, 1819, 1833).

Heckman, John. Gunstocker, rear of 18 Cherry St., Philadelphia (DeSD, 1823, 1835–36; MD, 1837).

Heed, Abraham. Gunsmith, 1796–1805, Bucks Co. (t).

Heed, Joseph. Gunsmith, 1856–61, Lumberville, Solebury Twp., Bucks Co. (t; BPaD, 1861; Registration of Births, no. 16 [1856].

Hegamin, Samuel. 186 N. 2d St., Philadelphia (PXD, 1808, 1810, 1813). Gunsmith.

Hehn, Peter. Gunsmith, 1805, Maxatawny Twp., Berks Co. (KKR, 252).

Heilen, C. V. (d. 1890). 1870–90, Carlisle, Cumberland Co. (c; t) Gunsmith. Maker of late-model half-stock rifles, percussion.

Heimburger, Thomas. 1776, powder maker of Philadelphia. Contracted with Committee of Safety to make gunpowder at Windsor Twp., Chester Co. (1 Pa. Arch. 4 at 710, 744; 1 Pa. Arch. 6 at 540).

Heinholder, Simeon. 1779, gunsmith, Strasburg Twp., Lancaster Co. (t; SED, 56).

Heisley, John. Gunsmith, Williamsport, Lycoming Co. He was the fourth gunsmith of that town, locating there in 1818. His shop was at the corner of Canal St. and Market St. (NAT, 214).

Hellings, Fincher. Artillery artificer, buckle filer and finisher, 1799 (2 Pa. Arch. 15 at 385).

Helwig, W. J. 24 Jackson, Allegheny City (Pittsburgh), gunsmith (PAD, 1884, 1887).

Hemminger, John. Gunsmith, 1814, Somerset boro (t).

Henderson, Charles. Gunsmith, 1829–31, Taylorstown, Buffalo Twp., Washington Co. (t).

Hendricks, Ahasures. 1640–50, Whorekill, Dutch colony (RAG, 222). Gunsmith.

Hendricks, James. Cutler, 1778–80, Philadelphia (t).

Hendricks, John. 1783–90, cutler, made trade tomahawks and scalping knives. Philadelphia (t; c).

Henninger, Daniel. 1841, gunsmith, Augusta Twp., Northumberland Co. (KKR, 252; t).

Henninger, John. Gunsmith, 1840–41, Augusta Twp., Northumberland Co. (t; c; KKR, 252).

Henritz, Balser. Artillery artificer, 1777, Capt. Will's Co. (BCR, 115).

Henry, Abraham. Gunsmith, 1781, Nazareth, Northampton Co.; 1797–1811, Lancaster boro. (t; SED, 56; HFAML, 336) Worked with Henry DeHuff and Peter Brong in 1801 on Virginia Militia weapons. Made militia weapons in 1798 with John Graeff (JKJ, 94–95).

Henry, Benjamin West. Active 1790–1802, Lancaster. Youngest son of William Henry, Jr. An "artist of considerable ability and reputation" (HFAML, 339). Gunsmith.

Henry, George. 1779, oversaw the sales of tools at the Pa. State Gun Factory (1 Pa. Arch. 7 at 366); 1779, gave final accounting of sales of equipment from that factory: £22,022/10/11 (3 Pa. Arch. 5 at 32). 1785, gunsmith at Philadelphia (t).

Henry, Granville (1835–1912). 1860–90, Bushkill Center, Northampton Co. Ran the family gun manufactory (RAG, 89). Patent 60,188 of 12/4/1866 for a firearm accessory.

Henry, J., & Son. Bushkill Center, Northampton Co. (BPaD, 1861); partnership of James Henry and his son Granville.

Henry, James (1809–95). Prop., Boulton Iron and Gun Manufactory, Bushkill Center, Northampton Co., 1835–1861. (t; c; BPaD, 1861) In 1850 they bought 1,000 gunlocks for $1,500 and produced 1,000 rifles valued at $10,000. Made guns, 1836–45, for John Jacob Astor's fur co. (RAG, 89; KKR, 253ff.; IC).

Henry, J. J. (1786–1836). Gunsmith, Philadelphia (DeSD, 1810, 1816, 1819, 1821). Partner of William Henry, III.

Henry, John. Gunsmith, 1747, Salisbury Twp., Lancaster Co. (SED, 56).

Henry, John Joseph (1758–1811). Son and apprentice of William Henry, Sr. He was a member of the ill-fated expedition against Canada and was wounded and never fully covered. He wrote "An Accurate and Interesting Account of the Hardships and Sufferings of that Band of Heroes Who Traversed the Wilderness in the Campaign Against Quebec in 1775." (reprinted in 2 Pa. Arch. 15 at 59–192). Died 4/22/1811 and buried in the Moravian Churchyard (PIR, 1:40; HFAML, 339).

Henry, John Joseph. 1788–1828, ran the manufactory at Bushkill Center, Northampton Co. (t).

Henry, John. 1876–80, Gunsmith Woodbury Twp., Bedford (t).

Henry, Michael. Gunsmith, 1828, Moore Twp., Northampton Co. (t).

Henry, Moses. Brother of William Henry, Sr. Indian trader, explorer, blacksmith, and gunsmith. Before 1765, Lancaster Co.; 1766–67, Pittsburgh; 1770–73, Ohio territory (DH, 340).

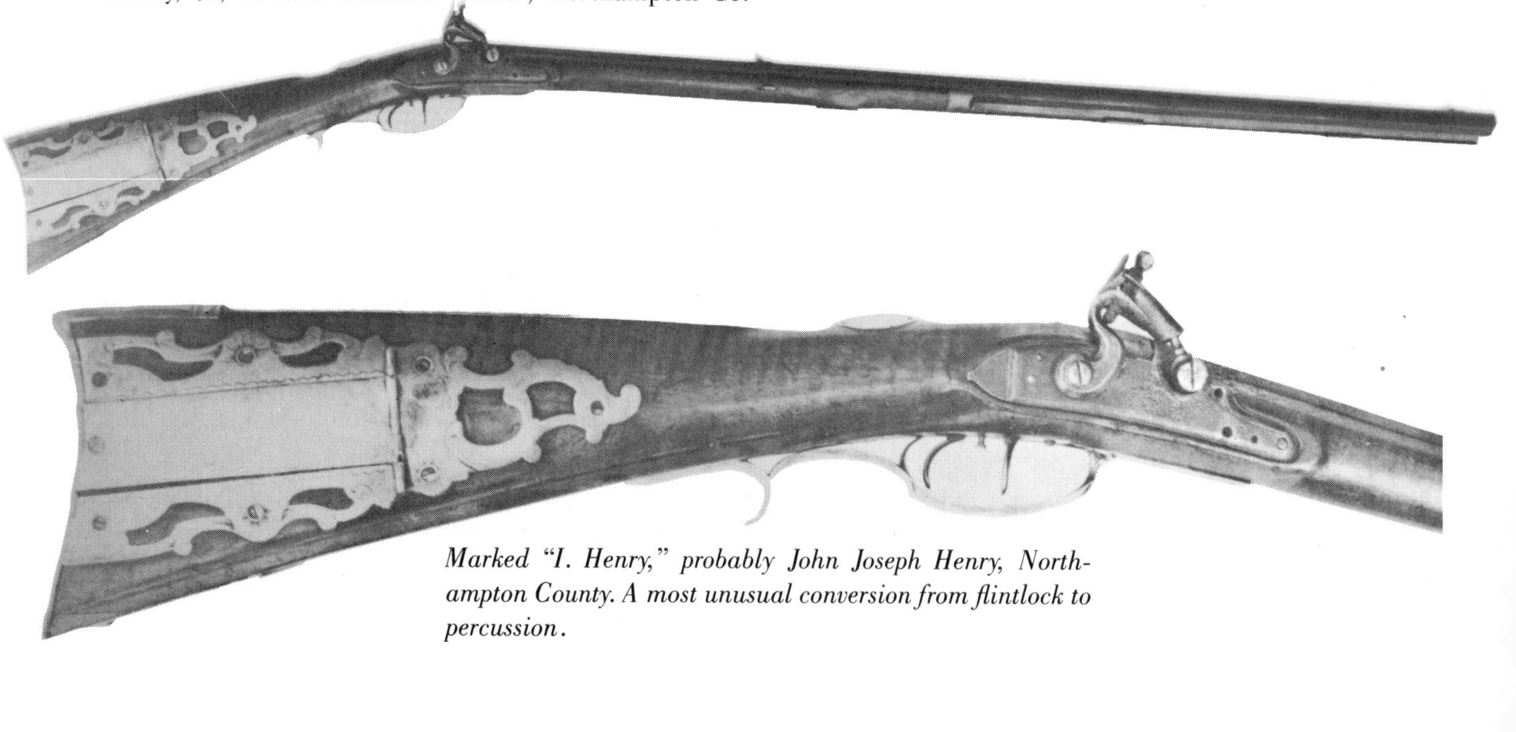

Marked "I. Henry," probably John Joseph Henry, Northampton County. A most unusual conversion from flintlock to percussion.

William Henry, I (1729–1886), Lancaster County.

Henry, William, Sr. (1729–1886). Apprenticed to Martin and Peter Roeser, 1758, in Lancaster. Armorer to Braddock expedition, 1755. Indian trader, with Joseph Simon, doing business as Simons & Henry. (2 Pa. Arch. 2 at 699; 1 Pa. Arch. 5 at 355) His apprentices, according to S. E. Dyke, included: Henry Gingerich, John Eberly, Christian Oberholtzer, Adam Deterer, Michael Dereiner, George Radfang, and Henry Myer (KRAN 11 no. 3 [1985]: 7). From 1776–1781 he was superintendent of arms for the Continental Army (1 Pa. Arch. 4 at 238; 1 Pa. Arch. 6 at 576). His account for the repair and maintenance of arms from September 1778 to September 1779 was £12,493/12/11 (3 Pa. Arch. 5 at 65).

Henry, William, Jr. (1757–1812). Apprenticed to Andreas Albrecht (ca. 1773. 1799–1807, Northampton Co. (t); 1780, Warwick Twp., Lancaster Co. (t). (SED, 56).

Henry, William, III (b. 1796). 1810–22, partner of J. J. Henry at Boulton Iron and Gun Works, near Nazareth. In 1822, he sold out to J. J. Henry (RAG, 89).

Hentz, Nicholas. Gunsmith, 1859–60, Pittsburgh (c; t; PAD, 1859–60).

Heriger, Solomon (1829–92). 1847–52, Clarion Co.; 1853–92, Knox Twp., Jefferson Co. (T; REH, 50–62).

Hero, Christian. Pewterer, 1769, Mulberry Ward, Philadelphia (3 Pa. Arch. 15 at 202).

Herron, John. Cannon borer, 1815, McClurg's Boring Mill, Pittsburgh. (PAD, 1815); gunsmith, Philadelphia (DeSD, 1816).

Hersh, Jacob. Gunsmith, Pennsburg, Montgomery Co. (BPaD, 1861; *Harris' Directories*, 1861).

Hertzog, Andrew (d. 1794). 1776, Lancaster (t); 1777–80, York Co. (t); 1790–94, Springhill Twp., Fayette Co. (t). Was paid £8/19/9, 1/1/6, 1/13/6, 0/11/6, and 9/17/9, 1777–79 at York, for the repair of various militia guns (3 Pa. Arch. 7 at 107–08).

Hertzog, Andrew, Jr. (b. 1804). 1834–39, Springhill Twp., Fayette Co. (t) Gunsmith, blacksmith, carpenter.

Hertzog, David (b. 1799). 1823–35, Springhill Twp., Fayette Co. Gunsmith. Son of George (t).

Hertzog, Frederick. Gunsmith, 1805–11, Georges Twp., Fayette Co. (t).

Hertzog, George (1776–1849). 1797–49, Springhill Twp., Fayette Co. (t). Gunsmith. Son of Andrew (t).

Hertzog, Jacob (d. 1823). Gunsmith and blacksmith, 1803–23, son of Andrew, Georges Twp., Fayette Co. (t).

Hess, Christian. Gunsmith, 1794, Lancaster boro (SED, 56; t).

Hess, David. Gunsmith, 1780–1828, Heidelberg Twp., Northampton Co. (t).

Hess, David II (1824–98). 1850–71, New Tripoli, Lehigh Co. (t; c; BPaD, 1861); Lynnport, Lynn Twp., Lehigh Co. (PD, 1882, 1890). Gunsmith.

Hess, Jonas. Gunsmith, 1790, Lynn Twp., Northampton Co. (t)

Hess, Jonas, II (b. 1799). 1832–61, Germansville, Lehigh Co., gunsmith. (t; c; BPaD, 1861) Son of, successor to, Philip Hess (t).

Hess, Philip (b. 1766). Gunsmith, 1780–1801, Lynn Twp., Northampton Co. (t).

Hess, Philip, II. 1832, Lehighton, Carbon Co. (t). Gunsmith.

Hess, Solomon (b. 1806). 1850, Heidelberg Twp., Lehigh Co. (c); Overton, Bradford Co. (*Harris' Directories*, 1861).

Hess, Thomas. Gunsmith, West Penn Twp., Schuylkill Co. (PD, 1882); Eldred Twp., Schuylkill Co. (BRD, 1887; PD, 1890).

Hewes, Josiah. 12/14/1776: "pay Josiah Hewes £33/10/0 for 9 muskets," (Col. Rec. vol. 11).

Heyberger, G. W. Gunsmith, 50 N. 3d St., Philadelphia (MD, 1865).

Heyberger, J. M. Gunsmith, 47 N. 2d St., Philadelphia (MD, 1865).

Heysinger, Isaac W. Patents 194,679 and 194,680, for a breech-loading gun and gunlock, Philadelphia, 1887.

Hibernia Furnace (built 1765). Owned jointly by Lord Stirling, Benj. Cooper, and Samuel Ford. 1776–77, managed by John Huff, made cannon on Pa. contract. By 1771 it was owned by Lord Stirling, (Col. Rec. 5:72).

Hiestand, Abraham. Cutler, 1786–88, Milford Twp., Northampton Co. (t).

Hildenbeitel, John. 1835–50, Cocalico Twp., Lancaster Co., gunsmith (t; SED, 56).

Hill, George. 1799–1800, West Bethlehem Twp., Washington Co. (t); 1805–8, Finley Twp., Washington Co. Gunsmith (t).

Hillegas, Henry. 151 Vine St., Harrisburg (BRD, 1871–72; GCD 1863–64; *Patriot Directory*, 1874–75). Gunsmith.

Hillegas, Joseph. 1838, Center St., Pottsville (t); 1841, Orwigsburg, Schuylkill Co. (t). Gunsmith.

Hillegas, Michael (4/22/1729–8/9/1804). Member of the firm of Winey & Hillegas, interested in manufacture of iron; also sugar refiners and merchants. Made cannon balls for the Revolution, (Col. Rec. 10:381).

Hindermyer, Phillip. Gunsmith, 210 W. Walnut St., Lancaster boro (BLD, 1875–76).

Hine, Jacob (1782–1865). 1825–32, Washington Twp., Westmoreland Co. (t); 1850–65, Loyal Hanna Twp., Westmoreland Co. (t; c). Gunsmith. His estate had a value of $1,220.92 and included gunsmithing tools. (estate appraisal of 24 May 1865, Register of Deeds Office, Westmoreland Co.).

Hiney, Jacob. Artillery artificer, 1782, Pa. Artillery (2 Pa. Arch. 11 at 236).

Hinkle, George J. Gunsmith, 408 S. Queen St., Lancaster (BLD, 1873–74, 1875–76; BLD, 1871–72).

Hinkle, William. Gunsmith, boarded at 408 S. Queen St., Lancaster (BLD, 1875–76).

Hinkles, Daniel. 1810–16, Philadelphia; gunsmith, swordsmith, and cutler (GAAAM, 71; DeSd, 1816).

Hinton, George. Cutler, Southern District, Spruce St., Philadelphia (c, 1790).

Hirth, August. Pittsburgh, partner in Bown & Hirth, gun manufactory; Enterprise Gun Works, ca. 1855–60 (FSP, 147).

Hixon, Daniel (b. 1818). Gunsmith, 1840–49, North Strabane Twp., Washington Co. (t); 1850, Washington boro (c); 1857–61, North Strabane Twp., Washington Co. (t; c).

Hixon, Joseph. Gunsmith and carpenter, 1816–22, South Huntingdon Twp., Westmoreland Co. (t) on 5/18/1793 C. Betts sold land to Joseph Hixon in Huntingdon Twp. for 1

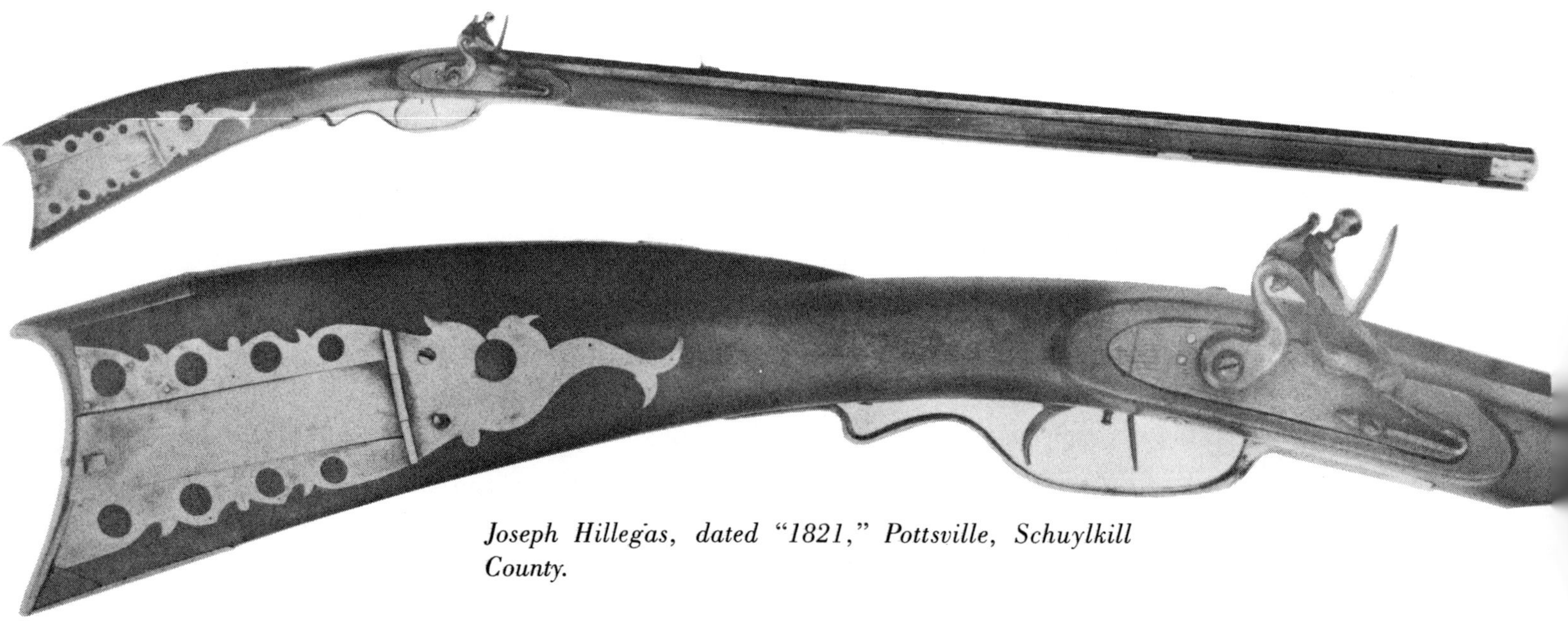

Joseph Hillegas, dated "1821," Pottsville, Schuylkill County.

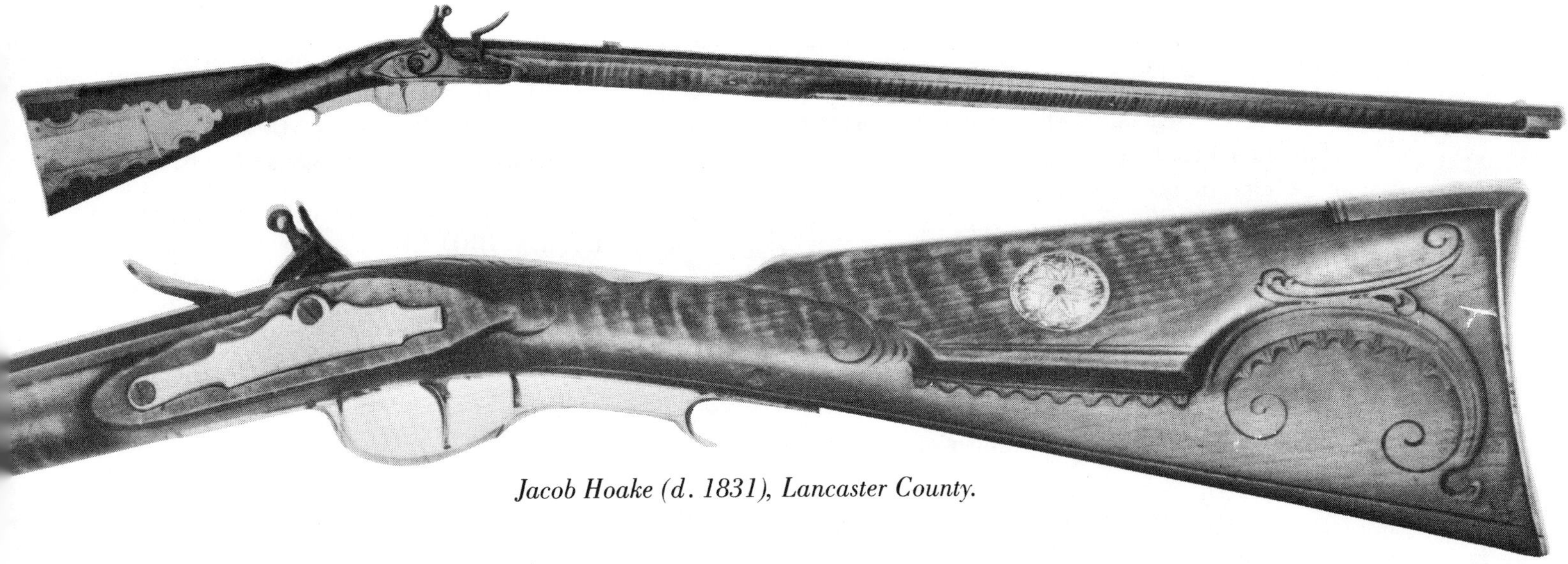

Jacob Hoake (d. 1831), Lancaster County.

barrel of whiskey (DB 1:453).

Hoak, Mathias. 1802–14, gunsmith, Strasburg Twp., Lancaster Co. (t; SED, 57).

Hoake, Jacob (d. 1831). Gunsmith, 1800–1831, Strasburg Twp., Lancaster Co. (SED, 57; t).

Hobbes, James. Gunsmith, rear of 128 Dillwyn, Philadelphia (DeSD, 1823).

Hofer, John G. Gunsmith, Front St. at Poplar St., Philadelphia (MD, 1846).

Hoff, Peter. Gunsmith, York St., Hanover boro. (c, 1860; BPaD, 1861).

Hoffman, Christian. Gunsmith, 43 Charlotte St., Philadelphia (PXD, 1813; DeSD, 1823, 1833); Lancaster boro., 1807–11 (SED, 57; t); 1835–55, St. Louis, Mo. (RAG, 93).

Hoffman, George (1828–97). 1850–97, Unity Twp., Westmoreland Co. (t; c; BPaD, 1861). Gunsmith. Appraisement (WB 10, 85–86): Gunsmithing tools $5; 6 gun barrels 50¢.

Hoffman, Gilian. Gunsmith, 1827, Quemahong Twp., Somerset Co. (t); 1840, Richard Twp., Cambria Co. (t); 1861, Jenners Crossroads, Somerset Co. (BPaD, 1861).

Hoffman, John. Gunsmith, 1828, Lykens Twp., Dauphin Co. (t).

Hoffman, Val. Artillery artificer, 1777–79, cannon factory at Philadelphia (2 Pa. Arch. 15 at 384; 2 Pa. Arch. 11 at 250).

Holdcraft, Sylvester. Artillery artificer, 1779, Capt. John Jordan's Co. at Carlisle (2 Pa. Arch. 11 at 260).

Holeman, John. Gunsmith, Lancaster, 1807–19 (t; SED, 57).

Holingshead, William. Gunsmith, 1780, High St. Ward, Philadelphia (t). Valuation £5,600; tax £16/6/0.

Holiday, Thomas Berlin. In the will of Samuel Holliday, 10/17/1792, Pittsburgh, he ordered that his oldest son, Thomas Berlin Holliday, be trained as a gunsmith.

Hollier, Thomas. Armorer on the Braddock expedition: "having ground and glazed 600 Dutch bayonets" (MIL-4 [microfilm], Survey Report 5820).

Holmes, Reuben. Gunsmith, 1837, Bridgewater, Beaver Co. (t); 1839, Petersburg, Ohio (t).

Holmes, Robert D. (d. 1928). 1890–1928, gunsmith, Cherry Hill Twp., Indiana Co. (t; c; f).

Holsapple, Phillip. Center, above High St., Pottsville, Schuylkill Co., Gunsmith (BPaD, 1861; c. 1860).

Holtry, Hiram. Gun-barrel maker; gunsmith at Mohn's Store, Berks Co. Among his apprentices was John White, who worked later at Angelica, Cumru Twp., Berks Co. (c, 1860; BPaD, 1861).

Holtry, Joseph. Gun-barrel maker; gunsmith at Mohn's Store. In 1850 he made 1,250 gun barrels. (IC) Active, ca. 1845–75 (t; c; BPaD, 1861).

Holtzscheider, J. Gunsmith, Philadelphia (PD, 1882; BPD, 1890).

Holtzscheider, L. 1421 N. 10th St., Philadelphia (BPD, 1890). Gunsmith.

Holtzworth, William. 1815–32, gunsmith, Lancaster (SED, 57; t).

Hoofman, Jacob. Gunsmith, Lancaster (DB C-38, 1750).

Hook, Jacob. Gunsmith, 1788, East Earl Twp., Lancaster Co. (SED, 57).

Hooper, Robert Lettis, Jr. Water St. above Chestnut, Philadelphia (*Pennslyvania Journal*, 8/18/1763; KKR, 268). Sold muskets, hardware, and the like.

Hoover Brothers. "Headed by Elijah Hoover, soon after the close of the Rebellion [Civil War], they began the manufacture of powder in this western portion of Packer Twp. Having had an explosion or two, they, in 1873, sold out to the Laflin Powder Manufacturing Co., which rebuilt the mills and continued this business until 1878 when another explosion resulted in the removal of the enterprise to the vicinity of Wilkes-Barre" (Brenckman, *History of Carbon County*, 299).

Hoover, Jacob. Waynesburg, Greene Co. (PD, 1882). Gunsmith.

Hopson, John. Lancaster Co. "shall have his powder . . . at the rate of £15/cwt." (2 Pa. Arch. 13 at 293).

Hormell, John, Jr. Gunsmith, 1788–96, Fredericktown, Washington Co. (t).

Horn, Conrad. 1820–62, gunsmith, Hazleton, Luzerne Co. (t; BPaD, 1861; c).

Horn, Stephen. 1785–90, gunpowder mill, Easton (GAAAM, 73).

Horn, Thomas (d. 1921). Gunsmith, East Hazleton, Luzerne Co. (BRD, 1879). Active about 1870–1900. Also sold real estate.

Horn, William. Gunsmith, 1837–39, Hazleton, Luzerne Co. (t).

Hornberger, Cyrus. Gun-barrel maker, Mohntown, Berks Co. (c, 1850).

Horner, Frederick. Gunsmith, 1805–11, Quemahoning Twp., Somerset Co. (t); 1815–34, Jenner Twp., Somerset Co. (t).

Horner, Jacob (b. 1805). 1825–70, Jenners Crossroad, Quemahoning Twp., Somerset Co. (c; t; BPaD, 1861) "Master gunsmith" (c, 1860).

Horner, John, Sr. (b. 1784). 1802–6, Conemaugh Twp., Somerset Co. (t); 1808–14, Conemaugh Twp., Cambria Co. (t); 1830–60, Quemahoning Twp., Somerset Co., gunsmith.

Horner, John, Jr. (b. 1804). 1825–63, Jenners Crossroads, Quemahoning Twp., Somerset Co., gunsmith (t; c; BPaD, 1861).

Horner, Lewis (b. 1825). Gunsmith 1860–61, Jenners Crossroads, Somerset Co. (t; BPaD, 1861); 1860, Davidsville, Quemahoning Twp., Somerset Co. (t; c).

Horning, Lewis. 1835–61, gunsmith, Schuylkill Haven, Schuylkill Co. (t; c; BRD, 1861).

Horstmann, W. H. Armorers, military suppliers, Philadelphia. In 1818 a lace and fringe factory was founded, the basis of the business. From 1828 to 1830, the firm was a military goods store and supplier. From 1843 to 1858, it did business as W. H. Horstmann & Sons, 51 N. 3d St. From 1845 to 1849 a New York store was operated, Horstmann Sons & Drucker. In 1859 John G. Franklin was added to the firm, which became Horstmann Sons/ Horstmann Brothers, 723 Chestnut St. In 1830 the firm purchased the Wildtman sword factory (RAG, 222).

Horton, William. Gunsmith, 104 Plumb St., Philadelphia (PXD, 1810).

Horverter, Henry. Gunsmith, Lemon St. between N. Queen St. and Prince St., Lancaster (GCD, 1863–64). See also Howerter, Henry.

Houston, James. Gunsmith, 1780, Mulberry Ward, east part, Philadelphia (t).

Howalter, Charles. Smallman near Clymer, Pittsburgh (PAD, 1864–65). Gunsmith.

Howard, Caleb. 2d St. near Federal, Philadelphia, gunsmith (DeSD, 1816).

Howard, George. Artillery artificer, 1779, Pa. Continental Line (2 Pa. Arch. 15 at 386).

Howell, Samuel. Supplied 11 cutlasses to Capt. Robert Towers, Commissary, on 10/20/1775 (Col. Rec. 10:375).

Howell, Thomas. Lock maker, importer; hardware items and accoutrements (DeSD, 1824).

Howell, William T. & Co. 1830–47, 190 Market St., Philadelphia. Gunlock importers, manufacturers. Successors to Thomas Howell Co. (MD, 1847, 1835; DeSD, 1830, 1837).

Howerter, Charles. Gunsmith, 455 Lafayette St., Lancaster. (GCD, 1871–72) Gun-barrel forger, Upper Leacock

Jacob Horner, Somerset County.

Twp., Lancaster Co. (c, 1850).

Howerter, George. Gunsmith, 1820–27, Lampeter Twp., Lancaster Co. (t).

Howerter, Henry. Gun-barrel forger, Upper Leacock Twp., Lancaster Co. (c, 1850; SED, 57) Gunsmith, 340 Middle, Lancaster (LCD, 1871–72).

Howarter, J. Gun-barrel maker, 1860–61, on Yellow Breeches Creek, Lower Allen Twp., Cumberland Co. (WSB, 50; c, 1860; BPaD, 1861).

Howland, Rufus J. Gunsmith, 1840–74, Binghamton, N.Y. (t; c); 1887–90, Coudersport, Potter Co. (PD, 1887, 1890).

Hsueder, Henry. Gunsmith, 1848–49, Manheim Twp., Lancaster Co. (t).

Huback, Henry. On 2/7/1776, Huback proposed building a powder mill on his property, Lower Milford Twp., Bucks Co. He had a race already built, and asked for financial aid in setting up the factory (1 Pa. Arch. 4 at 709).

Hubbell, William. Patents 3649 (1844) and 65,812 (1867) for his breech-loading Hubbell rifle. Philadelphia.

Huber, Abraham. Gunsmith, 1799, Manchester Twp., York Co. (KEA, 54).

Huber, George (1760–1813). Bethlehem. Moravian. Blacksmith and postmaster (Heckewelder's *Narratives*, 416).

Huber, Henry. 12/13/1776: "pay Henry Huber £30 toward manufacturing powder" (Col. Rec., vol. 11).

Hubner, Capt. George. gunpowder manufacturer, 1779 (2 Pa. Arch. 14 at 604, 610).

Hudd, Thomas. Artillery artificer and blacksmith, 1778, U.S. Ordnance Yard, Philadelphia. (2 Pa. Arch. 11 at 258).

Huggins, Thomas. Artillery artificer, 1778, U.S. Armory, Philadelphia (2 Pa. Arch. 11 at 258).

Hughes, John (b. 1818). 1849–50, Franklin Twp., Greene Co. (t; c). Gunsmith.

Hughes, Thomas (1789–1849). 1808–9, Waynesburg, Greene Co. (t); 1813–14, Uniontown, Fayette Co. (t); 1817–49, Wheeling, W.V. (t).

Huifener, Christian (b. 1780). Gunsmith, Carol Twp., Perry Co. (c, 1850).

Hull, Benjamin. Gunsmith, Sansoon's Alley above Noble St., Philadelphia (DeSD, 1819, 1834–35; KEA, 54).

Humes, John. Gunsmith, 1799–1800, N. 3d St., near Poplar St., Philadelphia (KEA, 54).

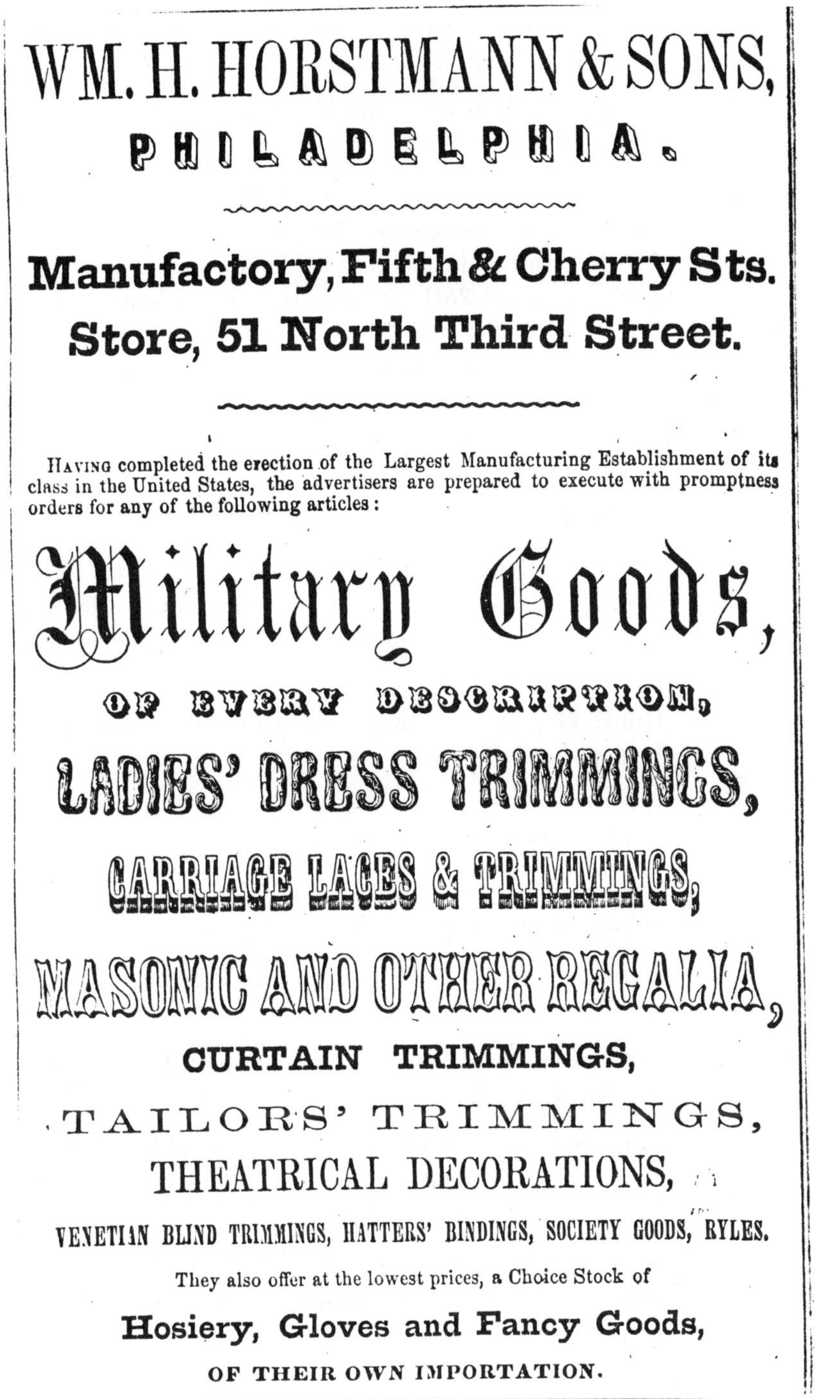

McElroy's Directory of Philadelphia, 1855

Hummel, John Jacob (b. 1756). 1797, Snyder Co., gunsmith. He is said to have once repaired the gun of an Indian who then shot him with it (*Snyder Company Historical Society Bulletin* 1, no. 243).

Hummell, Adam J. Gunsmith, 26 Napoleon St., Johnstown. (*Johnstown Directory*, 1876–77, 1889; PD, 1882), 1860, Market and Water Sts., Lebanon (c, 1860; BRD, 1860).

Hummell, Frederick. 1777: "in charge of the manufacture

of guns at Hummelstown, [promoted] to the rank of captain" (3 Pa. Arch. 23 at 462).

Humphrey, Rebecca. Smith's shop, 1779–80, Lower Darby, Chester Co. (t).

Humphreys (Humphries), Richard. On 10/20/1772 George Bardeck was indentured to Humphreys for 2 years, sold for £16/18/0 to pay for his passage to America to work as a goldsmith. On 1/26/1773 John Myers was indentured for 5 years, 5 months and 25 days to learn silversmithing. (PIDS) 1772–96, silversmith and swordmaker, Philadelphia (RAG, 222).

Humphreys, Whitehead. Made steel from iron for the use of the Continental army, 1777–78, 7th St., Philadelphia (SHIS, 382).

Hunlock, Bouman. Advertisement from *Pennsylvania Packet*, 1/28/1752: "gold and silver smith's tools, late the property of Bouman Hunlock, taken at the suit of John Ord, will be sold."

Hunsicker, Henry. 1821, gunsmith, Macungie Twp., Lehigh Co. (KEA, 54; t).

Hunt, Charles B. Springville, Venango Co. Patent 185,539 of 12/19/1876 for a magazine gun.

Hunter, John. Artillery artificer, U.S. Armory, Philadelphia (2 Pa. Arch. 11 at 258).

Hurlbut, William. Artillery artificer, 1778, Capt. Thomas Wylie's Co. at Carlisle (Pir, 2:248).

Hutchinson, R. J. Gunsmith, Williamsport, Lycoming Co. (BRD, 1878).

Hutt, Thomas. Swordsmith in the employ of Lewis Prahl, gunsmith; 1777, Northern Liberties, Philadelphia Co. (2 Pa. Arch. 3 at 104).

Hutton, John Strangeways (1683–1792). Died 12/20/1792 at the age of 109 at Philadelphia. He knew Teach (Blackbeard) the Pirate and was a privateer in Queen Anne's War. He was a self-taught silversmith and whitesmith. His hollow work was considered excellent. He made one tumbler in silver at the age of 94. At the age of 51 he married 18-year-old Ann VanLear (1716–1788). He used a heavy English musket in hunting (Hazard's *Annals*, 2:62–63).

Hutz, Benjamin (d. 1823), Heidelberg Twp., Lancaster Co. Gunsmith. He was one of those who lobbied Congress in 1803 to retain the impost on foreign-made guns (SED, 57; t).

Hyzer, Samuel. Gunsmith, Derringer's Ct., Philadelphia (MD, 1841, 1846).

I

Ickes, Jacob. 1867, Union Twp., Bedford Co. (t); 1870–76, King Twp., Bedford Co. (t; c); 1877–80, East St., Clair Twp., Bedford Co. (t; c; PD, 1882). Gunsmith.

Iler, Philip. Artillery artificer, blacksmith, 1778–82, U.S. Ordnance Yard, Philadelphia; reassigned 11/1/1782 to Southern Army (2 Pa. Arch. 11 at 258, 262).

Imhoff, Benedict. Gunsmith and physician, 1759, Heidelberg Twp., Lancaster Co. (SED, 57; t); 1781, Heidelberg Twp., Berks Co. (t); 1784, Middletown, Berks Co. (t; KEA, 55).

Indermaur, J. Armorer, Allegheny Arsenal, Pittsburgh (PAD, 1819).

Ingram, Joseph (1858–1914). Son of Eleazor Ingram, nephew and apprentice of William Ingram. 1880–1914, Mt. Morris, Greene Co. (t; c; f). Gunsmith.

Ingram, William (1828–1913). 1854–1913, gunsmith, Dunkard Ridge, near Mt. Morris, Greene Co. (t; c; f).

Innes, Robert. On 6/18/1772, apprenticed to Valentine Brown, whitesmith and silversmith, Philadelphia (PIDS).

Irish, Nathaniel (1737–1816). Artillery officer, cannon

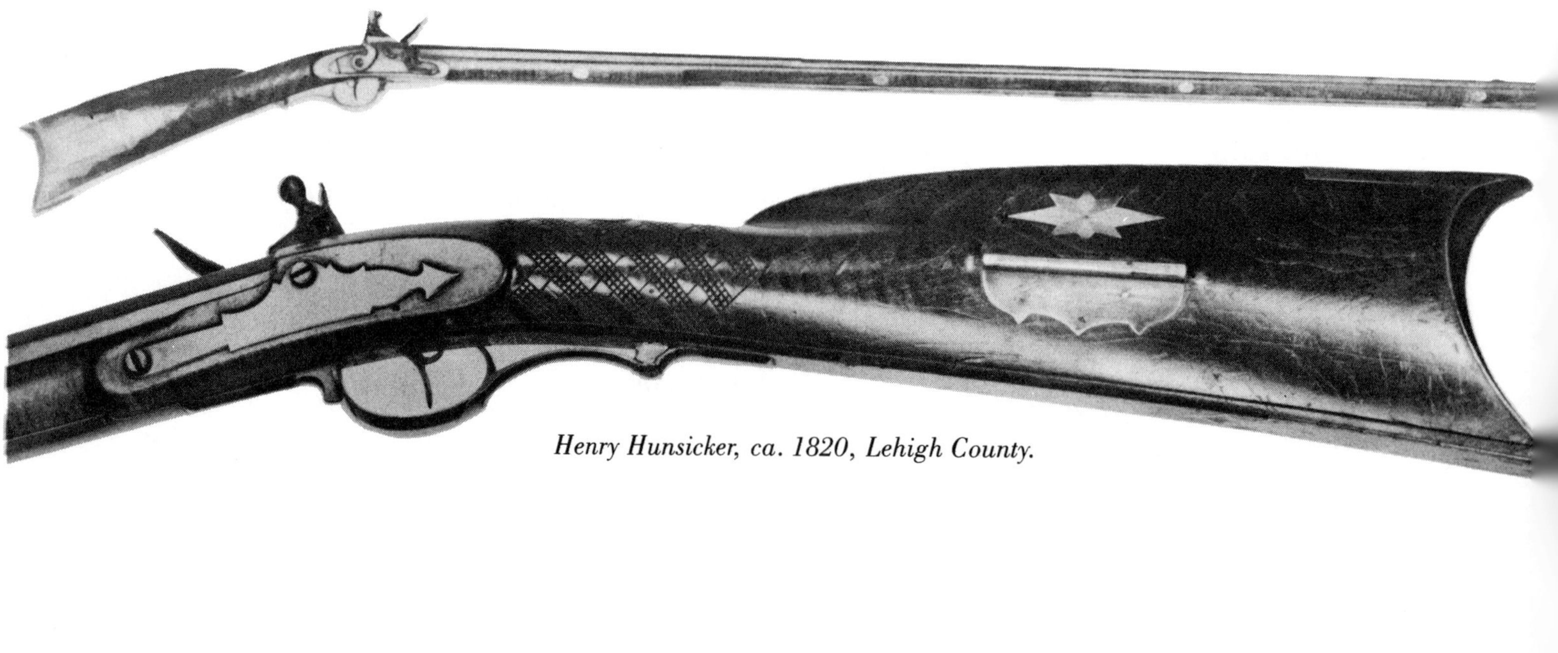

Henry Hunsicker, ca. 1820, Lehigh County.

maker, 1777. Died in Pittsburgh, 9/11/1816, at the age of 79 years, 4 months (2 Pa. Arch. 11 at 250; 2 Pa. Arch. 15 at 382).

Iron City Gun Works. 1856–77, Pittsburgh. Originally, 1856–57, William Craig and Daniel J. McDonald; 1858–64, McDonald only; 1864–77, H. H. Schutte. Gun manufactory (FSP, 158).

Irvin, George. Opened a store, 1760, Winchester St. and Cromwell St., Orbisonia, Huntingdon Co. (AFR, 261).

Irwin, James. Artificer, 1797, at Presqu' Isle, Erie Co. (Ella Chalfont, *A Goodly Heritage*, 25).

Irwin, Robert. Armorer, 1794, convicted of treason during the Whiskey Rebellion of 1794. Sentenced to £100 fine and 6 months in prison; pardoned after 20 days in jail. He had been a leader in running up the liberty pole (J. F. Meginness, *Otzinachson* [1857], 327).

Isch, Christian. 1775–83, Lancaster (t; SED, 57).

> *Resolved*, That in case any of the gun-smiths, in the county of Lancaster, upon application made to them by the members of the committees of the respective townships to which they belong, shall refuse to go to work and make their proportion of the firelocks and bayonets of this country, by the honorable House of Assembly, within two weeks from such application agreeable to the patterns, at the Philadelphia prices, such gun-smiths shall have their names inserted in the minutes of this committee as enemies to their country, and published as such, and the tools of the said gun-smiths so refusing shall be taken from them, and moreover the said gun-smiths shall not be permitted to carry on their trades until they shall engage to go to work as aforesaid, nor shall leave their respective places of residence until the arms are completed.
>
> *And it is further resolved*, That the Committee of Correspondence and Observation do take especial care that their resolves be carried into execution. . . .
>
> Christian Isch and Peter Reigart appeared in committee, and agreed to set to work on Monday, the twentieth day of November instant, and make muskets and bayonets for this country, (part of the number required from this county, by the honorable House of Assembly,) at the Philadelphia prices; and that they will confine themselves to that work entirely from that time to the first day of March next, and furnish as many as they can possibly complete in the time, and deliver the same to the commissioners of the county or this committee. [2 Pa. Arch. 13 at 299].

Isch, Peter. 1759–73, gunsmith and blacksmith, Lancaster (t. SED, 57)

Israel, Samuel. Gunsmith, 1808–12, Franklin Twp., Greene Co. (t).

J

Jackson, Cyrenus.; Heath Twp., Gunsmith Jefferson Co. (c, 1850; REH, 49).

Jackson, Isaac (1734–1807). Silversmith and clock maker, London Grove, Chester Co. (t; f).

Jackson, Ralph. 1702, Philadelphia (2 Pa. Arch. 19 at 388). Gunsmith.

Jackson, Silas T. Patent 45,830 of 1/10/1865 for a cartridge. Philadelphia.

Jacob, Joseph. 1861–99, Philadelphia. Double-barreled rifles and shotguns of the highest quality (RAG, 100).

Jacobs, J. H. Gunsmith and clock repairman, Curwensville, Clearfield Co. (*Rifleman's Journal*, 12/10/1856); Benner, Centre Co. (BPaD, 1861).

Jacobs, Michael. 1807–14, gunsmith, Warrington Twp., York Co. (KKR, 270; t).

Jacoby, Michael. Gunstock manufacturer, 10th St. below Parrish St., Philadelphia (MD, 1846).

Jacoby, Peter E. Gunsmith, Uniontown boro (PD, 1887, 1890).

Jaffet, Henry. Artillery artificer, 1779, at the laboratory at Philadelphia (2 Pa. Arch. 15 at 386).

Jakob, Joseph. 1156 Psyk Ave., Philadelphia (GPD, 1892; BSED, 1880; BPD, 1890; PD, 1882). Gunsmith.

James, H. Gunsmith, Nazareth (BRD, 1887).

Jameson, William. Gunsmith, 126 Broom St., Philadelphia (PXD, 1810).

Jarecki, Henry (b. 1826). 1861–72, Erie. Born in Posen, Prussia. 1872, Jarecki Manufacturing Co., brass foundry and gunshop. (BPaD, 1861; BRD, 1872; t; c) Between 1850 and 1855 he taught the gunsmithing trade to his brother-in-law, Oswald Peck (t, f).

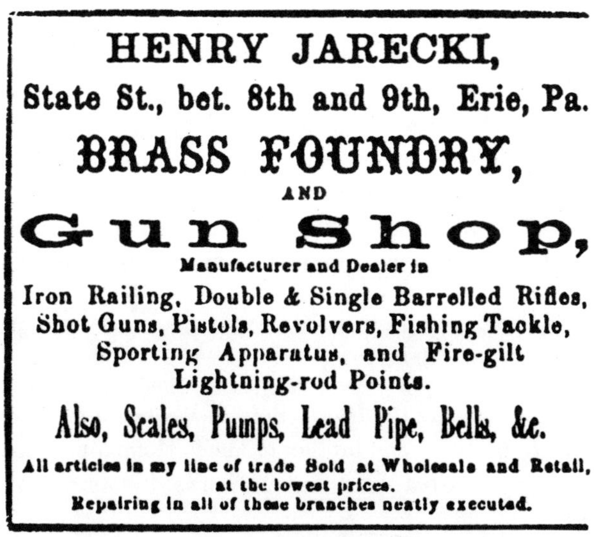

Directory of Erie and Erie County, 1877

Jeffrey, John. 1835–40, Lawrenceville, Tioga Co. (t). Armorer.

Jenkins, James. 5/30/1776; patriots reported that Jenkins "has 6 new gun barrels and will not deliver them to anybody." Listed as a nonassociator (2 Pa. Arch. 1 at 600).

Jenkins, Josiah. Artillery artificer, 2/19/1777 to 3/22/1781, Capt. Jesse Rowe's Co. at the cannon factory at Philadelphia (PIR, 2:244).

Jenkins, John. Swordsmith, 1778, Philadelphia (KEA, 56).

Jenks, Alfred & Son. Alfred Jenks died in 1854. He ran Bridesburg Machine Works, Philadelphia Co. (GR 5 [1962] 8).

Jenks, Barton (d. 1896). Patent 74,760 for his breech-loading gun. Bridesburg Machine Works, prop. Made contact in 1861 for Springfield rifled muskets. Son of Alfred Jenks (FSP, 160).

Johns, Isaac. Gunstocker, 1777, at French Creek Arsenal; 1780, Mulberry Ward, Philadelphia (t).

Johnson, Alfred. Gunlock patent of 10/12/1880, with Samuel Mier, Somerset Co. Johnson lived in New Geneva, Fayette Co.

Johnson, Joseph. Gunsmith, 59 Budd St., Philadelphia (DeSD, 1814, 1816).

Johnson, O. D. Boarded at Keystone House, Lancaster, gunsmith (BLD, 1873–74).

Johnson, Richard. Shot maker, 11 S. Water St., Philadelphia (DeSD, 1830).

Johnson, William. Gunsmith, 20 Green St., Philadelphia (DeSD, 1835–36); 4th at Brown (MD, 1841); 2 Rose Alley (MD, 1846–47).

Johnson, William. Gunsmith, 1838, Lancaster (SED, 57; t).

Johnston, Francis. 4/7/1785: "Paid Francis Johnston for two elegant rifles made, one for Aaron Hill of the Mohawks and the other for John Obeal of the Seneca Nation." (3 Pa. Arch. 7 at 485). Cost: £22/10/0. The Commonwealth also paid Johnston for cases for the two guns, and for engraving the names of the two chiefs thereon. The guns were sent, but were damaged on the trip, and Johnston was paid to repair them on 9/23/1785.

Johnston, George (b. 1840). Born in Waynesburg, a son of John H. Johnston and brother of James Hampton Johnston. Employed by Great Western Gun Works, Pittsburgh (WSB, 38).

Johnston, J. C. Gunsmith, McConnellstown, Huntingdon Co. (PD, 1890).

Johnston, James Hampton (1836–1916). 1856–63, Waynesboro, Franklin Co.; 1864–1916, Pittsburgh, prop., Great Western Gun Works. This was a major manufactory, primarily making percussion guns for the western trade (WSB, 35–39).

Johnston, John (b. 1768). Gunsmith, 1790, Listie, Somerset Co. (t).

Johnston, John A. (d. 1923). 1888–1923, prop., Great Western Gun Works, Pittsburgh. Son of, successor to, James Hampton Johnston (WSB, 35–39).

Johnston, John H. (1811–89). Gunsmith, apprenticed to Henry Carlile. 1832–89, Waynesboro, Franklin Co. (WSB, 35–39).

Johnston, John S. 1842, Washington Twp., Franklin Co. (WSB, 38); 1850, McConnellstown, Huntingdon Co. (t; AFR, 384).

Johnston, Joseph (b. 1833). Waynesboro, Franklin Co. (c, 1850). Gunsmith.

Johnston, Richard B. Associated with Robert McCormick on a contract of 5/4/1801 for M-1795 muskets; defaulted (RAG, 102). Philadelphia Co.

Johnston, Samuel. gunsmith, 56 Wayne St., Pittsburgh (PAD, 1850).

Johnston, William. Cannon borer and pattern maker, McClurg's Mill, Pittsburgh (PAD, 1815).

Johnston, William (1796–1872). Born in Butler Co. 1810–41, Indiana Co.; 1841–72, Winslow Twp., Jefferson Co. (REH, 39). Gunsmith.

Jonassen, Joost. Was paid 42 florins for cartouche boxes, 11/23/1661, at the Dutch colony on the Delaware (2 Pa. Arch. 5 at 457).

Jones, Albert A. Gunsmith, Aleppo Twp., Greene Co. (PD, 1876, 1882).

Jones, Aquilla. Cutler, 1765–74, New Town Twp., Philadelphia Co. (t; DB K-226) Henry Scoup was assigned as Jones's apprentice on 12/5/1772 (PIDS).

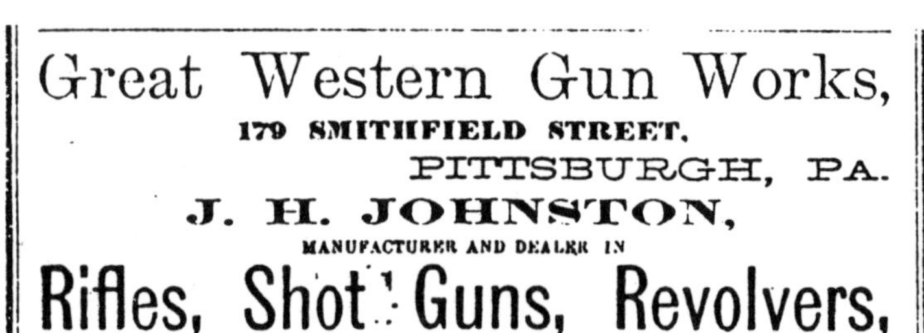

Directory of Pittsburgh and Allegheny City, 1865

GUNS! GUNS! GUNS!

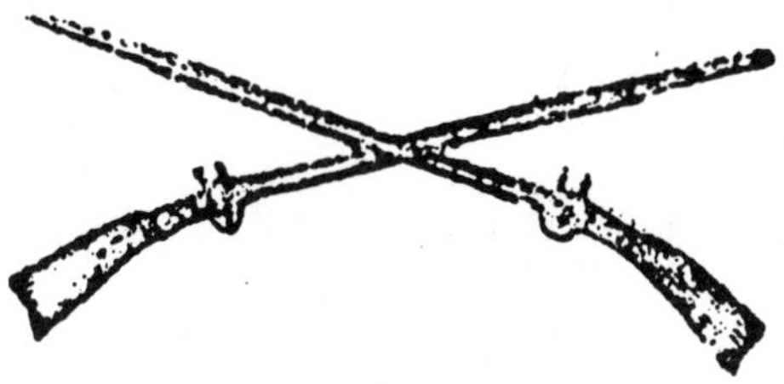

J. H. JOHNSTON,
Waynesboro', Franklin County, Penn'a.,

RESPECTFULLY informs the public in general, that he still continues his business at the old stand, and is carrying on more extensively than ever. He has now on hand and for sale, an unusually large assortment of RIFLES and other Guns of every description and price; so that no one giving him a call can fail to have himself accommodated. He is prepared to repair old Guns, and to make new ones according to order, at the shortest notice, having always on hand the largest and best assortment of all the materials used in the business, to be found at any manufactory in this or the neighboring counties.

His old customers, marksmen, and all who use the deadly weapon are requested to give him a call, and see for themselves May 22—6m.

HAGERSTOWN HERALD copy 3 weeks. 1856

Waynesboro Village Record, May 22 through November 20, 1856

Jones, Benjamin. 1780–82, cutler, Tredyffrin, Chester Co. (t) He has been listed as a maker of "Bowie" knives some 20 years before Bowie's birth in 1796 (RAG, 222).

Jones, Charles. 7/16/1776, apprentice gunsmith, Capt. Andrew Graff's Lancaster Militia (2 Pa. Arch. 13 at 357; 5 Pa. Arch. 7 at 1073); 1778, gunner in Capt. Thomas Wylie's artillery artificers at Carlisle (PIR, 2:248).

Jones, Daniel. "Daniel Jones, by consent of Mary Jones, by consent of his mother, indents himself apprentice to Richard Allen of Philadelphia, brass founder, for 6 years, from 20th November 1745, to be taught the trade of a brass founder" (PIDS).

Jones, Hugh. Gunsmith, 216 S. 5th St., Philadelphia (PXD, 1813).

Jones, I. N. Gilmore Twp., Greene Co. (PD, 1890). Gunsmith.

Jones, John. "smith's store," 1769, Radnor, Chester Co. (t).

Jones, John Bert (1866–1947). Born in Pine Creek Twp., Jefferson Co. and lived there until 1887; 1919 Main St., Brookville, 1887 gunsmith and clock repairman (REH, 140–41).

Jones, Joseph. Blacksmith, 1687–1719, Philadelphia Co. (2 Pa. Arch. 19 at 455, 620).

Jones, Owen. Small-time manufacturer, inventor, Philadelphia. Patents: 189,360 (1877) and 200,794 (1878). Made tip-up pistols.

Jones, Philip. Indian trader and gun repairman, Clarion Co. (REH, 147, 151) His activity is undated, but there were no native Americans in his area after 1830.

Jones, Robert. Apprentice gunsmith, 1776, in Capt. Andrew Graff's Lancaster Militia Co. (1 Pa. Arch. 13 at 347; 5 Pa. Arch. 7 at 1073).

Jones, Thomas. Artillery artificer, 1776–81, transferred to Capt. N. Irish's Co. from Capt. Greene's artillery (PIR, 1:93; PIR, 2:248).

Jones, William. 1775, ran away from Widow Jones, near Philadelphia (KEA, 56); by 1777, gunsmith, Bedford Co. Active until 1800 (t).

Jonsserant, Reimr. Artillery artificer, 1779, (2 Pa. Arch. 15 at 383).

Jordan, Capt.-Lt. John. Artillery artificer, 1777–83, at the cannon factory at Philadelphia. Listed on 1/1/1783 as deranged. 1829, living in Botetourt Co., Va. (2 Pa. Arch. 11 at 250, 251).

Jorg, Jacob. 1818–20, Greenwich Twp., Berks Co. (KEA, 56). Gunsmith.

Jorgenson, Frederick. Gunsmith, Howe Twp., Forest Co. (PD, 1882).

Kent County, on Delaware, July 10, 1775.

FIVE DOLLARS REWARD.

RAN AWAY from the ſubſcriber, living on the road leading from Dover to Queen's Town, and near the place called the Horſe-head, on Sunday night, the 2d inſt. a ſervant man named WILLIAM JONES, by trade a gun-lock maker, and very ingenious at any kind of whiteſmiths buſineſs, and even at blackſmiths work; aged upwards of thirty years, of a middle ſtature; he went off very well cloathed with ſundry ſorts of cloaths, and is ſuppoſed to have taken plenty of money with him. It is believed he is gone in company with another ſervant man who is alſo eloped from a neighbour, (name unknown). Whoever ſecures ſaid ſervant in any public gaol, ſo that his miſtreſs may have him again, or brings him home to her, ſhall have the above Reward and reaſonable charges, paid by

SARAH JONES, Widow.

Pennsylvania Packet, December 25, 1775

Joseph, William. Gunsmith, 1846–48, Upper Saucon Twp., Lehigh Co. (t).

Joshua (1741–1806). Mohican Amerindian. Spoke German and English as well as his own language. Born in Conn., lived at Moravian community at Bethlehem, and moved with the Brethren to Gnadenhutten, Ohio (Heckewelder's *Narratives;* DH, 381. Gunsmith.

Jost, Casper. 1785, gunsmith, Dauphin Co. (RAG, 104; KEA, 56).

Joy, Andrew S. (1802–53). 1830–31, Florence, Washington Co.; 1838–47, Market St. and 5th St., Pittsburgh. (PAD, 1847; t) 1850–53, Kittanning, Armstrong Co. (t; c). His inventory included: 17 gunstocks $2.37; 1 box patterns, $1.50; 1 set stocks & dies $7 (Armstrong Co. file 1014).

Joy, Daniel. Appointed by Committee of Safety to make and prove cannon at Reading Furnace, 1776. In early 1778 he cast the first cannon at Reading (2 Pa. Arch. 1 at 418; 3 Pa. Arch. 5 at 66).

Judd, C. W. Gunsmith, Woodcock Twp., Crawford Co. (CCD, 1874).

Judd, G. Meadville, Crawford Co. (CCD, 1873). Gunsmith.

Judson, W. H. Union City, Erie Co. (*Erie County Directory,* 1873–74). Dealer in all kinds of firearms; gunsmith; worked on sewing machines and locks.

Judy, Jacob (1740–1807). Born at Basel, Switzerland. Master gunsmith at Ft. Pitt during the Revolution (Va. appointment). 1788–92, Kaskasia. Ill.; 1792–1807, Monroe Co., Ill. (Johnson, *Illinois Gunsmiths,* 2, 17).

Justice, Philip S. Made edged weapons, guns, and swords, 6th Ward, Philadelphia. In 1860 he employed 10 men, made 2,500 guns valued at $15,000, and had a plant valued at $10,000. (IC; c, 1860) In 1862, he described himself as having had 20 years of experience as an importer and manufacturer. Made substandard weapons under military contract (W. R. Schaefer, MB 8 [1964] 9).

K

Kahn, J. Gunsmith, DBA Hartman & Kahn, Erie (c, 1860).

Kahn, W. T., & Co. Importers and gunsmiths, Philadelphia (MD, 1840).

Kantz, Gabriel. Gunsmith, Lafayette near Strawberry, Lancaster (GCD, 1863–64).

Kappal, J. 257 Poplar St., Philadelphia (MD, 1855, 1858). Gunsmith.

Kaup, Levi. Gunsmith, West Buffalo Twp., Union Co. (c, 1860).

Kaupert, John F. 500 Brook St., Philadelphia (BPaD, 1861). Gunsmith.

Kaye, S. R. 204 Green St., Philadelphia, gunsmith (MD, 1859, 1865).

Kaye, William W. Gunsmith, Philadelphia (MD, 1856, 1864–65).

Keaffer, Peter. Gunsmith, 1759, Lancaster boro (t).

Keefer, Isaiah (1829–61). 1829–61, Warren Twp., Franklin Co. (BPaD, 1861; WSB, 28).

Keefer, Jacob (b. 1821). Gunsmith, Shippensburg, Cumberland Co. (c, 1850).

Keefer, Michael (1803–89). Gunsmith, Sylvan, Warren Twp., Franklin Co. "Michael Keefer invented a homemade threshing machine and lost his arm in it. This accident curtailed his gun making" (Foreman, *History of the Little Cove* 52, 124).

Keeley, Matthias. Gunsmith to the Committee of Safety. Delivered 31 new muskets on 3/2/1776; was paid £189 for 42 new arms on 11/8/1776 (4 Amer. Arch. 5 at 718; 5 Amer. Arch. 1 at 186; Col. Rec., 10:535).

Keely, John. Gunsmith, rear of 105 Budd St., Philadelphia (PXD, 1810).

Keely, Sebastian. 11/9/1775: "Resolved that Mr Samuel Miles engage Sebastian Keely to make 100 fire locks for the use of the Province . . . [Keely] has engaged to deliver 6 firelocks per week until he compleats 100" (Col. Rec., 10:399).

Keenert, Lewis. Gunsmith, Stroudsburg, Monroe Co. (BRD, 1868).

Kehler, John. 1802–3, Lancaster. (SED, 57) 1823–44, Carlisle, Cumberland Co. (KKR, 274). Gunsmith.

Keim, John (1754–1815). Gun-barrel maker, Hanover Twp., Lehigh Co. (Mathews and Hungerford, *History of Lehigh and Carbon Counties* [Chicago: Beers 1884], 264) The basic factory begun by Keim operated as late as 1890.

Keiser, John. 1785, Locksmith, 5′ 7½″, born in Germany (5 Pa. Arch. 4 at 841.) (Militia enrollment list.

Kelker & Bros. Gunsmiths, 1861–74, Harrisburg (RAG, 105).

Keller, Elias. Gunsmith, Yardleyville, Bucks Co. (BPaD, 1861).

Keller, John. 1835–73, 27 W. Louther St., Carlisle, gunsmith (WSB, 38; KEA, 57; BRD, 1873).

Keller, Joseph (1821–79). 1844, McConnellsburg, Bedford Co. (t); 1850, Meadville, Crawford Co. (c, 1850); 1860–79, Clark Co., Ill. (research by Curtis Johnson).

Keller, Michael. 1777–79, delivered bayonet scabbards to the York Co. Militia (3 Pa. Arch. 7 at 108).

Kelly, James. Artillery artificer, 1779, at the cannon factory at Philadelphia (2 Pa. Arch. 15 at 385).

Kelly, Patrick. Artillery artificer, 1779, at the cannon factory at Philadelphia (2 Pa. Arch. 15 at 386).

Kemmerer, David, Sr. Gunsmith, 1830–40, Lehighton, Carbon Co. (t; c).

Kemmerer, David, Jr. Gunsmith, Lehighton, Carbon Co. (c, 1850).

Kempf, A. Youngstown, Gunsmith Westmoreland Co. (Westmoreland *Republican*, 10/9/1840).

Kempf (Camp, Kemp), Benjamin (1795–1879). 1819–23, Hempfield Twp., Westmoreland Co. (t); 1825-96, Mt. Pleasant, Westmoreland Co. (t; c).

Kempf, Garret (d. 1837). A native of Germany, Kempf settled in Westmoreland Co. before 1800. He also owned land in Columbiana Co., Ohio, 1803–4 (*History of Columbiana County* [Philadelphia: D. W. Ensign Co., 1879] 13). He died on 4/28/1837 in Hempfield Twp., Westmoreland Co. (t; death notice in Westmoreland *Republican*, 4/30/1837.

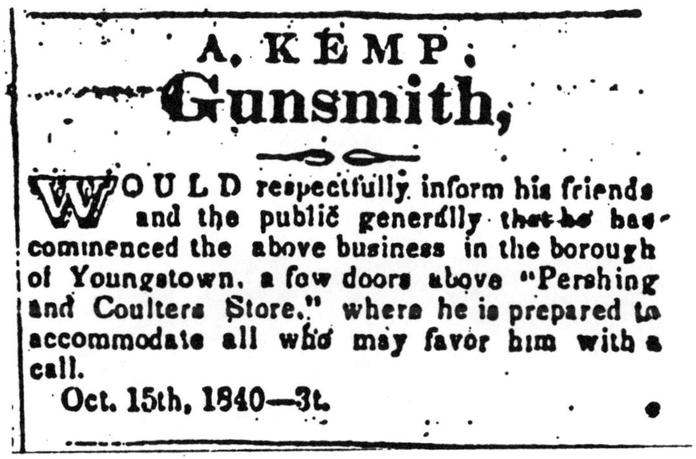

A. KEMP;

Gunsmith,

WOULD respectfully inform his friends and the public generally that he has commenced the above business in the borough of Youngstown, a few doors above "Pershing and Coulters Store," where he is prepared to accommodate all who may favor him with a call.

Oct. 15th, 1840—3t.

Westmoreland Farmer's Repository, October 15, 1840

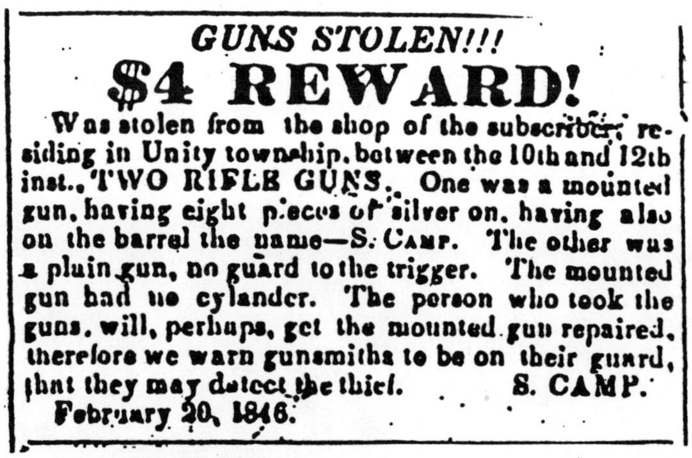

GUNS STOLEN!!!

$4 REWARD!

Was stolen from the shop of the subscriber, residing in Unity township, between the 10th and 12th inst., TWO RIFLE GUNS. One was a mounted gun, having eight pieces of silver on, having also on the barrel the name—S. CAMP. The other was a plain gun, no guard to the trigger. The mounted gun had no cylander. The person who took the guns, will, perhaps, get the mounted gun repaired, therefore we warn gunsmiths to be on their guard, that they may detect the thief. S. CAMP.

February 20, 1846.

Westmoreland Republican, February 20, 1846

Kempf (Camp, Kemp), Solomon (b. 1787). Gunsmith, 1808–25, Hempfield Twp. Westmoreland Co. (t); 1826–62, Unity Twp., Westmoreland Co. (t; c).

Kennedy & Curtis. Gunsmiths, 706 Market St., Philadelphia (GPD, 1892; BSED, 1888, 1895).

Kennedy, George S. Philadelphia. (BRD, 1899; BSED, 1888) Doing gunsmithing business as Kennedy & Curtis.

Kennedy, John. Gunsmith, Pittsburgh (PAD, 1868).

Kepelering, Lewis, Gunsmith, Berwick Twp., Adams Co. (c, 1850).

Kerchburgh, William. Gunsmith, 82 Race St., Philadelphia (MD, 1837).

Kerlin (Kirlin), John. Gunsmith and armorer to Committee of Safety. On 7/18/1776, delivered 50 muskets at £4/05/0 each (Col. Rec., 10:650; 5 Amer. Arch. 1 at 1297).

Kerlin (Kirlin), John, Jr. In 1779–81, he was paid £20 to repair Westmoreland Co. arms and noted as being from Chester Co. (3 Pa. Arch. 7 at 121) In 1780–81, John Kerlin, Jr. was a blacksmith, Amity Twp., Berks Co (t). There was a John Kerlin, gunsmith, in Perry Co., Ohio, 1846–53 (DH, 313). He took over John Miles, Jr's unfilled contract for M-1808 muskets; he was living then in Chester Co. He worked with Samuel Kerlin (RAG, 106).

Kerlin, Samuel. 1775–79, Amity Twp., Berks Co. (t), He made contracts of 5/5/1801 and 6/30/1801 for liberty cap muskets. 1780, blacksmith, Colebrookedale Twp., Berks Co. (t) (RAG, 106–7).

Kerling, Thomas. Gunsmith, Philadelphia. 1/4/1777: "pay Thomas Kerling £45 for 10 guns and bayonets." (Col. Rec., vol. 11) Gunsmith, 1805, Northumberland Co. (KKR, 275).

Kern, D. W. Gunsmith, Upper Milford Twp., Lehigh Co. (PD, 1882).

Kern, Frederick R. Gunsmith, James St. near Queen St., Lancaster boro (BLD, 1857).

Kern, Reinhart. Gunsmith, James St. near Queen St., Lancaster boro (BLD, 1857). Partner and brother of Frederick Kern.

Kerr, Archibald. 4/12/1813, supplied 100 cartridge boxes at $300 per 1,000 to Pa. militia (2 Pa. Arch. 12 at 652).

Kerr, John. Cutler and smith, 1779, Carlisle, Cumberland

Co. Used 1 black worker (3 Pa. Arch. 20 at 135).

Kerr, John L. 11/17/1874, Pittsburgh, patent 157,008 for gun barrels.

Kerr, Michael. Gunsmith, 448 N. Front St., Philadelphia (KKR, 274). 1799–1800.

Kessler, Israel. Gunsmith, Earl Twp., Lancaster Co. (PD, 1882).

Kettering, George (1803–66). Major gunsmith. Apprenticed to George Biddle. 1821, Bullskin Twp., Fayette Co. (t); 1825–66, Hempfield Twp., Westmoreland Co. (t; c).

Keys, John. Armorer, 1779, Pa. Navy, on armed ship *General Green* (2 Pa. Arch. 1 at 302).

Keyser, William W. 559 N. 2d St., Philadelphia (BPaD, 1861; MD, 1841, 1846, 1859, 1861, 1865). Gunsmith.

Kider, John. Gunsmith, Washington St., Philadelphia (MD, 1837).

Kienle, Jacob (b. 1718). Cutler and tool maker, from Doffingen, Germany; by 1750, Exeter Twp., Berks Co. (PaE, 73).

Kilburn, Benjamin. Artillery artificer, 1778, Capt. Thomas Wylie's Co. at Carlisle (PIR, 2:248).

Killets, Charles. 1849–50, gunstock maker, West Earl Twp., Lancaster Co. (SED, 57; t).

Killian, George. Gunsmith, Mulberry St., Lancaster (BLD, 1857).

Killin, William. Artillery artificer, 1778, Capt. Thomas Wylie's Co. at Carlisle (PIR, 2:248).

Kinaston, John. Artillery artificer, 1779, Capt. Nathaniel Irish's Co. (PIR, 2:247).

Kincaid, John. Artillery artificer, 1779–82, Capt. Nathaniel Irish's Co. (PIR, 2:247).

Kinder, Samuel. Gunlock maker, 1775–76, Philadelphia (GAAAM, 84). Signed the petition in 1801 asking Congress to retain the impost tax on foreign-made guns.

King, Chambers (1838–1920). Born on 3/24/1838 in Armstrong Co., King was already a gunsmith when he enlisted in the Union army on 2/24/1864 at Kittanning, where he was described as being 5′ 7″ tall with brown hair and eyes. He was discharged at Cloud's Mills, Va. on 7/13/1865. He was an armorer in Co. M, 2d Pa. Cavalry, and, later, Co. C, 1st Regt., Pa. Provisional Cavalry. In 1868 he moved to Rochester Mills, Canoe Twp., Indiana Co., where he lived until his death (research by Col. Russell Harriger).

King, Daniel. Germantown, Philadelphia Co. Operated a cannon factory, 1775. In 1795 he cast 2.85-inch bronze howitzers. (Col. Rec., 10:29; Col. Rec., 11:641) Before the Revolution he was a brass founder, specializing in

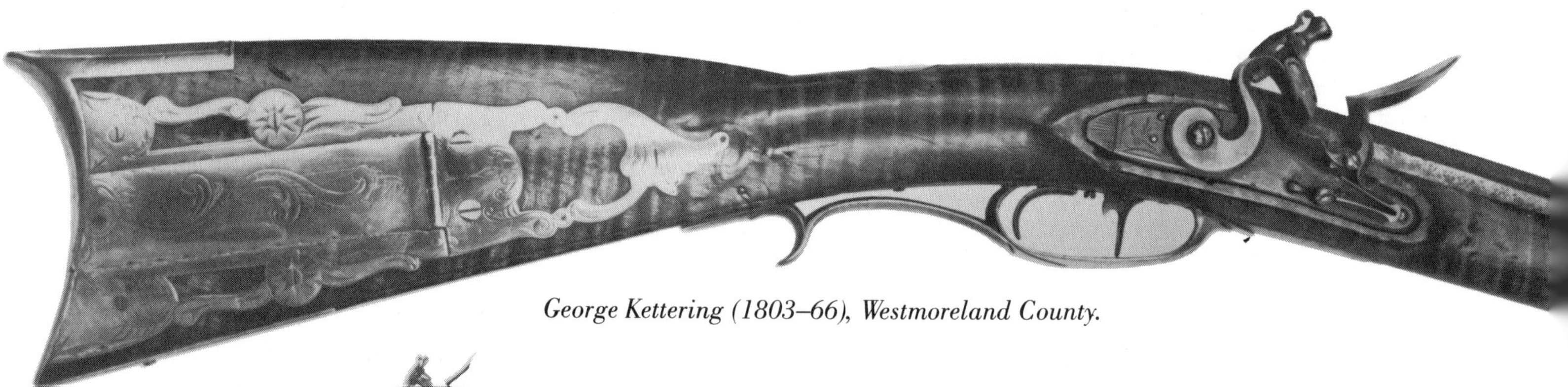

George Kettering (1803–66), Westmoreland County.

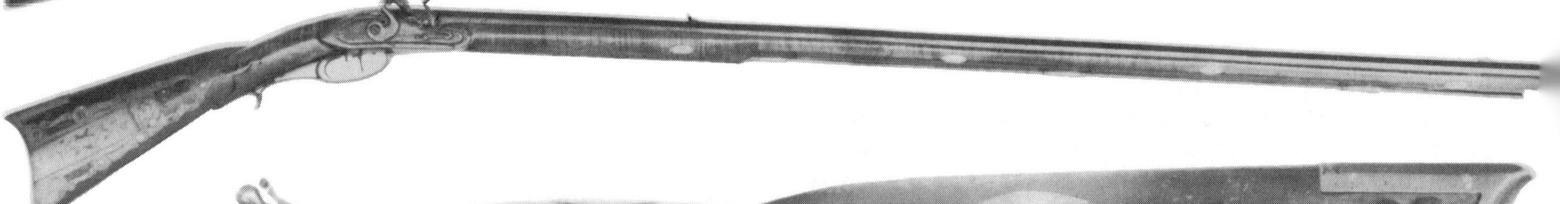

George Kettering (1803–66), Westmoreland County.

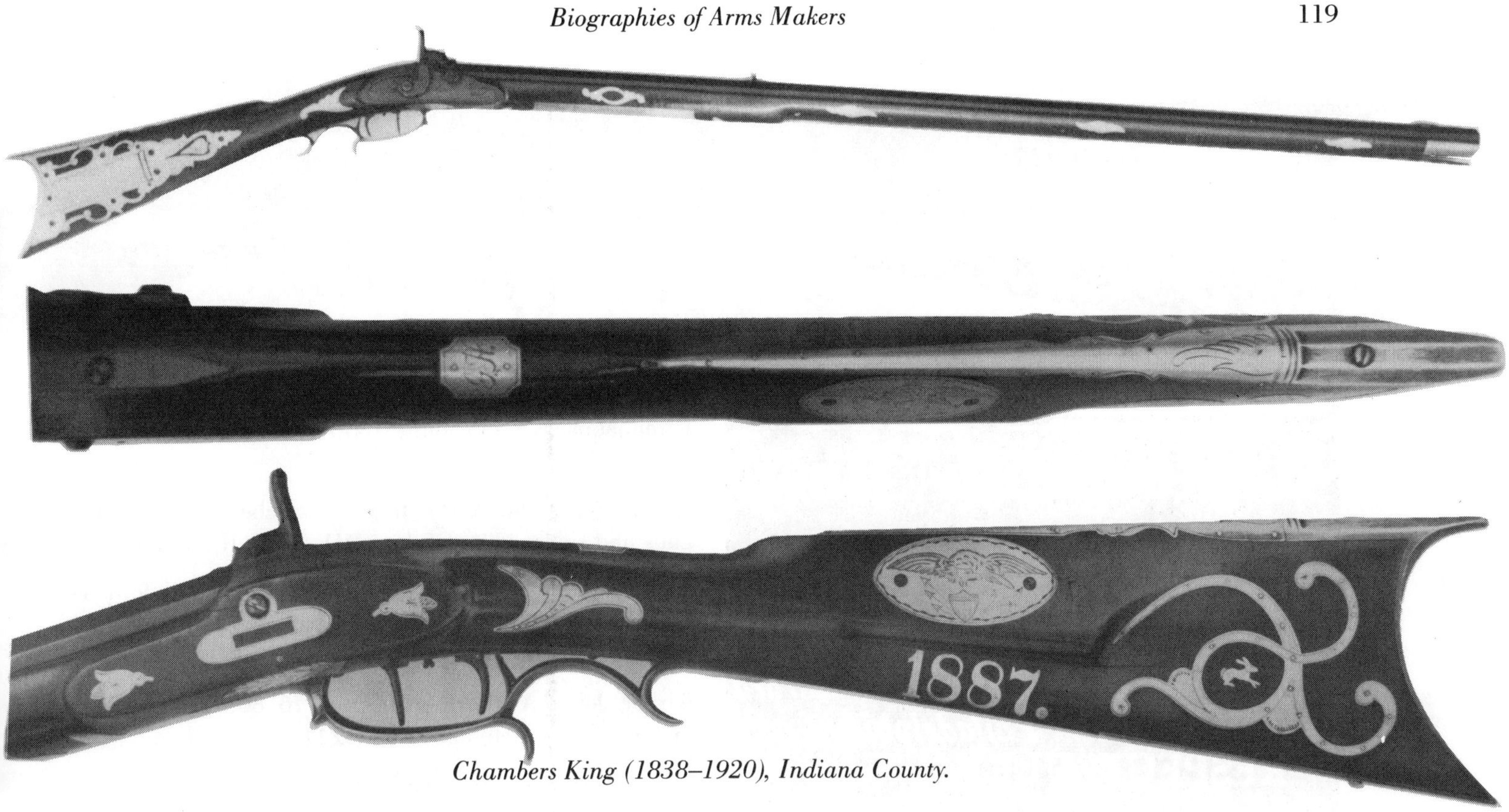

Chambers King (1838–1920), Indiana County.

fireplace instruments (Bridenbaugh, *The Colonial Craftsman*, 148).

King, Elmer (1869–1959). Son of Chambers King, Canoe Twp., Indiana Co. (MB 1 [1967] 8). Gunsmith.

King, Frederick. Gunmaker, Queen St. near James St., Lancaster (BLD, 1857; GCD, 1863–64); 163 Mary St., Lancaster (BLD, 1873–74).

King, Frederick. Gunstock maker. 1863, West Earl Twp., Lancaster Co. (t; SED, 57).

King, George (b. 1809). 1831–35, St. Clair Twp., Bedford Co., gunsmith; 1850–57, Sugar Twp., Armstrong Co. (t); 1860–1870, Washington Twp., Armstrong Co. (c).

King, Isaac. Somerset *Whig*, 1/8/1818. Gunsmith, trained in Philadelphia. Somerset boro (KKR, 275).

King, William. Artillery artificer, 1776, McClain's Co., 6th Battalion (PIR, 1:184).

King, William. Gunsmith, 1867–73, Washington Twp., Armstrong Co. (t).

Kinkinger, David. 1797–98, Northampton Twp., Northampton Co. (t).

Kinnard, A. M. & Brother. Gunsmiths, West Chester, Chester Co. (PD, 1882).

Kinnard, William. Artillery artificer (date unknown), Capt. Jesse Rowe's Co.; "additional private" (PIR, 2:245).

Chambers King (1838–1920), Indiana County.

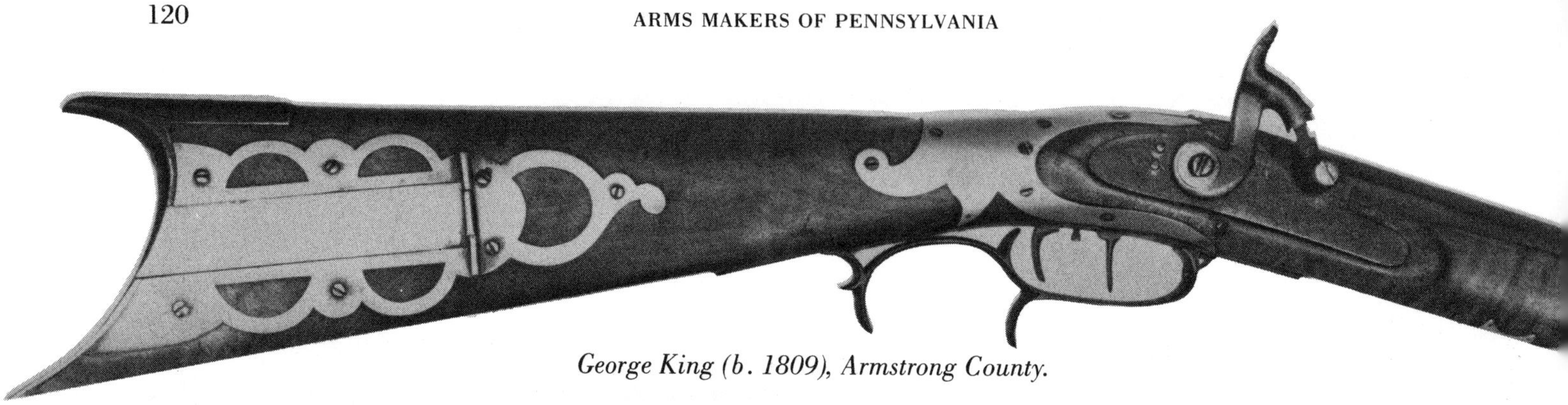

George King (b. 1809), Armstrong County.

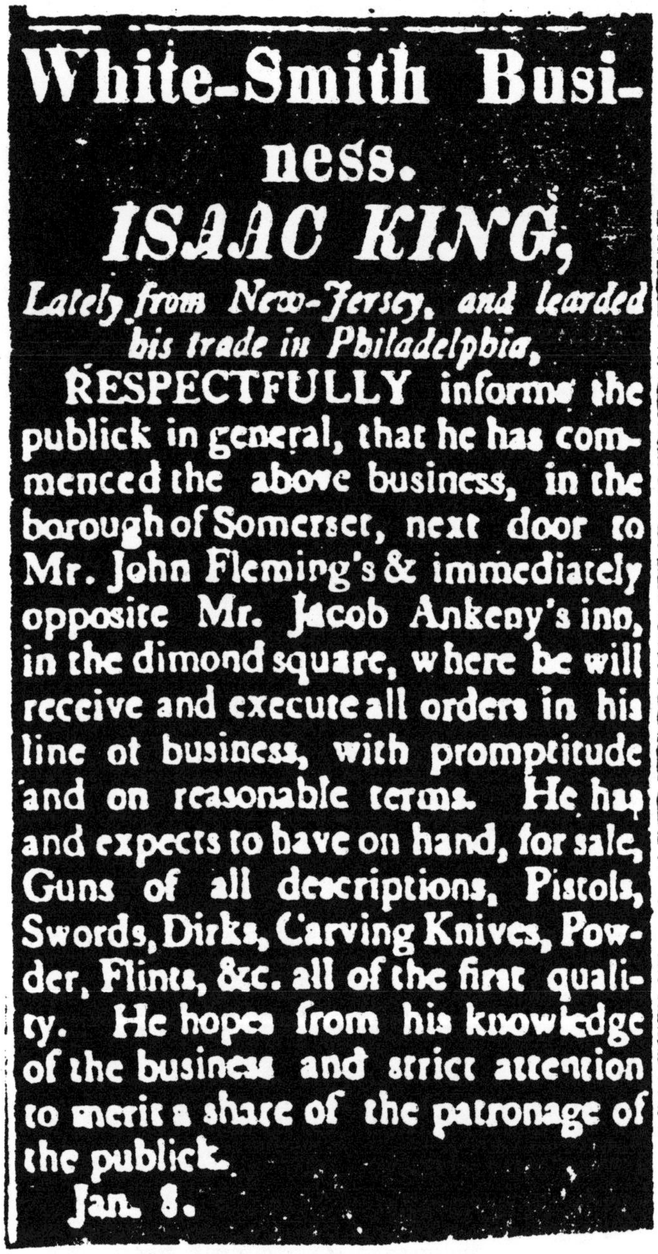

White-Smith Business.

ISAAC KING,

Lately from New-Jersey, and learded his trade in Philadelphia,

RESPECTFULLY informs the publick in general, that he has commenced the above business, in the borough of Somerset, next door to Mr. John Fleming's & immediately opposite Mr. Jacob Ankeny's inn, in the dimond square, where he will receive and execute all orders in his line of business, with promptitude and on reasonable terms. He has and expects to have on hand, for sale, Guns of all descriptions, Pistols, Swords, Dirks, Carving Knives, Powder, Flints, &c. all of the first quality. He hopes from his knowledge of the business and strict attention to merit a share of the patronage of the publick.

Jan. 8.

Somerset Whig, January 18, 1818

Kinnear, G. W. Gunsmith, Kane, McLean Co. (PD, 1887).

Kinsey, Edward. 10/19/1814: "I am now employed by Edward Kinsey in making knapsacks. [signed] John B. Weir." (MPHS 8 [1867])

Kirchbaum, E. Gunsmith, 1830–36, Danville, Montour Co. (t).

Kirchberg, W. M. & Co. Importer of cheap foreign shotguns and rifles, Philadelphia (MD, 1845; RAG, 108).

Kirk, Nathaniel. Blacksmith, 1779, in Capt. John Jordan's Co. of artillery artificers, making cannon (2 Pa. Arch. 11 at 260).

Kirker, John. Artillery artificer, 1776–80, at the cannon factory at Philadelphia (PIR, 2:244).

Kisselbach, Oswald. Artillery artificer, 1778, in Capt. Thomas Wylie's Co. (PIR, 2:249). Stationed at Carlisle.

Kistler, George. 1800, Gunsmith. Berks Co. (KEA, 58).

Kleine, Christian. Gunstock maker, 1792–93, West Earl Twp., Lancaster Co. (SED, 57).

Kleinfelter, George. Gunsmith, 1820, Bushkill Twp., Northampton Co. (t).

Kleist, Daniel (1716–92). Born at Frankfort, Germany. Arrived in America in 1749, a Moravian. (PaE, 42) 1750–92, Bethlehem Twp., Northampton Co., locksmith (t; Heckewelder's *Narratives*, 419).

Kline, A. Gunsmith, 1817–18, Dover, York Co. (t).

Kline, Conrad (d. 1842). He was the last surviving revolutionary war soldier in York Co. Gunsmith, Dover, York Co. (t; KEA, 58; RAG, 109).

Kline, Christian. Gunsmith, Harrisburg, 1811–17 (t; KKR, 275).

Kline, Jacob (b. 1792). 1828–63, Frankfort Twp., Cumberland Co. (WSB, 57; BPaD, 1861; KEA, 58).

Kline, Jacob (b. 1820). 1850, Loyalhanna Twp., Westmoreland Co. (c); 1860–68, Bell Twp., Westmoreland Co. (address: Saltsburg, Indiana Co.). In 11/1868, a certificate of removal was granted to Jacob Kline to join the Lutheran church at Knoxville, Iowa. Gunsmith.

Kline (Klyne), Michael. 1757, Conestoga Twp., Lancaster Co. (t); 1769–70, Ft. Johnson, N.Y. There he made a gun for Thomas King, an Oneida sachem and friend of Sir William Johnson (information provided by A. J. Swinney).

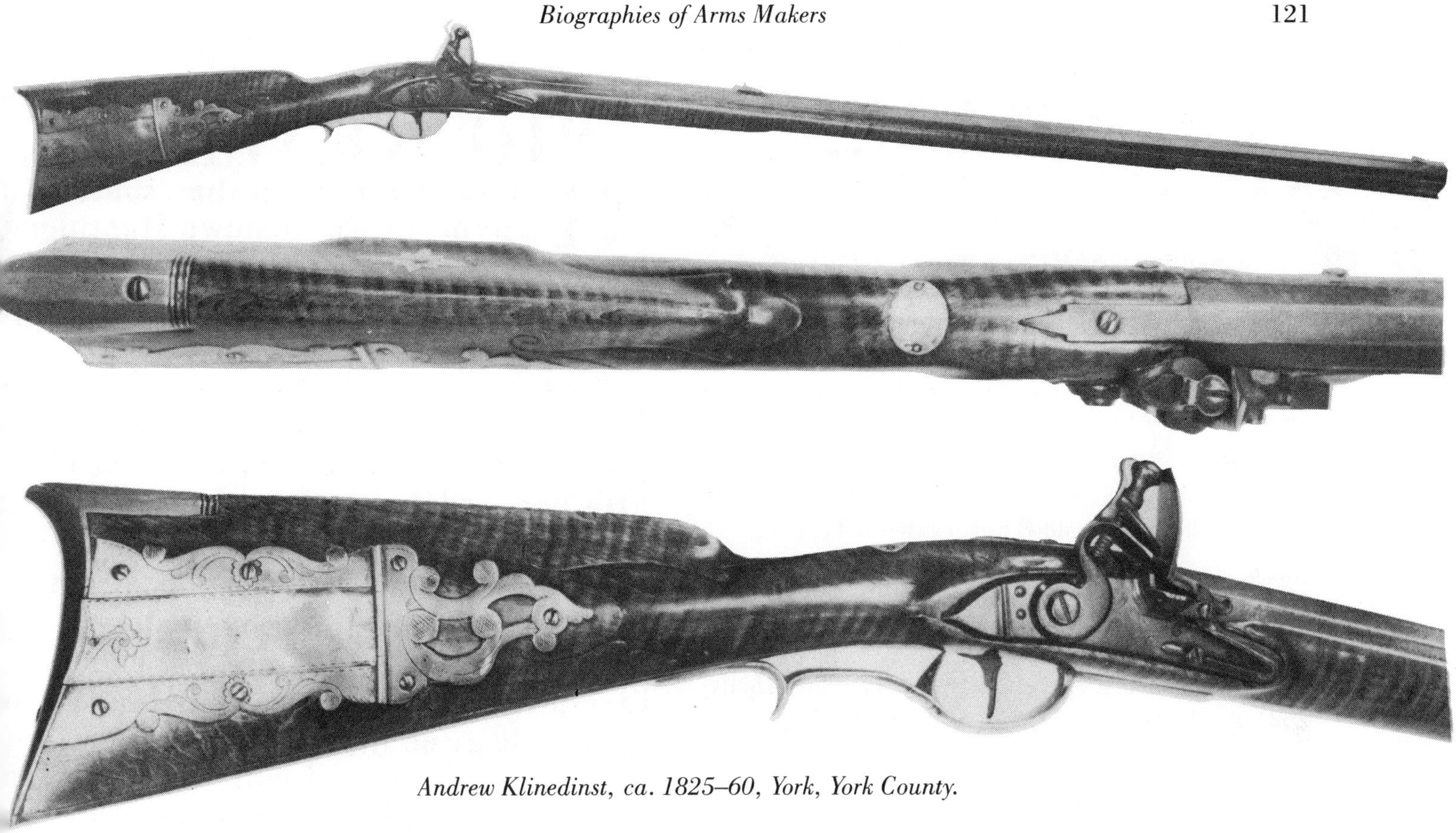

Andrew Klinedinst, ca. 1825–60, York, York County.

Kline, Philip. Halifax, Dauphin Co., 1814–62 (t; BPaD, 1861). Gunsmith.

Klinedinst, Andrew. 1825–60, Gunsmith, S. Beaver St., York (JKJ, 318–19; KEA, 58).

Knappenberger, Henry. 1821, Upper Milford Twp., Lehigh Co. (t) Gunsmith (KEA, 1818–21).

Knappinberger, Philip. Gunsmith, 1807–16, Franklin Twp., Westmoreland Co. (t).

Knauf, Henry (d. 1879). Nippenose, Lycoming Co. (BPaD, 1861); Collomville, Lycoming Co. (BRD, 1879). Gunsmith.

Knepper, Joseph. Gunsmith, 1833–47, Summerhill Twp., Cambria Co. (t).

Knous, Ludwig. Sicklesmith, 1788, Bethlehem Twp., Northampton Co. (t).

Knouse, William. Gunsmith, Marlborough St. and West St., Philadelphia (MD, 1846).

Knowles, Charles. Armorer, 1775, Pa. Navy, on armed schooner *Delaware* (2 Pa. Arch. 1 at 276).

Knox, Robert. Gunsmith, 1800–1801, Franklin Twp., Franklin Co. (WSB, 28).

Knupp, Charles Monroe (1863–1939). Gunsmith, Bakersville, Somerset Co. Son of Levi Knupp (t; c; f).

Knupp, Elias (b. 1837). 1860, Somerset Twp., Somerset Co., apprenticed to Jonathan Dunmyer; 1870, Cook Twp., Westmoreland Co.; 1880, Donegal Twp., Westmoreland Co.; 1900, Somerset boro (c).

Koch, A. Gunsmith, Henrysville, Monroe Co. (BPaD, 1861).

Koch, Charles. 1850, gun-barrel forger, Manheim Twp., Lancaster Co. (t; c; SED, 57).

Koch, Henry (b. 1780). 1796, apprenticed to John DeMuth, Frederick Co., Md., bound for 4 years, 4 months, and 16 days, at the age of 16. 1820, Tyrone Twp., Cumberland Co. (WSB, 58). Gunsmith.

Koch, L. Gunsmith, Henrysville, Monroe Co. (BPaD, 1861).

Kochler, Philip. Gunsmith, 1868–82, Lewisburg, Union Co. (BRD, 1868; PD, 1882).

Koenig & Warner. Gunsmiths. Frederick Koenig and John J. Werner, 110 N. Queen St., Lancaster (GCD, 1863–64).

Koenig, Frederick. Gunsmith, 130 N. Queen St., Lancaster (BLD, 1857; WCD, 1888).

Kohl, Conrad. Gun-barrel makers on Wyomissing Creek, Berks Co. (RAG, 110). Gave up gun-barrel making and turned the shop into a sawmill.

Kohn, J. J. Gunsmith, West Franklin Twp., Armstrong Co.

KŒNIG & WERNER'S

Gun, Gun Locks, and Rifle Manufactory,

110 North Queen Street, Lancaster, Pa.

☞ All orders for new Guns, Rifles, Pistols, &c., promptly filled, and at moderate terms.
☞ Gun-locks made and warranted to be good.

REPAIRING OF ALL KINDS DONE AT THE SHORTEST NOTICE.

Gopsill's Lancaster County Directory, 1863–64

(PD, 1887, 1890).

Kohr, William. 1837–38, gun-barrel forger, Manheim Twp., Lancaster Co. (SED, 57; t).

Kolb, Henry M. 2311 N. 16th St., Philadelphia. Made cheap pistols and guns. Active from 1890 to 1910, when firm was sold to Reginald F. Sedgley (RAG, 110).

Konce, D. (b. 1813). Gunsmith, Jackson Twp., Venango Co. (c, 1850).

Konigmacher & Fahnestock. Suppliers. Albert Konigmacher and Samuel Fahnestock, 1801–14, Philadelphia; 1821–28, Front St. and Queen St., Chambersburg. Lock makers and importers (WSB, 5–6).

Koons, F. G. Gunsmith, 1792–1820, Macungie Twp., Northampton Co. (t).

Kopp, Andrew (1781 or 1782–1875). 1810–11, Paradise Twp., York Co.; 1823–75, Frankstown Twp., Blair Co. Major gunsmith (JKJ, 438–39).

Kopp, George L. (1810–90). Gunsmith Son of Andrew Kopp. Worked in Frankstown Twp., Blair Co., except for a few years in Ill. ca. 1870 (t; c).

Kough, Samuel (1800–1870). Died 3/7/1870 at the age of 70; buried in Saltillo Cemetery, Huntingdon Co. No public record listed him as a gunsmith, but there is a splendid gun in the county's style bearing his name.

$10 REWARD.

RANAWAY from the subscriber living in Frankstown Huntingdon County on Surday night June 28th. an indented apprentice to the Mill Wright business named WILLIAM DIRK.—The said William had on when he left home a brown cloth coat, dark cassinet pants, black fur hat and monroe shoes. He had red hair and a fair complexion and was about 18 years of age.

He took with him when he left a short cut rifle, manufactured by Andrew Kopp belonging to the subscriber, which it is supposed he would sell near home as he had little or no money with him.

The above reward will be given to any person returning said William Dirk.

The public are hereby cautioned against trusting or harboring said William on my account as I am determined to pay no debts of his contracting after this date.

ANDREW KOPP.

Frankstown July 2. 1841.

Democratic Standard [Hollidaysburg], July 2, 1841

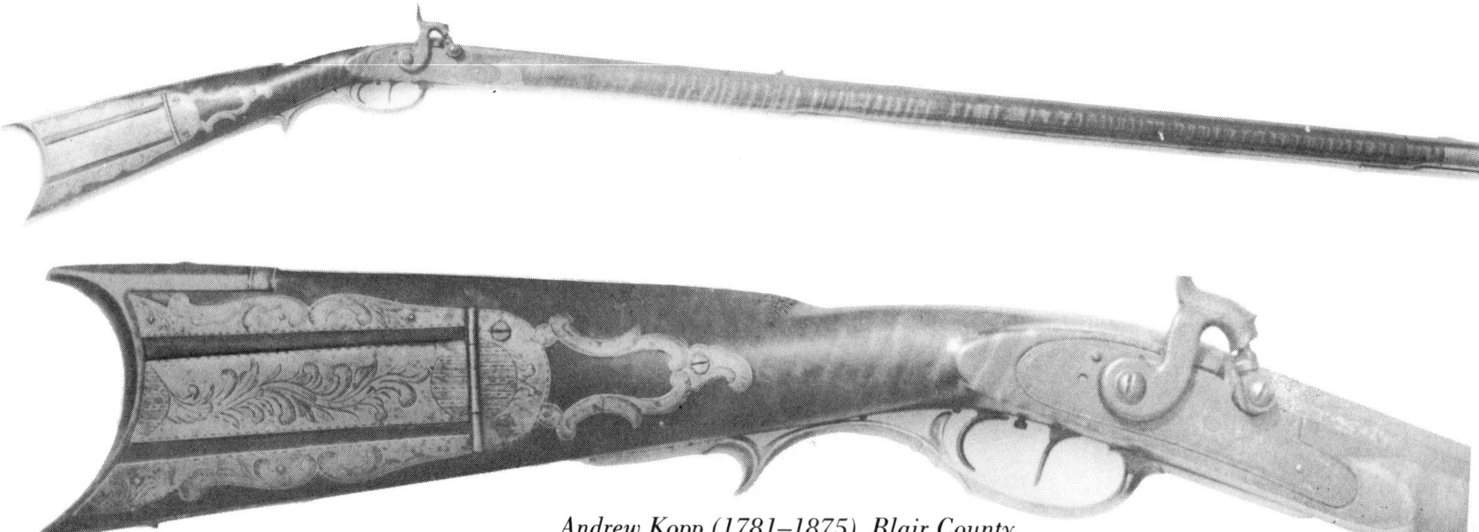

Andrew Kopp (1781–1875), Blair County.

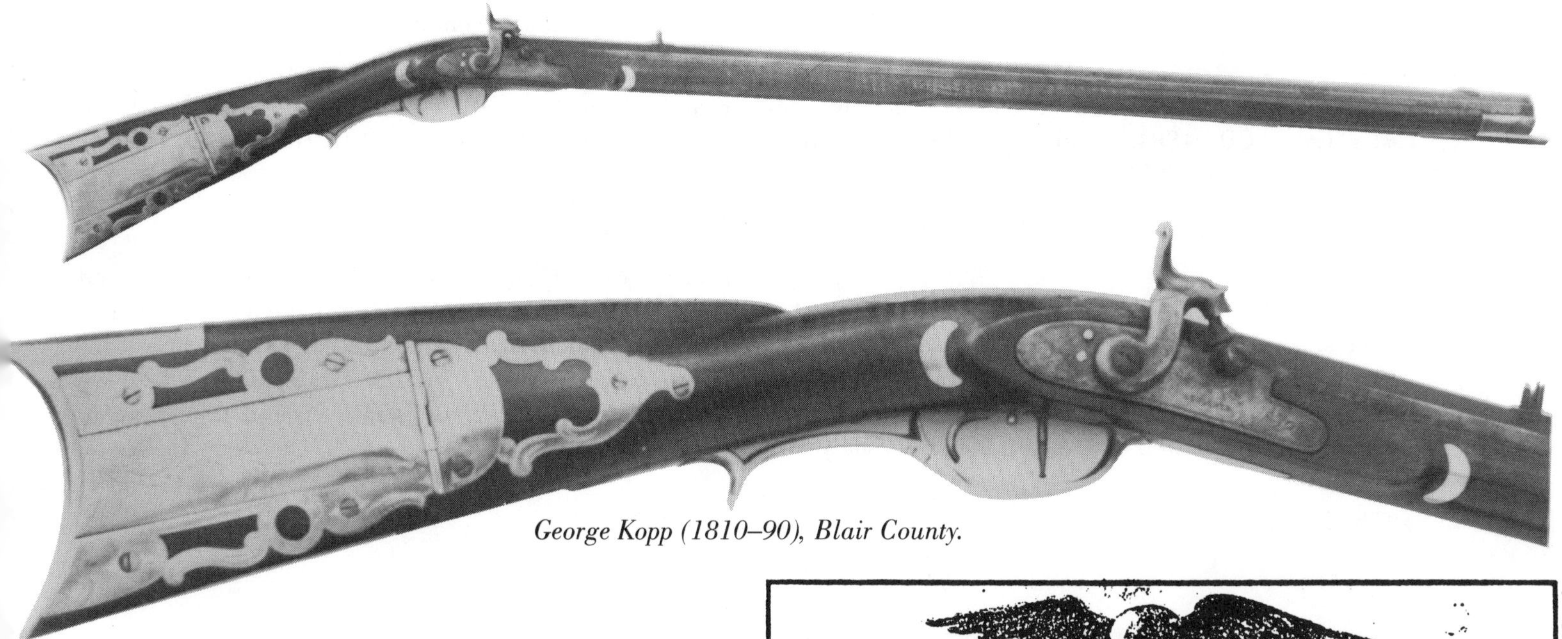

George Kopp (1810–90), Blair County.

Kraft, Jacob. Gunsmith, 1769–82, West King St. at Water St., Lancaster Co. In 1786 Kraft moved to Hagerstown, Md. In 1780 there was a Jacob Kraft, whitesmith, Cumru Twp., Berks Co. In 1781, "paid Jacob Kraft for repairing arms, £187/10/0" (3 Pa. Arch. 6 at 442). He married Catherine, sister of Peter Resor (SED, 57; t; JKJ, 91).

Krause, J. S. (d. 1815). 1777–1815, Bethlehem (t). Silversmith.

Kreuling, Mathias. Gunsmith, 1805, Hanover Twp., Northampton Co. (t).

Krider, John. Gun manufacturer, 2d St. and Walnut St., Philadelphia (GCD, 1892; BSED, 1880; PD, 1882; BPD, 1890).

Krider, John H. Gun manufacturer, 2d St. and Walnut St., Philadelphia (MD, 1837, 1846, 1849; BPaD, 1861).

Kriegbaum, George. Gunsmith, 1832–41, Northumberland Co. (t).

Krotz, Siegmund. Gunsmith, Sawmill Alley near Spruce, Duquesne, Allegheny Co. (PAD, 1860–61, 1864–65).

Krouse, William. Gunsmith, Philadelphia (MD, 1847).

Kryner, William. Gunsmith, Montgomery Twp., Franklin Co., only in these years: 1814, 1829, 1846 (t; WSB, 23–25).

Kuhn, Frank. Gunsmith, 502 Everett Place, Philadelphia

McElroy's Philadelphia Directory, 1855

Civil War rifle used by German Turner regiments. Lock marked "Krider."

(BSED, 1880; PD, 1880); 476 N. 8th St. (BPD, 1890; GPD, 1892).

Kuhn, William. Gunsmith, 22 W. Main St., Mt. Joy, Lancaster Co. (LCD, 1884, 1896; BLD, 1875–76, 1887, 1890).

Kuhns, Peter. Gunsmith, 1821, North Whitehall Twp., Lehigh Co. (t).

Kunkle, [?]. 8/30/1814, Philadelphia: "a sample parcel of muskets, I beg you to observe that the quantity is about 3000 and the price $14.50. The whole of them are made at Philadelphia by Mr Kunkle, whose name is stamped on the locks of them and who has, since a number of years, manufactured, under contract, for the Government of the U.S" MPHS 8:47–48).

Kunkle, Leonard G. 1831–72, gunsmith, 207–209 N. 2d St., Harrisburg (BRD, 1869; BPaD, 1861; t; c).

Kuntz, George. Gunsmith, 1816–30, East Bethlehem Twp., Washington Co. (t).

Kuntz, Jacob (1780 or 1781–1875). Major gunsmith and pistol maker. 1807–11, Allentown, Lehigh Co. (t); 1814–75, Philadelphia, at these addresses: 1814–17, Germantown Rd. near 2d St.; 1819–29, Germantown Rd. above Green; 1847–55, Cadwaller above Phoenix (Shumway, in MB 11 [1969]: 2–3, 25; t; c).

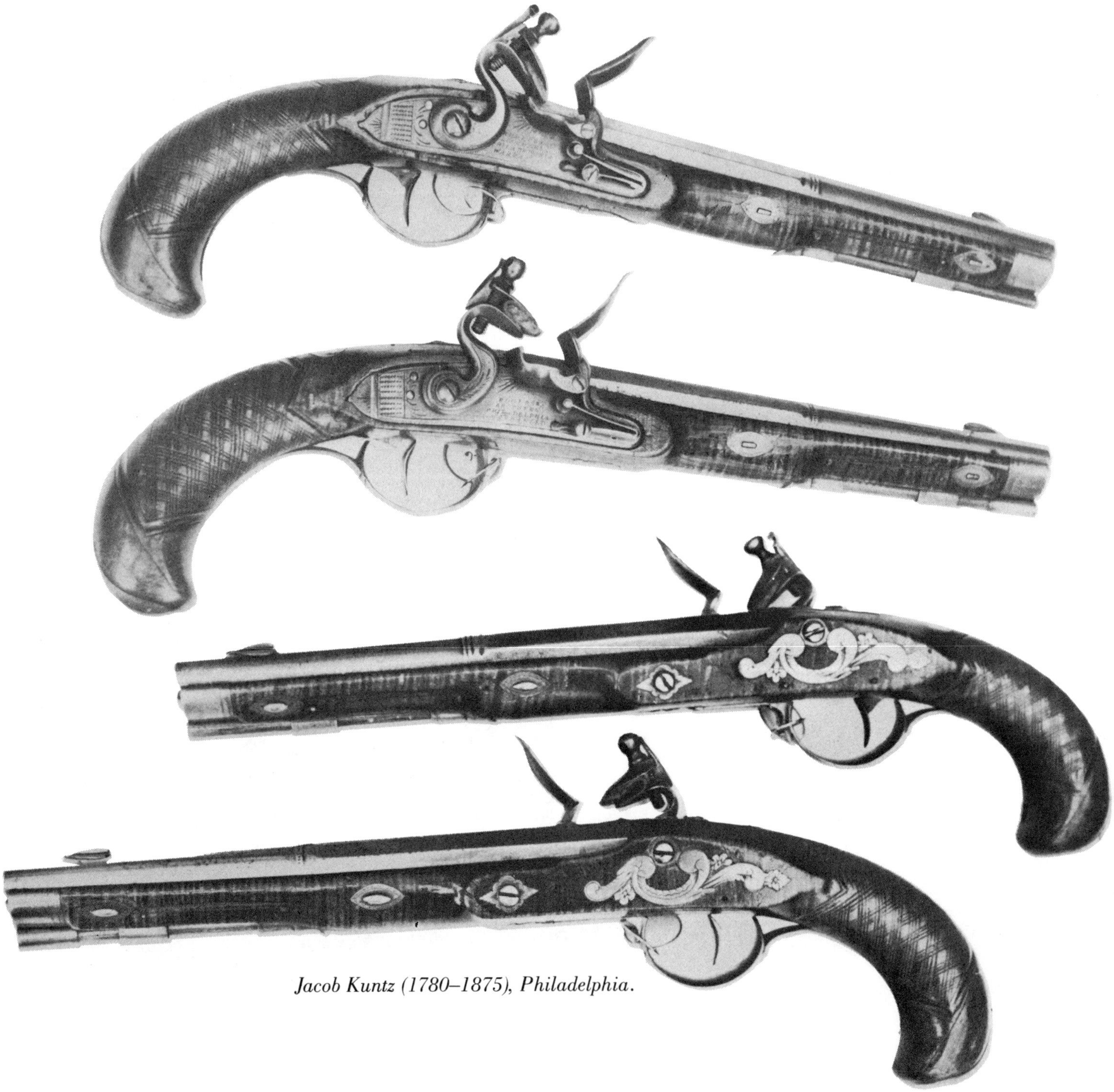

Jacob Kuntz (1780–1875), Philadelphia.

Kuntz, Michael. 1799–1807, gunsmith, Lancaster (SED, 57; t).

Kuntz, Peter (1795–1862). Gunsmith, 1817–62, North Whitehall Twp., Lehigh Co. (t; c; BPaD, 1861).

Kurton, Julius. 1871–90, 113 Norwegian St., Pottsville, Schuylkill Co. (t; c; PD, 1882; BRD, 1875). Gunsmith.

Kurtz, George. Gunlock maker, 1759, Lancaster (SED, 57).

Kurtz, Samuel. Gunsmith, 1837–43, Lancaster (SED, 57; t).

L

Lachland, Robert. Employed by the quartermaster general as a smith at Ft. Ligonier, 1758 (5 Pa. Arch. 1 at 350).

Lachler, Andrew. Artillery artificer, 1779, at the cannon factory at Philadelphia (2 Pa. Arch. 15 at 384).

Laflin. Powder makers. 1869, Smith & Laflin; 1871, Laflin & Rand; 1873–84, Laflin Powder Co. Had 2 magazines, each with a capacity of 5,000 kegs. Packer Twp., Carbon Co. (Brenckman, *History of Carbon County,* 751) Also at Philadelphia at 49 S. Front St. (BSED, 1880). In 1869, Laflin absorbed the Moosic Powder Works, Wyoming Co. (*Collection* [Wyoming Historical and Genealogical Society, 1909] 490).

Laib, Charles. 1848–55, Borough Twp., Beaver Co. (t; c); 1857, 1865–67, Madison, Wis. (FSP, 177).

Lake, Ira. 1832–38, Shamokin Twp., Northumberland Co. (t) Gunsmith.

Lake, Jesse, Gunsmith, Thompson Twp., Fulton Co. (PD, 1887, 1890).

Lake, Lewis. Gunsmith, Forsville, Sullivan Co. (c, 1860).

Lally, John. Gunsmith, 1811, German Twp., Fayette Co. (t).

Lama, Michael. 1861–71, gunsmith, Millheim, Centre Co. (GR, 11, 2/1961; BPaD, 1861; t; c).

Lambert, George. Gunsmith, 10 Green St., Philadelphia (DeSD, 1829, 1830, 1831).

Landis, Benjamin F. (b. 1800). In 1819, Jacob Roop (Rupp), Halifax Twp., Dauphin Co., advertised that his apprentice, Michael Landis, had run away. 1833, Water St., north of Main St., Cincinnati (DH, 175).

Lane, Christopher. Artillery artificer, 1780, U.S. Ordnance Yard, Philadelphia (2 Pa. Arch. 11 at 258).

Lane, William. Armorer, 1776–83, Lancaster Co. Revolutionary war contractor, Pa. State Gun Factory at French Creek (GAAM, p 112; 2 Pa. Arch. 15 at 384).

Lapkeehler, Henry. Gunsmith, Mifflinburg, Union Co. (c, 1850).

Lappington, William. Gunsmith, 1839, Lawrenceville, Tioga Co. (t).

Lard, David. Artillery artificer, 1778, U.S. Ordnance Yard, Philadelphia (2 Pa. Arch. 11 at 258).

Larosh, Jesse. Gunsmith, 1821, Macungie Twp., Lehigh Co. (t).

Larson, William. Gunsmith, 1820–35, Harrisburg (t; c).

Latsha, John. Silversmith, 1781, Reading Twp., York Co. (3 Pa. Arch. 21 at 417).

Lauck, Simon. 1783–85, Lebanon Twp., Lancaster Co. (SED, 57). There is a famous gunsmith from Winchester, Va. by this name; born 1760, died 1815. It is uncertain whether these two entries concern the same man (WSB, 155–58).

Laudenslayer, Henry (1839–1912). Gunsmith, Fair Oak, Snyder Co. (*Bulletin of the Snyder County Historical Society* 2).

Laufman, P. H. 1850–76, 28 Wood St., Pittsburgh. Sold gunlocks, gun barrels, and hardware (PAD, 1854–55, 1867, 1871–72; BPaD, 1861; c; t).

Lavelle, George (1745–94). 1779–86, cast cannon at McCulloch Iron Co., Springhill Twp., Fayette Co.; 1767–71, frontier ranger, Indian trader, and ran a commissary for Zack Morgan on the frontier (*Horn Papers,* 2:697–98).

Law, William (b. 1816). 1838–50, Shenango Twp., Lawrence Co. (t; c). Gunsmith.

Lawrence, John. Gunsmith, 465 N. 2d St., Philadelphia (PXD, 1810).

Lawrence, Richard (d. 1809). 1783–86, gunsmith, Antrim Twp., Franklin Co. (WSB, 29; t).

Lawrence, Thomas. Artillery artificer, 1779 (2 Pa. Arch. 15 at 383).

Lawrence, Thomas. Gunsmith, S. Queen St. near Middle St., Lancaster (BLD, 1857).

Lawrence, Walter. 1815–27, Williamsport, Lycoming Co., gunsmith (NAT, 214; t).

Lawser, John G. Gunsmith, 430½ Market St., Harrisburg (Patriot Directory, 1871, 1874–75, PD, 1882, 1887).

Lawser, W. H. 411½ N. Market St., Harrisburg (*Patriot Directory,* 1869); 131 Vine St. (*Patriot Directory,* 1874–75); DBA Lawser & Brother, 430 Market St. (PD, 1882;

Patriot Directory, 1887).

Layer, Adam. Swordsmith in the employ of Lewis Prahl, gunsmith, Northern Liberties, Philadelphia Co., 1777 (2 Pa. Arch. 3 at 104).

Leabo, Andrew (b. 1767). Born at Bremen, Germany. Arrived in Philadelphia on schooner *Betsky*, 1807, at age 40 (2 Pa. Arch. 17 at 660). Gunsmith, rear of 249 N. 2d St., Philadelphia (PXD, 1808).

Leach, Samuel Sheridan. 1886–1904, Everett, Bedford Co. (t; c) Patent 772,809 of 10/18/1904 for a single-trigger mechanism. He also modified bolt-action rifles to a straight-pull system (f).

Leacock, John. *Pennsylvania Packet*, 6/21/1753: "Goldsmith, removed to Walnut St., at the sign of the golden cup, Philadelphia."

Leaming & Co. Philadelphia, 1829–48, importers and parts suppliers (DeSD, 1829; MD, 1847).

Learn, G. Gunsmith and wagon maker, 1871, District 4, Green Twp., Indiana Co. (F. W. Beers, *Atlas of Indiana County*, 22).

Leary, Cornelius. Artillery artificer, 1778, Capt. Thomas Wylie's Co.; cannon maker at Carlisle (2 PIR, 2:249).

Leather, Jacob, Sr. On 9/16/1756 Martin Fry signed apprenticeship papers with Jacob Leather of York Co. to learn the art and mystery of gunsmithing. 1757, Jacob Leather, locksmith, York (t). In 1799, his eldest daughter was married in York (JKJ, 292).

Leather, Jacob, Jr. (1755–1835). Major gunsmith, government contractor. 1779–80, gunsmith, York (t). Probably active 1775–1820. Died 4/20/1835. Worked on government contracts for M-1795 and M-1808 muskets; he worked with Conrad Welshans on a contract dated 4/11/1798 (JKJ, 292).

Lechler, Henry. Prior to 1797, apprenticed in Lancaster. 1797–1843, Carlisle, Cumberland Co. Major gunsmith (JKJ, 472–73; WSB, 59–61).

Lechler, Henry, II. Gunsmith, 1829–55, 133 N. Front St., Philadelphia (DeSD, 1829, 1830, 1835–36, 1841; MD, 1846, 1847, 1855); Lafayette St. and Strawberry St., Lancaster (BLD, 1857).

Lechler, Jacob. Gunsmith, Lilly Alley, Philadelphia (DeSD, 1830, 1831).

Lechler (Leckler), John A. 1800, tin plater at Henry Lechler's, York St., Carlisle (WSB, 59); 35 Railroad at Yeager's, Philadelphia (MD, 1846). Gunsmith.

Leckner, John (d. 1835). 1751, John George Lechner naturalized in Lancaster Co. (2 Pa. Arch. 2 at 385) 1778,

Gunsmith Business.

HENRY LEEHLER,

from Lancaster, has commenced the GUN & RIFLE MAKING BUSINESS, *in the house of Mr William Blair, in the main street in the borough of Carlisle; he likewise makes new Gun and Rifle Locks and repairs old ones; as he has served a regular apprenticeship to the above branches, he hopes to give satisfaction to all who may employ him—And warrants all Rifles which are made by him, to be equal to any made in the state, and repairing done in the neatest manner. He wishes to purchase a quantity of seasoned curled Maple plank, of two and an half or three inches thick.*

Carlisle, May 1st, 1797.

Kline's Carlisle Gazette, May 18, 1797

gunsmith, York (t); 1782, Lancaster boro. (3 Pa. Arch. 17 at 759); 1802–5, Georges Twp., Fayette Co. (t). He then disappeared, but returned 1823–30. From 1832 to 1834 he was listed as a nonresident landowner (t).

Ledley, Gottlieb. Elizabeth Twp., Allegheny Co. (PD, 1887, 1890). Gunsmith.

Lee, James Paris. Prop., Lee Arms Co. Primarily worked in Bridgeport, Conn., but ca. 1880, Lee had a plant in Wilkes-Barre. Lee was a major inventor (RAG, 113).

Leech, Joseph. Artillery artificer, 1778, Capt. Nathaniel Irish's Co. at Carlisle (PIR, 2:247).

Lefevre, Isaac, Sr. Came to America about 1708; moved to Lancaster Co. before 1731. 1731–36, Lampeter Twp., Lancaster Co. (SED, 57) In 1712 a Maria Warenbur, widow, had two sons, Daniel Ffiere and Isaac Leffevre, and a patent for 2,000 acres in Philadelphia Co. dated 10/10/1712 (2 Pa. Arch. 19 at 529).

Lefevre, Philip. Gunsmith, 1756–70, Lampeter Twp., Lancaster Co. (SED, 57). Son of Isaac and Catherine Ferree.

Leffel, David. Gunsmith, Shanesville, Berks Co. (BPaD, 1861).

Lehman, Samuel. Gunlock maker, rear of Coates at 6th St., Philadelphia (MD, 1841).

Lehr, Joseph. Gunsmith, 1848, Allentown, Lehigh Co. (t).

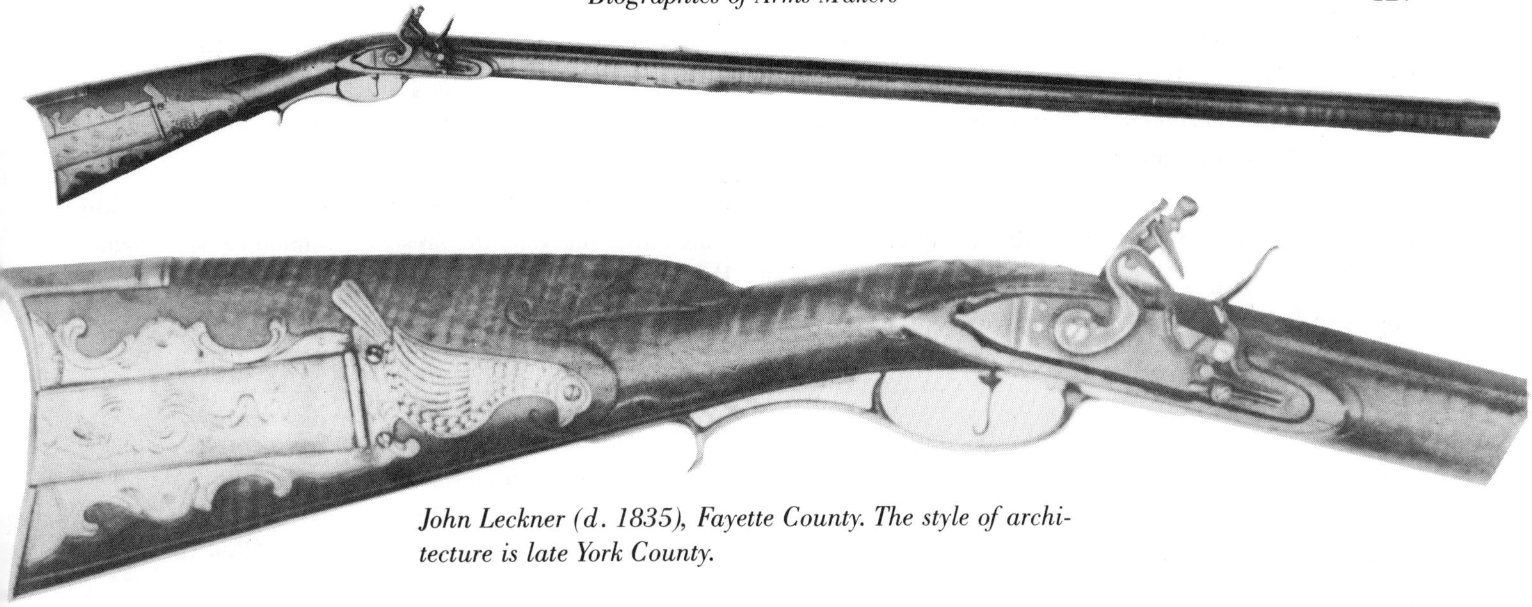

John Leckner (d. 1835), Fayette County. The style of architecture is late York County.

Leib, Alpheus. 1849–50, gun-barrel forger, Manheim Twp., Lancaster Co. (t; SED, 57).

Leib, John. 1842–50, gun-barrel forger, Manheim twp., Lancaster Co. (t; SED, 57).

Leibert, Henry (b. 1810). Montgomery Co. hatter, clock maker, and watchmaker. Killed experimenting with gunpowder (J. W. Gibbs, *Pennsylvania Clocks and Watches* [State College: Pennsylvania State University Press, 1984] 135–36).

Leinbach, Benjamin. Artillery artificer, 1777, Capt. Will's Co., Berks Co. Militia (BCR, 115).

Leinbach, Elias. 1843–46, Brecknock Twp., Lancaster Co., gunsmith. (t; SED, 57).

Leiper, Thomas. Cutler. 1824: "erected a blade mill on the site of Dr Robert Harris' powder mill on Crum Creek, Strath Haven, Delaware Co., a few miles above Chester. . . . In 1826 he made 200 dozen scythes and other items." After 1830 the building was used as a paper mill (Palmer, *A History of Delaware County* 1:134–35.).

Leister, Franklin and Jacob. Operators of the mechanized gunpowder mill on Perkiomen Creek, Montgomery Co., after 1810 (*Proceedings of the Pennsylvania German Society* 39, no. 34.).

Leitner (Lightner), Adam. 1779–1800, gunsmith, York. (t; JKJ, 322; KEA, 61) In 1783 he made guns valued at £253 and employed 1 man in his shop (*Gibson's History of York County,* 518). He advertised his services in the York *Reporter,* 5/2/1800.

Leitner, Ignatius. 1780–86, York, gunsmith. In 1783 he had 3 men employed in his shop, and his valuation was £143. (Gibson's *History of York County* [1886], 518.) In 1784–86, he worked on public arms (GAAAM, 91); 1/10/1787: "to Ignatius Lightener, for repairing public arms," York Co., £18/5/4 (3 Pa. Arch. 7 at 102).

Leitner, Jacob. 1779–83, York, gunsmith (t; KEA, 61).

Leman, Frank. Gunsmith, 588 Middle, Lancaster (BLD, 1873–74; LCD, 1871–72).

Leman, H. E. (1796–1835). Gunsmith, Lancaster (t).

Leman, Henry Eicholtz (1812–87). Walnut St. and Duke St., Lancaster; major manufactory at 23 E. James St. at Duke. (Prof. H. H. Beck, "Henry Leman, Riflemaker," *Papers of the Lancaster County Historical Society* 40, no. 3 [1936]) Apprenticed to Melchior Fordney, 1831–34; worked with George Tryon at Philadelphia, 1834 (FSP, 183).

Henry Leman (1812–87), Lancaster, Lancaster County.

Leman, Peter (1670?–1741). 1721–41, Lancaster, gunsmith (SED, 57; t).

Lennox, Andrew (1788–1842). Gunsmith, 1825–36, Pittsburgh (t); 1837–42, Fairview (Springdale), Allegheny Co. (t).

Leonard, Godfrey. Silversmith, 1780, York (3 Pa. Arch. 21 at 171).

Lepley, Adam (1774–1853). Smith, 1800, Rockland Twp., Somerset Co. (t); 1842–53, Southampton Twp., Somerset Co. (t; c). Gunsmith.

Lepley, Joseph. 1842–50, 1863–6, Southampton Twp., Somerset Co. (t; c).

Lepper, Lewis. Rifle maker, Alley below Chestnut St. and Walnut St., above Queen St., Lancaster (BLD, 1857).

Lerrett, Peter. Silversmith, 1782, Carlisle (3 Pa. Arch. 20 at 549).

Lesher, William L. Gunsmith, Clark St. at 3d St., Philadelphia (MD, 1846).

Lessley, Peter. Artillery artificer, 1779, Pa. Continental Line (2 Pa. Arch. 15 at 383).

Leu, Valerian (b. 1836). Born in Prussia. Gunsmith. 1860, naturalized in Philadelphia. 1875, Vallejo, Ca.; 1889, Watsonville, Ca (LSC, 160).

Levering, Joseph. 1781–84, Bethlehem (t); master of the gunshop at Christian's Spring set up by Andreas Albrecht (GAS-1, 173).

Lewes, Curtis. Smith, 1768, East Cain, Chester Co. (t).

Lewis, John. 1779, smith's shop, Nether Providence, Chester Co. (t).

Lewis, Joseph. Gunsmith, Pike Co. (c, 1870).

Lewis, Theophelus. Gunsmith, S. 8th St., between Chestnut St. and Walnut St., Philadelphia (*Stephen's Directory,* 1796).

Lewis, Dr. Webster. "until 1830 the flintlock gun was made; after that date the present invention came in vogue. Dr Lewis and Isaac Lloyd made the new patent here" (HYC, 632).

Ley, Frederick. Gunsmith, Rose St., Philadelphia (DeSD, 1829, 1830, 1831).

Libeau, Valentine. His naturalization papers, dated 1819, show that he came from Castle Hesse, Prussia; had worked in Russia; and came to Bristol, R.I. in 1816. 1819–20, Bedford boro. (t); 1825–57, Cincinnati, Ohio (DH, 125, 176).

Lichtenberger, Conrad. Smith, 1814, Somerset boro (t).

Lifred, John. Gun-barrel maker, Centre Twp., Union Co. (c, 1850).

Liggins, Thomas. Gunsmith, 6th St. at Greet St., Philadelphia (DeSD, 1835–36).

Lilly, David H. (b. 1826). Born in Va. 1848–49, Cumberland Twp., Greene Co. (t); 1850, Marion Co., W.V. (c). Gunsmith.

Lilly, James. 1845–50, German Twp., Fayette Co., gunsmith (t; c).

Lilly, John. Gunsmith, 1811, German Twp., Fayette Co. (t); 1815–17, Springhill Twp., Fayette Co. (t).

Lilly, John Patterson. Gunsmith, 1840–52, German Twp., Fayette Co. (t).

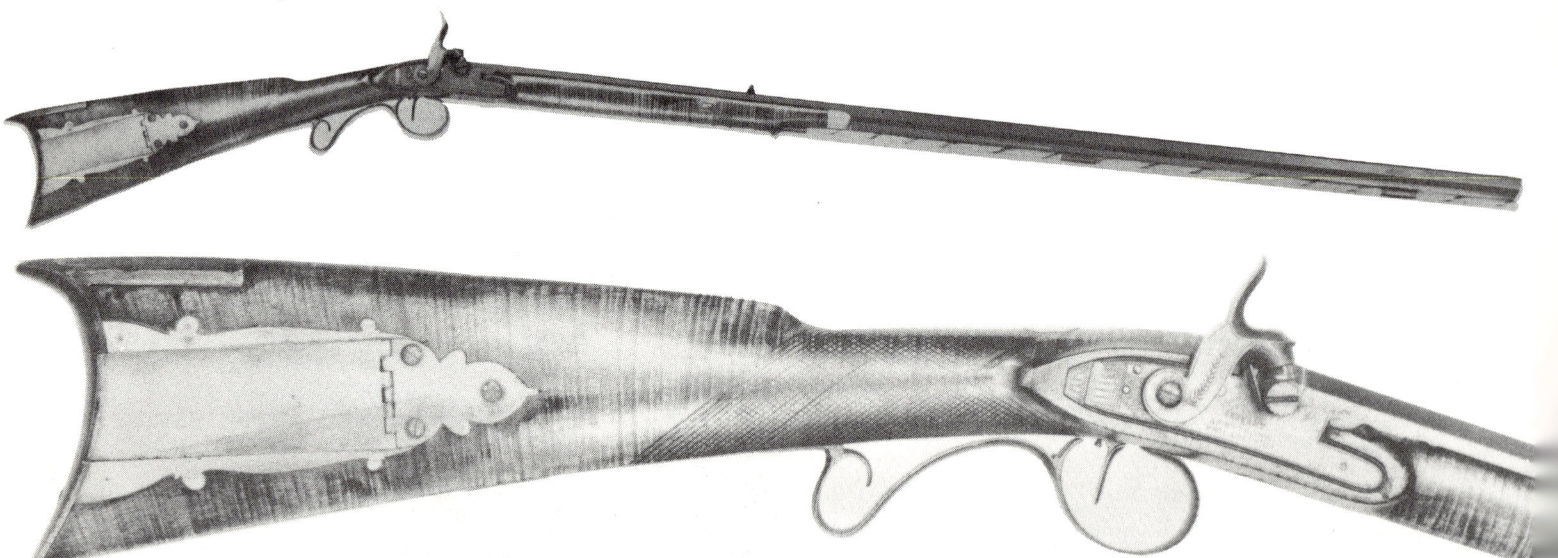

Valentine Libeau, ca. 1819, Bedford County. This gun was probably made in Libeau's active period, ca. 1825–57, in Cincinnati, Ohio.

Jonathan Dunmyer, Somerset County

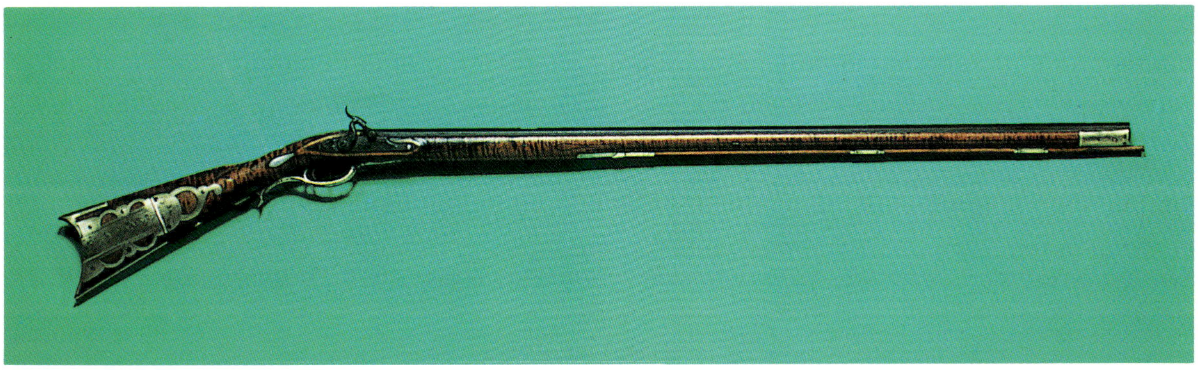

Daniel B. Border, Bedford County

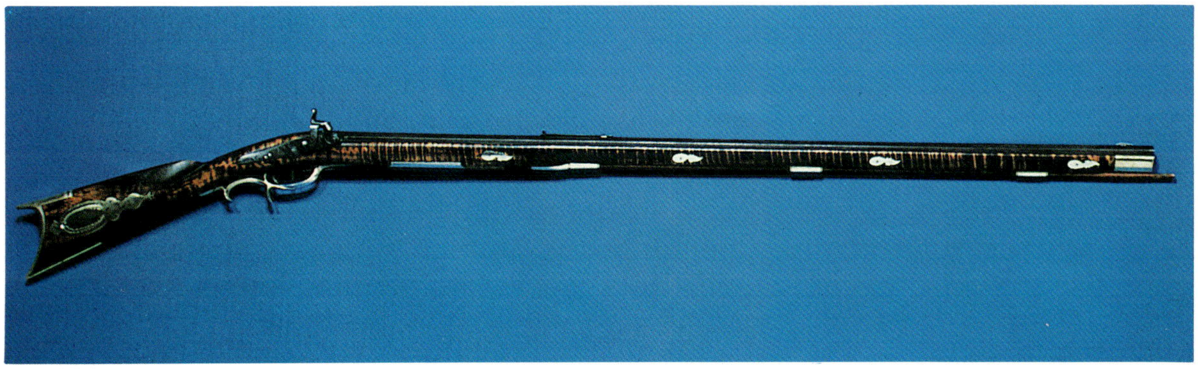

Elisha Pancoast, Westmoreland County

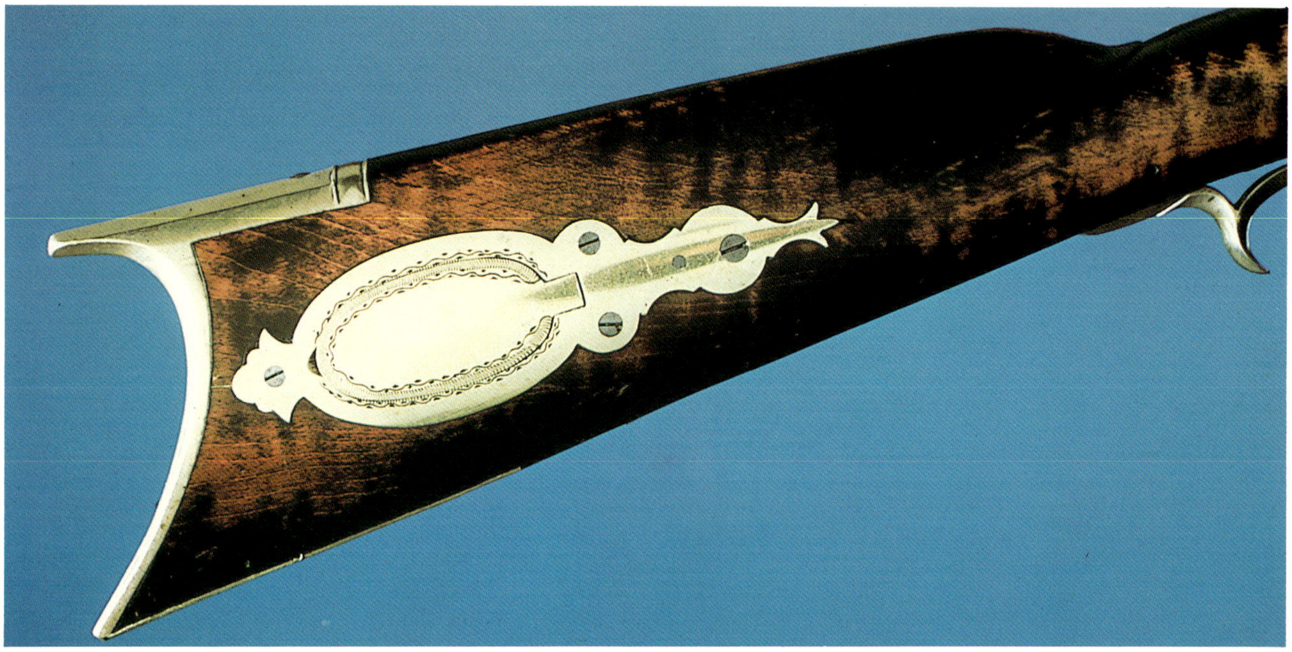

Elisha Pancoast, Westmoreland County

Adam Anstat, Berks County, with original powder horn and hunting pouch.

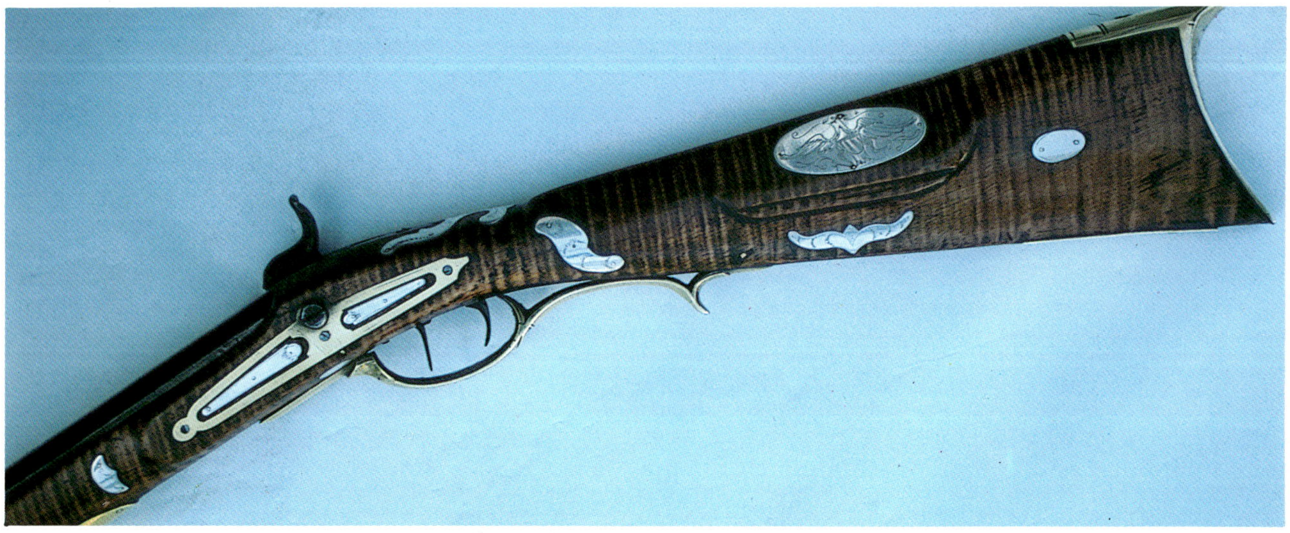

Samuel Mier, Somerset County

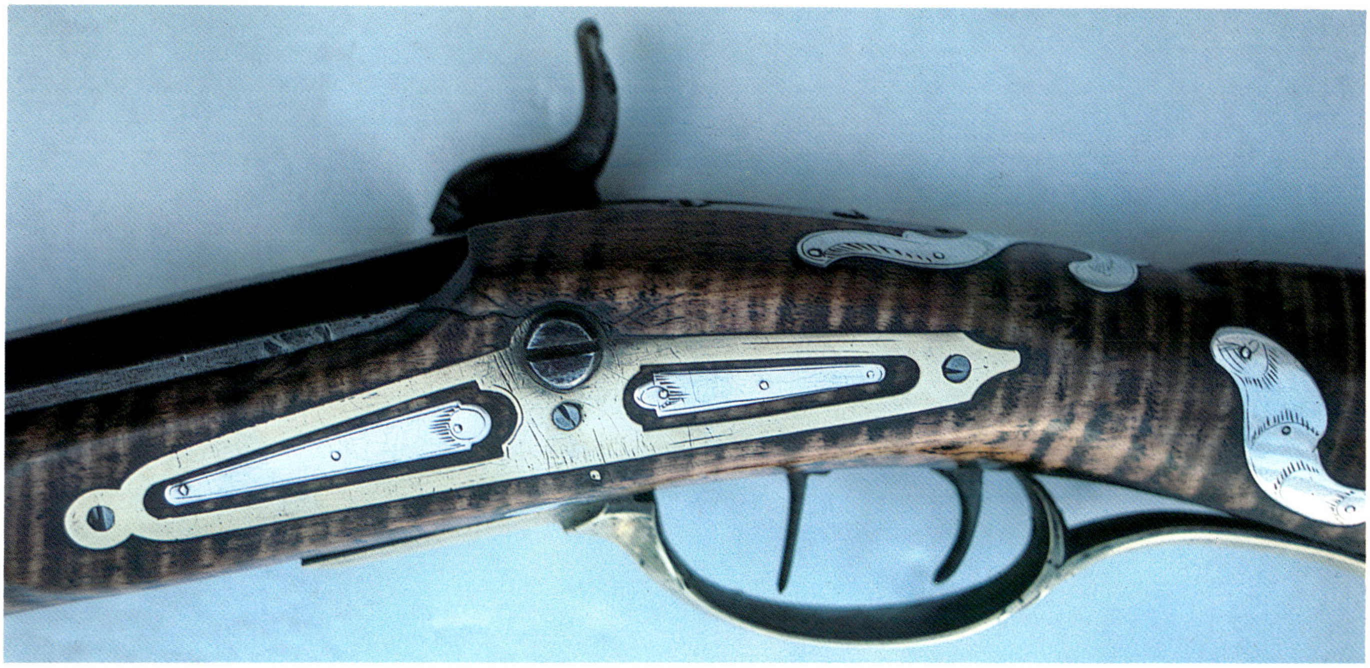

Samuel Mier, Somerset County

Samuel Mier, Somerset County

Benjamin F. Vore, Bedford County

Samuel Mier, Somerset County

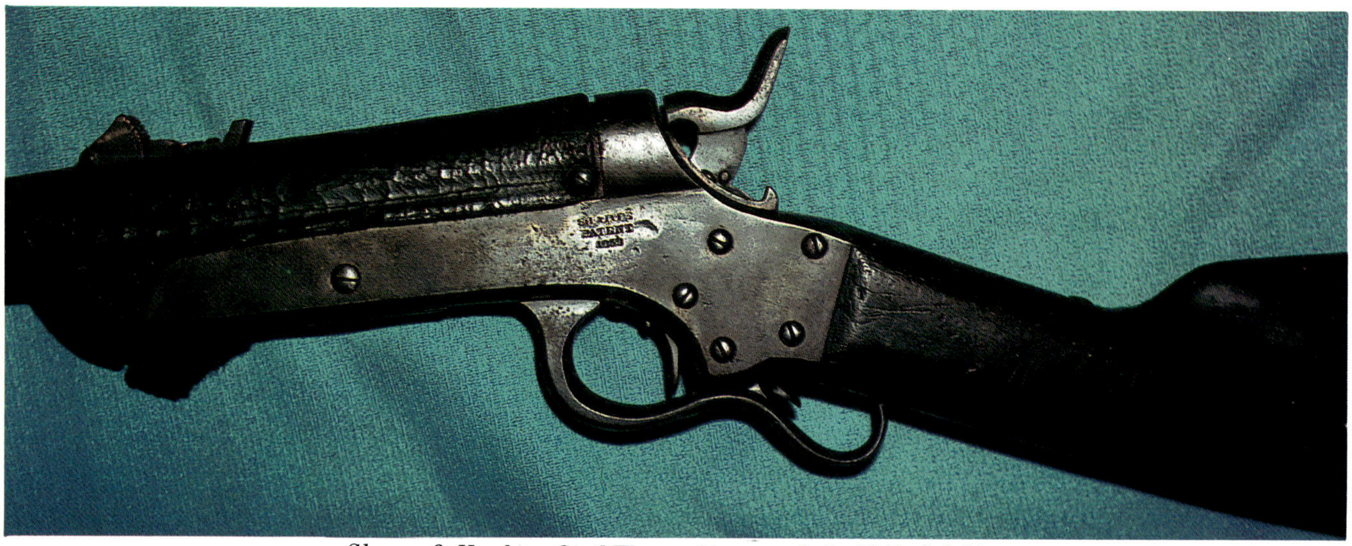

Sharps & Hankins Civil War carbine, made in Philadelphia

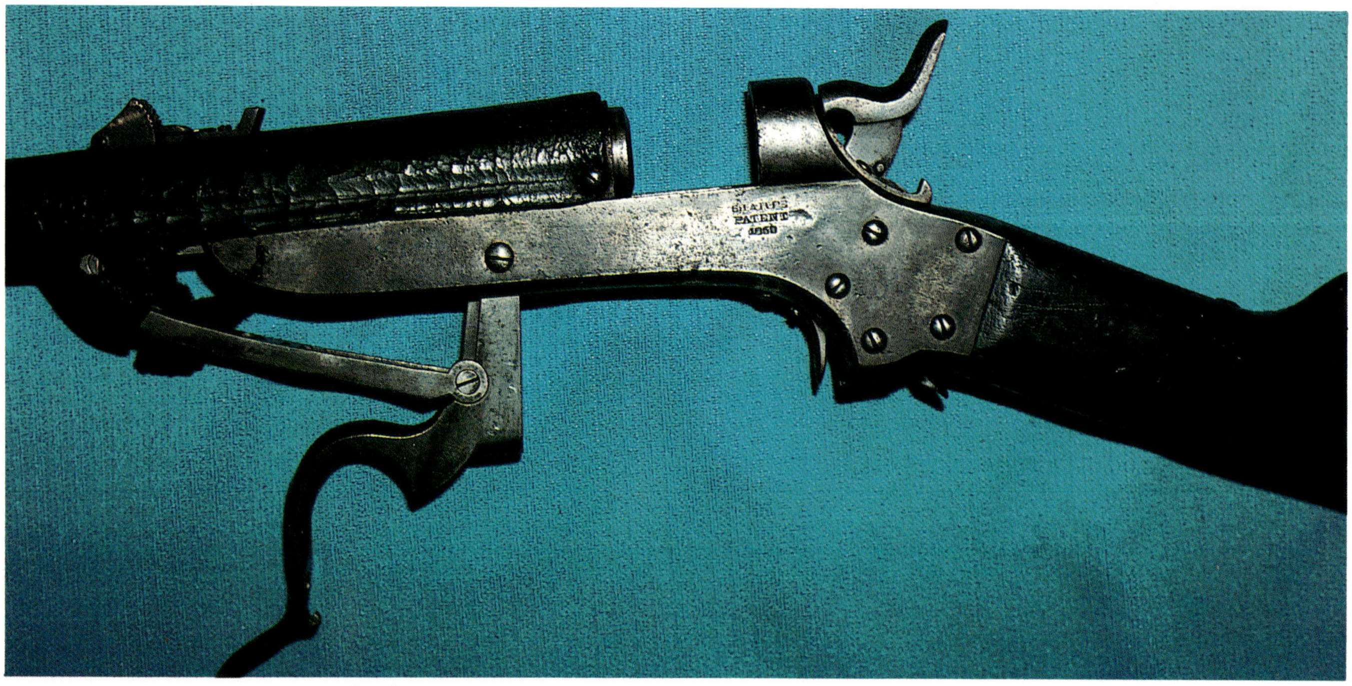

Sharps & Hankins Civil War carbine, made in Philadelphia

Lilly, Marzeeno C. (b. 1828). 1850, Menallen Twp., Fayette Co. (c); 1851, East Ward, Uniontown boro. (t); 1859–77, Elmwood, Peoria Co., Ill.; 1888, Macon Co., Ill. (research in Ill. by Curt Johnson; c; t).

Lilly, Robert A. (b. 1831). Gunsmith, Carmichael, Greene Co. (c, 1850).

Lilly, Thomas P. (b. 1818). Born in Va. 1846–52, Carmichaels, Greene Co. (t; c); 1859–60, Morgantown, W.Va. (t; c). Gunsmith.

Limerick, James. Gunsmith, 2 Coates Ct., Philadelphia (KKR, 290).

Limmervan, John. Gunsmith, 548 N. 3d St., Philadelphia (DeSD, 1823).

Lindley, John, Jr. 1811–35, gunsmith and blacksmith, Morris Twp. Washington Co. (t).

Lingard, Joseph. Gunlock maker, Sugar Alley, Philadelphia (MD, 1841).

Lingard, Samuel. Gunsmith, Philadelphia (c, 1850).

Linnard, Col. William. 1818, Philadelphia, patent for a breech-loading arm.

Lins, Adam Frederick. Manufacturer, 851 S. 2d St., Philadelphia, 1855–1895. Made and/or imported rifles, shotguns, and "Saturday Night Special" pistols (GAAAM, 92; PD, 1882; BPD, 1890, 1880).

Lins, Franz A. Gunsmith, 1853, Frederick & Franz Lins, partnership; 1855, 563 N. Front St., Philadelphia (MD, 1855); 116 Girard Ave. (BPD, 1861).

Lipp, Abraham. Gunsmith, Lancaster (c, 1850).

Little, Daniel. Barrel maker and gunsmith, 1848, Bellefonte, Centre Co. (t); 1852–60, J. & D. Little (PAD, 1852–53; 1859–60).

Little, James (b. 1809). 1830–34, Donegal Twp., Washington Co. (t); 1843–46, Franklin Co., Ohio (DH, 126); 1850–60, 9th Ward, Pittsburgh; 1827–30, Buffalo Twp., Washington Co. (t). See Little, J. & D., Rifle Works Pittsburgh.

Little, J. & D., Rifle Works. 1850, James Little, 4 Carson St., Pittsburgh (c); 1854, Allegheny St. and Penn St., Pittsburgh (PAD, 1852–53, 1859–60). Manufactured twist-rifle and shotgun barrels (KEA, 290).

Livegood, Jacob. Gunsmith, 1855–56, Lock Haven, Clinton Co. (t).

Livers, Lucas. 1824, apprentice gunsmith to Tower Adams. Adams was a Negro, so it is probable that Livers was also. Washington Twp., Washington Co. (t).

Livingston, James. Quartermaster, 1777, artillery artificers; killed at Brandywine (PIR, 2:245).

Llewellyn, Matthew. Gunsmith, 1792–1802, Lancaster boro. Gov. McKean gave him a contract for liberty cap muskets, 4/17/1801 (SED, 57).

Lloyd, Isaac. Lewisberry, York Co.: "until 1830 the flintlock was made; after that date the present invention came in vogue. Dr Lewis and Isaac Lloyd first made the new patent here" (HYC, 632).

Lloyd, James. Gunsmith, Minersville, Schuylkill Co. (c, 1850).

Lock, James (1790–1870). 1822–70, Wellsboro, Tioga Co. (t; BPaD, 1861). Gunsmith.

Lockhart, Josiah and Robert. "the county shall have their powder, being 5 quarter casks, at £15/cwt . . . and their lead at 45 shillings/cwt." (2 Pa. Arch. 13 at 293).

Lodge, Joseph. Gunsmith, 1855–65, Exchange, Montour Co. (BPaD, 1861; c, 1860; t).

Loesch, Jacob, Jr. 1780–81, Master gunsmith at the gunshop at Christian's Spring originally set up by Andreas Albrecht (GAS-1, 173).

Logan & Kennedy. Pittsburgh, 1831–47 (KKR, 291; t; c). Gunsmiths and parts suppliers.

Long, Daniel (1838–1823). Nephew and apprentice of William McCullough, 1845–1923, Brookville, Jefferson Co. (REH, 98). Gunsmith.

Long, George (1821–1915). New Berlin, Snyder Co. (t; c; f). Gunsmith.

Long, Jacob. Gunsmith, 1825, Penn Twp., Synder Co. (t)

Long, Jacob (b. 1808). 1835–67, Allegheny Twp., Westmoreland Co. (t; c). Gunsmith.

Long, James (1825–87). Beaver Springs, Snyder Co., gunsmith. (t; c; f).

Long, Joseph (1799–1872). 1820, White Deer Twp., Union Co. (t); 1826–1872, Beaver Springs, Snyder Co. (*Susquehanna and Juniata Valleys*, 2:1587; t; c; f). Major gunsmith.

Long, Peter (b. 1834). Gunsmith, 1870, 1880 (c) Greensburg boro.

Long, Ralph. Gunsmith, Beaver Twp., Union Co. (c, 1850).

Long, S. S. Gunsmith, Mill Creek Twp., Lebanon Co. (PD, 1882).

Long, William H. Gunsmith, Lewistown, Mifflin Co. (c, 1870).

Longer, John. 1833–38, Milton, Northumberland Co.(t) Gunsmith.

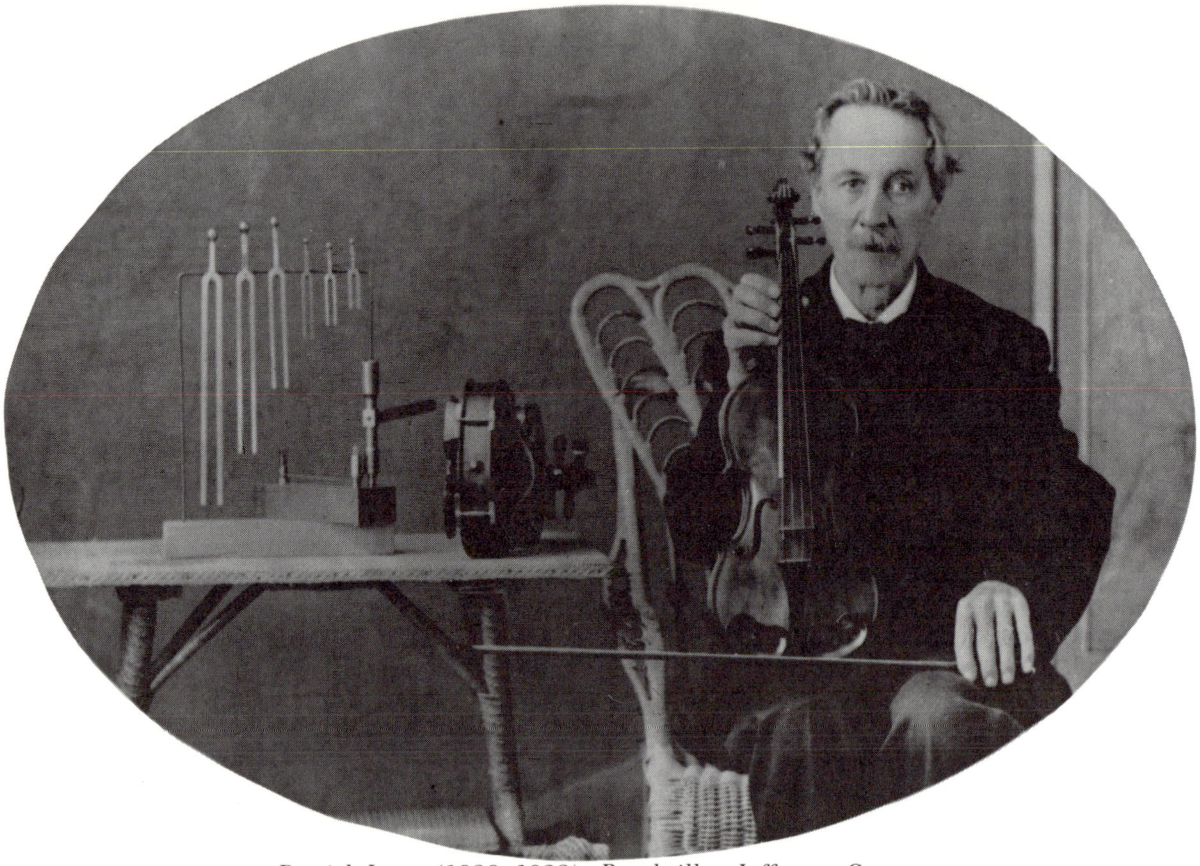

Daniel Long (1838–1923), Brookville, Jefferson County. Long made violins as well as guns.

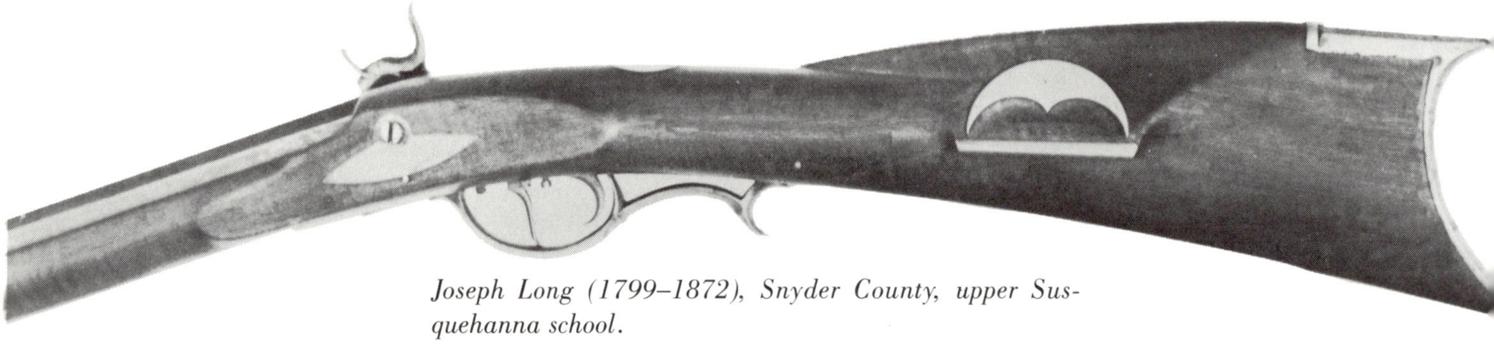

Joseph Long (1799–1872), Snyder County, upper Susquehanna school.

Lord, James. 1838–55, Orwigsburg, Schuylkill Co. (t; c) Patent 28,677 of 6/12/1860 for a firearm.

Lord, James, Jr. Tamaqua, Schuylkill Co.(PD, 1887, 1890). Gunsmith.

Losch, Georg. Powder maker, "Germantown, at the sign of the Conestoga Wagon, where Georg Losch, powdermaker, formerly had an inn" (from *Wochentlicher Pennsylvanischer Staatsbote*, 3/31/1779).

Losh, Jacob. 1777–78, powder maker for Continental army (2 Pa. Arch. 1 at 42; 3 Pa. Arch. 5 at 66). Was paid £188/1/4 for powder during March 1778.

Lotz, Peter. Gunsmith, 136 Middle, Lancaster (GCD, 1863–64; BLD, 1857).

Loudenslayer, Samuel. Gunsmith, Mexico, Juniata Co. (BPaD, 1861).

Lovell, James. Gunsmith, Green St. above 3rd St., Philadelphia (PXD, 1819).

Lovell, John P. (1820–97). 1840–73, gunsmith and small arms manufacturer, Philadelphia. By 1873 the firm, then J. P. Lovell & Sons, had been taken over by Iver Johnson. Made cheap pistols, rifles, and shotguns (RAG, 120).

Low, Nicholas. 1776, supplied 158 cutlasses to the Supreme Executive Council; was paid £75/5/0 (Col. Rec., vol. 11).

Lowe, William (b. 1755). Blacksmith, 1781, Philadelphia Militia; gray eyes, dark brown hair (2 Pa. Arch. 13 at 791).

Lower, John P. (1833–1915). Apprenticed to Joseph Grubb in Philadelphia. 1850–52, made Indian trade guns on government contract. He was active in Philadelphia until about 1870. 1872–1915, Denver, Colorado (RAG, 120).

Lowmaster, John. Major gunsmith. There is precious little information available on Lowmaster, although several extremely fine guns made by him are known. Single man, York tax list (KKR, 292); 1830, census, Fairfield Co., Ohio; died there in 1832 (DH, 113–14). Kindig showed several of his guns, and even in the second edition of JKJ, Dr. Shumway's annotation adds nothing to Kindig's original (which gives no solid facts) (KKR, 330–36).

Lownes, George. 1781–90, smith and cutler, Springfield, Chester Co. (t; RAG, 223).

Lownes, Hugh. 1781–90, smith and cutler, Springfield, Chester Co. (t; RAG, 223).

Lownes, Slater. 1781–90, smith and cutler, Springfield, Chester Co. (t; RAG, 223).

Loxley, Benjamin. Armorer, cannon maker, 1775–81, at Bustead's Furnace (1 Pa. Arch. 4 at 731–32; 4 Amer. Arch. 5 at 717; Col. Rec. 10:641).

DANIEL JOY TO COMMITTEE OF SAFETY, 1776

Gentlemen:

Agreeable to your instructions, I have prooved the two Brass cannon cast by Major Loxley, the first after a Scalleing charge, I tried it with four pounds of Powder, which burst it. The Second, after Scalleing, Loaded it with Three pounds, two Shot & three wadds, after which I gave it the full Proof of four pounds, Shot & wadds as before. Then gave it a third proof like the first, which it stood with out any other visible Damage than the blowing off apice of the Muzzle. Since which have got it sawed off square, Dressed the outside, fixed the Elivating Screw, mounted it on a traveling carriage & sent it to the State House yard. But as the carriag doth not Hold it I have described Mr Nevil to make one more Suitable. The Cylinder of the gun is Three feet five inches long, which is about ten and a half inches shorter than if cast perffect.

DANIEL JOY

(1 Pa. Arch. 5 at 121)

January 8th, 1777.

To the Honourable the Council of Safety:

GENTLEMEN:—I have lent Mr. Bias all the following Tools & Utensils belonging to the following Persons for your use, vizt:

My own:

One Tackle, fall & Blocks, one drawing knife, one square, two Cast Iron Plates under the Mortar Moulds, one Wooden Boreer & four knives, one round Tray, one Carpenter's work Bench & two Shooting Benches.

Hired of Benjamin Armitage's:

One Wood screw & Nut, one Iron Brace & one Drill for the Smith's Shops, one pair of Smith's Bellows & Tackle, one Vice & Block, one Beak Iron, one set Hammer & one Rivetting Tool.

Borrow'd of William Clifton's:

One wheel rope for Mr. Bursted's large Turning Lathe, two Poppet heads & Wedges, one Iron Wheel and Eight Screws, two Iron Roots, two Turning Tools and Three Handles.

Borrow'd of Charles Woolfall's:

One Anvil, one large Sledge & two Slitting Chizels.

Borrow'd of Nathaniel Brown's:

One small Sledge & one small Hammer for Smith's work.

Borrow'd of Morgan Burstead's:

One large wheel & Lathe for Turning, one Larg Iron Ladle & one Iron Kettle bound with wrought Iron.

And I have used of Morgan Burstead's Effects at the Furnaces & Forge, 6 Bushels of Isinglass Clay, one Cart Load of Moulding Loom, Nine Bushels of Lime & two Hundred Bricks.

By me,
Your Humble Servant,
BENJAMIN LOXLEY.

[2 Pa. Arch. 1 at 682]

Ludwig, Ann. Allentown, Lehigh Co., 1847. It is uncertain if the census entry meant to convey that *she* was a gunsmith or the widow of a gunsmith. In the census of 1840 she was listed as between 60 and 70 years of age.

Ludwig, Paul (d. 1846). 1827–46, Allentown, Lehigh Co. He was the husband of Ann Ludwig, above (t).

Lukens, Charles. York, Major, 1777–82, artillery artificer (2 Pa. Arch. 11 at 249).

Lukens, Isaiah (d. 1846). Air-gun maker, Philadelphia. He made the air rifle carried on the Lewis and Clark expedition (A. Halsey, AR 8 [1984]: 36–37, 80–82).

Lull, Orrin D. 1880, Tionesta, Forest Co. Brother and partner of Abraham. While in Watkins, N.Y., Lull patented a breech-loading firearm (38903 of 6/16/1863) and a form of minié ball (40761 of 12/1/1863). Their father, Charles, was a gunsmith in Ithaca, N.Y. (H. J. Swinney).

Lumas, L. Gunsmith, 6th St. near Green St., Philadelphia (DeSD, 1816).

Lund, Niels. Locksmith, 1761, arrived on the ship *Hope;* a Moravian, he went to Bethlehem (PaE, 52).

Lundy, Ebenezer. 1/9/1777; "pay Ebenezer Lundy £151/10 for 202 cartouche boxes and 200 bayonet belts" (Col. Rec. 11:82).

Lush, George. Powder maker, 1776–78. (Col. Rec., 10:382) On 2/7/1776, he "informed the Committee that he had rented 2 mill seats, the one on a stream called Stoney Run in Philadelphia Co., about 15 miles from town; the other on a stream called Mill Creek, in the possession of John Roberts, about 10 miles from town [Philadelphia] and that he will immediately begin to erect said mills. . . . He will endeavor to deliver 2400 pounds a week of good powder" (1 Pa. Arch. 4 at 709).

Lutz, C. & S. Gun-barrel makers, Reinholdsville, Lancaster Co. (BPaD, 1861) There was a Daniel Z. Lutz working with them (BPaD, 1861; t; c, 1860).

Lutz, George. 1842, gun-barrel maker, Cocalico Twp., Lancaster Co. (SED, 58; t).

Lutz, Jesse. Gun-barrel maker, 1850–61, Cocalico Twp., Lancaster Co. (t; c; BPaD, 1861).

Lydendecker, George. Gunsmith, 1770, Lehigh boro (t).

Lyne, John. 1751–1806, Conestoga Twp., Lancaster Co. (SED, 58). Gunsmith.

Lyons, Ignatius. Gunsmith, Philadelphia (c, 1850).

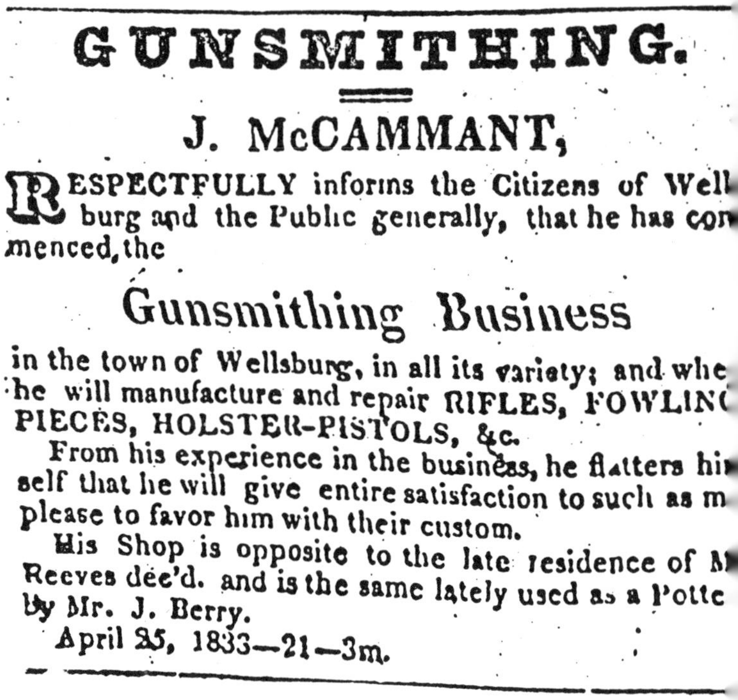

GUNSMITHING.

J. McCAMMANT,

RESPECTFULLY informs the Citizens of Well burg and the Public generally, that he has com menced, the

Gunsmithing Business

in the town of Wellsburg, in all its variety; and whe he will manufacture and repair RIFLES, FOWLIN PIECES, HOLSTER-PISTOLS, &c.

From his experience in the business, he flatters hi self that he will give entire satisfaction to such as m please to favor him with their custom.

His Shop is opposite to the late residence of M Reeves dec'd. and is the same lately used as a Potte by Mr. J. Berry.

April 25, 1833—21—3m.

Wellsburg, [West] Virginia, Gazette, April 25, 1833

Mc

McAlister, Coll. Gunsmith and whitesmith, SW corner, Redoubt Alley and 3d St., Pittsburgh. (PAD, 1815) Probably active in Pittsburgh, 1815–17; after 1818, Pickaway, Ohio (DH, 118, 316). Gunsmith.

McAvoy, Patrick. Artillery artificer, 1778, Capt. Nathaniel Irish's Co. (PIR 2:247).

McBride, John. Gunsmith, 1809, Mifflin Co. (t).

McCaffrey, Neal. Artillery artificer, 1778, Capt. Thomas Wylie's Co. at Carlisle (PIR, 2:248).

McCain, Hugh. Gunsmith, Versailles Twp., Allegheny Co. (t; c).

McCalla, John. Gunsmith, Walnut St. and 2d St., Philadelphia (DeSD, 1835–6).

McCallister, John. Gunsmith, 1764–65, Carlisle boro (WSB, 55).

McCaly, James. Gunsmith, 1824–25, South Huntingdon Twp., Westmoreland Co. (t); 1826–42, Monongahela Twp., Washington Co. (t).

McCamant (McCammant), James (b. 1777). 1797, Washington boro. (t); 1807–25, Wellsburg[h], Brooke Co., W.Va. (t; c); 1850, Perry Co., Ohio (c). Major gunsmith. His sons were also active in Wellsburg (t).

McCartney, William (1820–72). 1841–43, Uniontown boro. (t); 1846–72, Pittsburgh (t; c). Gunsmith.

McCauley, James. Gunsmith, 1824–25, South Huntingdon Twp., Wetmoreland Co. (t); 1825–42, Monongahela Twp., Washington Co. (t).

McClelland, Andrew. Gunsmith, 1809–13, German Twp., Fayette Co.; 1817–23, Greene Twp., Greene Co. (t); 1824–25, Monongahela Twp., Greene Co. (t).

McClelland, Asa. Gunsmith, 1798–1809, Waynesburg, Greene Co. (t); 1810–21, Franklin Twp., Greene Co. (t).

McClelland, David. Gunsmith, 1799, Waynesburg boro (t).

McClelland, Enos. Gunsmith, 1802–11, German Twp., Fayette Co. (t); 1812–13, Luzerne Twp., Fayette Co. (t).

McClelland, Ephraim. Gunsmith, 1828–29, German Twp., Fayette Co. (t).

McClelland, Hugh. Gunstock maker, Julian St. and Green St., Philadelphia (DeSD, 1829, 1830, 1831).

McClelland, James. Gunsmith, 1795–1816, German Twp., Fayette Co. (t).

McClelland, Jno. Smith, 1759, Ft. Ligonier (5 Pa. Arch. 1 at 350).

McClelland, John. Gunsmith, 1795–1800, German Twp., Fayette Co. (t).

McClelland, William (d. 1815). 1785–98, Uniontown, Fayette Co.; 1815, German Twp., Fayette Co. (t). He also ran an inn (t).

McClintock, Daniel. Cutler, 1779–81, Letterkenny Twp., Cumberland Co.; 1782–90, Guilford Twp., Fayette Co. (t). Made bayonets and edged weapons (3 Pa. Arch. 20 at 190, 326, 429).

McClintock, James. Sgt. in artillery artificers, 1778,

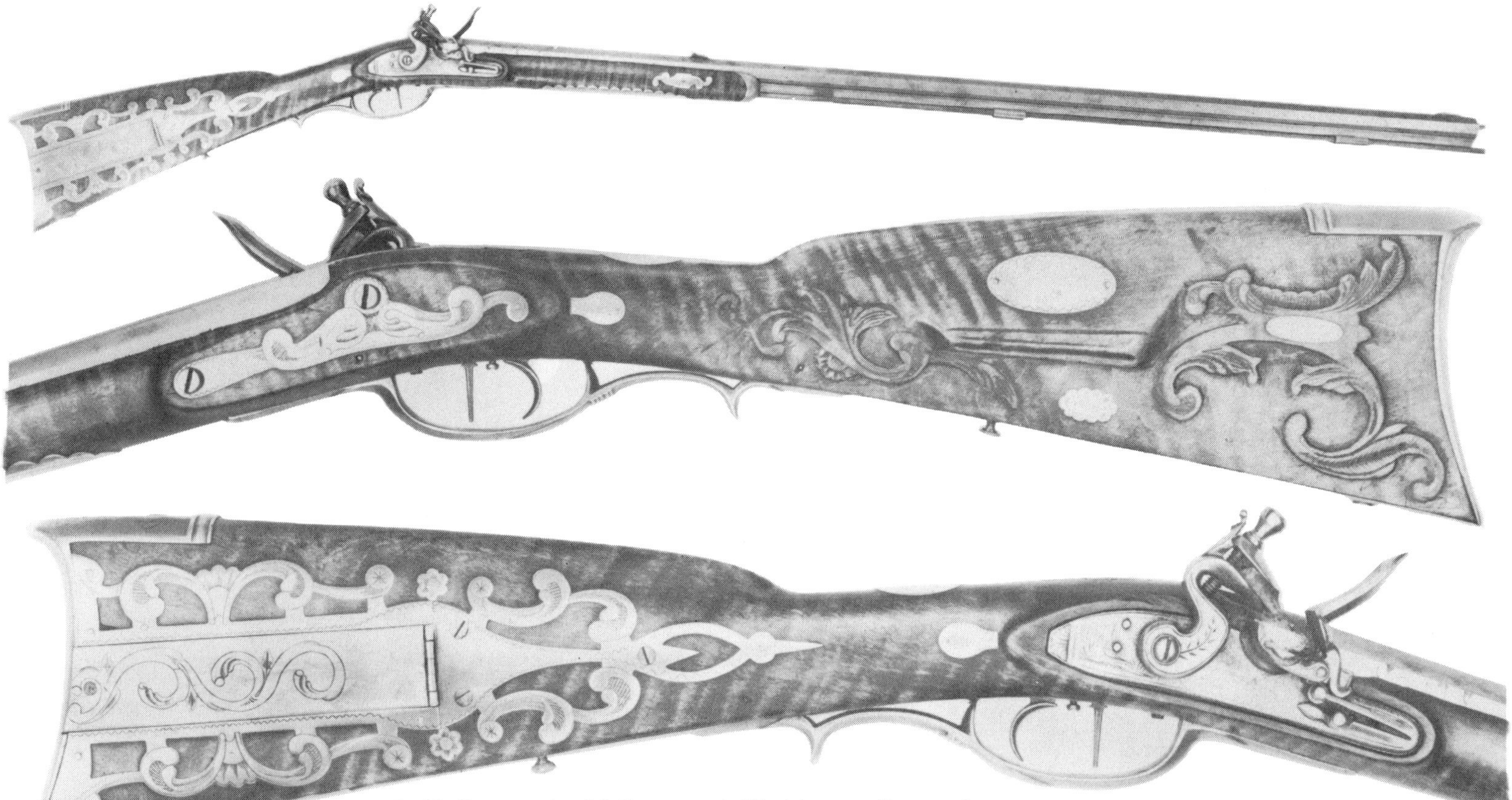

J. McCamant (or McCammant), Washington County, later of Brooke County, West Virginia.

Capt. Nathaniel Irish's Co. (PIR, 246).

McClure, William. 1822, gunsmith, West Hempfield Twp., Lancaster Co. (SED, 58; t).

McClurg & McKnight. NW corner of 5th between Wood St. and Smithfield St., Smithfield section of Pittsburgh. Air foundry, cannon makers. See Fort Pitt Cannon Co. (PAD, 1815).

McClurg, Alexander. Set up a foundry, air furnace, and, later, a cannon factory. "The iron foundry was established by Mr McClurg. . . . Joseph Smith and John Gormly were associated with Mr McClurg [who] took his son in partnership." The foundry was completed on 2/12/1806 (SHIS, 226). See Fort Pitt Cannon Co.

McClurg, Joseph. founder of McClurg Iron Foundry, later a cannon factory (see Fort Pitt Cannon Co.). In 1812 it was "converted . . . into a cannon factory and supplied the general government with cannon, howitzers, shells and balls. Commodore Perry's fleet on Lake Erie and General Jackson's army at New Orleans received their supplies of these articles in part from this foundry" (SHIS, 226).

McClurg, Joseph, Jr. A son of Joseph McClurg, he was active in the conversion of the iron foundry to a cannon factory.

McCook, Alexander. Artillery artificer, 1779, Pa. Continental Line (2 Pa. Arch. 15 at 384).

McCormick, James. 2/10/1777: "pay James McCormick, gunsmith, £2/11/0 for repairing guns." (Col. Rec., 10:) Leased Globe Mills, Philadelphia Co., to make 3,000 M-1795 muskets on contract at $13.40 each (RAG, 122).

McCormick, Robert. 1798–1801, Globe Mills, Northern Liberties, Philadelphia Co. He went bankrupt while trying to produce M-1795 muskets on contract and was sent to debtor's prison. His foreman, James Haslett, took over the contract and delivered 600 of the 3,000 muskets contracted (RAG, 122).

McCosh, Samuel (1796–1868). 1816–17, Peters Twp., Washington Co. (t); 1826–30, Pittsburgh (t); 1831–46, Nottingham Twp., Washington Co. (t); 1850–68, Collins Twp., Allegheny Co. (t; c). Gunsmith.

McCosh, Samuel, II (b. 1814). 1835–36, 1841–45, Nottingham Twp., Washington Co. (t); 1849–60, 22 Diamond St., Pittsburgh (t; c); 1862–90, Gastonville, Washington Co. (t; c). Gunsmith.

McCowing, William. Artillery artificer, 1778–80, Capt. Jesse Rowe's Co. Sent to the laboratory at Philadelphia Ordnance Yard (PIR, 2:244; 2 Pa. Arch. 11 at 252).

McCoy, Alexander. Gunsmith, 1779, Dock Ward, southern part, Philadelphia Co. (t).

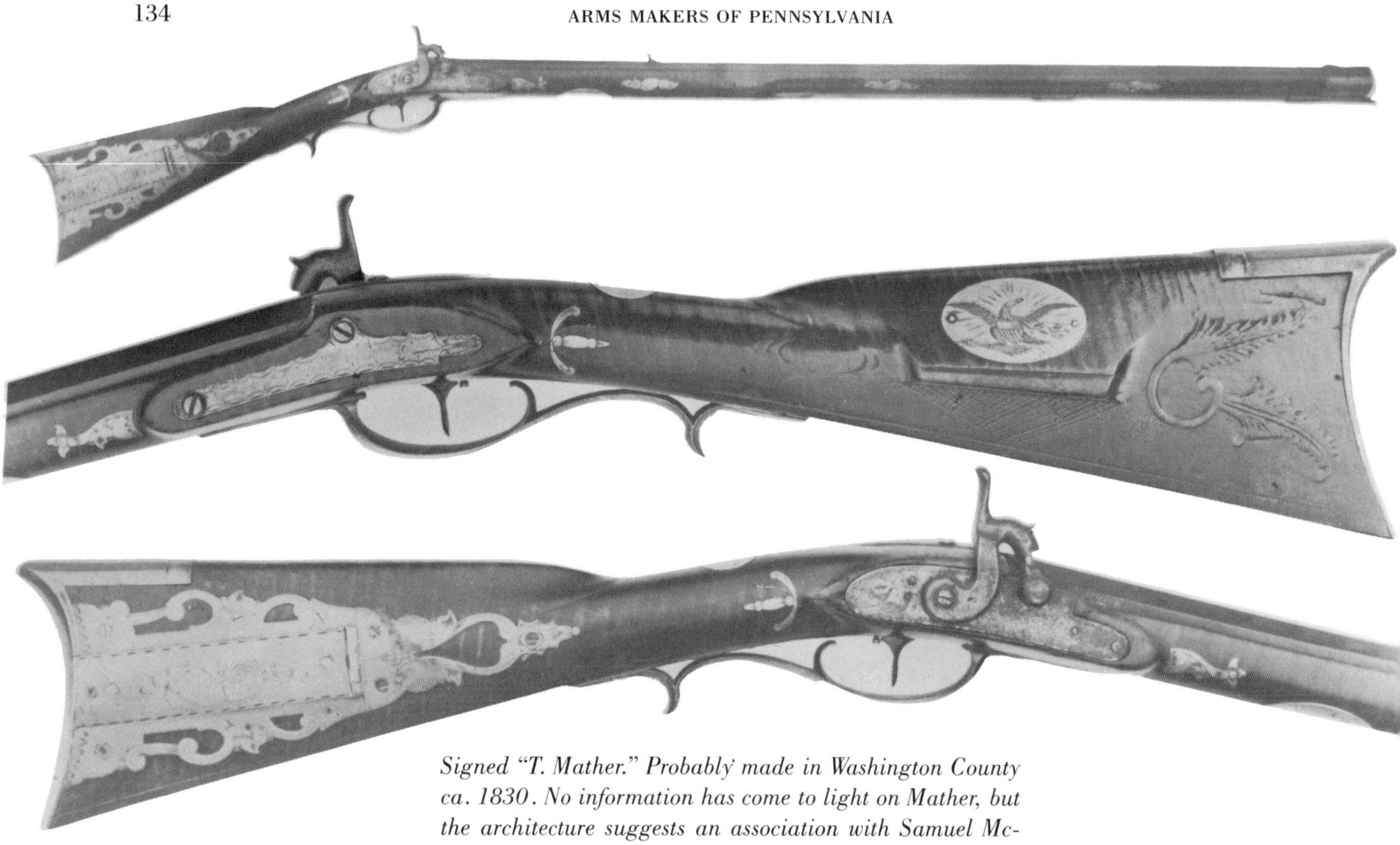

Signed "T. Mather." Probably made in Washington County ca. 1830. No information has come to light on Mather, but the architecture suggests an association with Samuel McCosh.

McCoy, George (b. 1820). 1785–90, Union Twp., Fayette Co. (t); 1813–20, Urbana, Ohio (DH, 51).

McCracken, William G. Gunsmith, Pittsburgh (c, 1850).

McCullough, William (1810–84). 1847–84, Brookville, Jefferson Co., except during the Civil War, when he was a member of Berdan's sharpshooters (REH, 40–48). Gunsmith.

McCune, William. Colerain Twp., Lancaster Co., 1826–46, gunsmith (t; SED, 58).

McDonald, D. J. (d. 1864). Gunsmith, 118 3d St., Pittsburgh. 1856–64, partner of William Craig in the Iron City Gun Works. Pistol maker (BPaD, 1861; t; c; PAD, 1860).

McDowell, Barr (b. 1822). 1843–60, Fairfield Twp., Westmoreland Co., gunsmith and innkeeper (t; c).

McEntosh, John. Goldsmith, 1759, Capt. William Bile's Co.; "age 45, born in Scotland, thin face, light hair" (5 Pa. Arch. 1 at 352).

McEwen, John (b. 1794). 1813, Pitt Twp., Allegheny Co.; runaway gunsmith's apprentice, working with Samuel

William McCullough (1810–84), Brookville, Jefferson County.

Galbraith (*Commonwealth*, 2/17/1813; KKR, 235).

McFadden, Charles (b. 1805). 1827–40, Fayette Co. (t); 1827–32, 1834, 1840, Bullskin Twp., Fayette Co. (t); by 1860, Portage, Wood Co., Ohio (DH, 405).

McFadden, J. B. Waynesburg *Messenger*, 5/27/1830: "Military Goods by J. B. Mcfadden, No. 83 Market St., Pittsburgh, announces that he has just received an assortment of military articles, consisting of swords, sashes, wings, epauletts, shoulder knotts, pompoons, plates, eagles, buttons &c., all of which will be offered at reduced prices."

McFarlin, David. Gunsmith, 4 Pa. Ave., Philadelphia (DeSD, 1831).

McGawghran, Patrick. Gunsmith, Troy, Bradford Co. (BPaD, 1861).

McGill, James. Artillery artificer, 1779, at the laboratory, Philadelphia (2 Pa. Arch. 15 at 386).

McGinley, John. 1779, supervised a blacksmith shop, made artillery, at Philadelphia (2 Pa. Arch. 15 at 384).

McGinnis, Lawrence. Artillery artificer, 1778, Capt. John Jordan's Co. at Carlisle (2 Pa. Arch. 11 at 260).

McGunnegle, George (1735–1820). Pittsburgh. Advertisements from the Pittsburgh *Gazette:* "The subscriber respectfully informs the public that he has removed his shop to the house formerly occupied by Mr Marmaduke Curtis, in Market St., which he carries on the black and white smith business. He has furnished himself with a very good horseshoer and country smiths; and also likewise makes locks, keys, hinges of all sorts, pipe tomahawks, scalping knives, boxes and pimns for vizes, grates, polished and unpolished and irons, shovels, tongs, pokers, chafing dishes, bread, toasters, ladles, skimmers, flesh forks & skewers, with all kinds of iron work for a kitchen, currying combs, plates, saddle trees, makes crapeing, curbing & pincing tongs, rupture belts, grinds swords, razors, scissors, and pen knives, cleans and polishes guns & pistols [8/1/1789]"; "He will give encouragement of a journeyman blacksmith & also wants an apprentice to said business [7/17/1796]"; "Runaway from George M'Gunnegle, James Nary, blacksmith apprentice, 6¢ reward, 17 years, red hair, sole clothes [5/12/1812]."

McKahan, John D. (1824–64). 1850–61, Washington boro., gunsmith. He had a long and distinguished Civil War record, and died of wounds received at the Battle of Peach Tree, 7/20/1864 (t; f; c). His grandson was the actor Alan Hale.

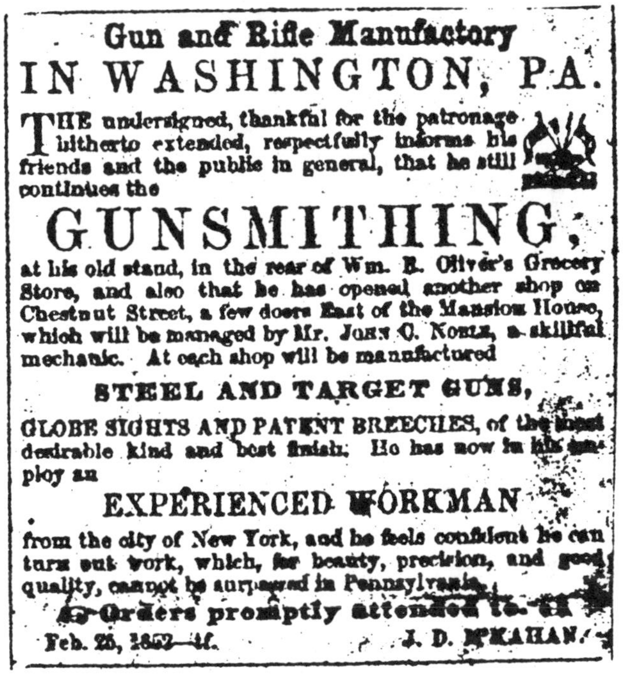

Gun and Rifle Manufactory

IN WASHINGTON, PA.

THE undersigned, thankful for the patronage hitherto extended, respectfully informs his friends and the public in general, that he still continues the

GUNSMITHING,

at his old stand, in the rear of Wm. B. Oliver's Grocery Store, and also that he has opened another shop on Chestnut Street, a few doors East of the Mansion House, which will be managed by Mr. John C. Noble, a skillful mechanic. At each shop will be manufactured

STEEL AND TARGET GUNS,

GLOBE SIGHTS AND PATENT BREECHES, of the most desirable kind and best finish. He has now in his employ an

EXPERIENCED WORKMAN

from the city of New York, and he feels confident he can turn out work, which, for beauty, precision, and good quality, cannot be surpassed in Pennsylvania.

Orders promptly attended to.

Feb. 25, 1852—4f. J. D. McKAHAN.

Washington Reporter, February 25, 1852

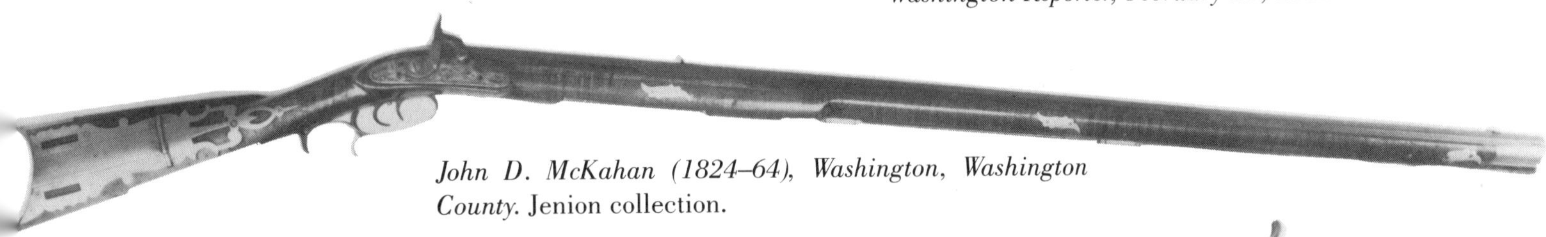

John D. McKahan (1824–64), Washington, Washington County. Jenion collection.

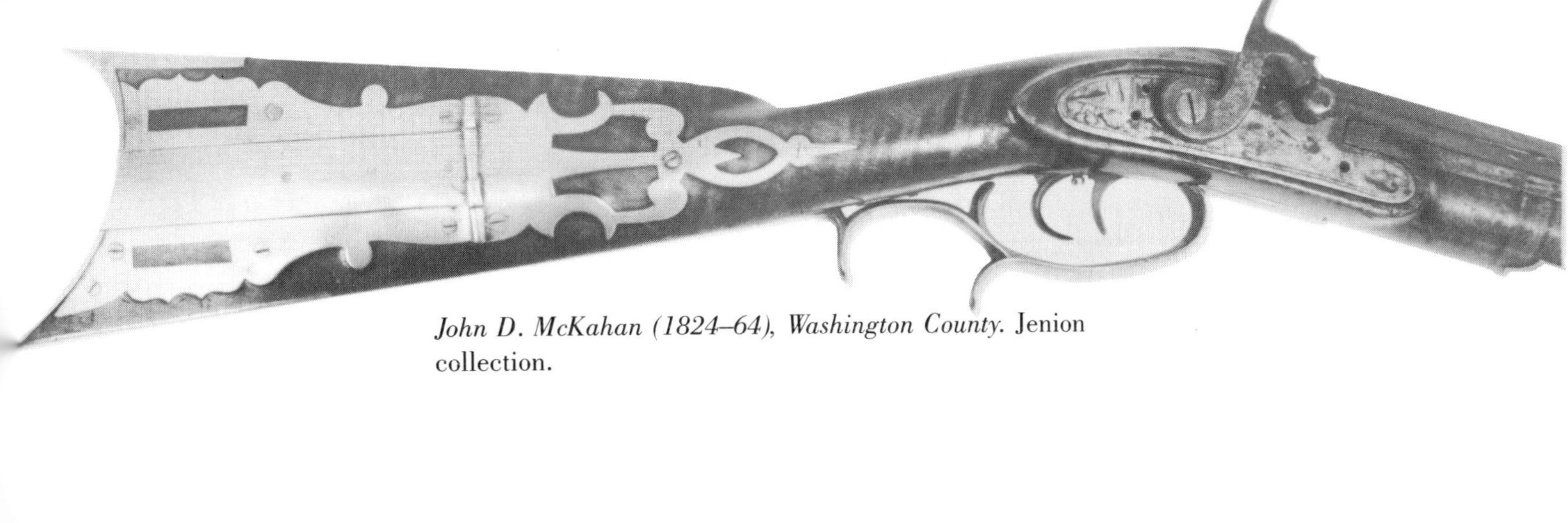

John D. McKahan (1824–64), Washington County. Jenion collection.

John D. McKahan (1824–64) and wife, Washington boro. McKahan was killed in Union service in the Civil War. He was the grandfather of actor Alan Hale, Sr.

would then have been part of "West Agusta" Virginia.) He served in the early part of the Revolutionary War, the first two years in the Pennsylvania militia. He was a locksmith and a gunsmith by trade, and, as such, his services were in great demand. About 1778 while he was burning a coal pit near his home, he was captured by the Indians. He left his coal pit to go past some brush and timber to his potato patch to gather potatoes. He gathered the potatoes and was returning to his coal pit when a party of 15 Indians secreted behind a fallen tree top took him prisoner. They took him down the Ohio River to the mouth of the Wabash River, and up that same river for some distance. They kept him with them about a year, and then took him to Detroit, and sold him to the British. He was kept a prisoner about Detroit some time, and one day some man wanted a gun lock repaired. There was a person at Detroit who pretended to do that work, but was not able to do it well. McKinney saw him at work and offered to help. It was then discovered that McKinney was a gunsmith; and the British then required him to make guns for the Indians, their allies in fighting the Americans. They gave him 2 pounds a day to make gun barrels and to finish guns; but he was shut in prison every night. He made the guns, but he spoiled every barrel so that they could not be relied upon to shoot his countrymen. It is supposed that he made about 250 guns and spoiled them all but one. There was one Indian who knew what a good gun was, and he got onto McKinney's scheme; he told him that he would not betray him if he would make him a perfect gun, which McKinney did. It is said that the Indians used one of McKinney's guns and shot 17 times at General Washington, but did not hit him once. McKinney remained a prisoner at Detroit until about 1783, when he was released. He went back to Pennsylvania and was married to Millie Doutheet." In 1788 Daniel McKinney, blacksmith, and his wife, Parmelia, sold land in Washington County. It is believed that Daniel was again captured by Indians in 1790, and this time his wife believing (or hearing reported) that he was dead, left, went to Kentucky, and remarried (actually bigamy). Parmelia died in Kentucky in 1834. The revolutionary war records merely listed him as an "artificer" (Nelson W. Evans, *A History of Scioto County* [Portsmouth, Ohio: Evans, 1903] 130–31).

McKean, Hugh (1763–1850). 1788–1801, Versailles Twp., Allegheny Co. (t); 1802–50, Cool Spring Twp., Mercer Co. (t; c). Gunsmith.

McKean, James (1799–1851). 1802–51, Franklin Twp., Westmoreland Co. (t) His estate inventory included: gun lumber 7¢; bellows & tools $2.75; 3 gunstocks 5¢; 1 pistol $1; 1 pistol 40¢; 1 small cannon 40¢ screwplate 50¢; 3 boxes gunsmithing tools $2; boreing apparatus $10 1 rifle gun $4.50; 1 rifle gun $12; 1 shotgun $5.10 (WB 3, 313).

McKean, James B. (b. 1832). Mercer boro., gunsmith (BPaD, 1861; c, 1860).

McKean, William (d. 1846), 1839–46), Franklin Twp., Westmoreland Co. (t; c; WB 3, 170). Gunsmith.

McKenzie, Duncan. Philadelphia, gunsmith. (Advertisements in Pa. *Gazette* of 12/28/1731 and 1/4/1732; KKR, 295; KEA, 64).

McKibben, William. Union Twp., Fulton Co. Patent 39,941 of 9/15/1863 for a gunsight.

McKim, Robert. 4/15/1777: "pay Robert McKim £100 for 25 stands of arms" (2 Pa. Arch. 1 at 32).

McKinney, Andrew. Artillery artificer, 1779 (2 Pa. Arch. 11 at 236); 1782, stationed at Ft. Pitt (2 Pa. Arch. 11 at 261).

McKinney, Barney. Artillery artificer, 1779 (2 Pa. Arch. 15 at 384).

McKinney, Daniel. "Daniel McKinney, Sr., was born in Washington County, Pennsylvania, in the year 1744. (It

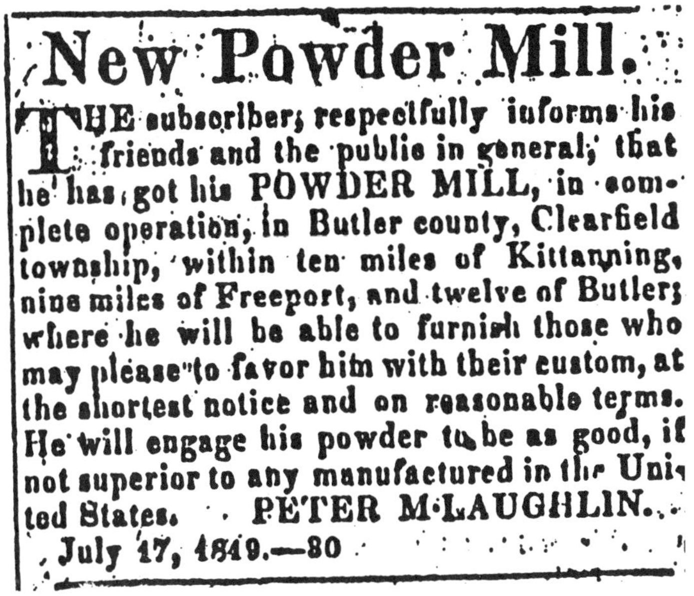

New Powder Mill.

THE subscriber, respectfully informs his friends and the public in general, that he has got his POWDER MILL, in complete operation, in Butler county, Clearfield township, within ten miles of Kittanning, nine miles of Freeport, and twelve of Butler; where he will be able to furnish those who may please to favor him with their custom, at the shortest notice and on reasonable terms. He will engage his powder to be as good, if not superior to any manufactured in the United States. PETER M'LAUGHLIN.

July 17, 1849.—80

Kittanning Gazette, July 17, 1849

McKinney, William. Artillery artificer, 1799. (2 Pa. Arch. 15 at 384).

McKnight, William. Edged-tool maker and swordmaker, south side, 4th between Wood St. and Smithfield St., Pittsburgh (PAD, 1815).

McLaughlin, Peter. "new powder mill," Clearfield Twp., Butler Co. (Westmoreland *Republican*, 7/17/1819 and 7/31/1819).

McLean, John. Gunsmith, 1821, West Pennsboro Twp., Cumberland Co. (t).

McLine, James. 1776, Cumberland Co. The Committee of Safety paid him £152/19/3 for 20 rifles and 43 muskets for the militia (1 Pa. Arch. 10 at 610).

McMahon, Samuel. Chester Co., 1783: "paid Samuel McMahon for a gun delivered in 1776 and for 3 muskets" (3 Pa. Arch. 6 at 188).

McNeely, William. Gunsmith, 1815, Union Twp., Fayette Co. (t).

McNitt (McNutt), Robert. Gunpowder maker, 1802–6, Beaver Twp., Beaver Co. (t).

McQuaid, James R. (1818–97). 1818–44, Clearfield Co.; 1844–97, Fulton Co. Ill. (t; c; research by Curtis Johnson) In 1850 he had $825 invested, bought 220 pounds of iron, had a payroll of $100 a month, and made 60 guns valued at $1,000 (IC).

McSwain, Hugh. Artillery artificer, 1779, at the laboratory at Philadelphia (2 Pa. Arch. 15 at 386).

M

Magrum, David. r118 Dillyn Ct., Philadelphia (DeSD, 1823, 1824). Gunsmith.

Mahaffy, James (b. 1794). In 1808 Mahaffy ran away from Henry Wolf, a Pittsburgh gunsmith. Wolf offered $2 for his return (KKR, 364).

Mahla, Philip. Gunsmith, 24 Diamond St., Pittsburgh (PAD, 1856–57).

Maine, George. Gunsmith, Perry Co. (c, 1850).

Mallison (Mallisen), Jacob. 1800–1802, Buffalo Twp., Allegheny Co. (c; t). Gunsmith.

Man, Robert. Swordsmith, 1777, in the employ of Lewis Prahl, Northern Liberties, Philadelphia Co. (2 Pa. Arch. 3 at 104).

Manger, Michael. Gunsmith, Honesdale, Wayne Co. (BPaD, 1861).

Mann, George. Gunsmith, and whitesmith, 1818–39, Pittsburgh (t).

Mann, William, Jr. 1835, ran the gun-barrel factory formerly run by George James and Jonas Spangler, a small, stone building in the narrows of Jack's Mountain on Kishacoquilla's Creek, Brown and Derby Twps., Mifflin Co. (t).

Manner (Miner, Mener), Henry. 1795–1800, Morris Twp., Huntingdon Co. (t) Gunsmith.

Manning, Joseph S. Gunsmith, Philadelphia (MD, 1857; GPD, 1869).

Markel, Samuel (1810–88). 1850, Venango Twp., Butler Co. (c); 1860–88, Allegheny Twp., Butler Co. (c). Gunsmith and blacksmith.

Markley, O. Gunsmith, Penn Twp., Lancaster Co. (PD, 1887).

Mars, William. Gunsmith, 1817, Pittsburgh (t).

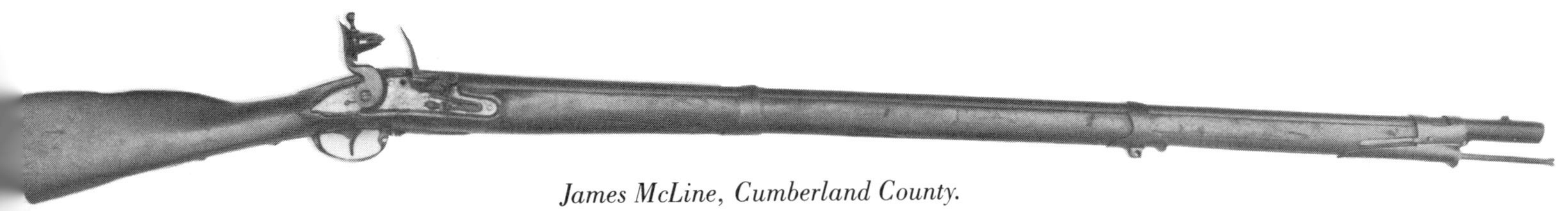

James McLine, Cumberland County.

Marsh & Stone. Cutlers, 1823, Fallston, Beaver Co. From 1824 to 1836 the firm was operated by William Blanchard (J. F. Richard, *History of Beaver County* [Chicago: Warner, 1888] 473).

Marshall, George. Artillery artificer, Capt. Jesse Rowe's Co. (PIR, 2:245).

Marshall, James B. 28 Wayne St., Pittsburgh (PAD, 1860–61); 249 Spring Alley (PAD, 1864–65). Gunsmith.

Marshall, Simeon. Philadelphia. Patents: 24,372 of (6/15/1859) for a self-priming gunlock, and 14,850 of (5/13/1856) for a cartridge.

Marshall, William. 8/23/1788: "an order was drawn . . . in favor of . . . William Marshall for £50/19/7 payable out of the militia fines of the county of Washington, in full . . . for repairing the public arms belonging to said county" (Creigh, *History of Washington County,* 75).

Marston (Marsdon), David. 1808, 106 8th St., Philadelphia; 1810, 111 Brown; 1833, 179 N. 4th St. (PXD, 1808, 1810; DeSD, 1833; KEA, 64).

Marston, Frank W. Philadelphia. Patents: 198,745 of 1/1/1878 and 199,717 of 1/29/1878 for a gun system.

Marston, John. 179 N. 4th St., Philadelphia (DeSD, 1829, 1830, 1831). Gunsmith.

Martic Forge. Pequa Creek, near Colemanville, Lancaster Co. "During the Revolution, round iron was drawn under the hammer at the forge and bored out for musket barrels at a boring mill, on a small stream, doubtless with a view to prevent discovery by the enemy . . . Negro slaves were employed from the beginning in hammering iron at the forge and it is a curious fact that Negroes continued to be the principal workmen down to the abandonment of active operations in 1883. . . . A long row of stone houses was occupied by the Negro workmen (SHIS, 183).

Martin & Smith. Ca. 1830–98, made cheap pistols and guns, Philadelphia (RAG, 126).

Martin, David. 1778, artillery artificer, U.S. Ordnance Yard, Philadelphia, blacksmith (2 Pa. Arch. 11 at 258).

Martin, David. Gunsmith, 1850, Brecknock Twp., Lancaster Co. (c).

Martin, Henry. Gunsmith, 1802–6, Georges Twp., Fayette Co. (t).

Martin, James. Gunsmith, 20 Duke St., Philadelphia (MD, 1837).

Martin, Joseph. Gunsmith, rear of 179 Coates, Philadelphia (MD, 1841).

Martin, Michael. Gunsmith, 1811–50, Brecknock Twp., Lancaster Co. (t).

WANTED,

A Journeymen Gun Smith, of steady habits, will meet with employ, and the highest wages, by applying to Thompson Martin, Elizabeth, Allegheny Co., Pa. None except a first rate workman need apply.

THOMPSON MARTIN.

june 10—dlw

Pittsburgh Gazette, June 10, 1839

Martin, Samuel. Gunsmith, Brecknock Twp., Lancaster Co., 1850 (SED, 58).

Martin, Thompson (1806–80). 1839–80, Elizabeth Twp., Allegheny Co. (t; c; BPaD, 1861; KKR, 296) Advertised in Pittsburgh *Gazette* for a "first rate workman," 6/10/1839.

Mary Ann Furnace (active 1762–1800). Owned by George Ross of Lancaster and Mark Bird of Philadelphia. "During the Revolution, it cast many shot and shells" (SHIS, 184). Ross was a signer of the Declaration of Independence.

Mason, Alonzo. Gunsmith, Monroetown, Bradford Co. (BPaD, 1861).

Mason, Charles. Artillery Artificer, 1778, Capt. Thomas Wylie's Co. at Carlisle (PIR, 2:249).

Mason, Henry. Gunlock maker, 1759, Lancaster (SED, 58).

Massa, George. Gunsmith, N. Queen St. above Frederick St., Lancaster (BLD, 1857).

Masterman, Peter S. Gunsmith, Philadelphia (MD, 1867).

Mathasie, Christian (b. 1797). 1838–49, Lancaster boro (SED, 58); 1850–60, Evansville, Indiana (c). Gunsmith.

Matthesiee, John Nicholas (1805–80). Born in Germany, 1863–80, Louiseville (Pavia), Union Twp., Bedford Co. (t; c).

Mathews & Moore. Bush Hill Works, Philadelphia. During the Civil War, they cast cannon weighing between 7,000 and 10,000 pounds (Scharf and Wescott, *History of Philadelphia* [Philadelphia: Barr, 1895], 311).

Mauger, Henry (1750–1826). 1781–1826, Douglass Twp., Berks Co. Major gunsmith. "Henry Mauger was an early resident of Douglass Twp., Berks Co., where he was engaged as a flintlock maker and a maker of guns. He lived on Rattlesnake Hill, and owned land there. He also worked as a stone mason, following his trades in season." (Montgomery, *History of Berks County* 2:1678) His will was probated on 3/12/1827 and showed gunsmithing tools. His total estate valuation was $4,786,82 (Shumway, MB 5 [1976]: 16–18).

Maurer, Jacob. In 1778, was "paid £1/10/0 for repairing Lancaster public arms" (3 Pa. Arch. 6 at 376).

Laurer, John. Gunsmith, 1779–1803, (SED, 58); Philadelphia (PXD, 1807).

Maus, Philip. 1776–98, Berks Co. (RAG, 128; DIL, 203). Also listed in are: F. E. Mause; Jacob Maus[e], and Philip Maus[e], Jr. Other, later researchers, myself included, have failed to locate these names, although there may be evidence we have overlooked.

May, William. Gunsmith, 216 N. 7th St., Philadelphia (DeSD, 1829, 1830, 1831).

Mayberry William. Armorer, 1776–77, on the ship *Effingham*, Pa. Navy (3 Pa. Arch. 23 at 16; 5 Pa. Arch. 1 at 429).

Maybud, David. Artillery artificer, 1778, Capt. Thomas Wylie's Co. at Carlisle (PIR, 2:249).

Mayer, Ernest. Gunsmith, West Chester, Chester Co. DBA Mayer & Eachus (PD, 1884; BSED, 1899).

Mayer, George. Gunsmith and gunlock maker, 1819–38, Lancaster boro (GAAAM, 96).

Mayer, George. Gunsmith, 22 N. Queen St., Lancaster (BLD, 1861).

Mays, James. 1759–90, Rapho Twp., Lancaster Co. (SED, 58). Gunsmith.

Mayweg & Nippes. Daniel Nippes and John Mayweg, props. Made cutlasses for the War of 1812 at $3 each (GAAAM, 223).

Mayweg, John. Gunstockers and gunsmiths, Dillwyn St. and Green St., Philadelphia (DeSD, 1829, 1830, 1831).

Mayweg, William. Gunsmith and gunstocker, Dillwyn St. and Green St., Philadelphia (DeSD, 1829, 1830, 1831, 1836–36); 216 N. 7th St. (MD, 1837, 1841).

Meadows, William (b. 1731). Silversmith, 5′ 10″ tall, born in England. Enlisted in New Pa. 11th Regt. on 2/11/1777 (2 Pa. Arch. 11 at 58).

Meakin, George. 1841–45, New York, N.Y. (Swinney); Fulmerville, Pike Co. (BPaD, 1861). Gunsmith.

Meals, John (1789–1852). Major gunsmith, 1798, Tyrone Twp., York Co. (t); 1800–1846, Menallen Twp., Adams Co.(t); 1849–52, Butler Twp., Adams Co. (WSB, 66–92; JKJ, 521–23). Bowers reproduced Meals's daybook.

Means, Samuel (d. 1823). 1810–25, Smith Twp., Washington Co. (t); 1829–33, Bridgeport, Ohio (t). Gunsmith.

Mears, John. 1776, Philadelphia. Supplied guns to the Committee of Safety, (Col. Rec. 10:681).

Mechlius, Edward. Gunsmith, 1815, Halifax, Dauphin Co. (KKR, 297).

Medeer, Brice. Gunsmith, Brownsville, Fayette Co., (c, 1870). One of his late apprentices was Uriah Fisher, also maker of long rifles.

Megahan, Abraham (b. 1816). 1850–70, Penn Twp., Huntingdon Co., Gunsmith. (t; c) He lived in the hamlet of Marklesburg (t).

Megahan, Benjamin L. Gunsmith, Penn Twp., Huntingdon

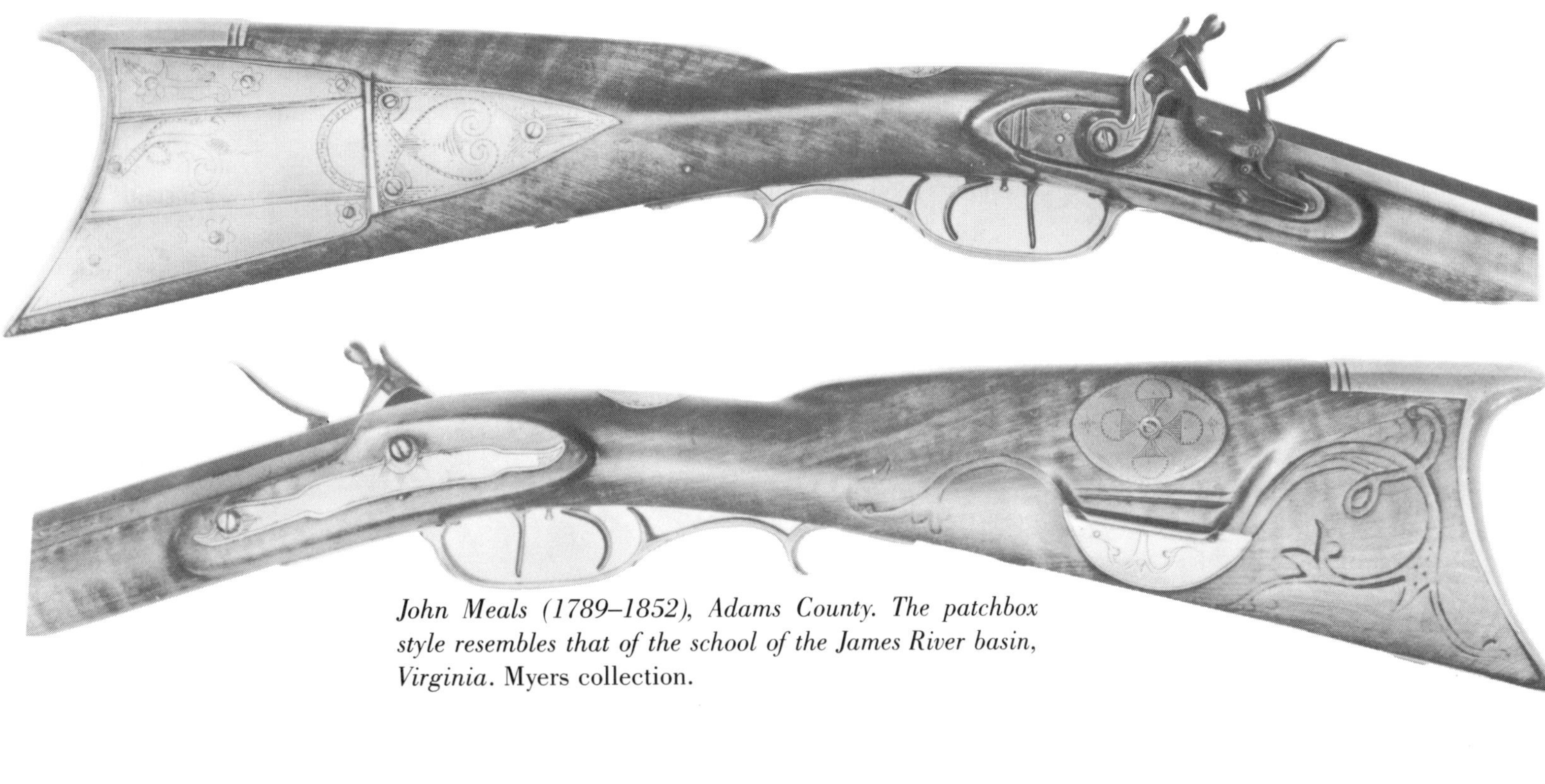

John Meals (1789–1852), Adams County. The patchbox style resembles that of the school of the James River basin, Virginia. Myers collection.

Co. Retired about 1854 (t; c).

Meggs, John. Artillery artificer, 1779 (2 Pa. Arch. 15 at 384).

Melloy, Hugh. Gunsmith, 1809, Lancaster (SED, 58); 1 Green St., Philadelphia (PXD, 1815).

Meng (Ming), Wolery. 11/29/1776: "pay Wolery Meng for 195 bayonet scabbards" (Col., Rec., vol. 22); 5/14/1777: "pay Woolery Meng £196/7/0 for bayonet belts, cartouche boxes &c." (2 Pa. Arch. 1 at 45); 7/21/1777: "pay Wilery Ming £174/17/0 for cartouche boxes & bombadiers' belts" (2 Pa. Arch. 1 at 67). Meng was evidently a soldier who was captured by the English and exchanged (Certificate of Exchange, 12/10/1779, 3 Pa. Arch. 23 at 393).

Merckley, Jacob. Gunsmith, 1781–90, Philadelphia (KEA, 65).

Merrill, William & T. 1108 Ridge Ave., Philadelphia (BSED, 1880). Gunsmiths.

Merz, Lewis. Gunsmith, 1865–80, Lock Haven, Clinton Co. (PD, 1882; t; c; BRD, 1865).

Messer, Joel. Gunsmith, 1837–43, Elizabeth Twp., Lancaster Co. (SED, 58). Gun-barrel smith.

Messersmith, George. Gunsmith, 1802, Lancaster (SED, 58).

Messersmith, Henry. Gunsmith, 1747, Lancaster (SED, 58).

Messersmith, Jacob. 1779–1810, gunsmith, W. King St. near Charlotte St., Lancaster (t; SED, 58; 3 Pa. Arch. 17 at 611).

Messner, Joel. Gunsmith, 1837–43, gun-barrel smith, Lancaster (SED, 58).

Mestrezat, William (b. 1809). Gunsmith, 1830–36, Mapletown, Greene Co. (t).

Metler, John E. (d. 1879). Gunsmith. 1850–79, Northampton Co. (t; c).

Mettler, Elias. Gunsmith, 36 Lehigh St., Easton (BRD, 1860).

Metzgar, Jacob. Gunsmith, 1754–77, Lancaster boro. (SED, 58); 1784, Frederick Co., Md., working with Jacob Groff (RAG, 131; JKJ, 465).

Metzgar, Jacob T. Gunsmith, N. Queen St., above Orange St., Lancaster (BLD, 1857); 1863, Hamilton, Ohio (DH, 57); 1879–85, Montgomery Co., Ind. (FSP, 207).

Metzgar, Philip. 1814–16, gunlock maker, Lancaster (SED, 58; JKJ, 465).

Mewhirter, John. Gunsmith, 1814–17, Shippensburg; 1833, Williamsport, Md; 1836–40, Chambersburg (WSB, 49).

Mewhirter, John, Jr. Gunsmith, Chambersburg. "on the corner of 2d and Washington, was an old two story brick dwelling with a gunsmith's shop. . . . John Mewhirter, a lively old chicken, who had a son, John, even livlier than himself, lived and made rifles here." (Cooper, *Recollections of Chambersburg*, 40) 1837–41, Chambersburg (WSB, 49). The shop was sold to Jeremiah Senseny, another gunsmith. (t).

Mewhirter, Robert. Revolutionary war soldier and gunsmith, 1783–1830, Chambersburg; 1833, Williamsport, Md.; 1836–41, Chambersburg (WSB, 39f.).

Mewhirter, William (d. 1814), 1770–90, Lower Swatara Twp., Lancaster Co.; 1811–14, Shippensburg (WSB, 39f.). Gunsmith.

Meyers, Charles A. Gunsmith, 212 Vine St., Philadelphia (GPD, 1892).

Meyer, George. Gunlock maker, 1814, Lancaster (SED, 58).

Meyers, James. Gunsmith, Stroud Twp., Monroe Co. (PD, 1882, 1890).

Meyers, Paul. 212 Vine St., Philadelphia, gunsmith (PD, 1882; BSED, 1880).

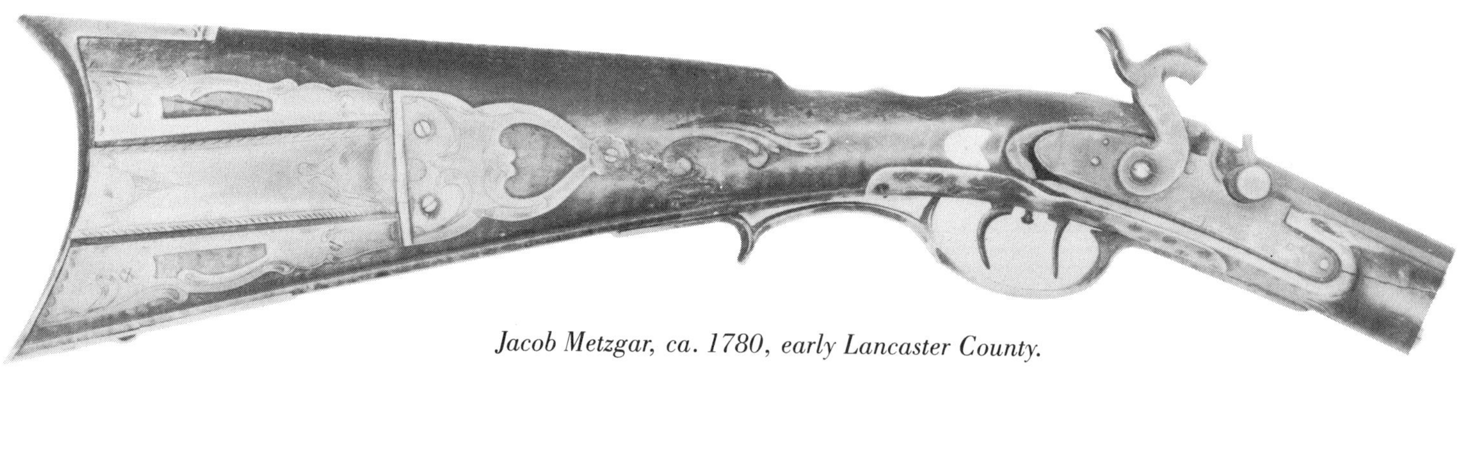

Jacob Metzgar, ca. 1780, early Lancaster County.

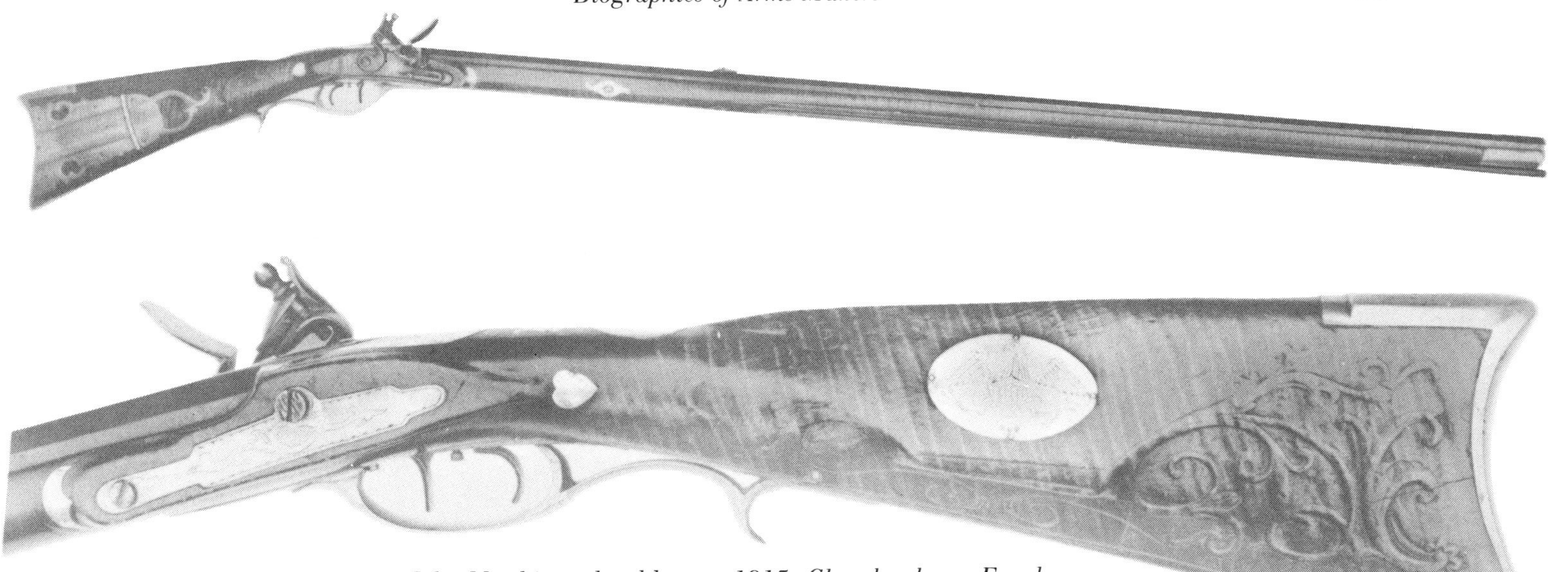

John Mewhirter the elder, ca. 1815, Chambersburg, Franklin County.

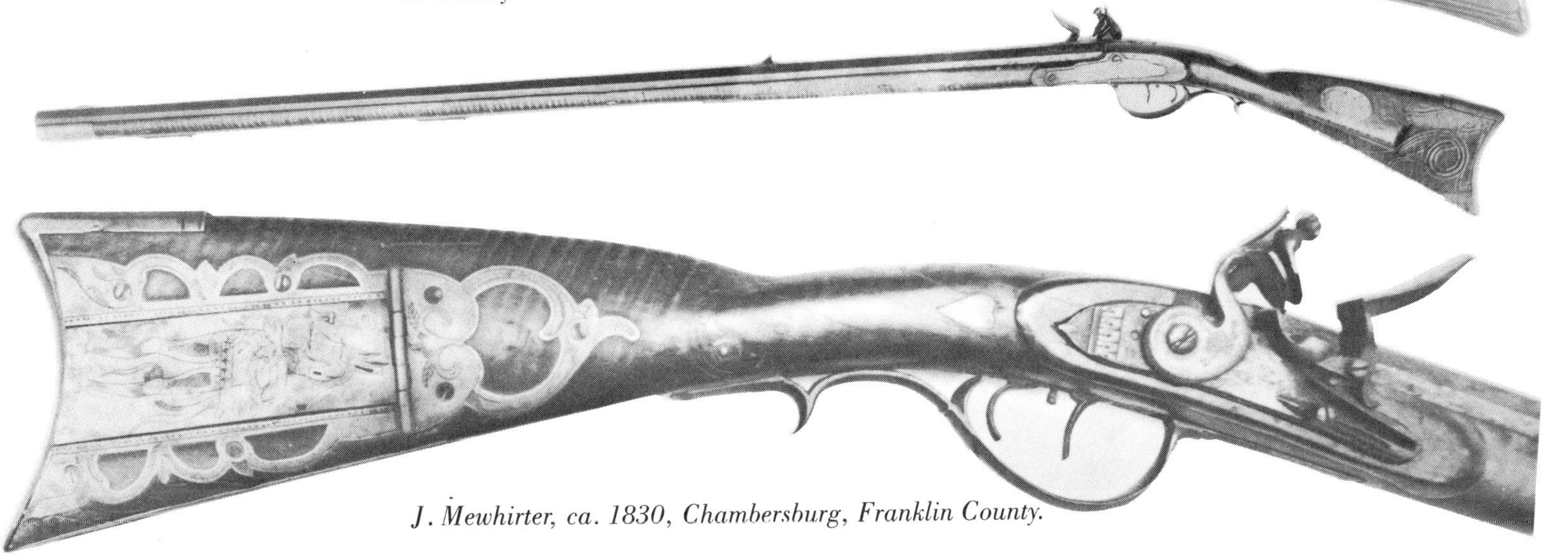

J. Mewhirter, ca. 1830, Chambersburg, Franklin County.

Meylan (Meylin), Martin (1690–1746). Few controversies in the gun collecting field have been more intense or longer in duration than that surrounding Martin Meylin. Rupp's *History of Lancaster County,* says that Meylin founded a gun-barrel boring mill in 1719. Beck's *History of Lancaster County,* says that "the gunsmiths, before 1745, over the entire county were few indeed." Presumably, this referred to Meylin, as there are few, if any, earlier ones. Herbert Beck (MB 6 [1960]: 4–5) believed that Meylin was the first gunsmith in Lancaster Co. Probably the best case for Meylin's having been a gunsmith was made by Richard Headley (KRAN, Winter 1978, 4–10). On the other hand, the major scholar of Lancaster Co. gunsmithing, S. E. Dyke, claimed that Meylin was not a gunsmith and that he had gunsmithing tools in his estate only because his son-in-law, Joshua Baker, owed him money and the tools were a guarantee of the debt. Dyke also said that the building marked by the Commonwealth as Meylin's gunshop, allegedly dating from 1719, was built only after Meylin's death. This was the position, also, of Clyde Groff (*Papers of the Lancaster County Historical Society* vol. 75, 1971). There are no public documents listing Meylin as a gunsmith. There are two rifles, one dated 1705, held by some to be the "first American long rifles" (Rupp, *History of Lancaster County,* 74–75; SED).

Michael, John. Gun-barrel forger, 1812, Manheim Twp., Lancaster Co. (SED, 58).

Michlin, Edward. Gunsmith, 1817, Halifax, Dauphin Co. (KKR, 208).

Mickey, Daniel (1812–1881). Confluence, Somerset Co. (t). Gunsmith.

Midvale Steel & Ordnance Co. 1880–83, Philadelphia, naval contractors for large cannon.

Mier, Jacob (1793–1873). 1851, Somerset boro.; 1852–73, Salisbury, Somerset Co. (t; c).

Jacob Mier, Somerset County.

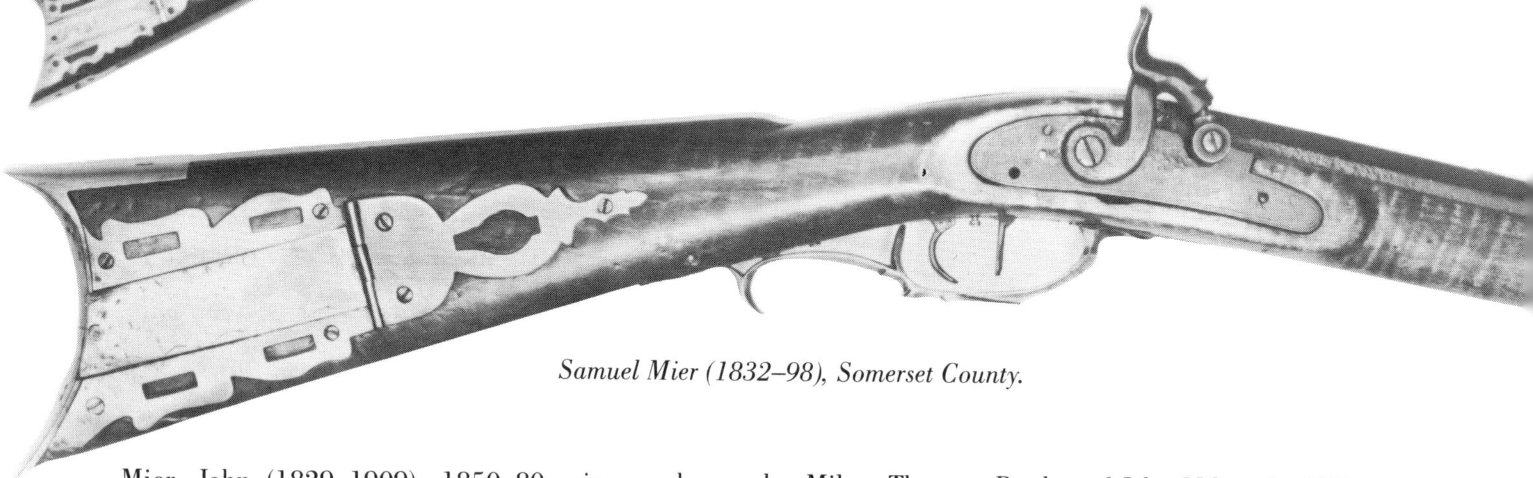

Samuel Mier (1832–98), Somerset County.

Mier, John (1829–1909). 1850–80, cigar maker and gunsmith, Somerset boro (t; c).

Mier, Samuel (1832–98). 1850–51, Somerset boro.; 1852–98, Salisbury, Somerset Co. (t; c). Major gunsmith.

Mier, William B. (1829–1909). Gunsmith, 1850–1909, Somerset boro (t; c).

Miles, John, Sr. (1752–1808). Gunsmith and military contractor, 1799–1808, 30 S. Third St., Philadelphia. Contract of 9/3/1798 for 2,000 liberty cap muskets; 4/16/1801, 2,000 liberty cap muskets; 7/20/1808, 9,200 M-1808 muskets (RAG, 132).

Miles, John, Jr. (d. 1852). Son of John Miles, partner of his father in government contracts, 1799–1808; 1808–52, Bordentown, N.J. (RAG, 132). Had his own contract, 10/7/1812, for 2,407 M-1808 muskets (RAG, 132).

Miles, Mary. Gunsmith, Philadelphia (DeSD, 1824).

Miles, Thomas. Brother of John Miles, Sr. Military contractor, 1775–76, Philadelphia (GAAAM, 100).

Miller, Benjamin. Gunsmith, 1841, Lancaster (SED, 58).

Miller, Daniel (b. 1801). Gunsmith, SE Ward, Reading (c, 1850).

Miller, David. Apprenticed to Christian Hawken, Washington Co. Md., 1796 (AMM, 225); gunsmith, 1800–14, Georges Twp., Fayette Co. (t).

Miller, Elizabeth. 10/20/1775, Committee of Safety purchased "of Elizabeth Miller, 52 cutlasses" (Col. Rec., 10:375).

Miller, Francis. 1777–81, artillery artificer, Capt. Jesse Rowe's Co., Lebanon (PIR, 2:246; 2 Pa. Arch. 11 at 252).

Miller, George. Gunsmith, 1811, Lancaster (SED, 58); 1821, Mahoning Twp., Columbia Co. (t); 203 Queen St., Philadelphia (MD, 1837, 1841, 1846; c, 1850).

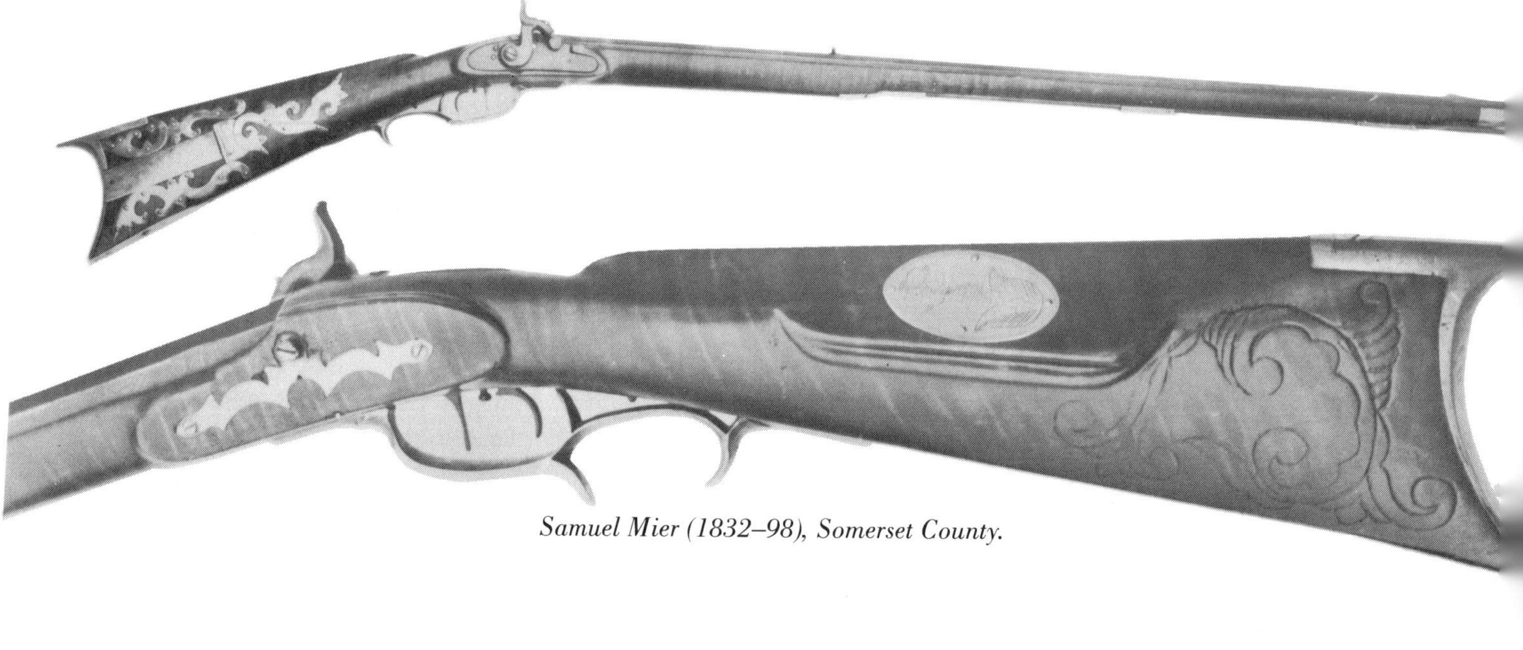

Samuel Mier (1832–98), Somerset County.

Miller, Henry. Gunsmith, 1842–50, Lancaster Co. (t; c; SED, 58).

Miller, Jacob. 1773, Conestoga Twp., Lancaster Co. (SED, 58); 1793–1800, West Bethlehem Twp., Washington Co. (t). Gunsmith.

Miller, John. Artillery artificer, 1779, at the laboratory, Philadelphia (2 Pa. Arch. 15 at 386).

Miller, John (1739–1810). 1761–89, gunsmith and blacksmith, Lancaster. (SED, 58; DB G-465 [1761], M-66 [1764]; McCord Ledger and Daybook, 36–37 [Pa. Dept. of Archives]; 3 Pa. Arch. 17 at 46, 10, 295, 460, 611) Was paid to repair public arms of Lancaster Co., £6/3/3 on 8/18/1777, and £25/7/7 on 11/22/1777 (3 Pa. Arch. 6 at 376).

Miller, John. Gunsmith, Alley below James St. and Frederick St., Lancaster boro (GCD, 1863–64).

Miller, John D. Gunsmith, 10 Gray's Ct., Philadelphia (MD, 1841); rear, Hope near Otter, Philadelphia (MD, 1861).

Miller, Joseph. Blacksmith, 1683, Philadelphia (PaE, 8).

Miller, Josiah. Gunsmith, 1814, Union Twp., Fayette Co. (t).

Miller, Mathias. 1775–1811, gunsmith, Easton (t; 3 Pa. Arch. 19 at 290).

Miller, Max. Gunsmith, 1810, Berks Co. (MB 1 [1962]: 5).

Miller, Philip (b. 1770). Smith, arrived at Philadelphia on brig *Mary*, 7/30/1796, age 26 (2 Pa. Arch. 17 at 601).

Miller, Robert. Gun-barrel finisher, 2d St. between Wood St. and Smithfield St., Pittsburgh (PAD, 1860–61, 1864–65).

Miller, Samuel. Gunsmith, 1823–49, Swatara Twp., Lebanon Co. (KEA, 54).

Miller, Simon. Gunsmith, 1820–55, Hamburg, Berks Co. (t; c).

Miller, William D. Pittsburgh. Patent 169,465 of 11/2/1875 for a recoil pad.

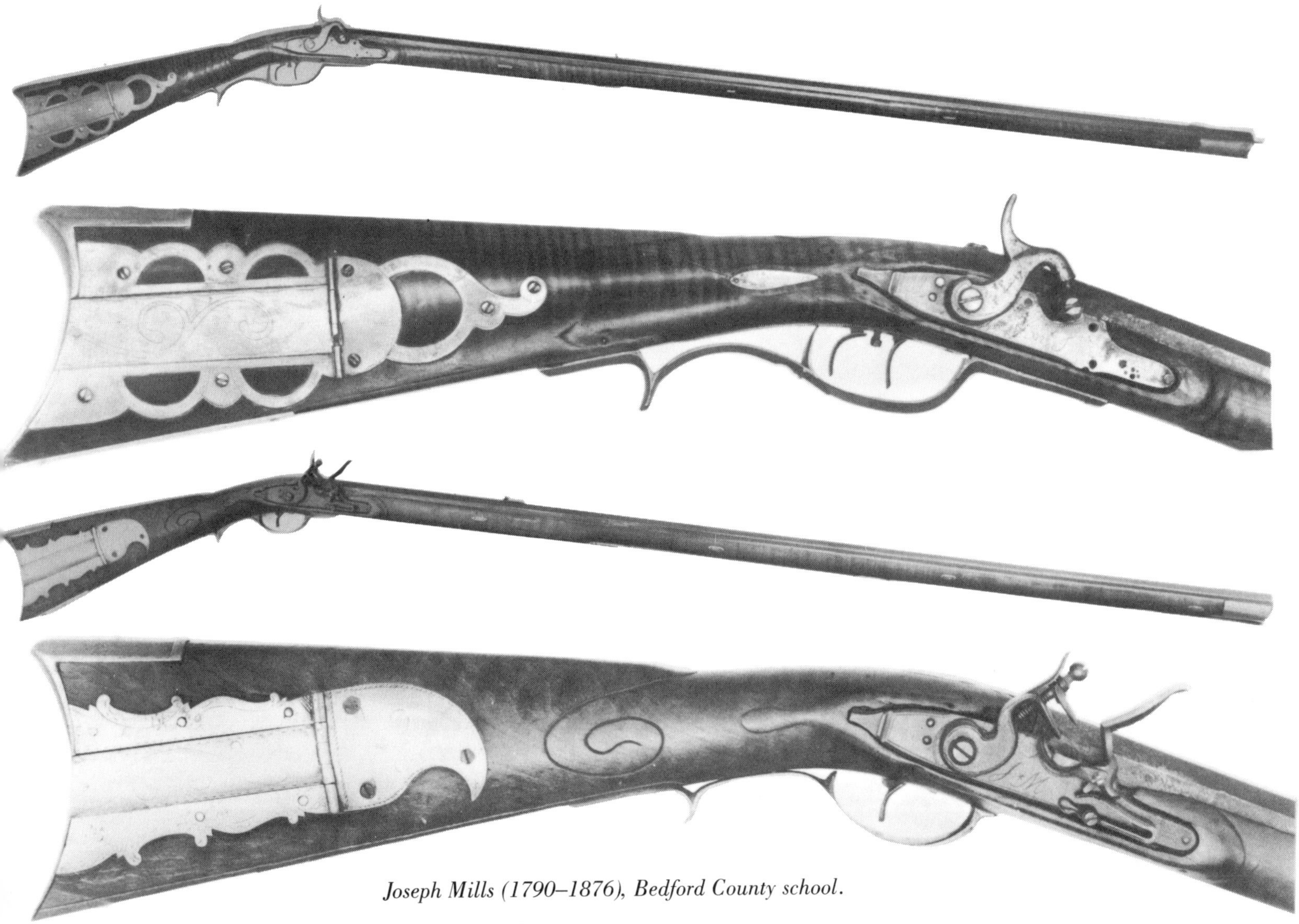

Joseph Mills (1790–1876), Bedford County school.

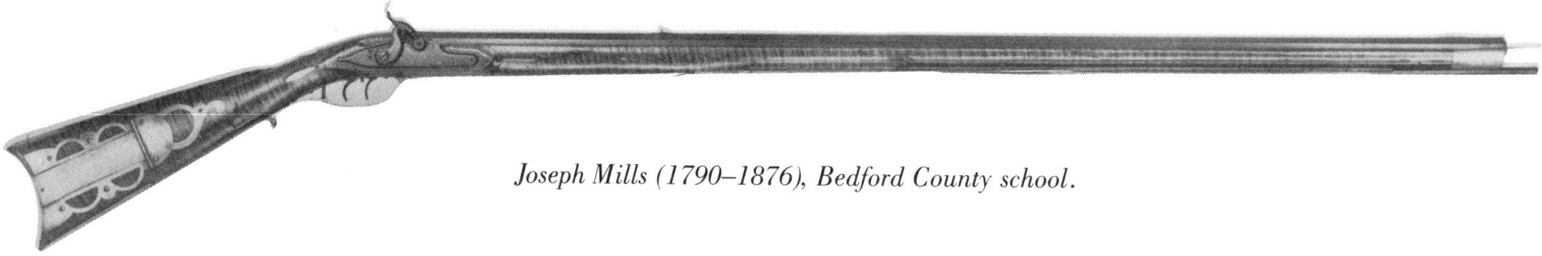

Joseph Mills (1790–1876), Bedford County school.

Miller, William T. Gunsmith, Danville, Montour Co. (BPaD, 1861).

Milley, Edward. Gunsmith, James St. near Mulberry St., Lancaster (GCD, 1863–64).

Milliron, Anthony (1774–1807). Gunsmith, 1805–7, Conemaugh Twp., Indiana Co. (t); 1800, Armstrong Twp., Westmoreland Co. (t; c).

Milliron, David (1805–90). 1850–63, Smicksburg, Indiana Co. (BPaD, 1861); 1864–80, Ringgold, Jefferson Co. (REH, 125–27); 1880–90, Armstrong Co. (t; c).

Milliron, John (1763–1808). Gunsmith, 1802–8, Hempfield Twp., Westmoreland Co. (t). Died intestate (WB I, 20).

Mills, Isaac (1801–36). Gunsmith, Bedford Co., 1822; 1828–30, Youngstown, Westmoreland Co; 1832–36, Rose Twp., Jefferson Co. (REH, 25).

Mills, Joseph (1790–1876). Major gunsmith. 1815–22, Colerain Twp., Bedford Co. (t); 1823–76, Perry Twp., Coshocton Co., Ohio (t; c). Mills made guns with 4 distinctively different lock types: flintlock, conventional percussion, percussion with an outside mainspring, and box lock. Additionally, a harmonica-lock gun is attributed to him.

Milne, Edmund. Goldsmith, 1773–82. In 1773 Milne had an apprentice in the silversmithing business. (PIDS) 10/4/1782: "towards completing an elegat sword for Capt. Barney, £91/4/4" (3 Pa. Arch. 5 at 334).

Milnor, Isaac. Gunsmith, 5 Norris Alley, Philadelphia (NTD, 1800).

Miltenberger, George. Gunsmith, 1818, Pittsburgh; also a tinsmith and coppersmith, and sold gun parts (PAD, 1818).

Miltz, Francis. Artillery artificer, filer, 1779, Pa. Continental Line (2 Pa. Arch. 15 at 384).

Minesinger (Mincsinger), David. On 2/27/1849 Minesinger patented a breech-loading gun, patent 6139, while living in Beaver Co. It was manufactured 1859–61 by John White. While still using a percussion cap for ignition, it used a steel-tapered cartridge.

Minfirt, Abraham. Gunsmith, Erie Co. (c, 1850).

Minrich, George. Gunsmith, Otter at Front St., Philadelphia (MD, 1849).

Mintzer, George. 1874–96, gunsmith, 320 Vine St., Philadelphia (BRD, 1874; BSED, 1880; GPD, 1892, 1896).

Mintzer, W. G. Swordsmith, Civil War contractor, Philadelphia (RAG, 223).

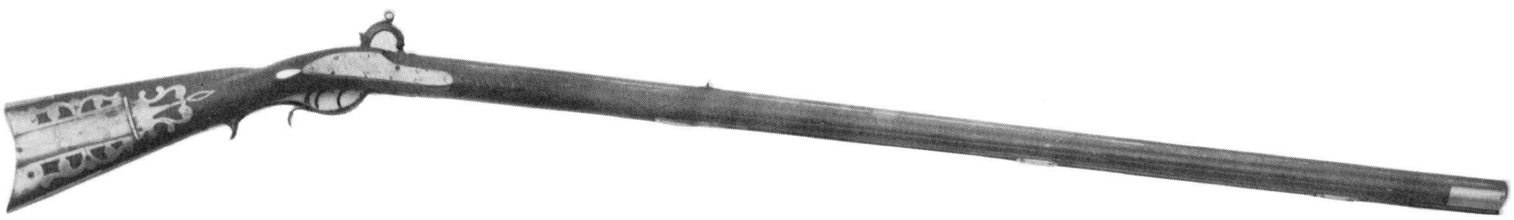

Joseph Mills (1790–1876), Bedford County. This box-lock, pill-lock rifle was probably made after Mills moved to Ohio.

Valentine Mohney (1824–1909), Jefferson County. Gunsmith and clockmaker.

Mitchell, Joseph. Gunsmith, 2 Pratt's Ct. (DeSD, 1823); Front St. at Otter, Philadelphia (DeSD, 1831); 591 N. Front St. (D, 1837, 1841; BPaD, 1861). He was an apprentice to Joseph Kunz and later worked for Brooke Evans at Valley Forge on government contracts (RAG, 223).

Mitchell, Richard. Cutler, 1781–90, Middletown Twp., Bucks Co. (t).

Mitchell, Simon. Artillery artificer, 1779, at the lead furnace (2 Pa. Arch. 15 at 385).

Mohney, Valentine (1824–1909). Gunsmith, Snyder Twp., Jefferson Co. (REH, 129, 222). Also a clockmaker.

Moll, Christian. Smith. Naturalized in Philadelphia Co., 1743, active thereafter (t; 2 Pa. Arch. 2 at 362).

Moll, David (d. 8/31/1853). 1820, Lower Saucon Twp., Northampton Co. Gunsmith. 1830–53, Hellertown, Northampton Co. (J. Flemish and D. Wendling, "Hellertown Molls," KRAN, Winter 1979, 2–6).

Moll, Edwin. Gunsmith, 1853–61, Hellertown, Northampton Co. (t; c).

Moll (Mull), Henry. Gunsmith, 1740–51, Lancaster Co. (DB C-251) He sold land in 1751 that he had bought in 1751 (KKR, 310; SED, 58; JKJ, 352–53). In 1777–78, Berks Co. he was on the list of those attainted by treason, and his property was seized and sold, including his smith's tools, bought by Elisha Readkars (6 Pa. Arch. 12 at 59). 1783, Huntingdon Twp., York Co. (t); gunsmith, 1788, Northamptontown (3 Pa. Arch. 19 at 314).

Moll, J. & W. H. Hellertown, Northampton Co., gunsmiths (BRD, 1873; BPaD, 1861; PD, 1882).

Moll, James G. Gunsmiths, DBA J. & W. H. Moll, Hellertown, Northampton Co. (t; c; BRD, 1860; PD, 1883).

Moll, John. On 9/16/1678, at Whorekill on the Delaware River, Capt. Billop delivered to John Moll, armorer, the following items belonging to the Dutch fort: "8 iron guns, 7 leaden aprons, 18 matchlocks, 6 firelocks, 24 muskets, 12 bandoliers, 66 iron shot, 465 musket balls, 1⅓ barrels of gunpowder, 3 quires of cartridge paper; 12 skeins of match" (Hazard's *Annals*, 1:457).

Moll (Mull), John, Sr. (d 1794). 1764–94, Northampton Co. in or near Allentown. (t; Roberts, *Anniversary History of Lehigh County*, 1:177–81) On 4/28/1772 Moll married Linda Rinker, daughter of a gunsmith. In 1776–80, he was employed by Ebenezer Cowell working on public arms. He was a Quaker, and was naturalized on 9/24/1759. (2 Pa. Arch. 2 at 401) He died 11/1794 leaving a widow and 2 sons (GAS-1, 254–55; FSP, 213).

Moll, John, Jr. (1773–1867). John, Jr. took over his father's business on 8/18/1793. In 1795 he married Elizabeth Neihard, a cousin of gunsmith Peter Neihard. On 4/11/1820 he sold his shop to his son, John, III. Listed as a gunsmith, Northampton Twp., Northampton Co., 1817–47 (t; c; GAS-1, 254; FSP, 213; JKJ,177).

Moll, John, III (1796–1883). "The John Moll who is mentioned as a gunsmith built the shop on 7th St., near the Lafayette Hotel where his son, John Moll, died at the age of 87 in the summer of 1883 and in which his grandson now carries on the same trade that was established here before the Revolution. The father of John Moll, Sr., was William, also a gunsmith and plied his trade as early as 1747. His greatgrandson has an heirloom descended from William, a device for cutting threads on screws, neatly made of iron, and bearing in plainly legible characters the inscription, 'April 10, 1747—W. M.'"" (Alfred Mathews, *History of Lehigh and Carbon Counties* [Philadelphia: Everts & Richards, 1884] 123–24). John, III ran the shop his grandfather had founded from 1820 to 1883 (FSP, 213; KKR, 305ff.; t; c).

Moll, John J. (d. 1900). Son of Edwin. 1850–1900, Hellertown, Northampton Co. (c; BRD, 1882).

Moll, Nathan. Born at Keelersville on the Old Bethlehem

Rd., Bucks Co. 1828–35, gunsmith, Hellertown, Northampton Co. (Paul Boebert, MB 2 [1945]: 13) Made many guns in large bores, some of them iron mounted.

Moll, Peter (1799–1879). Gunsmith, 1820–35, Salisbury, Northampton Co. (t); Shumway (GAS-2, 11) showed a gun by him dated 5/26/1826 (JKJ, 177, 184).

Moll, William. Gunsmith after 1747, Northampton Twp., Northampton Co. (see Moll, John, III; GAAAM, 101).

Moll, William. 1853–68, gunsmith, Hellertown, Northampton Co. (t; c).

Moll, William Henry (b. 1825). Gunsmith, 1863–83, Allentown. The shop was demolished in 1883, after which he did no more gunsmithing. His career as a gunsmith began in 1848 (t).

Monferrer, Peloux A. Gunsmith, 174 S. 2d St., Philadelphia (PXD, 1810).

Monroe, I. S. Powder maker and dealer, 1876, Catawissa, Columbia Co. (*Atlas of Columbia and Montour Counties*, Beers 1876).

Moody, Baltis. Gunpowder manufacturer, before 1777, York Co.; after 1777, Market St., Philadelphia. "compiled an impressive record in manufacturing gunpowder" (J. Kelley, *Pennsylvania: The Colonial Years* [New York: Scribner's, 1980] 741–42).

Moore, Abraham. 1770–80, Coventry Twp., Chester Co. Contractor to Committee of Safety (t; Col. Rec., 10:687).

Moore, Andrew P. (b. 1833). Gunsmith, Clearfield Co. (c, 1850); Lawrence Twp., Clearfield Co. (c, 1860, 1870).

Moore, George. Cartridge maker, 1778, York Co.; supplied Committee of Safety (3 Pa. Arch. 7 at 106, 107); artillery artificer, fife maj., 1779, at Carlisle (PIR, 2:249).

Moore, James. Smith, 1684, Philadelphia (2 Pa. Arch. 19 at 328).

Moore, James. Gunsmith, James St. near Front St., Chester, Delaware Co. (BPaD, 1861).

Moore, John (1806–72). Born in Va. 1850–72, Clearfield boro (c; t; BPaD, 1861). Gunsmith.

Moore, J. T. Gunsmith, Springhill Twp., Greene Co. (c, 1870, 1880).

Moor, Ralph. Artillery artificer 1778, at the cannon factory, Carlisle (PIR, 2:249).

Moore, Samuel (b. 1838). Lawrence Twp., Clearfield Co. Son and apprentice of John Moore (c, 1860, 1870).

Moore, William (b. 1833). Jacob Stoudenour's apprentice. Colerain Twp., Bedford Co., gunsmith (c, 1850; t, 1855).

Moosic Powder Works. Organized in 1864; merged with Laflin, Boies & Yurik in 1869. Capital investment, \$100,000; capacity; 1,000 kegs a day. Jermyn, Lackawanna Co. (*Wyoming Historical and Genealogical Society Collection* [New York: Munsell, 1880], 490).

Morey, Nathan. Gunsmith, 60 Apple St., Philadelphia (MD, 1837).

Morgan, Edward. Whitpain Twp., Montgomery Co., "a manufacturer of guns, some of which were used in the Revolutionary War . . . Quaker . . . His paternal grandfather for whom he was named was also a grandfather of Daniel Boone" (*History of the Wyoming and Lackawanna Valleys*, 1:522).

Morgan, George. Gunsmith, 1776, Philadelphia (KEA, 69).

Morgan, John. 1852–55, gunsmith, Freeport Twp., Armstrong Co. (t; Smith's *History of Armstrong County* [1883], 420).

Morgan, Joseph. Gunsmith, rear of 225 N. 3d St., Philadelphia (KEA, 69).

Morgan, Phillamon. Gunsmith, 1817–30, Springhill Twp., Fayette Co. (t).

Morganstern, William. Inventor, 1863–67, Philadelphia; 1868–90, Hartford, Conn. (RAG, 136).

Morineau, Philip. Inventor. Patent of 10/10/1832 for a breech-loading firearm; patent 32,460 of 5/28/1861 for a combination sword and gun. Philadelphia.

Morris, G. W. Gunsmith, Main St., Cory, Erie Co. (*Erie County Directory*, 1873–74).

Morris, John. 1768–83, Southwark Twp., Philadelphia (t). Gunsmith and flint cutter.

Morrison, Samuel. 1826–43, Milton, Union Co. Patent of 2/10/1836 for a superposed multi-charge gun system. Gunsmith (t).

Morrow, Abraham. 1775–98, Front St., east side, Southern District, Philadelphia. On 2/22/1777 he delivered 22 new guns to the Committee of Safety. In 1788 he was paid £15 for repairing "arms belonging to the city and liberties of Philadelphia," and he also repaired 150 stands for the city in 1789, for £30. (3 Pa. Arch. 7 at 222) He was perhaps the major military arms contractor and repairman in Philadelphia before 1800 (c, 1790; KEA, 69; t).

Morrow, William. Gunsmith, 1842–43, Uniontown, Fayette Co. (t).

Morton, David (1770–1823). 1810–23 West Newton, Westmoreland Co. He died on 3/5/1823 at the age of 53. His estate appraisal included: "1 pattern for gunstock; 1

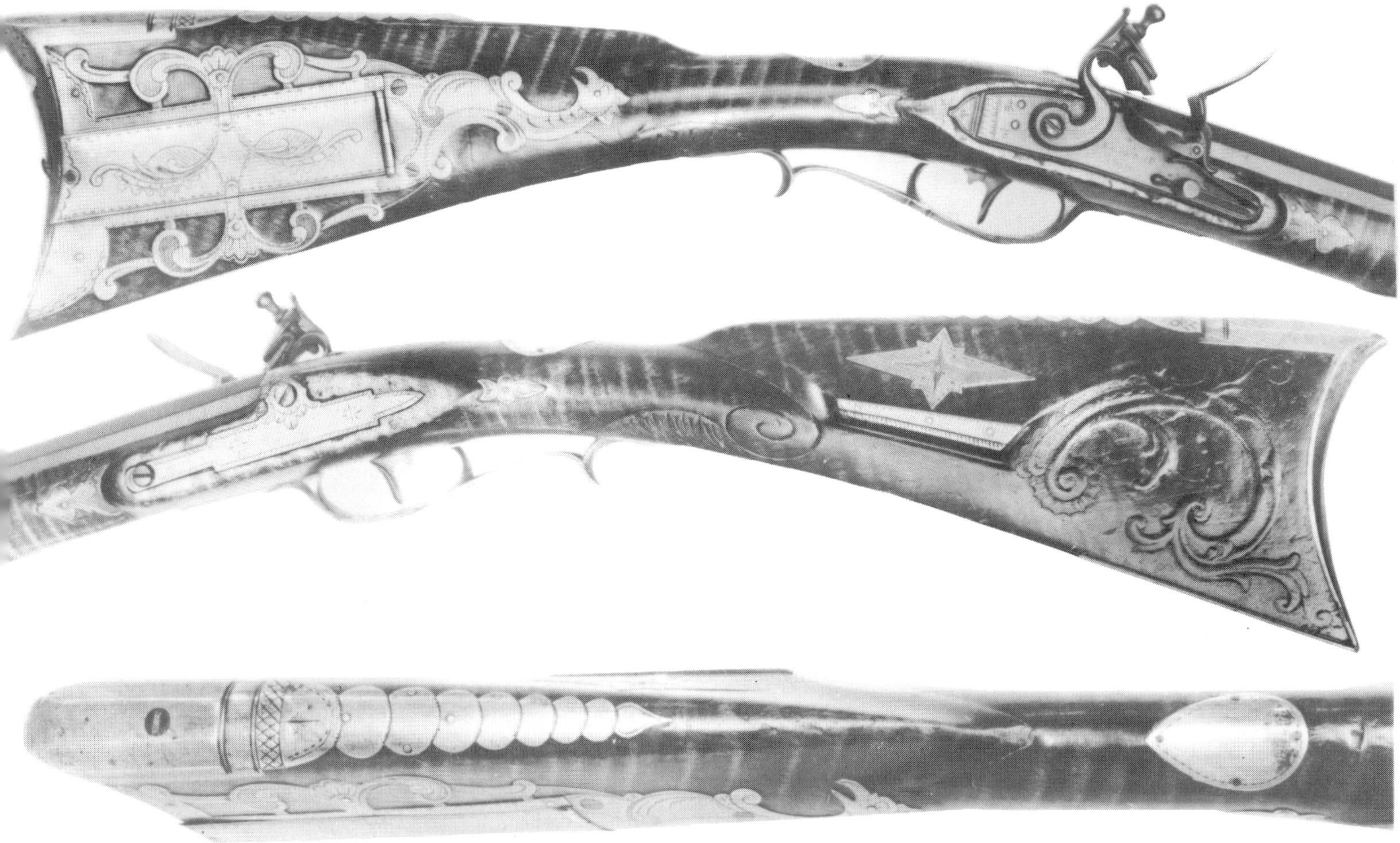

David Morton (1770–1823), Westmoreland County.

ramrod auger; 8 plains [planes] for stocking guns; and 2 small pistol barrels & 1 stock (t; C. Kaufmann, KRAN 10:2–7).

Morwitz, Edward. Inventor, Philadelphia; worked with William Morganstern. Patent 40,572, of 11/10/1863 for a breech-loading gun.

Moss, Elijah. Gunsmith, 1801–3, Georges Twp., Fayette Co. (t) In 1813, Experience Moss bound her son Josiah (b. 1798) to George Rizer of Cumberland, Md. to learn gunsmithing. Their home was in Bedford Co. (DH, 226).

Most, Henry. Gunsmith, 28 Kunckel, Philadelphia (PXD, 1813); 370 N. 6th St. (MD, 1841).

Montanye, Abraham. Artillery artificer, buckle filer and finisher, 1779 (2 Pa. Arch. 15 at 385).

Moulder, Joseph. In 1776, he was paid £50/5/9 for cartridge boxes (3 Pa. Arch. 5 at 9).

Mount Joy Forge (Valley Forge Armory). Active ca. 1742–1842. At the mouth of East Valley Creek and the Schuylkill River. It was destroyed by the British army in 1777 and rebuilt by David and Isaac Potts on the Montgomery Co. side, about a half mile from the original forge. Potts and Rutter cast cannon there and got iron for the various contracts of the Evans family (SHIS, 176).

Mower, John. Germantown Rd. below 5th St., Philadelphia, gunsmith (MD, 1846, 1847).

Moyer, Absalom. Aetna, Allegheny Co. (PAD, 1860–61); Mulberry Alley near Wilkins, Pittsburgh (PAD, 1864–65). Gunsmith.

Moyer, Daniel. Gunsmith, Kimberton, Chester Co. (BPaD, 1861).

Moyer, George. Gunlock maker, 1819–21, Lancaster boro. (SED, 58); 1810–19, Swamp Creek, Montgomery Co., gun-barrel manufactory with machinery driven by power from the Perkiomen Creek (t); 1821–40, gun-lock maker and gunsmith, Lancaster boro (SED 58).

Moyer, William. Gunsmith, Delmar Twp., Tioga Co. (PD, 1890).

Mulgrew, William. Gunsmith, 1841–47, Lancaster boro (SED, 58).

Mulholland, James. 1861–65, Reading, Berks Co. Contract for U.S. M-1861 rifled muskets. He made 5,502 (RAG, 137).

Mull (Moll), Ludwig. Gunsmith prior to 1779, York Springs, Huntingdon Twp., York Co. (*History of Cumberland and Adams Counties* [Chicago: Warner-beers, 1886] 288).

Mull, Nathan. Gunsmith. Penn Twp., Lancaster Co. (BPaD, 1861).

Mullen, Hugh. Gunsmith, 391 Front St., Philadelphia (DeSD, 1831).

Mullin, John Clay (1850–1905). 1872–1905, gunsmith, blacksmith, and tanner, Harrison Twp., Bedford Co. (t; c).

Munebach, William. Artillery artificer, 1779, Berks Co. (BCR, 142).

Munshore, Nicolas. Smith, 1769, Coventry, Chester Co. (3 Pa. Arch. 11 at 634).

Munson, H. Gunsmith and letter cutter, 2d St. and Liberty St., Pittsburgh (PAD, 1845, 1847).

Murphy, John. Gunsmith, 1817, Donegal Twp., Lancaster Co. (SED, 58).

Murphy, Joseph. Gunsmith, 1845–51, Washington boro. (t) He placed an advertisement in the Washington *Examiner* (4/21/1849) for an apprentice to gunsmithing.

Murray, Simon. Artillery artificer, 1779, Pa. Continental Line; shop foreman (2 Pa. Arch. 15 at 383).

Murrelle, J. W. Sold guns, ammunition, and the like, Wilkes-Barre, Luzerne Co. (PD, 1882).

Musgrave, Benjamin. In 1854 he secured a marriage license in Allegheny Co. (no. 267). He listed his occupation as gunsmith and said he was born in Beaver Co. His shop was on Leacock near Federal in Allegheny City (Pittsburgh) (PAD, 1860–61).

Musgrave, Harry. Artillery artificer, 1779, at the cannon factory at Philadelphia (2 Pa. Arch. 15 at 384).

Musketnuss, Peter. Gunsmith, Charlotte near W. King St., Lancaster boro (GCD, 1863–64, 1871–72).

Mussman, Frank. Gunsmith, 56 Marion, Pittsburgh (PAD, 1860–61, 1864–65).

Mussman, Robert G. Drum maker, 1827; was paid $84 for 12 drums (Pennsylvania Auditor General's Report, 1827).

Myers, Daniel. Gunsmith, 1847–50, Manor Twp., Lancaster Co. (SED, 58).

Meyer, George, Gunlock maker, 1814, Lancaster boro (SED, 58).

Myers, George S. Gunsmith, Apple Grove, York Co. (BPaD, 1861).

Myers, Henry. (b. 1807). Shirleysburg, Huntingdon Co. "Henry Myers, now living [1833], aged 76, was one of the pioneer gunsmiths of Shirleysburg" (AFR, 345).

Myer, Jacob. Naval master of arms, Pa. Navy, 1776. (2 Pa. Arch. 2 at 708) 3/27/1777: "1 pr smith's bellows delivered to Jacob Myers, £15" (2 Pa. Arch. 1 at 108). 3/8/1777, the Pa. naval board paid Jacob Myers, smith, £23/8/0 for work done (2 Pa. Arch. 1 at 91). 1779, Jacob Myers, smith, Windsor Twp., Berks Co. (3 Pa. Arch. 18 at 297).

Myers, Jacob. Gunsmith, 1818, Pittsburgh (t).

Myers, John. High-quality swordsmith, 1785–95, Philadelphia (RAG, 223).

Myer, John. Locksmith, 1779, York boro. (t) On 1/25/1773 he had been apprenticed to Richard Humphreys, silversmith, for 5 years, 5 months, 25 days (PIDS).

Myers, Philip. Gunsmith, 130 Fulton St., Pittsburgh (c, 1850).

N

Nace, John. Gunsmith, rear of 1 Mint's Ct., Philadelphia (MD, 1841).

Nagle, Charles. Orchard St. near Green's Row, Philadelphia. Gunsmith (MD, 1837).

Nagel, George. 1776, Reading, Berks Co. "such arms . . . as were left in this place since the last [French and Indian] war have been taken care of by Messrs Jonathan Potts and George Nagel, who, as they are informed, have been at considerable expence in repairing them" (1 Pa. Arch. 4 at 697).

Napier, Robert. 1865–70, Capitol Park, Harrisburg. A gun dated 1870 and repaired or made by Napier is in Arnold State Arsenal. (*Donations and Purchases*, 1865–70) Gunsmith (*Patriot Directory*, 1874–75).

Nary, James. In the Pittsburgh *Gazette*, 5/12/1812, Nary was represented as a runaway apprentice of George McGunnigle, a gun repairman and cutler. At the time he fled, Nary was described as being age 17 and 5′ 4″, with red hair.

Nauman, Jacob. Gunsmith, 1809–32, Lancaster boro (SED, 58).

Nave, Jacob. Gunsmith, 1804–14, Metal Twp., Franklin Co. (WSB, 22).

Naylor (Nailor), James. Gunsmith, 1811–12, Hempfield Twp., Westmoreland Co. (t).

Nebinger & Co. Gun-barrel makers, ca. 1858, Cumberland Co. (WSB, 50).

Neeper, John. Gunsmith, LeMoyne Twp., Cumberland Co. (KKR, 312; c, 1800).

Neff, Johannes (1773–1853). Gunsmith, Kutztown, Lebanon Co. (MB 9 [1967]; FSP, 211).

Neihard, Conrad. Gunsmith, 1820–35, Lehigh Co. (t).

Neihard (Neihart), Peter (1743–1813). Major gunsmith. 1775–1813, North Whitehall Twp., Lehigh Co. (t) Married Magdalena Kohler (d. 1777) and Catherine Miller (1758–1842). Member of Capt. John Moritz's Northampton Co. Militia, at the Battle of Brandywine. Son of Michael Neihart (1713–93) (R. Gabel, 9, no. 1 [KRAN]; GAS-1, 238–41; JKJ, 54).

Neil, James (b. 1727). Born in Bucks Co., son of Robert and Jane (Rice) Neil. Associate (1744) of John Harris on the Susquehanna River. During the Revolution he was a gunmaker in Greene Co. (*Horn Papers*, 2:689).

Neil, Richard. Gunsmith, 1817, Allegheny Arsenal, Pittsburgh (t).

Nelson, Alexander. 1776, Philadelphia, VA. Received contract for 600 muskets (RAG, 139).

Nelson, Francis. Gunstocker, 1776, Strawberry Alley, Philadelphia (t).

Nelson, Robert. Artillery artificer and blacksmith, 1778–81 (PIR, 2:247).

Nelson, Seth Iredell, Sr. (1809–1905). Gunsmith, Clinton Co. (t; c; f).

Nelson, Seth Iredell, Jr. (1838–1934). Gunsmith, East Keating Twp., Clinton Co. (t; c; f).

Neth, Jacob (b. 1804). 1860–74, West Newton, Westmoreland Co. (t; c); after 1876, Cass Co., Iowa (Westmoreland DB 87, 461f.). Gunsmith.

Netter, Joseph. Gunsmith, rear of 2 Rose Alley, Philadelphia (MD, 1846).

Neubecker (Newbecker), Philip. 1817–26, Halifax Twp., Dauphin Co. (t; SED, 58) Gunsmith.

Nevell, John. On 1/8/1773, Nevell was apprenticed to John Handlyn, an established Philadelphia gunsmith, but the trade Nevell was to learn was not specified (PIDS).

Nevins, John G. Gunsmith, Buckhorn, Columbia Co. (BPaD, 1861).

Newcomer, Abraham. Gunsmith, Mennonite, Lancaster boro. (Francis Parkman, *Conspiracy of Pontiac*, appendix). He was in Lancaster boro in 1761 (SED, 58) and 1764 (Rupp, *History of Lancaster County*, 165). He was probably active for a much longer period (GAS-1, 308; MB 8 [1965]: 18).

Newcomer, Christian. Gunsmith, 1802, Hempfield Twp., Lancaster Co. (SED, 58; GAS-1, 308).

Newcomer, John (d. 1782). 1761–82, Hempfield Twp., Lancaster Co. Gunsmith and blacksmith. His estate inventory showed the trade tools of a gunsmith, wheelwright, and blacksmith (DB M-160; GAS-1, 308–11; JKJ, 115–16; "Old John, Gunsmith," MB 8 [1965] 11).

Newcomer, John, Jr. Gunsmith, 1780–83, Hempfield Twp., Lancaster Co.; 1805–10, Hellam Twp., York Co. (SED, 59; GAS-1, 308–11; JKJ, 115–16; "Young John, Gunsmith," MB 8 [1965] 18).

Newingham, Daniel (b. 1754). Born in Ireland. Blacksmith, New Pa. 11th Regt., 1777. (2 Pa. Arch. 11 at 62).

Newton, Jonathan. Artillery artificer, 1778, at the cannon factory, Carlisle (PIR, 2:249).

Nicely, Jacob (b. 1810). 1837–47, Little Beaver Twp., Beaver Co.(t); 1848–70, Darlington Twp., Beaver Co.(t; c); 1880–90, Little Beaver Twp., Lawrence Co. (c; t). Gunsmith.

Nichman, John. Gunsmith, before 1829, Beavertown, York Co. (JKJ, 478–80).

Nicholls, Isaac. Brass founder and cannon caster, 1780, U.S. Arsenal, Philadelphia (2 Pa. Arch. 11 at 258).

Nichols, Marcus. Gunsmith, Erie (PD, 1890).

Nicholson, John. Gunsmith, Dock Ward, Philadelphia. On 2/28/1789 he was paid £119/19/0 for repairs to the city's arms. He was also paid for repairs on guns on 12/12/1787. (3 Pa. Arch. 7 at 226) Nicholson made the pattern guns sent to all Pa. counties by the Committee of Safety, in an attempt to standardize the arms used in the Revolution (Col. Rec. 10:335). On 6/11/1773 he purchased an indentured servant, Richard Trested, who had come from London, to employ in the gunsmithing business (PIDS). On 7/18/1789 there was a reported balance in Nicholson's account for arms repairs of £129/2/0 (Hazard's *Register*, 2:78). His plan for creating a state gunsmith shop is interesting.

PLAN FOR CARRYING ON A GUN FACTORY, 1776.

By John Nichalson, Philad[a]., June, 4th, 1776.

First.—There must be a convenient place for erecting a Mill for boaring and grinding barrells, to be under the direction of M[r]. Tomlinson, or some other barrell maker, & shops fit to forge barrells, in, with 3 or 4 forges.

Second.—A Shop to contain three forges for forgeing Locks, with a good sett of tools to each, and a shop or shops to contain forty Lock filers, with a good sett of tools to each, suitable to the part of the Lock they have to file, with a forge for every ten Lock filer to harden & temper the Springs, mend tools, & case harden, &c.

Third.—A shop or shops for ten Gun stockers, with each man a sett of tools.
Fourth.—A Casting Shop, & propper tools for a Brass founder to Cast the mounting, & a shop to finish ditto in.
Fifth.—A Shop with two forges to forge Bayonets and Steel Ramrods, & works erected at the Mill for grinding and polishing ditto.
Sixth.—A forge for makeing Swivals, trickers, Pins & Springs for the Ramrods, &c.
Seventh.—A Small Shop for putting the Guns together in, near the Stocking Shop.

N. B.—It will be necessary to get all the files & Brass that can be got, as them Articles are become very scarce, and to provide a file Cutter to cut the files over Again. It will be likewise Nessecary to have a Clark that understands Something of the hardware business, in order to provide things as they may be wanted. [1 Pa. Arch. 4 at 767–68]

Nichtman, John. Gunsmith, 1813–28, Tyrone Twp., Adams Co. (WSB, 95).

Ninemeyr, Charles (b. 1807). 1833–50, Greenville, Mercer Co. (t; c). Gunsmith.

Nippes, Abraham. Gunsmith, 1800, Longswamp Twp., Berks Co.; 1812–13, 262 St. John's, Philadelphia (KEA, 71).

Nippes, Abraham, II. Gunsmith, 449 N. 3d St., Philadelphia (PXD, 1812); 262 St. John's (PXD, 1812); 111 Dillwyn (DeSD, 1829, 1830, 1831, 1835–36; MD, 1841, 1846).

Nippes, Albert S. 1848–60, Mill Creek, Philadelphia Co. Made Sharps rifles, 1849–51 (FSP, 223).

Nippes, Charles. Gunsmith, 262 St. John's, Philadelphia (DeSD, 1823).

Nippes, Daniel (1796–1848). Major manufacturer and government contractor, Philadelphia (FSP, 223).

Nippes, William. Arrived from Holland in 1796, active in Philadelphia as a gunsmith and government contractor, 1796–1829 (FSP, 223).

Nitre Hall Mills. Gunpowder manufacturers, 1810–30, Haverford Twp., Delaware Co. In 1825 the mill made 10,000 casks of gunpowder and employed 25 men (Charles Palmer, ed., *A History of Delaware County,* 2 vols. [Harrisburg: National Historical Association, 1932] 1:134).

Nitschmann, Martin. Cutler, 1749, arrived from Moravia on the *Irene*, part of the John Nitschmann Moravian colony who went to Bethlehem (PaE, 42).

Noble & Brady. Gunsmiths. John Noble and Freeman Brady, Washington boro (PAD, 1863–64).

Noble & Little. Gun-barrel makers, Pittsburgh (PAD, 1856–57).

Noble, John. Claysville, Washington Co. Patent 34,126 of 1/28/1862 for the Noble magazine gun.

Noll, Henry (1791–1855). 1820–55, Washington Twp., Franklin Co., gunsmith (WSB, 29–35; JKJ, 516–17).

Noll, John (1747–1824). Major gunsmith. 1800–1824, Washington Twp., Franklin Co. (WSB, 29–35; JKJ, 500–9).

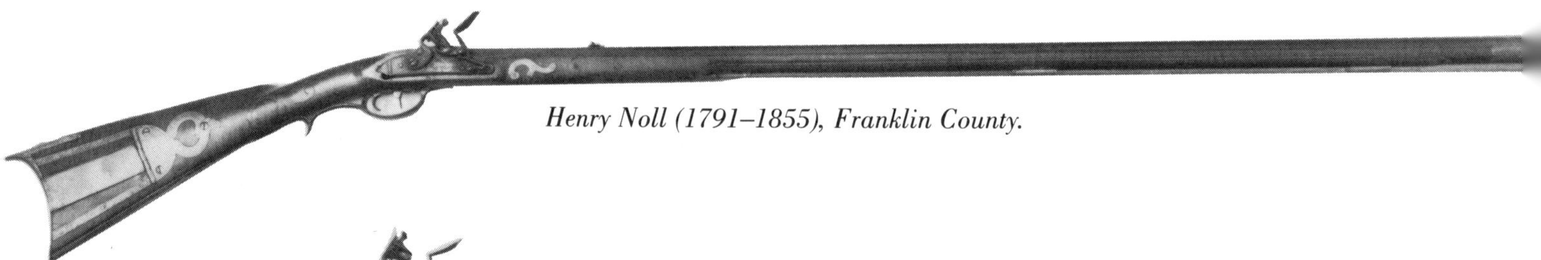

Henry Noll (1791–1855), Franklin County.

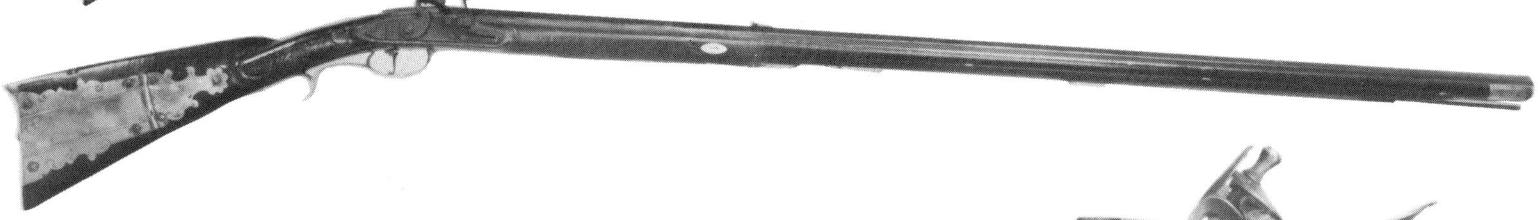

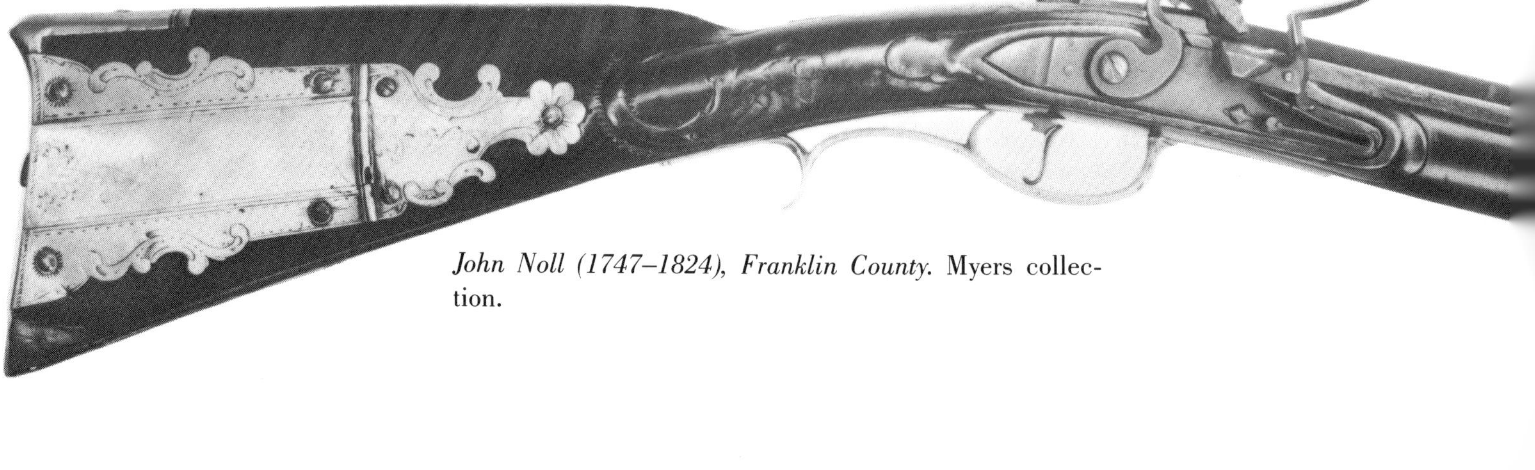

John Noll (1747–1824), Franklin County. Myers collection.

Noll, John, II. (b. 1784 or 1785). Gunsmith, before 1834, Antrim Twp., Franklin Co. (WSB, 29–35); 1850, Monroe Co., Ohio (c).

Nord, Jacob. Gunlock maker, 19th Ward, Philadelphia (c, 1860).

Norman, John. Gunsmith, Springfield, Bradford Co. (BPaD, 1861).

Norton, Jacob (b. 1767). Son of John and June (Noss) Norton of Lancaster Co. Gunsmith, 1800–34, Wayne Twp., Mifflin Co. (t).

Norton, John (b. 1735). 1765–95, Newton Hamilton, Wayne Twp., Mifflin Co., gunsmith (t; f; c).

Norton, Lyman. Gunsmith, Susquehanna boro (BRD, 1867).

Noss, J. A. Gunsmith, New Castle, Lawrence Co. (PD, 1882).

Nottridge, James. Armorer on the Braddock expedition. Petition from his wife of 7/10/1775 for back pay, £45/15/0 (Virginia Colonial Report, MIL-8, Survey Report 5822).

Novelty Iron Works. Main St., Union City, Erie Co. W. H. Judson, prop. Gunsmiths, locksmiths, bell hangers, and the like. (*Erie County Directory*, 1873–74).

Null, Nathan. Gunsmith, Seller's Tavern, Bucks Co. (BPaD, 1861). Gunsmith.

Numberal, D. 1880, Gunsmith Rogersville, Center Twp., Greene Co. (t; c).

Nungesser, George. Silversmith, 1786–88, Easton boro (t).

Nunnemacher, Abram. Gunsmith, 1779–91, York boro (t).

Nunnemacher, Andrew. Gunsmith, 1779–83, York boro (t; JKJ, 337–38).

Nunnemacher, George. Gunsmith, 1802–10, Huntingdon Twp., Huntingdon Co.

O

Oakes, Samuel. Gunsmith, ca. 1785–90, Northumberland Co.; ca. 1792, Ft. Washington (Pittsburgh); 1800, Philadelphia (KKR, 314; RAG, 143).

Oberholtzer, Christian, Sr. Gunsmith, Hempfield Twp., Lancaster Co. (t) In 1789 worked for William Henry, Sr. making guns for the Committee of Safety (RAG, 143).

Oberholtzer, Christian. Gunsmith, 1802, Hempfield Twp., Lancaster Co. (SED, 59).

Oberteuffer, Charles A. (b. 1811). 1842–50, Philadelphia (t; c); 1836–40, Baltimore, Md. (FSP, 226).

O'Brien, P. Gunsmith, Sugar Grove Twp., Warren Co. (BRD, 1867).

Ochs, Charles. Locksmith, Bushkill Center, Northampton Co. (c, 1860)

Oellers, James. 11/18/1776: "pay James Oellers £3364/4/0 for 128 casks of powder" (Col. Rec., vol. 11).

Oerter, John Christian (1747–77). 1771–77, Lower Nazareth Twp., Northampton Co. (t) Apprenticed to Andreas Albrecht at Christian's Spring gun shop. He became the shop manager at the age of 19, in 1776, and continued until his premature death at the age of 29, on 3/29/1777 (GAS-1, 173).

Old, James (1730–1809). Born in Wales, buried at Bangor Churchyard, Lancaster Co. In 1773 he leased Reading Forge on French Creek. In 1776 he was paid £1,500 for his services by the Committee of Safety, and in 1777, he was paid an additional £1,400 for cannon balls cast at his furnace. (2 Pa. Arch. 1 at 25) After initial difficulties, he began to cast cannon at the iron works (3 Pa. Arch. 5 at 36, 60; 2 Pa. Arch. 1 at 498; SHIS, 181).

Olden, Leigh. Gunsmith, Main below 7th, Carbondale, Lackawanna Co. (BPaD, 1861.

Oldham, G. Philadelphia. sold guns, ammunition, and the like. (PD, 1882).

Oldham, Thomas (1808–91). 1832, apprenticed to George Slonaker, Greenfield Twp., Bedford Co; 1832–54, St. Clair Twp., Bedford Co.; 1856–91, Mallory Twp., Clayton Co., Iowa (t; c; f). Major gunsmith.

Olenhousen, Jacob. Gunsmith, Pennsylvania Ave., near Dinwiddie, Pittsburgh (PAD, 1847); later, Wooster, Ohio (DH, 400).

Oliphant, Andrew and John. In 1798, with Nathaniel Breading, they purchased Fairfield Furnace on Georges Creek, Fayette Co. By 1800 the Oliphants became the sole owners, along with owning the Sylvan Forge on Georges Creek. "While the Oliphants operated Fairfield Furnace, they cast a quantity of shot which was used by General Andrew Jackson's artillery at New Orleans. It was shipped down the Monongahela, Ohio and Mississippi Rivers" (SHIS, 215).

Oliphant, John. Uniontown, Fayette Co. Patents 37,406 and 37,407 of 1/13/1863 for a safety nipple and a breech-loading firearm.

Oliver, Robert. Gunsmith, Pittston, Luzerne Co. (PD, 1890).

Oliver, Thomas. Artillery artificer, 1779, at the laboratory

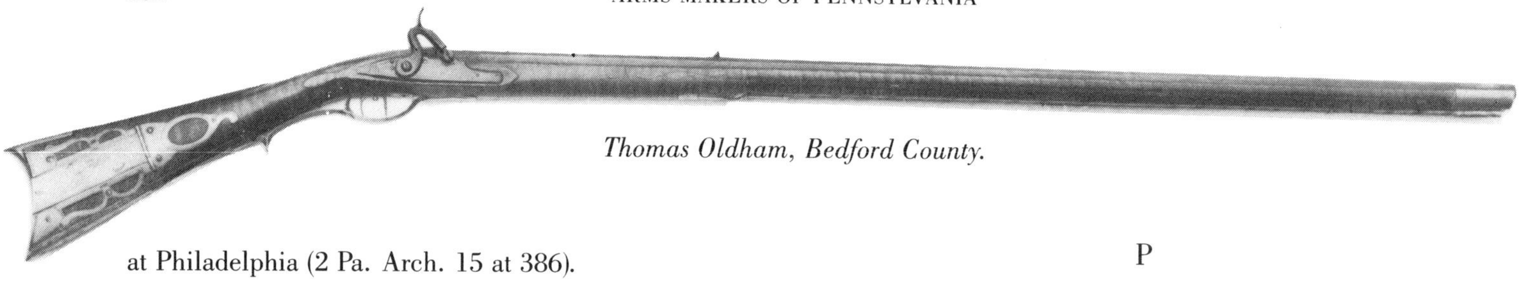

Thomas Oldham, Bedford County.

at Philadelphia (2 Pa. Arch. 15 at 386).

O'Neil, Charles. Gunsmith and cutler (advertisement in the *Franklin Repository,* 11/11/1803).

O'Neil J. Palmer. 66 5th Ave., Pittsburgh. Sold guns, ammunition, and the like. (PAD, 1885; PD, 1882).

Oppelt, Edwin A. (b. 1882). Gunsmith and physician. 1842–48, Clarion boro; 1849–50, Clay Twp., Tuscarawas Co., Ohio (REH, 193–95).

Orban, John. Locksmith, 1781, Dover Twp., York Co. (t).

Orman, John Michael. Gunlock maker, 1759, Lancaster (SED, 59).

Ormstead, Abraham. Artillery artificer, ca. 1779, Capt. Noah Nicholas's Co.; cannon maker at Philadelphia (PIR, 2:246).

Orre, James (1796–1887). Before 1830, Westmoreland Co; after 1830, Belmont Co., Ohio (DH, 49). Gunsmith.

Orton, John (b. 1729). Gunsmith, near Mulberry St., Philadelphia, 1790–95 (t; c; KEA, 73), but probably active much longer.

Orwan, Frederick (b. 1776). Patent of 2/25/1815 for a multi-shot gun. New Bloomfield, Perry Co. (c, 1850).

Orwan, John. 1793–1800, Carlisle, Cumberland Co. (WSB, 58). Gunsmith.

Overbaugh, C. E. 1870–75, Philadelphia; 1879–87, 300 Broadway, New York City. (RAG, 144) Late, quality percussion target guns.

Overdorf, John. Gunsmith, Brush Valley, Indiana Co. (BPaD, 1861).

Owens, E. G. 1850–63, employed as a gunsmith by Henry Deringer at Philadelphia; 1877–80, Denver, Colo. (RAG, 144).

Owens, John. Indian trader and gunsmith. 1747, unlicensed trader; 1751 at Logstown, Ohio; 1758, guide for Gen. Bouquet. He set his losses in the French and Indian Wars at £1,040/6/9, including iron, steel, gun parts and gunsmithing tools valued at £30, (Col. Rec. 5:532, 6:536; Col. Rec. 6:160; S. K. Bailey, *Ohio County Papers,* 150f; 1 Pa. Arch. 2 at 14; Bausman, *History of Beaver County,* 978).

P

Padden, Pashem. Artillery artificer, 1777, Capt. Nathaniel Irish's Co. (PIR, 2:247).

Page, Thomas. Artillery Artificer, 1779, at the laboratory at Philadelphia (2 Pa. Arch. 15 at 386).

Page, William. Artillery artificer and blacksmith, 1779, at the cannon factory at Philadelphia (2 Pa. Arch. 15 at 384).

Painter, Jacob (1765–1824). Born in Mecklenberg, Germany. Came first to Berks Co. 1805–6, gunsmith, Allegheny Twp., Somerset Co.; 1808–24, Harrold settlement, Greensburg (t; DAR).

Palm, Jacob (d. 1777). 1759–68, Lancaster; 1768–77, Ulster Co., N.Y. (RAG, 145). Gunsmith.

Palm, John. Gunsmith, Womelsdorf, Berks Co. (BPaD, 1861).

Palmer, Frederick. Gunsmith, Cedar at 8th St., Philadelphia (MD, 1837).

Palmer, George. Gunsmith, 109 Shippen above 8th St., Philadelphia (MD, 1846).

Palmer, Joseph. On 5/24/1773, Samuel Wigfal purchased Palmer, who came from Ireland as an indentured servant to learn the cutler's trade. The indenture was for 7 years. In the 5th year Palmer was to receive £5, in the 6th year, £6, and in the 7th year, £7. Wigfal paid £15 for Palmer. (PIDS) 1778, Pa. 9th Regt. working at the armorer's camp (2 Pa. Arch. 15 at 490).

Palmer, Stephen. Gunsmith, Franklin Lane, Philadelphia. (MD, 1855).

Palmer, Theodore. Gunsmith, 1770–85, north side of Market St., Philadelphia (t).

Palmer, Thomas. 1769–1818, Philadelphia. Revolutionary war contractor and arms supplier. On 7/6/1776, guns made by Palmer ordered to be proofed. (Col. Rec. 10:648, 653) Committee of Safety sought to purchase gunlocks from Palmer, 1/18/1776 (2 Pa. Arch. 1 at 470). U.S. Arms Inspector at Philadelphia, 1808–18 (RAG, 73; KEA, 73).

Pancoast, David. Capt., 1777, artillery artificers (2 Pa. Arch. 11 at 250).

Pancoast, Elisha L. (1822–96). Gunsmith, 1850–51, Elizabeth, Allegheny Co. (t); 1852, West Newton, West-

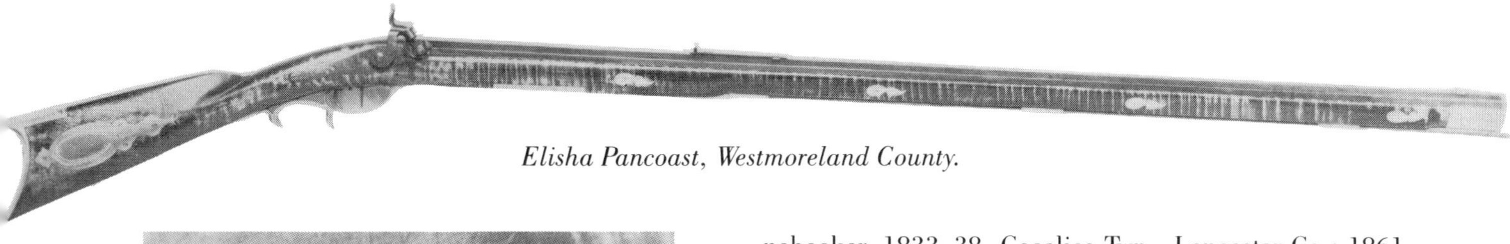

Elisha Pancoast, Westmoreland County.

Elisha Pancoast (1822–96) and his second wife, Mary Stewart. An old tintype taken on their wedding day. Pancoast was active in Westmoreland and Allegheny counties.

moreland Co. (t); 1853–56, Morgantown, W. Va. (t); 1858–59, Elizabeth boro. (t); 1860, West Elizabeth (c); 1861–80, Elizabeth boro. (t); 1882–90, Greensboro, Greene Co. (t; c). Died at Confluence, Somerset Co., 3/30/1896 (obituary in Elizabeth *Herald*, 4/3/1896). Twice married: first wife, Elizabeth; second wife, Mary, daughter of Alexander and Margaret (Drake) Stewart. Elisha was a son of Israel and Mary (Lindsey) Pancoast of N.J. (f).

Pannebecker, Daniel, Sr. (1761–1825). 1799–1825, Cumru Twp, Berks Co. A son of John and Magdalena Pannebacker; married Anna Maria, daughter of Johannes and Anna Barbara Mosser. Gun-barrel boring mill and forging hammer (Goodling, "Some Notes," KRAN, 3, no. 1 [1976]: 5–8).

Pannebecker (Pennypacker), Daniel, Jr. (1791–1872). 1828–61, gun-barrel maker and gunsmith, Cocalico Twp., Lancaster Co. (t; c; BPaD, 1861; Goodling, "Some Notes").

Pannebecker, Jesse. Son of Daniel and Anna Maria Pannebecker. 1833–38, Cocalico Twp., Lancaster Co.; 1861, Clay Twp., Lancaster Co. (t; SED, 59; c; Goodling, "Some Notes"; BPaD, 1861) Gun-barrel maker and gunsmith.

Pannebecker, David. "The pioneer gun maker on this [Wyoming] stream was David Pennypacker [Pannebecker], father of William, who worked near the head of the stream, as early as 1786. He made guns complete, including the woodwork, while William made barrels only" (Morton L. Montgomery, comp., *History of Berks County*, 2 vols. [Chicago: Beers, 1909] 714).

Pannebecker, John. 1826–34, Brecknock Twp., Lancaster Co. (SED, 59). Gun-barrel forger and gunsmith.

Pannebecker [Pennypacker], Samuel (1794–1876). Son of Daniel and Anna Maria Pannebecker. 1828–44, Brecknock Twp., Lancaster Co. (SED, 59; Goodling, "Some Notes"). Gunsmith.

Pannebecker [Pennypacker], William, Sr. (1785–1870). Gunsmith, Mohntown, Berks Co., ca. 1800–38 (t; c; Goodling, "Some Notes").

Pannebecker, William, III. Gun-barrel forger and gunsmith, Cocalico Twp., Lancaster Co. 1850–61 (c, 1850, 1860; BPaD, 1861).

Pap, Anthony. Artillery artificer, 1777 (BCR, 115). Capt. Will's Co.

Park, John, Sr. (1773–1851). 1789, Franklin Twp., Huntingdon Co. (t); 1814–19, Woodberry Twp., Bedford Co., gunsmith and tavern keeper (t); 1829, Dublin Twp., Bedford Co. (t). Died at Cassville, Huntingdon Co.; buried at Saltillo, Huntingdon Co. (died 12/10/1851, at the age of 78.

Parke, Theophilus. Artillery artificer, 1777; dismissed from service in 1780 (2 Pa. Arch. 11 at 250).

Parker, Andrew. Gunsmith, 1780, Dock Ward, south part, Philadelphia (3 Pa. Arch. 15 at 220).

Parker, E. R. Gunsmith, Scranton, Lackawanna Co. (PD, 1887).

Parker, Samuel. Brass founder, artillery artificer, at the cannon factory at Philadelphia (2 Pa. Arch. 15 at 385).

Parker, T. C. Sold guns, ammunition, and the like, Wilkes-Barre (PD, 1882).

Parker, William. Carpenter and sergeant, 1780, artillery

artificers, Capt. Thomas Wylie's Co. (PIR 2:248).

Parkhill, Andrew. Gunsmith, 1779, Dock Ward, south part, Philadelphia (t).

Parkinson, Benjamin F. Patent 48,201 of 6/13/1865, Washington.

Parkinson, Joseph (d. 1807). Settler at Conococheague settlement. 1877, Catfish Camp (near Washington boro.), Yohogania Co., Va. Appointed by Committee of Safety at West Augusta to make guns, tomahawks, scalping knives, and other arms. 1801, Dunnbar Twp., Fayette Co.; 1804, Somerset Twp., Washington Co.; 1807, died in Brooke Co., W.Va. (Crumrine, *Washington County*, 565).

Parkison (Parkinson), Thomas (d. 1808). One of 5 brothers from Conococheague Settlement, Washington Co., Md. On 1/29/1777 the Council for West Augusta appointed Thomas a gunsmith and scalping-knife and tomahawk maker for Yohogania Co. (Greene, *Washington and Fayette Counties of Pennsylvania*) 1773–75, Tygary Valley, Marion Co., W.Va.; 1777–83, Monongahela City. Will dated 4/1808, Brooke Co., W.Va. (Boyd Crumrine, *History of Washington County* [Philadelphia: Everts, 1882] 186).

Parkison, William. Appointed 1/2/1777 by Council for West Augusta as a gunsmith and tomahawk and scalping-knife maker, Yohogania Co. (present-day Monongahela City, Pa.) (Crumrine, *Washington County*, 186).

Parks, John, Jr., (1816–94). 1821, Henderson Twp., Huntingdon Co. (t); 1837–94, Selinsgrove, Snyder Co. (t; c; f).

Parks, John K. 1819–20; 1822–45, Penn Twp., Snyder Co. (t); 1821, Henderson Twp., Huntingdon Co. (t); 1847–50, Bradford Twp., Clearfield Co. (t; c). On 1/18/1816 he married Jane Parkinson in Huntingdon Co. (f).

Paschall, Stephen. Cutler, 1771–72, Philadelphia (t).

Paschall, Thomas. Pewterer, 1681–82, Penn's Grant (PSSP, 198); 1701, pewterer, "formerly from City of Bristol, now of Philadelphia" (3 Pa. Arch. 3 at 338).

Pase, Jake. Gunsmith, Cooper Twp., Clearfield Co. (BRD, 1872, 1880).

Pass, John. "as early as 1734 . . . shot for cannon were cast by John Pass . . . the same [man] who recast the statehouse bell" (Hazard's *Annals*, 1:325).

Patchen, Ira. Gunsmith, Covington, Tioga Co. (PD, 1882).

Patterson, Andrew. Gunner, 1779–82, Capt. Thomas Wylie's co. of artillery artificers; wounded in the wrist. 1807, lived in York Co. (PIR, 2:249).

Patterson, Jacob. Gunsmith, Harrisville, Butler Co. (BPaD, 1861).

Patterson, James. Locksmith and gunsmith, Birmingham, Allegheny Co. (PAD, 1815).

Patterson, James S. 1873–90, McCoysville, Juniata Co. Gunsmith. Patent 346,941 of 8/10/1886 for a gunlock (t; c; RAG, 147).

Patterson, Robert. Gunsmith, 1827, Smith Twp., Washington Co. (t)

Patterson, Samuel J. Gunsmith, Bank St. and Tremont St., Pittsburgh (PAD, 1858–59).

Patton, Anthony (b. 1761). Blacksmith, 1779, New Pa. 11th Regt.; 5′ 11″ tall (2 Pa. Arch. 11 at 67).

Patton, John. Armorer on the *Argus*, 10/1/1776, Pa. Navy. (2 Pa. Arch. 1 at 244); 12/11/1776, was paid £35 for cannon shot, Berks Co. (BCR, 178).

Paul, James. Gunsmith, 1800, Horsham Twp., Montgomery Co. (KEA, 147).

Paul, William (b. 1825). 1850–52, Summerhill Twp., Cambria Co., gunsmith. (t; c) In the census of 1850, there was a William Paul, son of William Paul, both listed as gunsmiths, father age 48, son age 17, in Centreville, Wayne Co., Ind.; both were born in Pa.

Paulhill, James. Sgt., 1778, in Capt. Nathaniel Irish's Co. of artillery artificers (PIR, 2:246).

Payne, S. L. Gunsmith, 1846–51, Erie boro (BRD, 1846; c, 1850).

Peale, Charles Wilson. The great American artist experimented with David Rittenhouse on early telescopic gunsights, 1776–80 (Kelley, *Pennsylvania: The Colonial Years*, 742).

Pearson, James. Philadelphia, Committee of Safety contractor, musket-barrel maker (RAG, 148).

Pearson, Joseph. Blacksmith shop and cutler, 1767–80, Lower Darby, Chester Co. (t).

Peck, Levi (Eli). 137 Green St., Philadelphia. Gunsmith (DeSD, 1830, 1823, 1829, 1833).

Peden, John, Armorer, 1776–77, Pa. Navy, the Arnold Battery (5 Pa. Arch. 1 at 425; 3 Pa. Arch. 23 at 11).

Peebles, Robert. Cumberland Co. Committee of Safety musket maker, 1777–80. (Col. Rec., 10:700).

Pelck, Rudolph Oswald (1827–85). Born in Posen, Prussia. 1850–55, Erie boro., brother-in-law and apprentice to Henry Jarecki; 1856–85, Freeport, Stephenson Co., Ill. (t; c; Curtis Johnson).

Peloux (Feloux, Pelaix), Peter. 190 South St., Philadelphia (DeSD, 1829); 190 Cedar St. (DeSD, 1829, 1830); 60 Cedar St. near 5th St. and German St., Philadelphia (MD, 1841, 1846). Armorer and gunsmith.

Pembrook, Thomas. 1778, Capt. Nathaniel Irish's Co. of artillery artificers (PIR, 2:247).

Pendleton, John. Cannon maker, artillery artificer, 1779, Pa. Cannon Factory at Philadelphia (2 Pa. Arch. 15 at 385).

Pennel, Joshua. Gunsmith, miller, and farmer, 1835–62, Southampton Twp., Bedford Co. (t; f; c).

Pennsylvania Rifle Works. George Dunlap, prop. Philadelphia, post–Civil War. Made and imported cheap guns of many kinds (RAG, 149).

Pennsylvania State Gun Factory. Established 3/6/1776 on Cherry St., Philadelphia. 12/1776, moved to French Creek, near Valley Forge; 9/1777, moved to Hummelstown, Dauphin Co. Liquidated by William Henry, Sr., armorer, 12/17/1778 (RAG, 149).

Pennsylvania State Gun Repair Shop. Established 9/1777 at Allentown. By 5/1778 it had repaired and assembled 800 guns, and 150 more were in repair (RAG, 149).

Pennypacker, Henry. Smith, 1774, Perkiomen Twp., Philadelphia Co. (t).

Pennypacker, John (1850–1926). Adamstown, East Cocalico Twp., Lancaster Co. Gunsmith and barrel maker (t; c; Goodling, "Some Notes").

Pennypacker, William, Jr. (1818–78). Gun-barrel maker, Gouglerville, Berks Co. (Goodling, "Some Notes"; t; c) In 1845 he took over his father's shop; in 1860–65 he was active in Trenton, N.J.; 1870–78, Mohntown (Cyrus T. Fox, *History of Reading and Berks Counties*, 3 vols. [New York: Lewis, 1925] 351).

Perkin & Coutty. Samuel Coutty and James Perkins, 1784, Philadelphia (KEA, 74). Gunsmiths, government contractors (3 Pa. Arch. 5 at 341).

Perkin, Joseph. Gunsmith, 1779–90, Dock Ward, north part, Philadelphia. 1779, artillery artificer. (2 Pa. Arch. 15 at 383). 5/18/1784, Perkin & Coutty, gunsmiths, were paid £81/1/0 for repair work (3 Pa. Arch. 5 at 341). 4/12/1788, repaired city arms, Philadelphia, and were paid £100; 9/11/1788, were paid £200; 3/18/1789, were paid £318/14/5 (3 Pa. Arch. 7 at 306); 12/20/1787 (2 Pa. Arch. 2 at 710).

Perkins, Reuben (1767–1816). 1767–70, Susquehanna Co. (t); 1788–90, Greene Co.; 1798–1816, Belmont Co., Ohio (DH, 50). Gunsmith.

Perkins, William. Smith, 1779, Philadelphia (3 Pa. Arch. 15 at 472).

Perrein, Dennis. 1796–1800, Philadelphia (KEA, 74). Gunsmith.

Pershing, Christian (1751–1825). Born in York Co. By 1785, moved to Greensburg and worked as a weaver. 1792–1810, ran a gunpowder mill in Westmoreland Co., which he sold to Jacob Rugh in 1810 (Edgar J. Pershing, *The Pershing Family in America* [Philadelphia: Ferguson, 1924] 103–4).

Peterman, Abraham. 483 S. Front St., Philadelphia. (MD, 1846; BPaD, 1861); 131 Walnut St. (MD, 1870, 1875); 708 Market St. (GPD, 1892). A. Peterman & Sons. Gunsmiths.

Peterman, Lewis C. Gunsmith, Germantown Rd. at Phoenix, Philadelphia (MD, 1837, 1841, 1846).

Peterman, A., & Sons. 708 Market St., Philadelphia (BPD, 1890). Gunsmiths.

Peters, Jacob. Artillery artificer, 1779–82. (2 Pa. Arch. 11 at 236); armorer at Ft. Pitt, 1782 (2 Pa. Arch. 11 at 262).

Peters, Jonas. Gunsmith. 1795, Heidelberg Twp., Northampton Co. (t).

Peters, William. Artillery artificer, 1779, at the laboratory at Philadelphia (2 Pa. Arch. 15 at 386).

Pettibone, Daniel. 1799–1814, Philadelphia. Patent of 1802 for a method of welding using borax; patent of 2/12/1814 for a method of boring gun barrels. On 9/15/1814 he was given a public arms contract for 200 pikes. (MPHS, 8:142) 1808–9, U.S. arms inspector, Philadelphia. 1811–14, made belt axes and scalping knives (RAG, 223).

Peusch, Frederick. Powder-horn maker, Lancaster. (c, 1850, 1860; BPaD, 1861) In 1860 he made 6,000 horns (IC).

Pfeiffer, Charles. Gunsmith, alley between Walnut St. and Chestnut St., above Queen St., Lancaster (BLD, 1857).

Pharres, William. Gunsmith, Broadheadsville, Monroe Co. (BPaD, 1861).

Pheaster, Samuel (b. 1812). Gunsmith, 1838–60, West Bethlehem Twp., Washington Co. (t).

Philipee, Jacob. Gunsmith, 1842, Heidelberg Twp., Lebanon Co. (KEA, 75).

Philippi, Samuel (1801–77). 1821–77, Easton, Northampton Co. (t; c; f); gunsmith (BRD, 1865, 1877).

Philips, Lavan. Gunsmith, 1791–92, Pittsburgh (t).

Phillippi & Son. 16 S. 3d St., Easton (PD, 1882; Uzal W. Conduit, *History of Easton* [Easton: West, 1889] 477). 56 N. 2d St. (BRD, 1860). Gunsmiths.

Phillippi. Solomon. 1850–77, gunsmith, Easton, son of Samuel (t; c).

Phillips, John. 10/20/1775: "Capt Robery Towers, Commissary, reports that he has purchased the following items . . . of John Phillips, 3 cutlasses at £0/25/6 each" (Col. Rec., 10:375).

Pickel, Henry. 1780–1809, York (JKJ, 284; t; KKR, 322); 1811, Manheim Twp., Lancaster Co. (SED, 59). Contract of 4/17/1801 for liberty cap muskets; work to be done with Conrad Welshans and Jacob Doll, York.

Pickets & Wier. James Weir (Wier), Church Alley and 2d St., Philadelphia. "1500 lock brushes from Pickers & Wier" on 11/8/1775 (Col. Rec., 10:398).

Pieper (Piper), Abraham. 1798–1803, Lancaster. (t) In 1803 he signed the petition asking Congress to retain the import duty on foreign-made guns.

Pierce, John (1792–1836). Born in Conn., he came to northeastern Pa and there married Susan Shupp (1792–1861), daughter of Col. Philip and Katherine Shupp of Northampton Co. Gunsmith and farmer, Plymouth Twp., Luzerne Co. (*Genealogical and Family History of the Wyoming and Lackawanna Valleys* [Chicago: Beers, 1906], 1:525).

Piper, Edwin. Springfield, Mass., patent 51,391 of 12/5/1865 for a gun part for a breech-loader. 1347 S. 2d St., Philadelphia. (PD, 1882; BSED, 1880. BRD, 1873–74) Gunsmith and inventor.

Piper, George (b. 1783). On 8/31/1801 Piper was apprenticed to John Armstrong, Emmitsburg, Md. In 1817, Shippensburg, Cumberland Co., gunsmith. (WSB, 59; AMM, 47, 119–20). 1821–24, 1842–43, Hopewell Twp., Bedford Co. (t).

Piper, Thomas. Gunsmith, 1829–43, Hopewell Twp., Bedford Co. (t) He was also listed as a watchmaker, surveyor, and clock maker.

Pirkle, Henry. "The second gunsmith in the city of Williamsport was Henry Pirkle who came here in 1812. He had his shop on the northeast corner of 4th and Market Sts." (NAT, 214).

Pittsburgh Fire Arms Co. 68 5th St., Pittsburgh; 1837–58, gunlock makers; 1860–84, gunsmiths, lock makers, and importers (PAD, 1837, 1884; c, 1850, 1860; IC, 1840, 1850, 1860).

Plank, William. 1821, Greenwood Twp., Columbia Co. (t). Gunsmith.

Plants, Christian. 1861–73, East Finley Twp., Washington Co. (t). Gunsmith (c, 1860, 1870).

Pleasant & Charnler. Gunlock makers and importers, 1850, Philadelphia (c. 1850; RAG, 52; IC, 1850).

Plumley, George. Cutler, Philadelphia (advertisement in *Pennsylvania Packet*, 3/28/1749).

Plumstead, Francis. Ironmonger, 1701, Philadelphia; of the Monories, London (3 Pa. Arch. 3 at 338).

Pole, Edward. 5/21/1777: "pay to Edward Pole £0/15/0 for cannon powder horns," Philadelphia (2 Pa. Arch. 1 at 160).

Pollard, John. 3/26/1776: "upon application of Jno Pollard and Jacob Acfield for the payment of their accounts for bayonets and repairing firearms . . . two orders were drawn . . . one in favor of John Pollard for £2/11/1" (Col. Rec., 10:525)

Pollit, John H. (1820–95). Born in England. Active in Pittsburgh, 1842–95 (t; c; f). Gunsmith.

Pollock, Benjamin Smith. Gunsmith, 1840, Neshannock, Lawrence Co.; 1841, New Castle boro. (t); 1849, Shenango Twp., Lawrence Co. Brother of David (f); signed his guns "B. Smith Pollock" or "Benj. Smith Pollock."

Pollock, David (1795–1876). 1820, Mifflin Twp., Allegheny Co. (t); 1829–30, Jefferson Twp., Allegheny Co. (t); 1840, Mifflin Twp., Allegheny Co. (t; c); 1850–60, Baldwin Twp., Allegheny Co. (c); 1870–76, Taylor Twp., Lawrence Co. (t; c). Gunsmith.

Polsz, Frederick. Philadelphia (MD, 1847). Gunsmith.

Ponsler, John. 5/10/1773. Apprenticed to Henry Sheets, smith, White Marsh Twp., Montgomery Co., (PIDS).

Poole, Edward. Parts dealer, 1795, Philadelphia (KEA, 75).

Poor, William. Artillery artificer, 1778, at the cannon factory at Carlisle, Capt. Thomas Wylie's Co. (PIR, 2:249).

Poorman, Michael. Gunsmith, 1772, Heidelberg Twp., Lancaster Co. (SED, 59); 1790, Dauphin Co. (c; JKJ, 279).

Pope, John W. On 7/17/1809 he received a patent for a method to make shot. Philadelphia.

Porter, James B. Received a patent of 8/4/1832 for a method of rifle manufacture. Girard, Erie Co.

Post, Samuel B. (1862–1947). Gunsmith. Apprenticed to George Craft of Morris Twp., Washington Co. Active most of his adult life on the National Road, west of Washington boro (Dr. Jim Herron).

Postley, Nelson & Co. 1852–75, 17 Market St., Pittsburgh. Gunsmiths (PAD, 1860–61; t; c). Also made shovels, vises, gun barrels.

Potts & Rutter. Cast cannon at Valley Forge, 1776–81. (See Potts, Thomas; Valley Forge Armory; Mount Joy Forge.)

Samuel B. Post (1862–1947), Washington County.

Potts, Jonathan. 1776, Reading. (See Nagel, George.) On 4/9/1776, was paid £19/12/0 for repairing 9 muskets (BCR, 176).

Potts, Thomas. Steelmaker, supplied steel to Committee of Safety as well as to many gunsmiths and cutlers. 1765–80, Coventry, Chester Co. Owned 1,065 acres and a steel furnace at East Nantmill, Chester Co. (t) "The cannon cast by Potts & Rutter to be annealed further, proved and mounted." 7/29/1776 (2 Pa. Arch. 1 at 487).

Powell, William. Coppersmith and parts supplier, 1782, Pittsburgh. Noted in George Croghan's will (HAN, 2:83).

Prahl, Lewis. 1774–97, Philadelphia. (t) 10/23/1775: "Mr Fox informed the Committee that the Commissions of Philadelphia Co. had contracted with a certain Lewis Prahl, gunsmith, for the making of 150 firelocks, that none were yet ready, but that he expected 24 of them would be delivered on Saturday next, and the gunsmith had promised to deliver for the future, 12 firelocks every week until the whole number of them were completed." He had another musket contract dated 11/9/1775. On 1/16/1776 he delivered "new arms." He also had a contract for 1,000 swords for Col. Benj. Flowers's Pa. Horsemen, 5/22/1777 (2 Pa. Arch. 3 at 104; 3 Pa. Arch. 15 at 392; t; c; BHAM, 1:572).

Pralsh (Pratsh), Charles A. Gunsmith, boarded at Blickenderfer's Hotel, Lancaster (BLD, 1857); gunsmith, Millersburg, Mercer Co., Il. (c, 1860).

Presoner, John. Smith, came from Maxfield, Cheshire Co., England, to Philadelphia, 1683 (PaE, 9).

Price, Benjamin. 1753–61, Lancaster Co., goldsmith (DB C-479, D-511, G-293) 1753, Conestoga Twp.; 1759–61, Lancaster boro (t).

Price, John. Gunsmith, 1805–6, Somerset boro (t).

Price, Joseph. Gunsmith, 498 N. 2d St., Philadelphia (KKR, 323).

Pringle, Jacob. Gunsmith, 1832–40, Cambria Twp., Cambria Co. (t) Philip Pringle's will, dated 11/21/1837 and probated 11/25/1841, named sons: Jacob and John.

Pringle, John. 1775–80, importer and armorer. 11/22/1775: "resolved, that . . . a committee purchase for this Board . . . 48 muskets, 12 cutlasses, and 540 pounds of gunpowder imported by Messrs. York and Pringle, (Col. Rec., 10:407; RAG, 155).

Pringle, John. Gunsmith, 1838, Cambria Twp., Cambria Co. (t) He may be the same man who made the first steam-powered paddle-wheel boats on the Ohio–Mississippi

J. Pringle, ca. 1835, Cambria County.

River complex at Brownsville (t, f, c).

Proctor, William. Gunsmith, 1815–16, German Twp., Fayette Co. (t); 1817–19, Menallen Twp., Fayette Co. (t).

Pryor, Silas D. (b. 1795). Born in Va. 1819–21, Allegheny Arsenal, Pittsburgh (t); 1826–33, Versailles Twp., Allegheny Co. (t); 1834, St. Clair Twp., Allegheny Co. (t); 1844–48, Baldwin Twp., Allegheny Co. (t); 1848–50, South Pittsburgh (c; t). Gunsmith.

Purdom, Calvin. Gunsmith, 236 Callow Hill, Philadelphia (PXD, 1816).

Pyle, Nathan. Gunsmith, 1827–36, East Bethlehem Twp., Washington Co. (t).

Q

Quick, J. M. Lackawaxen, Pike Co. (BPAD, 1861). Gunsmith.

Quigsall, Thomas. Swordsmith, in the employ of Lewis Prahl, gunsmith, 1777, Northern Liberties, Philadelphia Co. (2 Pa. Arch. 3 at 104).

Quinn, Kennedy. Gunsmith, 420 Lynd St., Philadelphia (BPAD, 1861); 1861–63, Quinn & Clark, Philadelphia. (t).

Quirk, William. Gunsmith, 111 W. 3d St., Chester (BRD, 1873, 1876).

R

Raaf, Jacob. Gunsmith, 1820–25, Halifax, Dauphin Co. (t; KEA, 77).

Rader, Benjamin. Gunsmith, Fraily Twp., Schuylkill Co. (c, 1850).

Radfang (Rathfong), George (1750–1819). 1770–80, Conestoga Twp., Lancaster Co. (SED, 59) Born in 1750, he had his own shop by 1770. In 1776 he was in Capt. N. Page's Co., Lancaster Co. Militia, and was reassigned in 1777 to make guns under William Henry. 1785–1819, Lancaster boro. (Alexander Harris, *A Biographical History of Lancaster County* [Lancaster, Barr, 1872] 465).

Radfang, Jacob (d. 1839). 1811–39, Donegal Twp., Lancaster Co., gunsmith (SED, 59; Harris, 465).

Raeser, Daniel. 1774, Strasburg Twp., Lancaster Co. (SED, 59). Gunsmith.

Raesor, David (d. 1774). Locksmith, 1759, Lancaster (SED, 59); 1759–74, Lancaster (JKJ, 91).

Raesor, Peter. Gunsmith, 1782–86, Lancaster boro. (SED, 59).

Raffsnyder (Reiffsnyder), George. Smith, 1780–81, Ruscomb Twp., Berks Co. (t).

Raffsnyder, John. Gunsmith, 1779–81, Reading, Berks Co. (t).

Rainey, James. Gunsmith, 1799–1818, Barree Twp., Huntingdon Co. (t).

Raker, Charles F. (1823–1907). Born in Holland. 1849–50, Tod Twp., Fulton Co., son of David and Caroline Raker. 1865–1907, Knox Co., Ill (t; c; Curtis Johnson). Gunsmith.

Raker, Conrad. Little Mahoney Twp., Northumberland Co. (PD, 1887, 1890). Gunsmith.

Ramsey, Robert. Gun-barrel forger, 1796–98, Martic Twp., Lancaster Co. (SED, 59).

Randall, Jacob. Gunsmith, 25 Tammany, Philaelphia (MD, 1837).

Randall, Joseph C. Gunlock 1847–61, makers and importers, Philadelphia (RAG, 157).

Randolph, Benjamin. Gunsmith, armorer, 1775–80, Philadelphia (KKR, 323; Pennsylvania *Evening Post*, 5/3/1777, 5/6/1777).

Rankin, John. 1783, gun-barrel factory along Bennett's Run, York Co. (HYC, 624, 632) Newberry Twp., York Co. (t).

Rate, Jacob. Gunsmith, White Deer Twp., Union Co. (BPaD, 1861).

Rauch, Benjamin. Gunsmith, Germansville, Lehigh Co. (BPaD, 1861).

Rauch, Johann Heinrich. Arrived in Philadelphia, 1761, on the *Hope;* a locksmith and Moravian; went to Bethlehem (PaE, 52) Came from Cologne on the Rhine River, Germany. 1769, Lititz, Lancaster Co., made gunlocks, edged tools, and coffee mills; 1776, developed an auger that would bore a hole and simultaneously expel the chips, a distinct improvement over the old pot-style auger. (Brenckman, *History of Carbon County,* 548) Lititz, Warwick Twp., Lancaster Co. (t, 1769–77).

Rawsell, John (b. 1813). Gunsmith, Jefferson Twp., Fayette Co. (c, 1850).

Ray, John. Employed 9/5/1785 to 12/10/1785, 87 days at £2/5/0 a day for gunsmithing work done among the Amerindians (3 Pa. Arch. 7 at 484). This may be the Indian trader John [W]ray, after whom Raystown, now Bedford, was named; or a relative of [W]Ray's.

Raysor (Reasor, Razor), Jacob (1764–1833). Reasor "was also a German and kept a gunshop. He came early in 1795, and purchased a lot the same spring. He was a little old man known to all the early residents as 'Daddy Raysor' and lived on Walnut St. until his death in 1833." (Brown, *History of Crawford County*, 344, 435) There was a son of Peter Reesor who arrived in Lancaster Co. in 1739, who had a son Jacob, born 1764, died 1835, whose dates obviously come very close to those of this gunsmith (Egle, *Notes and Queries*, 140).

Reading Furnace. Daniel Joy was sent to Reading Furnace on 6/12/1776 and 4/4/1777 to proof cannon made there (GAAAM, 110).

Readkars, Elisha. On 9/18/1778, Readkars bought the smithing tools sold by the Commonwealth after Henry Mull was declared "attainted by treason." Smith's tools sold for £39/5/0; sundry tools, £7/5/0. (6 Pa. Arch. 12 at 59). As I have not located any public record of Readkars otherwise, it may be that he planned to use the tools in some capacity other than gunsmithing.

Ream, Charles. Gunsmith, Ream's Crossing, near DuBois (t, 1846).

Reasor, David. 1748–60, Lancaster boro., gunsmith (SED, 59; DB C-38; t). Made gunlocks.

Reasure, John (b. 1739). 1779–81, Strasburg Twp., Lancaster Co. (SED, 59). Gunsmith.

Redacker, William. Gunsmith, Penn Twp., Berks Co. (c, 1850).

Redsecker, G. W. Gunsmith, S. Market St., Elizabethtown, Lancaster Co. (BRD, 1896).

Reed & Gardner sold guns, pistols, ammunition, and the like, Philadelphia (PD, 1882); 510 Girard Ave., Philadelphia (BPD, 1890).

Reed, Daniel. Artillery artificer and blacksmith, 1778, in Capt. John Jordan's Co. at Carlisle (2 Pa. Arch. 11 at 260).

Reed, James. Apprentice gunsmith, 1776, in Capt. Andrew Graff's Co., Lancaster Militia (2 Pa. Arch. 13 at 347; 5 Pa. Arch. 7 at 1073).

Reed, Philip. Gunsmith, Philadelphia (BRD, 1898).

Reede, Charles. Gunsmith, 1813–15, Springhill Twp., Fayette Co.(t); 1815–19, Washington Twp., Fayette Co. (t).

Reedy, B. E. (b. 1818). Finley Twp., Schuylkill Co. (KKR, 327; t, 1862). Gunsmith.

Reedy, E. Gunsmith, 1862, Lykens Twp., Dauphin Co. (t).

Reedy, Leonard (1774–1837). 1792–1837, Gratztown, Lykens Twp., Dauphin Co. Major gunsmith (JKJ, 218–20; t; c; KKR, 325–27).

Rees, David. Drafted into the artillery artificers. Member of Capt. Hamilton's Co. before 7/1777; afterward in Capt. Ziegler's 1st Pa. Continental Line (PIR 1:369).

Reichenbach, George. Gunsmith, Lancaster boro (c, 1850; SED, 59).

Reid, Samuel. Stock maker, 91 Dillwyn, Philadelphia (DeSD, 1829, 1830, 1835–36; MD, 1841).

Reifsnyder, John. Artillery artificer, 1777, Capt. Will's Co. (BCR, 115).

Reigart, Peter. Gunsmith, 1754–77, Paxton Twp., Lancaster Co.; 1777–80; Lancaster boro. (SED, 59; 2 Pa. Arch. 13 at 299) On 11/17/1775, he agreed to "make muskets and bayonets for this county at Philadelphia; and that they will confine themselves to that work entirely . . . and furnish as many as they can possibly complete."

Reily, Edward. Artillery artificer, 1778, Capt. Thomas Wiley's Co, making cannon at Carlisle (PIR, 2:249). Shoemaker by trade.

Reimer, David (b. 1753). Artillery artificer, 1778, Capt. Thomas Wylie's Co. at Carlisle. In 1833, age 80, living in Dearborn Co., Ind. (PIR, 2:249).

Reiner, Michael D. 1777–90, gunsmith, Conestoga Twp., Lancaster Co. (SED, 59; t).

Reinfried, Leo. Gunsmith, 1325 Frankford Ave., Philadelphia. (MD, 1861); 205 Frankford Rd., Philadelphia (MD, 1865).

Reinfried, P. C. Joint patent 23,224 of 3/15/1859 with Joseph Barber for a breech-loading gun. Bridesburg, Philadelphia Co.

Reitenbaugh, William. Gunsmith, 1838–47, Lancaster boro (SED, 59).

Reiter, E. (b. 1783). Locksmith, arrived at Philadelphia on 8/11/1804 on brig *Newton;* 5′ tall (2 Pa. Arch. 17 at 609).

Reiter, Leonard. Gunsmith, 1800, Heidelberg Twp., Berks Co. (KEA, 78).

Reitmyer, John. 2/1776; "for cartouch boxes, £90," Berks Co. (3 Pa. Arch. 5 at 9); 2/22/1776, cartouch box maker with John Deisler and Peter Brecht; on 5/8/1776, an addition order was issued for £108/15/0, for 150 cartouches (BCR, 175).

Remley, George J. Gunsmith, 647 W. King St., Lancaster (LCD, 1871–72; BLD, 1875–76).

Remley, John H. Gunsmith, W. King St. Near Mulberry St., Lancaster (BLD, 1857, 1871–72, 1873–74, 1875–76).

Remley, John Y. Gunsmith, 1836–50, Lancaster (SED, 59).

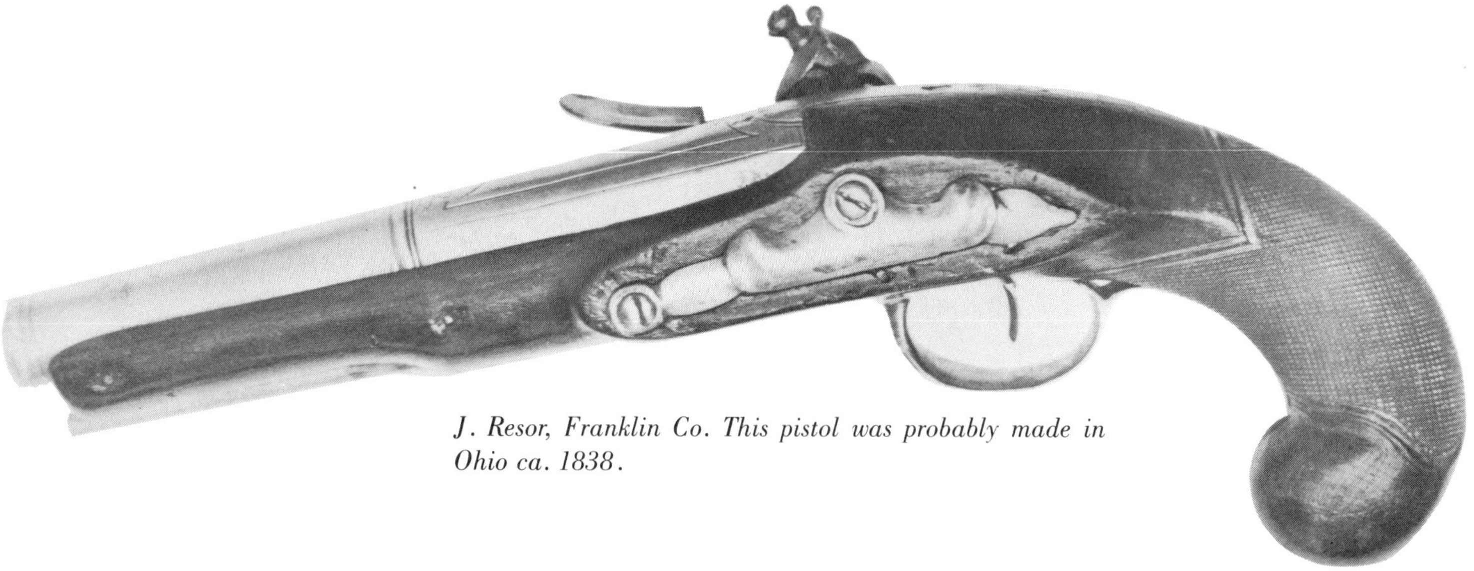

J. Resor, Franklin Co. This pistol was probably made in Ohio ca. 1838.

Renker, Rudolph. Gunsmith, boarded at Miller's Hotel, Lancaster (BLD, 1857).

Repeating Rifle Co. 1533 Bristol St., Philadelphia (PCD, 1915–16).

Resor (Reasor), Jacob (1786–1836). Gunsmith, born in Lancaster. 1807–8, Mercersburg, Franklin Co. (t); 1811–36, Cincinnati, Ohio (WSB, 21; DH, 185). The relationship between this gunsmith and the man in Meadville, Crawford Co. is unknown. Note the similarity in dates of birth and death. The guns with the Jacob Resor name on them that I have seen were in a style that suggests that they were all made by the Mercersburg and Ohio gunsmith, not the Crawford Co. gunsmith.

Resor, John. 1798–1803, German Twp., Fayette Co. (t). Gunsmith.

Reynall, Richard. Gunsmith, 1797, Philadelphia; 1801 and 1816–29, Baltimore, Md. (AMM, 237–38).

Rhodes, Frederick. Gunsmith, 1823, Allegheny Co. (KKR, 328).

Rice, Joseph. Gunsmith, 132 South St., Philadelphia (DeSD, 1823–24).

Richard, Harmon. Gunsmith, 1852, Somerset Twp., Somerset Co. (t).

Richards, John C. Patent 179,609 of 7/4/1876 for a breech-loading firearm. Pittsburgh.

Richards, Joseph. Armorer, 1777, Pa. Navy, armed boat *Chatham* (2 Pa. Arch. 1 at 262).

Richardson & Overman. Produced ordnance 1862–65; made Gallagher carbines under Mahlon J. Gallagher's supervision and patent (22,728).

Richardson, C. H. Patent 191,178 of 5/22/1877 for a revolver. Philadelphia.

Richardson, George. Artillery artificer, 1778, Capt. Nathaniel Irish's Co. (PIR, 2:247).

Richardson, George J. Patent 43,929 of 8/23/1864 for a breech-loading gun, Philadelphia.

Richardson, James. Advertised in *Pennsylvania Packet* on 12/24/1751 as a silversmith on Front St., Philadelphia.

Richardson, Joseph and Nathaniel. Silversmiths and cutlers, 1748–85, Philadelphia. (Bridenbaugh, *Colonial Craftsman*, 86) Advertised in *Pennsylvania Packet*, 10/26/1749, as silversmiths on Front St., Philadelphia; *Packet*, 9/20/1750, Joseph Richardson, goldsmith; 11/7/1754, Joseph Richardson, Front St., silversmith.

Richardson, William. Gunsmith, 1839–50, Elizabeth Twp., Lancaster Co., (SED, 59).

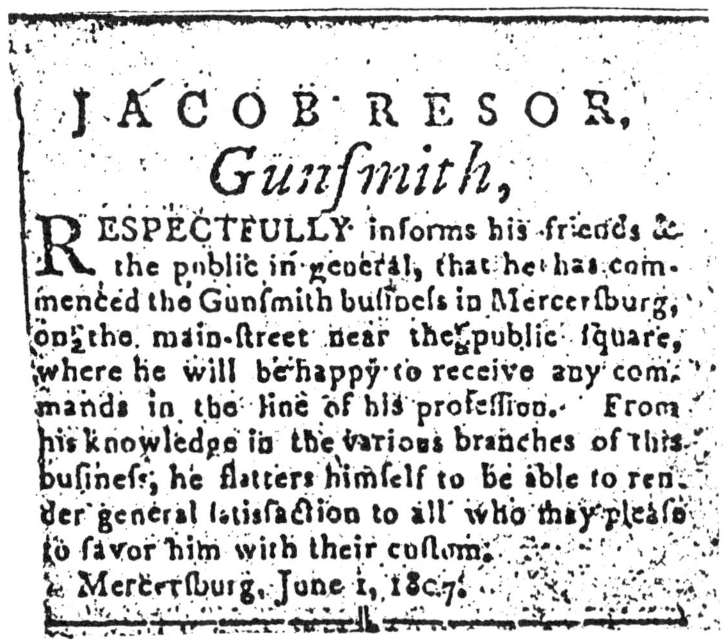

JACOB RESOR,
Gunſmith,
RESPECTFULLY informs his friends & the public in general, that he has commenced the Gunſmith buſineſs in Mercerſburg, on the main-ſtreet near the public ſquare, where he will be happy to receive any commands in the line of his profeſſion. From his knowledge in the various branches of this buſineſs, he flatters himſelf to be able to render general ſatisfaction to all who may pleaſe to favor him with their cuſtom.
Mercerſburg, June 1, 1807.

Franklin Repository, June 1, 1807

Riche, Henry. Gunsmith, 3d St. near Poplar St., Philadelphia (PXD, 1816).

Richwine (Reichwein), Cornelius. Gunsmith, 1838–40, Cumru Twp. Berks Co. (t).

Reid, Johan. (b. 1784). Silversmith, born in Württemberg, Germany; dark, age 21, arrived on ship *Margaret* at Philadelphia on 8/26/1805 (PaE, 2:169).

Rigden, Henry J. (b. 1795). Born in Md. in 1817, Georgetown, District of Columbia; 1818–50, Brownsville, Fayette Co. Gunlock maker, silversmith (t; c; FSP, 254).

Rigdon, William. Importers of arms, merchants, 1776, Dock Ward, Philadelphia. 5/30/1776: "Has arms but will not deliver them up." Listed as a nonassociator (2 Pa. Arch. 1 at 600).

Riggleman, J. Gunsmith and wagon maker, New Hamburg, Mercer Co. (t, 1867; c, 1870).

Riley, Edward. Gun-powder manufacturer, 208 Chestnut St. and 413 High St., Philadelphia (PXD, 1816).

Riley, John, Jr. On 12/27/1771 Riley was apprenticed to Stephen Paschall of Philadelphia, for 4 years, 10 months, to learn the cutler's trade (PIDS).

Riling, John, Jr. (1824–1902). 1850–92, gunsmith and cabinetmaker, 608 58th St. at Eldorado Ave., Eldorado, Blair Co. (F. E. Long, *Eldorado Saga* [1971], 11; t; c).

Ringle, Mathias (1801–72). 1835-72, Blairsville, Indiana Co. (t; c) He had the unusual habit of stamping his name on barrels twice.

Rippy, Hugh (d. 1790). Gunsmith, 1789–90, Pittsburgh (t).

Ritchie, Henry. Gunsmith, 487 N. 3d St., Philadelphia (PXD, 1810).

Rittenhouse, Benjamin. In 2/1776, the Committee of Safety appointed Rittenhouse to superintend the Pennsylvania Gunlock Factory, initially located on Cherry St., Philadelphia, and later moved to Allentown. He indicated that he could make 200 gunlocks a week. By 1776 he had become an independent contractor and agreed to make muskets at £4/5/0 each. When the factory was disbanded, Rittenhouse asked that the tools that he had brought with him be returned. He wished to buy them back at £25, the sum he had charged to the Commonwealth (1 Pa. Arch. 7 at 112; BHAM, 1:572; Col. Rec., 10:481).

Rittenhouse, David. In 1776–80, with Charles Wilson Peale, Rittenhouse had been experimenting with telescopic sights. He invented the crosswire used in most telescopic sights to this day. He was also a munitions maker, personally appointed by George Washington, and a musket subcontractor. Later, he was the first director of the U.S. Mint. Jefferson called him the greatest mind in science. He also made and designed brass clocks (Kelley, *Pennsylvania Colonial Years*, 742; *History of the Wyoming and Lackawanna Valleys*, 1:522).

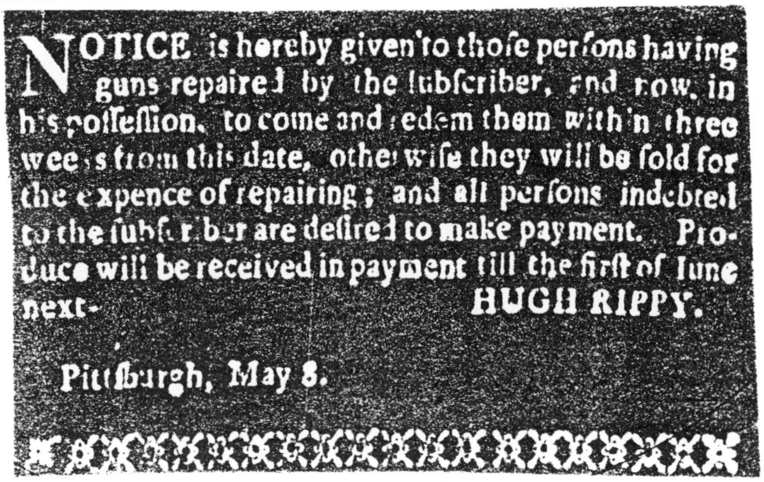
NOTICE is hereby given to thoſe perſons having guns repaired by the ſubſcriber, and now in his poſſeſſion, to come and redeem them within three weeks from this date, otherwiſe they will be ſold for the expence of repairing; and all perſons indebted to the ſubſcriber are deſired to make payment. Produce will be received in payment till the firſt of June next. HUGH RIPPY.

Pittſburgh, May 8.

Pittsburgh Gazette, May 23, 1789

Ritter, Jacob. Gunsmith, 1775–83, Philadelphia (GAAAM, 126).

Ritter, Peter Paul (1795–1876). Born in Bavaria, Ritter was the pioneer gunsmith of St. Mary's boro., 1847–76 (t; c; REH, 217–18).

Ritter, William. Locksmith, 1769–74, North Ward, Philadelphia (t).

Roads (Roth), Christopher. 1777–80, artillery artificer, Capt. Jesse Rowe's Co. at the cannon factory (2 Pa. Arch. 11 at 252; PIR 2:244).

Robbins, John. Artillery artificer, 1779, at the lead factory (2 Pa. Arch. 15 at 385).

Robbins, William E. Gunsmith, 1847–60, Charleston Twp., Tioga Co.; 1861, Sullivan Twp., Tioga Co. (c; t; BPaD, 1861).

Robbinson, John (1808–52). Gunsmith, Lancaster Co. (Registration of deaths, no. 16, shows he died 7/12/1852; occupation: gunsmith).

Roberts, Isaac. Artillery artificer, 1779. (2 Pa. Arch. 15 at 384).

Roberts, John. Smith and powder maker, 1784, Upper Dublin Twp., Philadelphia Co. On the list of persons attainted by treason. (1 Pa. Arch. 10 at 250ff.) On 3/10/1780 Roberts's vast holdings were confiscated and sold at public auction for his having been a Loyalist. His property included 378 acres and an oil mill, gunpowder mill, sawmill, paper mill and other buildings. (6 Pa. Arch. 13 at 313f.) Among his goods that were seized were: 40 tons of iron; 3 millsaws and handsaw; another lot of iron; 1 set of cooper's tools; and 2 draw knives (6 Pa. Arch. 12 at 715–16).

Roberts, Joseph. Armorer, 1776, armed boat *Hancock,* Pa. Navy; reenlisted on 5/1/1778 and assigned to *Hancock.* (2 Pa. Arch. 1 at 306, 309)

Roberts, Peter Gunsmith, 512 Church St., Lancaster (BLD, 1871–72, 1875–76).

Roberts, Thomas. On 3/10/1780 "the gunpowder mill of John Roberts which was sold at auction for John's treason was purchased and operated by Thomas Roberts." Lower Merion Twp., Philadelphia Co. (6 Pa. Arch. 13 at 313).

Roberts (Robinson), William. Contracts for cartridge boxes and related accoutrements: 4/1/1777 for £75/4/0; 5/7/1777 for £85/2/0; and 6/25/1777 for £114/12/0 (2 Pa. Arch. 1 at 22, 43, 60).

Robertson (Robison), John. Gunsmith, Lancaster (c, 1850).

Robertson, William. Gunlock maker and importer, gunsmith, S. 2d St. and Dock St., Philadelphia (MD, 1855); 131 Walnut St. (MD, 1846, 1859); 102 Carpenter St. (DeSD, 1829, 1830, 1831).

Robinson & Krider. Gunsmiths. John H. Krider and William Robinson, Washington St. below 6th St., Philadelphia (MD, 1837).

Robinson, William. Gunsmith, 90 S. 2d St. above Walnut St., Philadelphia (MD, 1837, 1841, 1849).

Rod, Thomas. Sickle and knife maker and swordmaker, west side of Penn St., between Pitt St. and St. Clair St., Pittsburgh (PAD, 1815).

Rodes, John. Gunsmith, Schaeffertown, Lebanon Co. (BPaD, 1861).

Rodman, Thomas Jefferson. In 1849, invented a way to cast cannon instead of having to bore them out of a block of steel. Used first at Ft. Pitt Cannon Co., Pittsburgh. Over 2,000 cannon of 20-inch bores were cast during the Civil War using his process (See Fort Pitt Cannon Co.)

Roe, Jesse., Capt. 1777–81, commanded the artillery artificers who made cannon at Carlisle and Philadelphia (2 Pa. Arch. 11 at 249).

Roe, John. Artillery artificer, 1777–80, Capt. Jesse Rowe's Co. at the Philadelphia cannon factory (2 Pa. Arch. 11 at 252).

Roedler, Maximillian. 1868–70, gunsmith, Ash Alley, Harrisburg (c; BRD, 1869).

Roeser, Mathias (1708–71). 1749–71, Lancaster boro. (t; gun dated 1749; SED, 59) Major gunsmith. William Henry, Sr. was Roeser's apprentice. Buried in Moravian graveyard. His daughter, Katherine, married Jacob Kraft, a gunsmith (JKJ, 91; GAS-1, 279).

Roesser, Peter (d. 1823). Major gunsmith. 1775–86, Lancaster (SED, 59; JKJ, 91–92); 1790, Elizabethtown (Hagerstown), Md.; 1796–1812, Mercersburg, Franklin Co.; 1823, died in Lancaster (WSB, 19–21, 137–38). In 1795 Roesser married Catherine, daughter of Joseph Welshans, a gunsmith. In 1774 Peter was executor of Mathias Roeser's estate.

Rogers Brothers, John and Charles. Gunmakers, 1809–46, Philadelphia (t; c). (See Mount Joy Forge, Valley Forge Armory.)

Rogers, Charles. Ironmonger, government contractor at Valley Forge, working with John Rogers and Brooke Evans, ca. 1800–1820.

Rogers, Charles H. Ironmonger and government contractor, ran Valley Forge ca. 1830. Nephew of John Rogers.

Rogers, John. Government contractor, ironmonger, 7 N. 2d St., Philadelphia (PXD, 1809); Valley Creek, Chester Co. (t, 1814). In 1817, he filled a government arms contract with/for A. McRae; in 1821, filled a contract for U.S. muskets (RAG, 165).

Rogers, Jonathan (b. 1781). Son of John Rogers, an ironmonger in Chester Co.; assisted his father in filling a contract of 7/28/1817 for 10,000 Va. Militia arms at $14 a stand; and of 1/1/1825 for 5,000 muskets at $12.25 a stand for Frankford Arsenal. (RAG, 165) By 1850 he was in Hurnon Co., Ohio (DH, 212).

Rogers, William, Sr. Gunpowder manufacturer, 413 High St., Philadelphia (PXD, 1813); 30 Browne St., Philadelphia, gunsmith (DeSD, 1829, 1830, 1831).

Rogers, William, Jr. Gunpowder maker; took over his father's business about 1825. His brand was "Nitre Hall." 3 Minor St., Philadelphia. (DeSD, 1829, 1830) "In Haverford Twp., Delaware Co., the Nitre Hall Mills were used for the manufacture of gun powder before the War of 1812. In 1825, William Rogers, Jr., manufactured 10,000 casks of gunpowder and employed 25 men in that work" (Palmer, *History of Delaware County,* 1:134).

Rohr, John Frederick. Gunmaker, toolmaker, saw and armor maker, 2d St., Philadelphia, before the Revolution (KEA, 80).

Rohrer, John. Gunsmith, 639 Columbia Ave., Lancaster (BLD, 1873–74).

Rohrer, Leopold (1851–1939). Born at Baden, Germany; came to America in 1871. 1871–73, worked at Great Western Gun Works, Pittsburgh; 1873, went to Chicago, working at salvaging and rebuilding guns found after the great fire; 1873–1939, New Castle, Lawrence Co., ran Rohrer's Gun Store (t; f, c).

Rolland, Henry E. 488 S. Wyoming St., Hazleton,

Luzerne Co. (BRD, 1882; PD, 1882).

Roller, & Sprecher, Gunsmiths, Lancaster (c, 1850).

Ropp, Adam. Gunsmith, alley between Walnut St. and Chestnut St., above Queen St., Lancaster (BLD, 1857).

Roop (Rupp), John. Blacksmith, 1777, Noxamixon Twp., Lehigh Co. (t).

Rose, John. Artillery artificer, 1778, Capt. Nathaniel Irish's Co. (PIR, 2:247).

Rose, William. Government contractor, 1805–13, Block;ey Twp., Philadelphia Co. Contracts: 12/9/1807, 2,000 horsemen's swords; 3/24/1812, 500 iron swords and scabbards at $6.50 each. Also known as William Rose & Sons. (RAG, 224).

Ross, Elijah (1785–1864). Gunsmith and swordsmith, born in Fayette Co. By 1804, Zanesville, Ohio, through 1864 (DH, 302).

Ross, James (d. 1816). 1799–1802, East Bethlehem Twp., Washington Co.; 1803–16, Steubenville, Ohio (t; c). Gunsmith.

Ross, Samuel. Gunsmith, 1848–50, 12 5th St., Pittsburgh (c; PAD, 1848).

Ross, William. 1795–96, Luzerne Twp., Fayette Co. (t); 1798–1818, Fredericktown, Washington Co. (t). Gunsmith.

Rossell, Job, Jr. (b. 1813). Born in Columbiana Co., Ohio; active there until 1839. 1848–50, Connellsville, Fayette Co. (c, 1850; t). Gunsmith.

Rossett, Robert (b. 1749). Born in England. Brass founder, 1779, New 11th Pa. Regt. (2 Pa. Arch. 11 at 65).

Roth, Charles (d. 1852). Born in New York City. Registration of Deaths, Luzerne Co., no. 10, 1852; occupation: gunsmith; lived at corner of Main St. and Union St., Wilkes-Barre.

Roth, Charles (1815–1903). Gunsmith, Main St. and Union St., Wilkes-Barre. Born in Darmstadt, Germany. Left Germany in 1837; arrived in New York, 4/19/1837. Came to Wilkes-Barre in 1840 and retired from gunsmithing in 1887 (t; c; f).

Roth, Ernest E. (1853–1917). 69 S. Main St., Wilkes-Barre. Took over his father's (Charles, Jr.) business in 1887. In 1891 he convinced Parry Firearms to relocate in the Wyoming Valley, doing business as Wilkes-Barre Gun Co. It went bankrupt in 10/1895 (D. C. Rowan, "Charles Roth, Wilkes-Barre Gunsmith," 8, no. 3; KRAN Registration of Births, no. 138 for the year 1853).

Roth, George. Gunsmith, Moore Twp., Northampton Co.; 1844–46, Heidelberg Twp., Lebanon Co. (KEA, 80).

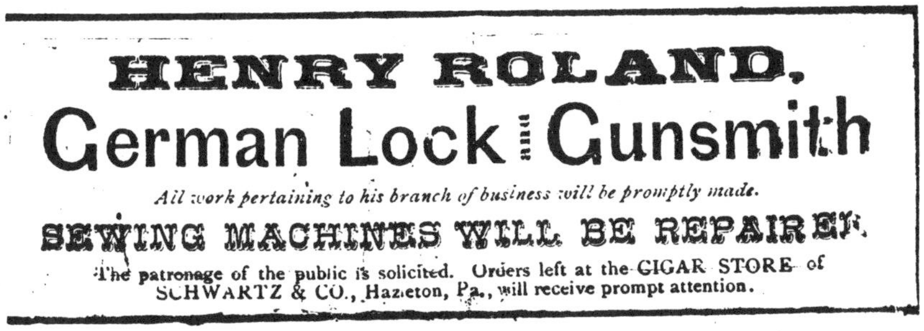

Polk's Pennsylvania Gazetteer, 1882

Roth, Henry, Sr. (1686–1774). Gunsmith and Dunkard minister, before 1771, Ephrata Twp., Lancaster Co.; 1771–74, Ursine, Brothers Valley Twp., Bedford Co. He may have been at Frederick, Md., 1749–67. "Made rifles although this was contrary to Dunker practices." His estate inventory does not reveal anything that suggests his trade (DH, 245–46).

Roth, Samuel. 1830–50, gun-barrel maker, Elizabeth Twp., Lancaster Co. (SED, 59; c, 1850; t).

Rothrock, Ed L. (d. 1880). Gunsmith, Middle Creek, Snyder Co. (c, 1870, 1880).

Rothrock, Edward, Jr. (1872–1934). Gunsmith, Middle Creek, Snyder Co. (t; f; c).

Rothrock, John. Locksmith, 1779, York (3 Pa. Arch. 21 at 8).

Charles Roth (1815–1903), Wilkes-Barre, Luzerne County.

Rothrock, Joseph. Silversmith, 1779, York (3 Pa. Arch. 21 at 8).

Rothrock, Joseph, Jr. (b. 1755). Silversmith, 1783, York Twp., York Co. (t).

Rowan, John and George. On 6/21/1778, were paid "in full on their account for repairing the city artillery, £0/16/0" (3 Pa. Arch. 7 at 219). Philadelphia.

Rowland, Thomas H. (1813–67). 1849–54, Brownsville, Fayette Co. (t); 1854–58, New Castle, Lawrence Co. (t); 1862–67, Pittsburgh (t; c; PAD, 1864–65). Gunsmith.

Royer, W. S. Gunsmith, Pottstown, Montgomery Co. (PD, 1890).

Royet, Louis. Born in France. Came to Reading in 1858; active there until 1889, at 58 S. 7th St. (BSED, 1873; PD, 1882; c; t; f).

Rubert, Peter. Gunsmith, 512 Church St., Lancaster (BLD, 1873–74).

Rudisill, Henry. Gunsmith, boarded at 510 N. Duke St., Lancaster (BLD, 1873–74, 1875–76).

Ruggles, Alexander. Gunsmith, 357 S. 11th St., Harrisburg. (*Patriot Directory,* 1880, 1882, 1887).

Rugh, Daniel (b. 1812). 1837–42, Blairsville, Indiana Co.; 1847–56, Galena, Ill. (t; c; Curtis Johnson).

Ruhland, Henry. Gunsmith, Philadelphia (PD, 1893; BPD, 1895).

Runfielt, S. C. Gunsmith, Lancaster (SED, 59; c, 1850).

Rupertus, Jacob. Gunsmith, 1924 N. 4th St., Philadelphia. (PD, 1882; BSED, 1880) Jacob Rupertus had a number of patents, mostly for cheap pistols he made, dated from 1859–1899. He ran the Rupertus Patented Pistol Manufacturing Co., ca. 1858–1900.

Rupp, A. Herman (1756–1831). Major gunsmith. 1784–1809, Lower Macungie Twp., Lehigh Co. (t; GAS-1, 242–47).

Rupp (Roop), Jacob. 1780–1800, Ruppville, Annville Twp., Lebanon Co. In 1819 he advertised for his runaway apprentice, Ben Landis (cf. DH, 175) (JKJ, 175–76; t; c). Gunsmith.

Rupp (Roop), Jacob (1787–1858). 1825–26, Roopsville, Centre Co.; 1827–44, Bellefonte. His wife, Susanna, died on 5/14/1847, and her obituary indicated that they had moved from Centre Co. in 1844. She died at Grannsville, Tippecanoe Co., Ind. (John B. Linn, *History of Centre and Clinton Counties* [Philadelphia: Everts, 1883] 311; t; c). Gunsmith, brewer, farmer.

Rupp, John (1762–1836). 1789–1807, Lower Macungie Twp., Northampton Co. (t). Uncle of John Rupp II. Major gunsmith.

Rupp, John, II (1786–1848). 1812–48, Weisenberg Twp., Northampton Co. (t). Major gunsmith.

Rush, Benjamin. 1778–81, Philadelphia, professor of chemistry and "authority on the making of gunpowder" (Kelley, *Colonial Years* 741–42).

Rush, James (1679–1727). Gunsmith, Byberry, Philadelphia Co. (MB 12 [1968]).

Rush, John, Sr. Gun repairman, ca. 1683, Byberry, Philadelphia Co. (RAG, 167).

Rush, John, II (1712–51). Son of James. (MB 12 [1968]) "Two men, supposedly theives, tried to buy pistols from Mr Rush, gunsmith" (*Pennsylvania Packet;* 8/31/1749); "accounts with the estate of John Rush, gunsmith, dec'd, to be settled with William Rush, blacksmith, and Hannah Rush" (*Pennsylvania Packet,* 10/24/1751).

Rush, William. Son of John Rush, II. William was a blacksmith and locksmith, if not a gunsmith. 1769, Mulberry Ward, Philadelphia. (3 Pa. Arch. 15 at 201) 4/10/1745: "John Troy assigns Richard Berry, his servant, to William Rush, blacksmith, for the remainder of his time." (PIDS) *Pennsylvania Packet,* 6/20/1754, advertised Rush's services as a locksmith and lockmaker, Front St., Philadelphia.

Rutt, Henry. Gunsmith, Earl Twp., Lancaster Co. (SED, 59; t, 1820).

S

Sackett, Jacob T. Gunsmith, 1870–74, Woodcock, Crawford Co. (c; CCD, 1874); 1880–85, Saegertown, Crawford Co. (c; PD, 1880, 1882; CCD, 1885).

Sager, John. Gunsmith, 1777, Lampeter Twp., Lancaster Co. (SED, 59).

Sailor, Henry. Gunsmith, 1792, Harrisburg (KKR, 333).

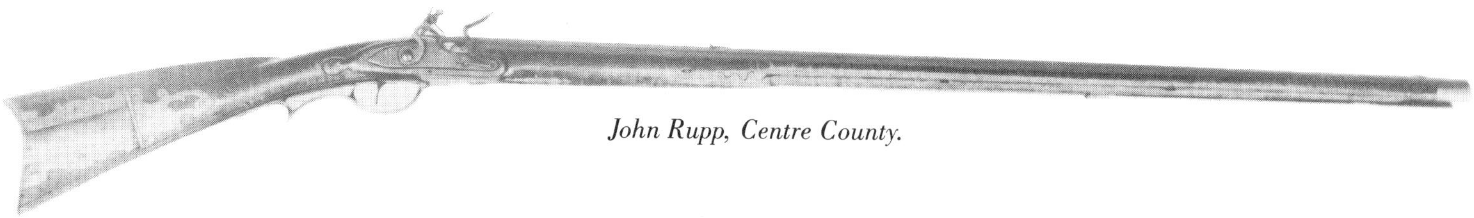

John Rupp, Centre County.

St. Johns, William. Gunsmith, 1839–54, Lancaster (SED, 59).

Salada, Melchior. Enlisted in 1st Pa. Rifle Regt., 1776; released by order of Congress, 12/1776, to work with D. & S. Hughes making cannon (2 Pa. Arch. 10 at 312).

Salter, John H. (1845–1923). Born in N.J.; died in Emporium, Cameron Co. Patents received in Elk. Co.: 208,696 (1878) for a magazine gun; 232,766 (1880) for a magazine firearm (REH, 245–52).

Sanders, William. Gunsmith, 44 Cedar St., Philadelphia (DeSD, 1819); 38 Cedar St. (MD, 1841, 1846).

Sankey, William. Gunsmith, Union Twp., Mifflin Co. (c, 1850).

Sansfield, John. Armorer, 1775, Pa. Navy, assigned to armed boat *Congress* (2 Pa. Arch. 1 at 268).

Sarvey, Daniel (b. 1829). Apprentice gunsmith, with Nicholas Shennefelt in Clarion boro (c, 1850).

Sauer, Urias. Gunsmith, boarded at 213 Church St., Lancaster (BLD, 1871–72, 1873–74).

John H. Salter (1845–1923), Elk County.

Saupp, Andrew (1805–55). 1833–55, Bedford boro. Gunsmith, tinner, postmaster, storekeeper. Married Elizabeth Sellers on 2/12/1835. Buried at Old Catholic Cemetery, Bedford boro (t; f; c).

Savery & Co. Gun-barrel makers, 1855–58, Philadelphia (RAG, 169).

Saxton, Joseph (1799–1873). 1824–73, Philadelphia, inventor, gunsmith. Built percussion cane guns; later worked in Isaiah Lukens's gunshop (GR, 7-1968; FSP, 265).

Saylor, Jacob, Sr. (1715–93). Gunsmith, 1772, Summit Twp., Somerset Co. (t); 1775, obtained land at Ft. Pitt by Va. warrant. 1773–90, Bedford boro. On 5/21/1780 was paid for militia arms repairs, £22/8/6 (3 Pa. Arch. 7 at 25). 1790–93, Summit Twp., Somerset Co. Buried on his farm, "3 Hills." His wife Magdalena died 1792. (t; c) He was the only gunsmith in Bedford Co. at the beginning of the Revolution.

Schaanfelder, John Daniel. Goldsmith. On 11/26/1771 he was apprenticed to William Taylor for 2½ years to learn the art and mysteries of the goldsmith. He was sold as an indentured servant for £18/14/0 (PIDS).

Schaeffer, James. Silversmith, 1800, Washington boro (t).

Schalck, Christopher Christian. 1825–75, Williamsport, Lycoming Co., gunsmith (NAT, 214; t; c).

Schalck, George (1821–92). Born in Weisbaden, Germany. 1854–92, Pottsville, Schuylkill Co. (t; c, MB 5 [1960]: 10; BPaD, 1860; BRD, 1860; PD, 1882). Worked with T. P. Cherrington, Jr. on the "Schalck rifling system." May have apprenticed with T. P. Cherrington, Sr.

Schalk, Andrew. 1847–58, Pottsville, Schuylkill Co. Gunsmith and inventor, working with T. P. Cherrington, Jr. (t; c).

Schaub, Adam. Gunsmith, 1857, Lancaster (BLD, 1857).

Schaub, Elias (1811–81). 1845–47 Philadelphia, gunsmith (MD, 1845, 1847); 1850–81, Bethania, N.C. (FSP, 266).

Schell, Martin, Sr. (1737–96). 1770–74, Heidelberg Twp., Lancaster Co. (SED, 60), gunsmith. 1780–96, Paxton Twp., Dauphin Co. Made muskets during the Revolution. Major gunsmith (c; t; JKJ, 274–75).

Schell, Martin, Jr. (1763–1817). 1783–1817, Lower Paxton Twp., Dauphin Co. (t; c; JKJ, 274–78). His estate inventory of 9/2/1817 showed smithing tools. Shell died 8/8/1817. (Daupain County wills and estates).

Schilling, Frank (1835–1916). Gunsmith, born in Baden, Germany; came to N.Y. in 1850, apprenticed in Philadelphia, 1854; 1860, San Francisco, Calif.; 1862, San Jose, Calif. (LSC, 199).

Schilling, Frederick. Gunsmith, boarded at Blickenderfer's Tavern, Lancaster (BLD, 1857).

Schilling, Peter. Gunsmith, boarded at Blickenderfer's Tavern, Lancaster (BLD, 1857).

Schimmel, Gerrit. On 9/1/1661 was paid 85 florins "for snaphaunce delivered" at Dutch Colony on the Delaware River (2 Pa. Arch. 5 at 455).

Schlegelmilch, Herman. Gunsmith, born in Saxony, Germany; came to U.S. in 1853 and worked at Bethlehem, late Moravian religious sect; moved to Iowa and Wisconsin; died in 1903 (RAG, 170).

Schlichter, Nicholas. Artillery artificer, 1777, Berks Co. (BCR, 115).

Schmidt, Heinrich. Gunsmith, Middle St., Lancaster (BLD, 1857).

Schnader, Franklin (1831–1906). Gun-barrel maker, Wyomissing Creek, Berks Co. In 1835 he apprenticed with Frank Miller; 1839, bought out John Keim's shop; 1852–84, gun-barrel maker, exclusively. He had an annual capacity of 2,500 to 3,000 gun barrels a year (RAG, 170; MB 12 [1963] 9.)

Schnader, James F. (b. 1869). Ran a gun-barrel factory in Cumru Twp., Berks Co. Son of Franklin Schnader (Reading *Eagle*, 9/23/1934).

Schneider, Franklin Alexander (1790–1864). Born in Schleswig-Holstein, Germany; migrated to U.S. to avoid conscription into the army. He came to Philadelphia in 1808 and worked at Chambersburg, moving in 1829 to Stark Co., Ohio (DH, 368).

Schneider, Jacob. Gunsmith, 1805, Heidelberg Twp., Northampton Co. (t).

Schneisitzer, Charles. Susquehanna St. below Race St., Mauch Chunk, Carbon Co. (BRD, 1873). Gunsmith.

Schneiter, John. Gunsmith, 1793, Lancaster Twp., Lancaster Co. (SED, 59; t).

Schoeneman, Frederick (b. 1833). Born in Berlin, Germany; apprenticed there as a gunsmith. 1851, journeyman gunsmith at Philadelphia; by 1852, gold miner in California; later, he worked in San Francisco as a gunsmith, 1873–76 (LSC, 208–09).

Schoener, Henry. 1850–83, gunsmith, N. 4th St. near Walnut St., Reading (KKR, 333; c, 1850; t; BRD, 1860; BPaD, 1861).

Schopp, Francis. Patent 51,225 for a harmonica gunlock, Philadelphia, 1865; 1875–76, New York City (FSP, 268).

Schorer, Andrew. 1775–80, gunsmith, Bethlehem Twp., Northampton Co. (GAAAM, 134; RAG, 131).

Schreckengost, Benjamin. Gunsmith, 1842–60, Armstrong Co. (t; c; 3 Pa. Arch. 26 at 657).

Schreckengost, Coonrod. Gunsmith, born in Germany, a son of Jacob. Came, first, to Bucks Co.; moved to Armstrong Co. "the later part of the 18th century." Coonrod had a grandson born in 1829. (S. T. Wiley, *Biographical and Historical Cyclopedia of Indiana and Armstrong Counties* 381) Lived in Kittanning Twp., Armstrong Co. Coonrod's father apprenticed out his son George for 9 years to the ship's captain to pay for the family's passage to America. The area in which Coonrod, "a gunsmith by trade," lived is now in Valley Twp. (*Armstrong County* [1982], 81) "He could use a gun as well as make one" (Beers, *History of Armstrong County* [1914], 677–78).

Schreckengost, Daniel. "Daniel Schreckengost was born in Valley Twp. [Armstrong Co.], where he died. He was a gunsmith by trade and by industry and thrift acquired a farm of 300 acres of land in his native township. He married Mary Crum." (Wiley, *Indiana and Armstrong Counties*, 560) 1820–29, Bullskin Twp., Fayette Co. (t); 1830–34, Connellsville, Fayette Co. (t).

Schreckengost, Henry John. "Henry John Schreckengost . . . was born in Cowanshannock Twp., Armstrong Co., and eventually moved to Valley Twp., where he bought a farm. . . . He then bought a farm in Wayne Twp., upon which he spent the remainder of his life. . . . His wife was Sarah, daughter of George Cline" (Wiley, *Indiana and Armstrong Counties*, 659). He obtained 300 acres by warrant in Armstrong Co. on 1/11/1828 (3 Pa. Arch. 26 at 657). Part-time gunsmith.

Schtreckengost, John (1792–1859). Gunsmith, Kittanning Twp., Armstrong Co. "He was a blacksmith and gunsmith and has the record of making the first plow manufactured in Armstrong Co." (Beers, *History of Armstrong County*, 678) Married Sarah, daughter of Peter Turney (1789–1870) (Wiley, *Indiana and Armstrong Counties*, 694).

Schreckengost, Lincoln (1865–1949). Gunsmith, operated a general store in Putneyville, Armstrong Co. Son of William (t; c; f).

Schreckengost, Peter. 1870–71, gunsmith, Kittanning, Armstrong Co. (t; c; f).

Schreckengost, William (1818–94). Putneyville, Armstrong Co. Died 9/11/1894 of a rupture, age 76, gunsmith. Born and married in Armstrong Co. Buried at Putneyville Cemetery. (Registration of Deaths) "William Schreckengost was an early settler in Mahoning Twp. He was a gunsmith and followed his trade at Putneyville for

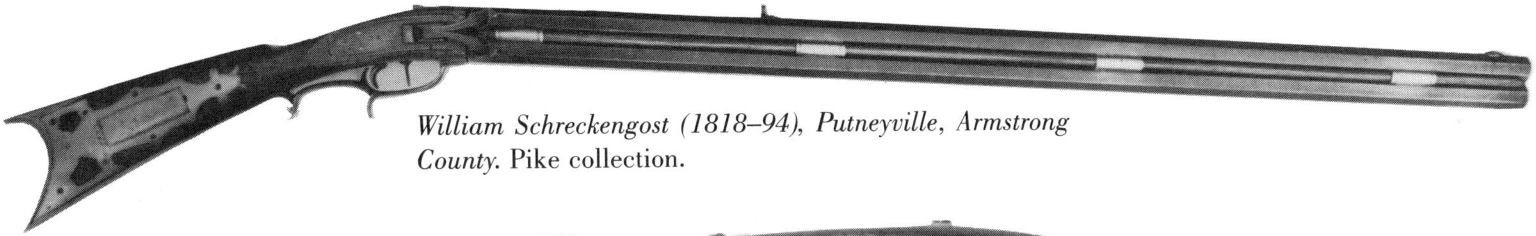

William Schreckengost (1818–94), Putneyville, Armstrong County. Pike collection.

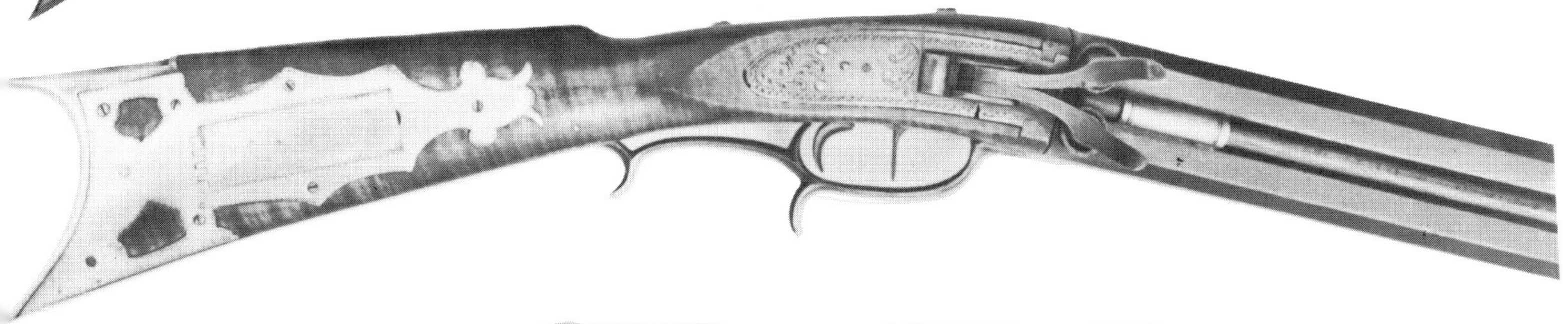

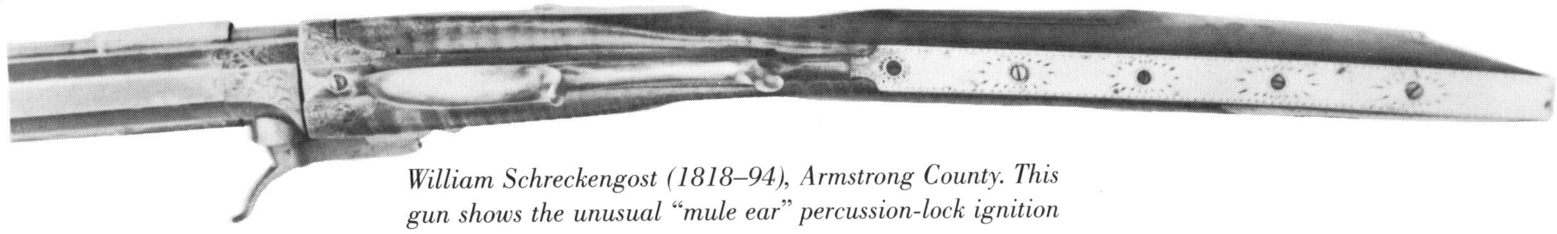

William Schreckengost (1818–94), Armstrong County. This gun shows the unusual "mule ear" percussion-lock ignition system.

many years, attaining wide renown as the manufacturer of the best muzzle-loading rifles in his day in the United States. He himself was a crack shot" (Beers, *History of Armstrong County: Her People Past and Present* [Chicago: Beers, 1914] 844).

Schreit, John. 1758–76, Reading, gunsmith; 1777–80, Cocalico Twp., Lancaster Co. (SED, 59; GAS-1, 84-87). Major gunsmith.

Schribener, John W. 1865–80, Locust St. near Arch St., Mechanicsburg, Cumberland Co. (t; c; BRD, 1869, 1871).

Schroeder, Valentine. The sole reference to him in any way in Bedford Co. is through an advertisement in the *Gazette*, 6/30/1843, in which he offers to make or repair guns, fix stoves, and the like. 1843–47, Baltimore, Md. (FSP, 269).

Schulte (Schutlz), Henry H. Gunsmith, 300 Liberty Ave., Pittsburgh. Ran Iron City Gun Works, 1864–77 (BRD, 1873; t; c).

Schuler, Hans (b. 1699). Born at Durrwangen, Germany. Children: Hans Ludwig (b. 1728); Christian (b. 1730); Andreas (b. 1740); Martin (b. 1745). Arrived on 10/16/1754 on schooner *Peggy* (PaE, 155). Smith.

Schuler, Jacob. Gunsmith, 1789, Macungie Twp., Northampton Co. (t).

Schuler, Leonard (b. 1804). Before 1830, Lewistown, Scuylkill Co. (CHAN, 50f.); 497 High St., Philadelphia (DeSD, 1829, 1830, 1831). Gunsmith.

Schuler, Valentine (b. 1767). Born in Tulpelhoken, Bucks Co.; after 1812, moved to New Philadelphia, Ohio. (DH, 382) Gunsmith. There was also a Valentine Schuler, born 1810 in Perry Co., Pa., in Tuscarawas Co., Ohio (c, 1850; DH, 382).

Schutz, Peter. Silversmith, 1771–99, York Twp., York Co. (t).

Schuylkill Arsenal. Gray's Ferry Rd. at Washington Ave., Philadelphia. Land was purchased by the federal government in 1799 and the first buildings were erected in 1800. The records actually began in 1781 when gifts for various Amerindian chiefs were made in that area. The arsenal outfitted the Lewis and Clark expedition and the forces that attacked the Barbary pirates. After 1818 it handled mostly clothing, for with the opening of the Frankford Arsenal, Schuylkill lost its character as an arsenal. In 1906 the name was changed to Philadelphia Depot, Quartermaster's Corps, U.S. Army.

Schwartz, Peter. Silversmith, 1758–90, York Twp., York Co. (t) Died 1790 (t).

Schweitzer, Abraham (1769–1831). Major gunsmith. 1794, Lancaster; 1795–1831, Chambersburg boro (WSB, 1–6).

Schweitzer, Jacob (1810–64). 1833–64, Chambersburg, gunsmith. Killed by Yankee soldiers by accident in a bizarre incident (described in detail in WSB, 6).

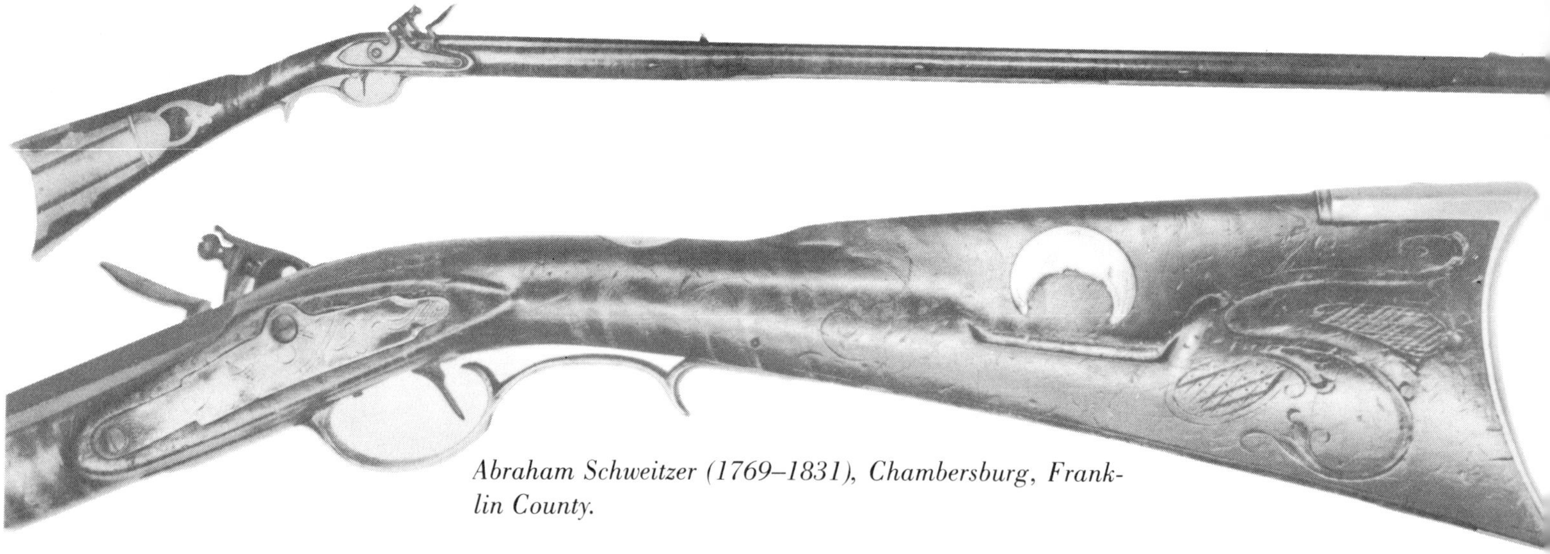

Abraham Schweitzer (1769–1831), Chambersburg, Franklin County.

Schweitzer, John (b. 1805). 1829–32, Chambersburg, gunsmith. (t); 1850, Greenville, Ohio (c, 1850; WSB, 1–6).

Sciek, Abraham. Artillery artificer, 1779. (2 Pa. Arch. 15 at 385).

Scott, Alexander. Silversmith, (d. 1821). 1800–1821, Chambersburg. (t).

Scott, John. Silversmith, 1781–96, Montgomery Twp., Cumberland Co. (t).

Scott, J. Q. A. Patent 36,174 of 8/12/1862 for a magazine firearm, Pittsburgh.

Scoup, Henry. On 12/5/1772, assigned to Aquilla Jones, cutler, New Town, Chester Co., as an apprentice (PIDS).

Scout, Jacobus "Cobe" (1736–1829). Gunsmith and silversmith, Warminster Twp., Bucks Co. Taught silversmithing to John Fitch, assisted Fitch in the manufacture of the first steam-powered boat. He carried a gun of his own manufacture during the Revolution and is reported to have shot a Hessian dead across the Delaware River in 1776. Buried at the VanSant Cemetery (MB, 8 [1951] 8; *History of Bucks County* [1876], 220).

Scull, William. Artillery Artificer, 1777, Capt. Nathaniel Irish's Co. (PIR, 2:247).

Sedgley, R. F. & Co. 2301 16th St., Philadelphia, 1897 and later. Successor to Henry M. Kolb. Made "baby hammerless" cheap revolvers, and, later, Springfield sporters.

Sedwick, George (1832–62). Gunsmith, 1860–61,

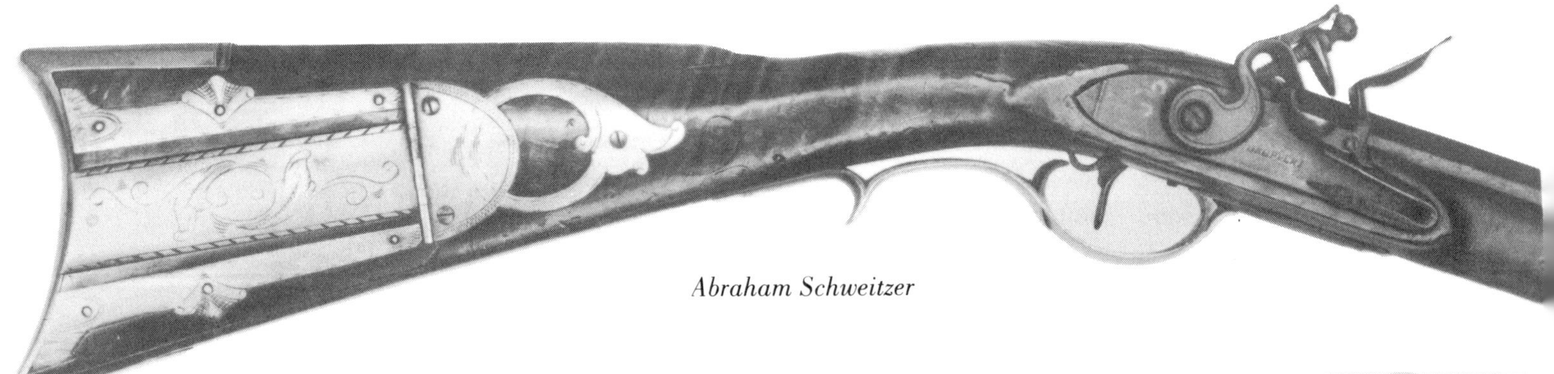

Abraham Schweitzer

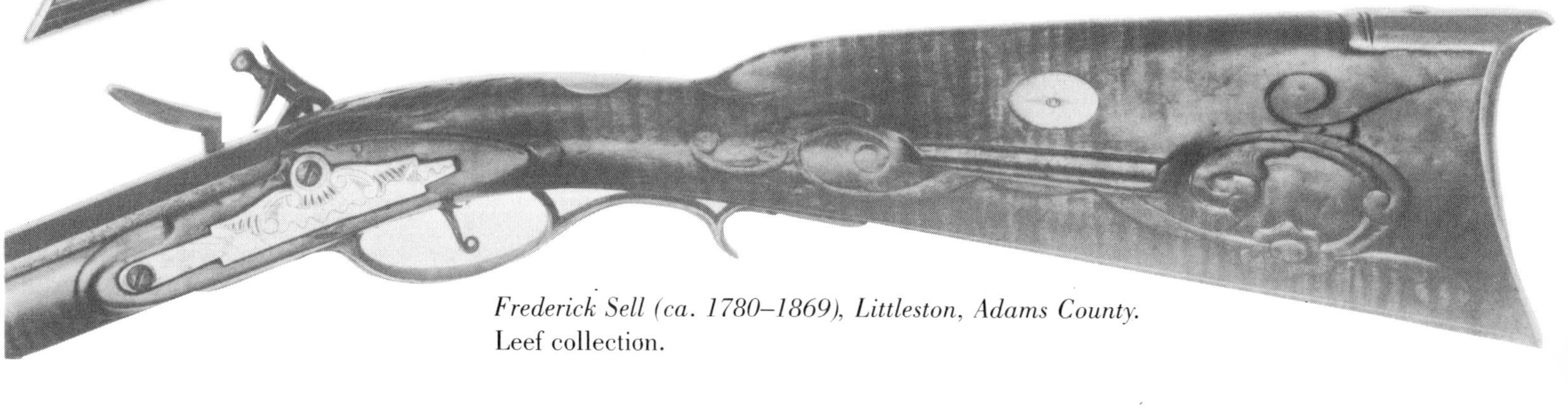

Frederick Sell (ca. 1780–1869), Littleston, Adams County. Leef collection.

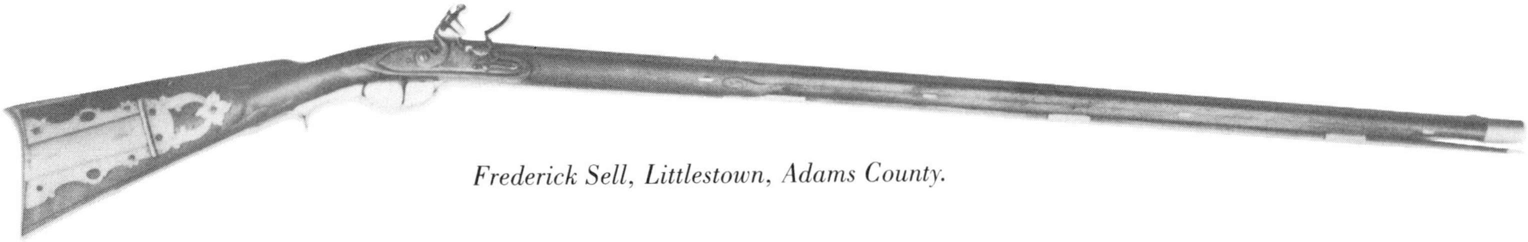

Frederick Sell, Littlestown, Adams County.

Rimersburg, Clarion Co. (BPaD, 1861); died in 1862 of typhoid in Civil War, Union army (REH, 199).

Seebrooks, William. Gunsmith, 1803–8, Martic Twp., Lancaster Co. (SED, 58); 1807, Providence Twp., Lancaster Co. (*History of Lancaster County*, 1015).

Seely, D. T. Gunsmith, Bradford, McKean Co. (PD, 1882, 1890).

Sees, Jacob (d. 1812). 1800–1812, gunsmith, Elizabeth Twp., Lancaster Co. (SED, 59; JKJ, 150–51).

Seibble (Seiple), Christian. Gunsmith, 1797–1803, Lancaster. (SED, 60); 1808, Middletown, Lancaster Co. (KKR, 344).

Seibert, George. Gunsmith, 1813–20, Montgomery Co., Franklin Co. (t); 1818–24, Peters Twp., Franklin Co. (t). In 1823 he was working with Samuel Shillito, gunsmith (WSB, 21–22).

Seidel, Nicholas. Artillery artificer 1777, Capt. Will's Co. (BCR, 115).

Seigler (Sigler), Amos. Gunsmith, Irish Lane, Philadelphia (DeSD, 1819); rear of 142 St. John's Philadelphia (MD, 1846).

Seinman, Frederick. Gunsmith, 12 Budd St., Philadelphia (DeSD, 1823).

Sell, Frederick (ca. 1780–1869). Major gunsmith, 1807, York; 1816–69, Littlestown, Adams Co. (WSB, 92–94; JKJ, 407–12; BPaD, 1861).

Sell, Jacob, the elder (1741–1825). 1793–99, Germany Twp., York Co.; 1810–25, Littlestown, Adams Co. Major gunsmith (WSB, 92–94; JKJ, 386–90; *History of Adams* [1886], 265).

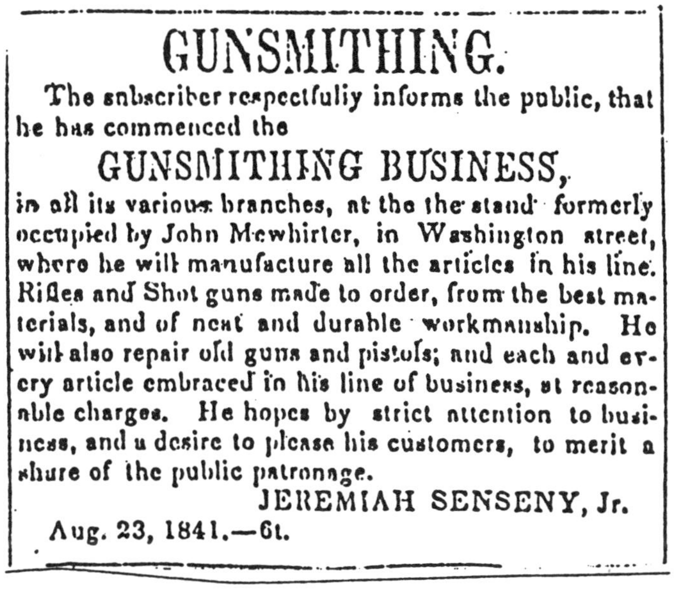

GUNSMITHING.

The subscriber respectfully informs the public, that he has commenced the

GUNSMITHING BUSINESS,

in all its various branches, at the the stand formerly occupied by John Mcwhirter, in Washington street, where he will manufacture all the articles in his line. Rifles and Shot guns made to order, from the best materials, and of neat and durable workmanship. He will also repair old guns and pistols; and each and every article embraced in his line of business, at reasonable charges. He hopes by strict attention to business, and a desire to please his customers, to merit a share of the public patronage.

JEREMIAH SENSENY, Jr.

Aug. 23, 1841.—6t.

Chambersburg Franklin Repository, August 23, 1841

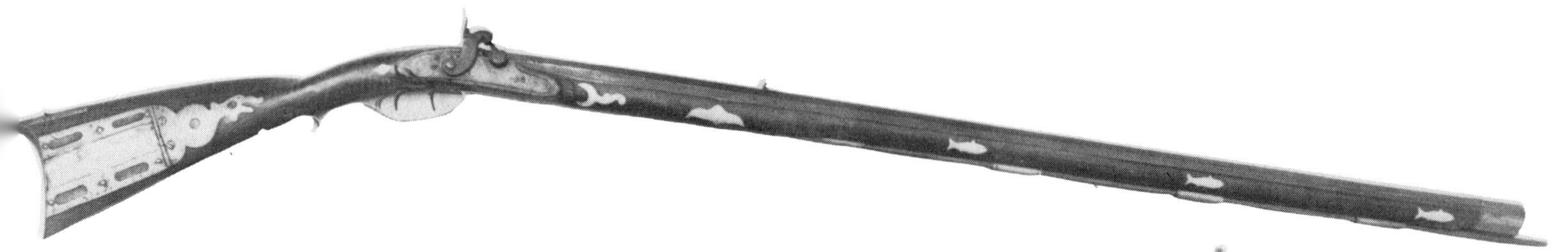

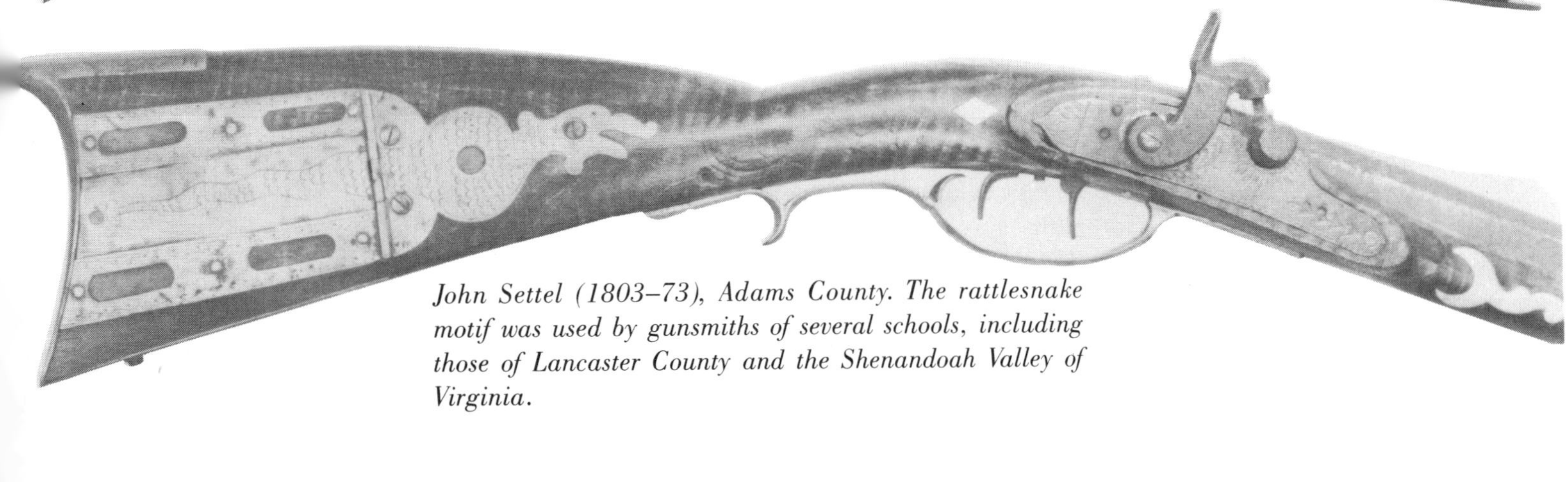

John Settel (1803–73), Adams County. The rattlesnake motif was used by gunsmiths of several schools, including those of Lancaster County and the Shenandoah Valley of Virginia.

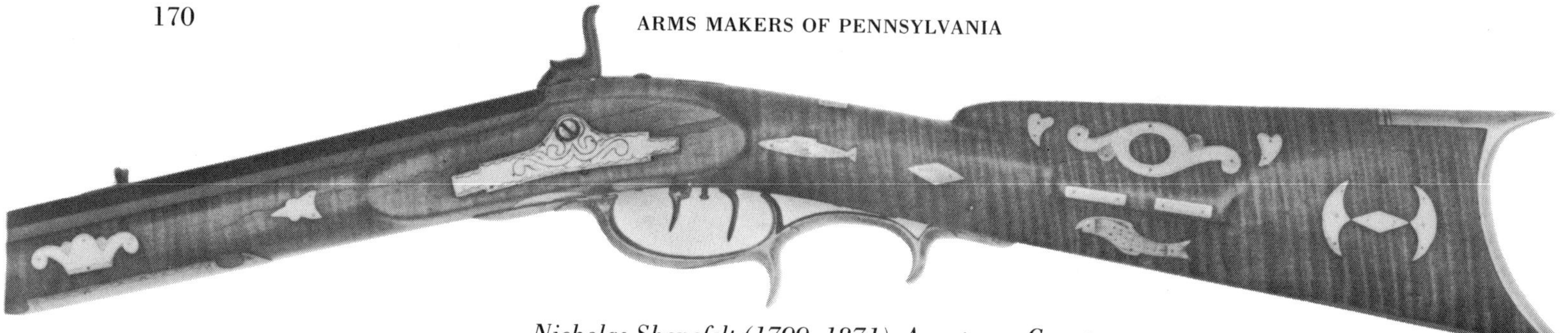

Nicholas Shenefelt (1799–1871), Armstrong County.

Sell, Jacob, the younger (1780–1855). 1815–55, Littlestown, Adams Co. Major gunsmith. Married Catherine, daughter of John Gobrecht (JKJ, 386–90; WSB, 92–94; t; c).

Semmence, Ed. Gunsmith, Erie boro (PD, 1890).

Senseny, Jeremiah (1821–73). Major gunsmith. 1841–73, Chambersburg, Franklin Co. Bought out the Mewhirter shop in 8/1841. One of his apprentices was J. H. Johnston (WSB, 19).

Settel, John (1803–73). 1831–73, Franklin Twp., Adams Co. (WSB, 62–66) Gunsmith, died 8/27/1873 at Cashtown.

Shaeffer, John R. 1822, gunpowder maker on Pine Run, Burrell Twp., Armstrong Co. Blew himself up in 1824 (Smith, *History of Armstrong County*, 369).

Shaffner, John. Gunsmith, 1788, Lancaster (SED, 60).

Shallenberger, Daniel. Gunsmith, 1820–29, Bullskin Twp., Fayette Co.; 1830–34, Connellsville boro., Fayette Co. (t).

Shanafelt, Andrew (1832–75). Son and apprentice of Nicholas Shenefelt. Graduate of Bucknell Univ. and Rochester Theological Sem. Gunsmith and Baptist preacher. Died 3/16/1875 (t; c; f).

Shanefelt (Shenefelt), Nicholas, Jr. (1799–1871). Born in Haines Twp., Centre Co. 1822–23, Union Twp.; Huntingdon Co. (t); 1825–35, McConnellstown, Huntingdon Co.(t); 1835–40, Redbank Twp., Armstrong Co.(t); 1840–50, Potter Twp., Clariòn Co. (t); 1850–71, Clarion boro. (t; c). His apprentices included J. S. Johnston, Joseph Douglass, Andrew Shanefelt, Henry Bonnet, and Daniel Sarvey. He sold his shop in Clarion to Henry Bonnet in 1868 (AFR, 284; REH, 179–93).

Shannon, W & H. 1808–16, 21 Passyunk St. and 293 S. 5th St., Philadelphia. Gunsmiths and cutlers. (PXD, 1808, 1811; DeSD, 1818) Militia arms contract of 10/7/1812 for M-1808 muskets: "25 rifles will today be completed by Shannons" (MPHS 8 [1867]) 10/1/1814.

Shannon, Hugh. Gunsmith, 1805–20, Philadelphia. 1805–7, 47 Sassafrass St. (PXD, 1805, 1807); 24 Passyunk St., near 5th St. (PXD, 1808, 1810, 1816); 293 S. 5th St. (DeSD, 1818, 1820).

Shannon, William (1745–1823). 1808–23, gunsmith, 224 Shippen, Philadelphia (PXD, 1808, 1810; DeSD, 1816, 1818; RAG, 174).

Sharp, Daniel. 1812–16, U.S. Arms Inspector, Pa. Militia (t).

Sharpless, Daniel. 1774–81, Nether Providence Twp., Chester Co. (t). 10/7/1775: "that Daniel Sharpless should be employed in making a number of firearms" (Col. Rec., 10:357–8).

Christian Sharps (1811–74), Philadelphia.

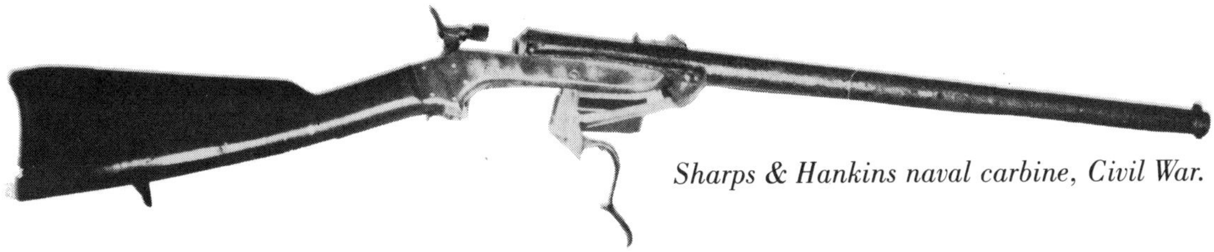

Sharps & Hankins naval carbine, Civil War.

Sharps, Christian (1811–74). Inventor and major manufacturer, Philadelphia, west end of the Wire Bridge, 336 Frankford Rd. Born in N.J.; died in Vernon, Conn.

Sharps, Christian & Co. Fairmount, on the Schuylkill River, across the wire bridge. Operated as Christian Sharps, 1858–63; Sharps & Hankins, 1863–72; C. Sharps & Co., 1872–75.

Shaub, Adam. Gunsmith, Chestnut St., between Prince St. and N. Queen St., Lancaster (BLD, 1857).

Shaubel, John (b. 1779). On 12/3/1807, arrived in Philadelphia from Germany on *William Johnston;* brown hair, blue eyes, born Marbach, Germany (2 Pa. Arch. 17 at 664). Locksmith.

Shaw, George. Artillery artificer, 1779. (2 Pa. Arch. 15 at 384).

Shaw, Joshua (1777–1860). Born in Lincolnshire, England; died in Burlington, N.J. From 1814 to 1818 he worked on finding a practical application of Rev. Forsyth's discovery of the explosive power of mercury-fulminate in a percussion cap. He tried iron and pewter before settling on brass as the medium to house the explosive charge. Shaw held more patents before the Civil War than anyone else in the firearms field: 5485X (5/7/1829) for a firearm; 7313X, 7314X, 7315X (12/3/1832) for cannon; 8090X (5/17/1834) for a combination gun; and patent of 1/30/1841 for a method of discharging a firearm. Congress paid him $25,000 in lieu of royalties denied him for his discovery of the percussion cap, as the patent was denied because Shaw was a British subject (Thompson Wescott and John T. Scharf, *History of Philadelphia*, 3 vols. [Philadelphia: Everts, 1884] 2:1054, 3:2270).

Shay, Michael. Locksmith, Lancaster (BLD, 1857).

Sheaff, Henry. Gunsmith, 1832–33, Lancaster (SED, 60).

Sheanor, William. Gunsmith, 1780, Reading (RAG, 169).

Sheble & Fisher. Swordsmiths and cutlers, Philadelphia. Produced swords and sabers for the Union army (GAAAM, 224).

Sheets, Henry (d. 1816). On 5/10/1773, John Ponsler, with the consent of his father, Ludwig, binds himself apprentice to Henry Sheatz of White Marsh Twp., Montgomery Co. (PIDS) There was a family of gunsmiths in Berkeley Co., W.Va., including Henry Sheetz. Henry and his brother Philip, the latter already established in Shepherdstown (then called Mechlenberg), advertised in the Philadelphia newspaper, *Pennsylvania Packet*, 12/25/1775, that they had a gunshop in Shepherdstown.

Sheffer, Adam. On 3/15/1777, "pay Adam Sheffer £33/5/0 for arms delivered" (2 Pa. Arch. 1 at 14).

Sheffer, George. Smith, 1767, Maxatawny Twp., Berks Co. (t).

Shefler, George. 1779, locksmith, Lancaster (t); 1817–31, Hempfield Twp., Westmoreland Co. (t). Gunsmith.

Sheiber, John. Gunsmith, Mechanicsburg, Cumberland Co. (BPaD, 1861).

Sheivley, Henry. Cutler, Philadelphia (c. 1790).

Sheldrake, David. Artillery artificer, 1779 (2 Pa. Arch. 15 at 386).

Shell, Daniel. Gunsmith, 1825, Paxton Twp., Dauphin Co. (KKR, 339).

Shell, John (1790–1875). 1812–75, West Hanover, boro. or immediate area, Dauphin Co. (BPaD, 1861; MB 9 [1970] 9; JKJ, 274). Gunsmith.

Shell, Samuel. Gunsmith, 1800–1810, Perry Co. (Chandler and Mitchell).

Shelly, Joseph (b. 1831). Gunstock maker, Rostraver Twp., Westmoreland Co. (c, 1850).

Shenefelt. See Shanefelt.

Shener, John. Gunsmith, Reading, 1810–10 (KKR, 341).

Shener, William. Gunsmith, 1779–85, Reading (3 Pa. Arch. 18 at 657, 793, 538, 396, 276).

Shener, William, Jr. Gunsmith, 1800–1810, Reading (KKR, 341).

Shepherd, John. Gunlock filer, 135 Green St., Philadelphia (DeSD, 1829, 1830, 1831; MD, 1841).

Sheppard, Thomas. Artificer, 1760–61, at Ft. Pitt (Bouquet Census).

Sheppard, William. 1779, married Ann McGinnis at St. Michael's Zion Lutheran Church, Philadelphia. His occupation was given as gunsmith on the church records. (2 Pa. Arch. 19 at 370) 1779, artillery artificer, Pa. Continental Line (2 Pa. Arch. 15 at 383).

Henry Sherry (1850–1908), Clarion County.

Sherer, John. Artillery artificer, 1778–79, Capt. Nathaniel Irish's Co. (PIR, 2:247).

Sherritt, Joseph. Gunsmith, 1780, North Ward, Philadelphia (t).

Sherry, Henry (1850–1908). Son and apprentice of Michael Sherry, Beaver Twp., Clarion Co. (REH, 169–70).

Sherry, John, Sr. (1773–1853). Born in Lehigh Twp., Northampton Co. (t); in 1799 his son Wilhelm was born in Westmoreland Co. 1800–1813, Hempfield Twp., Westmoreland Co. (t); 1814–23, Richland Twp., Venango Co. (t); 1824–39, Beaver Twp., Venango Co. (t); 1840–53, Beaver Twp., Clarion Co. Major gunsmith (REH, 151–57).

Sherry, John, Jr. (1808–93). 1830–93, Beaver Twp., Venango Co. (In 1839 the county of Clarion was created and thus John Sherry, Jr. and Sr., "moved" only in the political sense.) Gunsmith. (REH, 168–77).

Sherry, Michael (1814–84). Gunsmith. Younger son of John Sherry, Sr., he lived all his life in Beaver Twp., Clarion Co. (REH, 177–78).

Sherts, Christian. 1779–82, gunsmith, Paxton Twp., Lancaster Co. (SED, 60; t).

Shey, Timothy. Artillery artificer, 1780; deserted Capt. John Jordan's Co. (2 Pa. Arch. 11 at 260).

Shields, W. H. Gunsmith, Philadelphia (PD, 1882).

Shillinger, George. Gunsmith, 1824, Forks Twp., Northampton Co. (t).

Shillito, Samuel (1794–1852). 1819, Peters Twp., Franklin Co. (t); 1825–35, Ayr Twp., Bedford Co. (t); 1836–52, Chambersburg. Died 9/18/1852 of cholera (WSB, 18–19). Gunsmith.

Shillito, Samuel Miller (d. 1904). Gunsmith, son of Samuel Shillito. 1846, Chambersburg (t); 1850, Montgomery Twp., Franklin Co. (c); 1853–1904, Chambersburg. (t; c; WSB, 18–19) In 1864 fire burned out his gunshop. He died 3/10/1904 at his home, 171 Queen St., Chambersburg (WSB, 18–19).

Shirer (Schairer, Shearer), Joseph. Gunsmith, 1856–65, Greensburg, Westmoreland Co. (t).

Shiver (Shirer), Joshua (b. 1805). Gunsmith, Greensburg, born in Würtemberg, Germany. (c, 1860) I suspect that this gunsmith and the one named above are one and the same.

Shock, John. Probably an apprentice of Valentine Fondersmith at Lancaster, ca. 1780 (GAS-1, 316–17).

Shoemaker, Jacob (b. 1837). Gunsmith, South New Beth-

Samuel Shillito, Sr. (1794–1852), Fulton and Franklin counties.

lehem, Armstrong Co., 1858–68. Apprenticed to William Schreckengost (t; c).

Shoemaker, M. Gunsmith, 467 N. 3d St., Philadelphia (PXD, 1810).

Shoemaker, Thomas. On 11/14/1776 was paid £3/8/0 for 1,700 flints. Flint maker to Committee of Safety (Col. Rec., vol. 11).

Shoff (Shough), Jacob. 1800, Cavernon Twp., Berks Co. (KKR, 342); rear of Noble near 3d St., Philadelphia (PXD, 1813, 1814). Gunsmith.

Shomo, John. 1780, Richmond Twp., Berks Co. (RAG, 176). Gunsmith. His descendants became important gunsmiths in W. Va.

Shorer, Andrew. Gunsmith, 1775, Bethlehem Twp., Northampton Co. (t).

Shorer, John. Gunsmith, Perry Co. (c, 1850).

Shough, Joseph (b. 1761). Gunsmith, 1809–27, German Twp., Fayette Co. (t).

Shoup, George (b. 1798). 1830–35, Jackson Twp., Cambria Co.(t); 1842–60, Brush Valley Twp., Indiana Co. (c; t). Gunsmith.

Showalter, Isaac (1853–1905). 1870–1905, Brookville, Jefferson Co. Son of John. Gunsmith and sawmiller (REH, 32–33).

Showalter, John (1814–83). Son of Samuel (d. 1814) and Margaret (Rowland) Showalter (1781–1849). After Samuel died, Margaret remarried, to John Farver. Thus, John Showalter was a half-brother to William Farver (b. 1818), a gunsmith. John Showalter was active in Lancaster boro., 1838–40 (t), perhaps as an apprentice. He then set up shop as a gunsmith on Old Mill St. (Franklin St.), Brookville, Jefferson Co. (REH, 32–38; t; c).

Shreid, Casper. Gunsmith, Cocalico Twp., Lancaster Co. (DB H-173).

Shriver, John, the elder (1752–83). 1781, Mt. Pleasant Twp., York Co. Buried at Christ Reformed Church, Littlestown (JKJ, 425). Gunsmith.

Shriver, John, the younger (d. 1830). 1798, Mt. Pleasant Twp., York Co. (t); 1800–1830 Mt. Pleasant Twp., Adams Co. (t); 1831, "John Shriver's heirs" (WSB, 99; JKJ, 425; GAS-1, 156–59).

Shroyer (Schreyer, and other spellings), George Sr. (b. ca. 1739–1821). 1755–66, Hanover, York Co.; 1767–68, Reading, Berks Co. (t), gunsmith. 1770 and after, Hanover, York Co. There is today much speculation in the absence of useful documentary evidence (JKJ, 374–77; GAS-1, 394–419).

John Shriver's gunshop, York County.

Shroyer, George, Jr. 1792–93, Hanover, York Co.; 1804–5, Antrim Twp., Franklin Co. (t; WSB, 23); 1815, Hanover (as George Schreye) (Official Program of the Hanover Centennial, 17); Baltimore, Md. 1810, (FSP, 269). Gunsmith.

Shroyer, Lewis. Gunsmith, Georges Twp., Fayette Co. (t); 1807–27, Grafton, Taylor Co., W. Va. (t; c). His sons became important gunsmiths in both W. Va. and Ind.

Shryock, Christian. Artillery artificer, 1778, Capt. Nathaniel Irish's Co. (PIR, 2:247).

Shugart, John, & Co. Cutlers, before and after 1828, Chambersburg (Hazard's *Annals*, 2:264).

Shuler, Caleb. B. (b. 1825). Liverpool, Perry Co. Son of Samuel Shuler (CHAN, 53ff.).

Shuler, Henry (1751–1820). Smith. Pvt., Pa. German Regt., 1776–79, died 1/10/1820 in Mifflin Co. (2 Pa. Arch. 11 at 86).

Shuler, John (d. 1827). A gun at Ft. Ticonderoga Museum is marked, "This rifle made by John Shuler, Gunsmith, 1800." (GAS-1, 270–76) In 1802, purchased land in Milford Twp., Bucks Co.; 1800–1802 Milford Twp., Bucks Co. (t). By 1817, Perry Co., first at Millerstown and later at Liverpool A John Shuler was a teamster in Pulaski's Legion, and in 1820 was living in Lehigh Co. (PIR, 2:156). Several of his sons became gunsmiths: Samuel, Leonard, John, Jr., and Joseph. (CHAN, 44ff.).

Shuler, John, Jr. (1804–79). 1827–79, Liverpool, Perry

John Shuler, ca. 1780, Bucks County.

John Shuler, Bucks County.

Joseph Shuler (1811–79), Perry County.

Co. After 1834 he was a full-time gunsmith in the rear of his brother, Samuel's, store. (t; c; CHAN, 44ff.).

Shuler, John Ramsey (b. 1829). Son of Samuel. Before 1857, Liverpool; after 1857, Bloomfield boro (CHAN, 69f.).

Shuler, Joseph (1811–79). Although trained as a gunsmith, he became a politician, and was elected high sheriff of Perry Co. He later ran a boat yard and a monument works (CHAN, 47ff.).

Shuler, Joseph R., II (b. 1838). Perry Co. Employed before the Civil War as a gunsmith. Served in the 192d Pa. Volunteers, then 7th Pa. Reserves, 1863–65 (CHAN, 47ff.).

Shuler, Samuel (1799–1875). 1827–75, Liverpool, Perry Co. After 1834, when he bought out his brother's interest in their store, it seems probable that John Shuler, Jr. did the gunsmithing and Samuel ran the store (CHAN, 44ff.).

Shull, Thomas. Patent 23,505 for conversion of muzzle-loading Springfield rifles to breech-loaders. Millersburg, Dauphin Co., 1859.

Shultie, Hiram. Gunsmith, Wylie, Allegheny Co. (PAD, 1864–65).

Shultz, Henry. Gunsmith, Richland (Missemer's Mills), Lebanon Co. (BPaD, 1861; BRD, 1860).

Shunk, Isaac. 1810–11, Lancaster (t); gunsmith, 1819–30, Pitt Twp., Allegheny Co. (t); 1831–32, Pittsburgh (t).

Sible, Jacob. Gunsmith, 1788–1814, Lancaster (SED, 60).

Sible, John. Gunsmith, 1788–1814, Lancaster (SED, 60).

Sibthorp, Christopher, Braizier, 1656; arrived at Philadelphia from London (PaE, 18); braizier, 10/1681, Philadelphia (2 Pa. Arch. 19 at 152).

Sides, Henry (d. 1806). Gunsmith, 1788–1806, Bedford boro. (t).

Siegfried, Daniel. Two entries: gun-barrel maker, Mohn's Store; gun-barrel maker, Wernersville, both Berks Co. (BPaD, 1861).

Siegfried, Daniel. Gun-barrel maker, 1846–51, East Cocalico Twp., Lancaster Co. (SED, 60) In 1850 Siegfried bought 5 tons of iron at $450 and produced 750 gun barrels valued at $1,100 (IC); 1861, Mohn's Store, Berks Co. (BPaD, 1861); gunsmith, 435 E. Orange St., Lancaster boro (BLD, 1871–72).

Siegfried, Henry. Gun-barrel maker, 1846–51, East Cocalico Twp., Lancaster Co.; gunsmith, 435 E. Orange St., Lancaster boro (BLD, 1871–72).

Siegfried, Reuben. Gun-barrel smith, East Cocalico Twp., Lancaster Co., 1850 (c; SED, 60).

Sifford, Leonard. 1800–1830, whitesmith and gunsmith, Letterkenny Twp., Franklin Co. (t; WSB, 27).

Silfries, Jacob. Artillery artificer, 1775–77, Capt. Will's Co. (BCR, 115).

Silknitter, Solomon. Apprentice gunsmith, Barree Twp., Huntingdon Co. Apprenticed to Peter Smith (c, 1850).

Simcock, Joseph. Artillery artificer, 1779 (2 Pa. Arch. 15 at 383).

Simes, Richard. Negro. Bound himself apprentice to Leonard Snider, Warren Twp., Franklin Co., to learn the gun-barrel making business, May 1816 (WSB, 28).

Simpson, Oliver K. Gunsmith, Diamond Alley near Diamond St., Pittsburgh (PAD, 1864–65).

Simpson, William. Artillery artificer, 1778, Capt. Thomas Wylie's Co. (PIR, 2:249). Shoemaker by trade.

Sims, Joseph. 10/20/1775, supplied 7 cutlasses at £0/3/9 each. (Col. Rec., 10:375) Purchased by the Committee of Safety.

Sinclair, John. Gunsmith, 1794–1854, Lancaster (SED, 60).

Siner, John T. Gunsmith, 5th St. below George St., Philadelphia (MD, 1846); Lawrence below Franklin (MD, 1855).

Siner, L. K. Successor to John Krider, S. 2d St. and Walnut St., Philadelphia (MD, 1855); 5th St. below George St., Northern Liberties (MD, 1846). Gunsmith.

Sink, David (1836–1910). 1858–67, Cameron, Marshall Co., W. Va.; 1868–72, Marietta, Ohio, with A. C. McGirr. After 1875, Punxsutawney, Jefferson Co. (PD, 1882; REH, 121–24). Gunsmith.

Sink, Jacob (1842–1910). 1858–67, Cameron, Marshall Co. (t; c); 1875–80, Punxsutawney, Jefferson Co.; Indiana boro. (PD, 1882). Died at Sharon, Pa., 1910. (REH, 114–116). Gunsmith.

Sink, John (b. 1826). 1855–57 Punxsutawny, Jefferson Co.; 1858–67, Cameron, Marshall Co., W.V.; 1868–70, Georgeville, Indiana, Pa. gunsmith. (t)

Sink, Joseph (b. 1844). 1858–67, Cameron, Marshall Co., W. Va.; 1870, gunsmith Indiana Co.; 1910, Ford City, Armstrong Co. (REH, 110–24 passim).

Sink, William (b. 1803). 1830, Conemaugh Twp., Cambria Co. (t); 1835–37, Mahoning Twp., Indiana Co. (t); 1850, South Mahoning Twp., Indiana Co. (c); 1860, Marshall Co., W. Va.; 1870, South Mahoning Twp., Indiana Co. (c). In 1854 Sink bought many tools at Samuel Ghrett's sale. Gunsmith.

Sins, Fred. Gunsmith, 817 S. 2d St., Philadelphia (MD, 1859).

Sipes, Charles. 1796–1811, Buffalo Twp., Armstrong Co. (t; Smith, *History of Armstrong County*, 444).

Sipe, Charles (1812–76). 1850–76, Indiana boro (t; c). Gunsmith.

Skerl, Adolph. Patent 176,367 for a gunlock. Philadelphia, 1876.

Skevington, William. Artillery artificer, 1778–79, Capt. Thomas Wylie's Co. at Carlisle (PIR, 2:249).

Skillen, Charles. Gunsmith, 1800, Conaquanessing Twp., Allegheny Co. (t).

Slaybaugh, Henry. 1821, runaway apprentice of John Meals, Menallen Twp., Adams Co. 6½¢ reward offered. Apprentice gunsmith (WSB, 66).

Slaysman, Charles (1812–76). Gunsmith, 1850–76, Indiana boro (t; c).

Slaysman, George (1782–1862). Born in Northampton Co.; 1805–23, Lewistown, Mifflin Co.; 1825–32, Woodberry Twp., Huntingdon Co.; (t) 1832–62, Punxsutawney, Jefferson Co. (REH, 28–31). Gunsmith.

Slonaker, George (1797–1870). 1829–32, Greenfield Twp., Bedford Co. (t); 1835–43 St. Clair Twp., Bedford Co. (t); 1844–70, Union Twp., Bedford Co. (t; c; f). Gunsmith.

Slonaker, John George (1830–89). 1855–89, Alum Bank (Pleasantville), St. Clair Twp., Bedford Co. (t; c; f). Gunsmith.

Sloop (Sloope), Jacob. 1796, Bethel Twp., Bedford Co. On the Constable's list of tradesman and tavern keepers. Gunsmith at Warfordsburgh (t).

Slotter & Co., Philadelphia. Made "Deringer" pistols that were probably better than the real "Deringers." After Deringer forced the Slotterbeck brothers to abandon the Deringer name, they formed Slotter & Co. to sell these pistols in Calif. through A. J. Plate. To use the "Derringer" name, they rented the last name of a tailor, named John Derringer.

Slotterbeck, Charles (1831–86). By 1847, worked for James Golcher at Philadelphia, until 1852, when he began working for Henry Deringer. He managed Deringer's shop until 1858 when he left for Calif. Beginning in 1859, he made "fake" Deringer pistols in Calif., which were at least as good as, perhaps better than, those made in Philadelphia (LSC, 220–34).

Slotterbeck, Frederick (d. 1873). Brother of Charles and Henry, part of "Slotter & Co."; made fake Deringer pistols at Philadelphia, which were sold through A. J. Plate in Ca. In 1847, worked with James Golcher; 1847–59, worked for Henry Deringer; 1859–60, worked at Slotter & Co., all in Philadelphia (LSC, 230–44).

Slotterbeck, Henry (1836–88). Like his brothers, Charles and Frederick, Henry worked for James Golcher in 1847 in Philadelphia, then worked, ca. 1852, for Henry Deringer; he, too, was involved with the "Slotter & Co." fake Deringer pistols. By 1869 he was in Calif. He was accidentally shot in the chest and killed (LSC, 220–44).

Slough, Jacob. Gunsmith, 24 Duke St., Philadelphia (PXD, 1816).

Slough, Mathias. 5/4/1775, Lancaster Co. Allowed the Committee of Safety to have his powder at £15 per cwt. and his lead at 45 pence per cwt. (2 Pa. Arch. 13 at 293).

Small, David. 1800, South Beaver Twp., Allegheny Co. (t); 1804–7, New Lisbon, Ohio (t). Gunsmith.

Small, Jacob. Gunsmith, South Beaver Twp., Allegheny Co. (t).

Small, John N. (d. 1823). 1775, West Augusta, state arms repairman; 1780, Philadelphia, armorer; 1788–1821, Vincennes, Ind. (Lindsay, *Kentucky Rifle* [York, Pa.: York County Historical Society, 1976] plate 54; FSP, 282.

Small, John. Gunsmith, 1802–6, Georges Twp., Fayette Co. (t).

Smalt, John. Artillery artificer, 1779, at the laboratory, Philadelphia (2 Pa. Arch. 15 at 385).

Smith, Abias Butler (1818–1900). Gunsmith, 1838–66, Clinton, Allegheny Co. Apprenticed to Joel Ferree (KKR, 122, 145, 345–46).

Smith & Rand. Gun-powder makers, 1869–1884, Packer Twp., Carbon Co. In 1873 it became Laflin Powder Co.; in 1871 it was Laflin & Rand. It had 2 magazines with a capacity of 5,000 kegs each (Matthews, *History of Carbon County*, 751).

Smith, Anthony. 1772–75, Bethlehem, Northampton Co. (t). Gunsmith.

Smith, "Widow" Catherine. Gun-barrel maker, White Deer Twp., Union Co. Widow of Peter Smith (d. 1773). She ran a gun-barrel mill until 1779, when it was burnt by the Amerindians. Col. Hunter to Col. M. Smith, 7/23/1779: "Immediately after the evacuation of Ft. Muncy . . . they burned the Widow Smith's mills and killed one man, 17th inst" (1 Pa. Arch. 7 at 574; L. E. Theiss, NAT 9 [1950]: 179–85).

Smith, Charles. Gunsmith, 14 Bread St., Philadelphia (MD, 1846).

Smith, Conrad. Apprentice gunsmith, 1852, with William Defibaugh, Monroe Twp., Bedford Co. (t).

Smith, C. F. Gunsmith, Brokenstraw Twp., Warren Co. (PD, 1882, 1890).

Smith, David (1780–1868). 1801–22, Berks Co.; 1822, Columbiana Co., Ohio; 1832, Sandusky Co., Ohio (t; c; f). Gunsmith.

Smith, Henry (b. 1810). 1832–55, Hempfield Twp., Westmoreland Co. (t; c). Gunsmith.

Smith, Jacob (1763–1847). Died 3/23/1847, West Beaver Twp., Union Co., age 83 or 84. (death cert.) Gunsmith. Probably the son of Peter Smith and "Widow" Catherine; Jacob was born in Berks Co. (Death certs.).

Smith, Jacob. Gunsmith, Murraysville, Westmoreland Co. (BPaD, 1861).

Smith, James. Artillery artificer, 1779 (2 Pa. Arch. 15 at 383).

Smith, J. E. (b. 1834). Barree Twp., Huntingdon Co., after 1855; gunsmith (t; c).

Smith, J. H. 1850, powder-horn maker, Lancaster boro (t; c).

Smith, John (b. 1744). Smith and cutler, 1769–75, Coventry, Chester Co. 1777, artillery artificer, smith, New 11th Pa. Regt. (2 Pa. Arch. 11 at 72). He was born in England.

Smith, John. Gunsmith, 1802–10, Salem Twp., Westmoreland Co. (t).

Smith, John. Gunsmith, 1850, Lancaster (c; t).

Smith, Johnston. Gunsmith, 1776, Easton. (t) on 4/10/1776: "a certain Mr Fraily of Germantown has a number of arms for sale, which he has offered [to sell to] Johnston Smith" (1 Pa. Arch. 4 at 731). In 2/1776 he received a Va. contract for 1,000 strands of arms, which he had completed by 5/1/1776 (GAAAM, 140).

Smith, Lewis, Swordsmith, 1777, in the employ of Lewis Prahl, gunsmith, Northern Liberties, Philadelphia Co. (2 Pa. Arch. 3 at 104).

Smith, Lorens. Gunsmith, Penn Ave. near Lackawanna Ave., Scranton (BPaD, 1861).

Smith, Martin. Gunsmith, 1814–19, Lancaster (SED, 60).

Smith, Nat ca. 1880–90, Newport, Perry Co. (CHAN, 22). Gunsmith.

Smith, Peter (b. 1756). Born in Longswamp Twp., Berks Co., a son of Peter and "Widow" Catherine Smith. In 7/1777, Capt. Druckenmiller's Co.; 1778, pvt., Capt. Good's Co.; 1780–90, Longswamp Twp., Berks Co.; 1798–1813, gunsmith, Sunbury, Northumberland Co.; 1814–18, Lewistown, Mifflin Co.; 1818–33, Lake Twp., Wayne Co.,

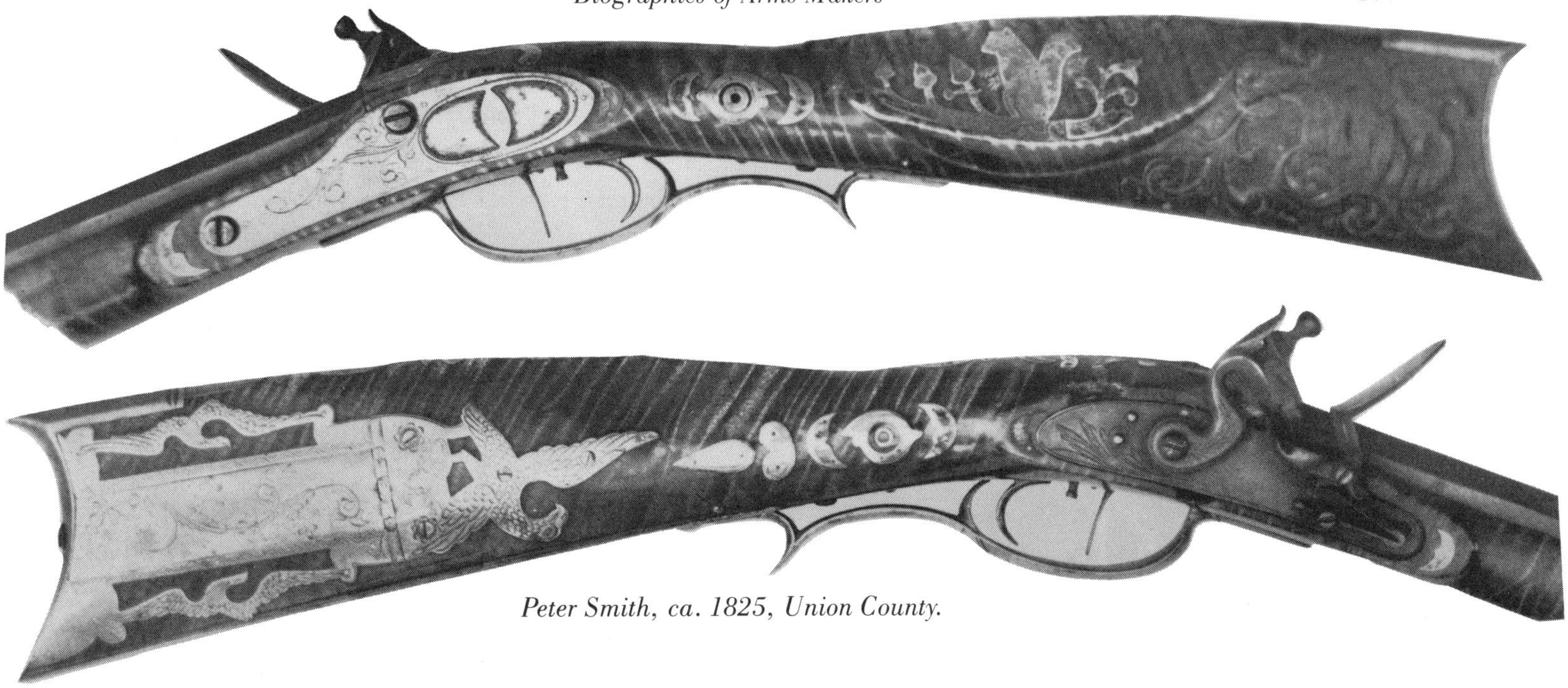

Peter Smith, ca. 1825, Union County.

Ohio. Military pension application dated 1833, Wayne Co. (3 Pa. Arch. 18 at 377; t; c; f; LIN, 370).

Smith, Peter (b. 1796). 1825–28, New Berlin, Union Co. (t); 1830–37, Union Twp., Mifflin Co. (t); 1838–39, Menno Twp., Mifflin Co. (t); 1844–45, Henderson Twp., Huntingdon Co. (t); 1845–50, Barree Twp., Huntingdon Co. (t; c); 1879, Union Twp., Mifflin Co. (t). Major Gunsmith. Ennisville, Huntingdon Co. (BPaD, 1861).

Smith, Peter (b. 1823). 1850, apprenticed to Peter Smith, Barree Twp., Huntingdon Co. (t; c); 1855, "Peter Smith was the only gunsmith that ever opened a shop to carry on that trade at Mill Creek, Brady Twp." (AFR, 225); 1860–61, Jackson Twp., Huntingdon Co. (t); 1876–83, Huntingdon boro (t; c). Gunsmith.

Smith, Philip. 3/29/1798, Carlisle *Gazette:* "Runaway, Thursday evening, 22 of March, from the subscriber in Carlisle, an apprentice named Philip Smith, 16 years old, fair complexion, stout made, left handed . . ." (WSB, 55–57).

Smith, Samuel (b. 1826). Gunsmith, son and apprentice of Peter Smith, Barree Twp., Huntingdon Co. (c, 1850).

Smith, Samuel (1802–72). Gunsmith, 1823–31, Redstone Twp., Fayette Co. (t); 1850–72, Astoria, Fulton Co., Ill. (Curt Johnson).

Smith, Stephen. Blacksmith, artillery artificer, 1779, at the cannon factory, Philadelphia (2 Pa. Arch. 15 at 384).

Smith, Stophel. Gunsmith. Undated militia enrollment list, pre-Revolutionary: "Stoeful Smith, 5′7″, age 40, Capt William Biles' Bucks Co. Regt. born in Scotland, thin face, light hair" (5 Pa. Arch. 1 at 358 and 361). 4/13/1781, Capt Andrew Smith's Lancaster Co. Militia, Stopel Smith, Upper Paxtang Twp., Lancaster Co. (2 Pa. Arch. 13 at 386). Capt. Dillin (*Kentucky Rifle*, 152) listed

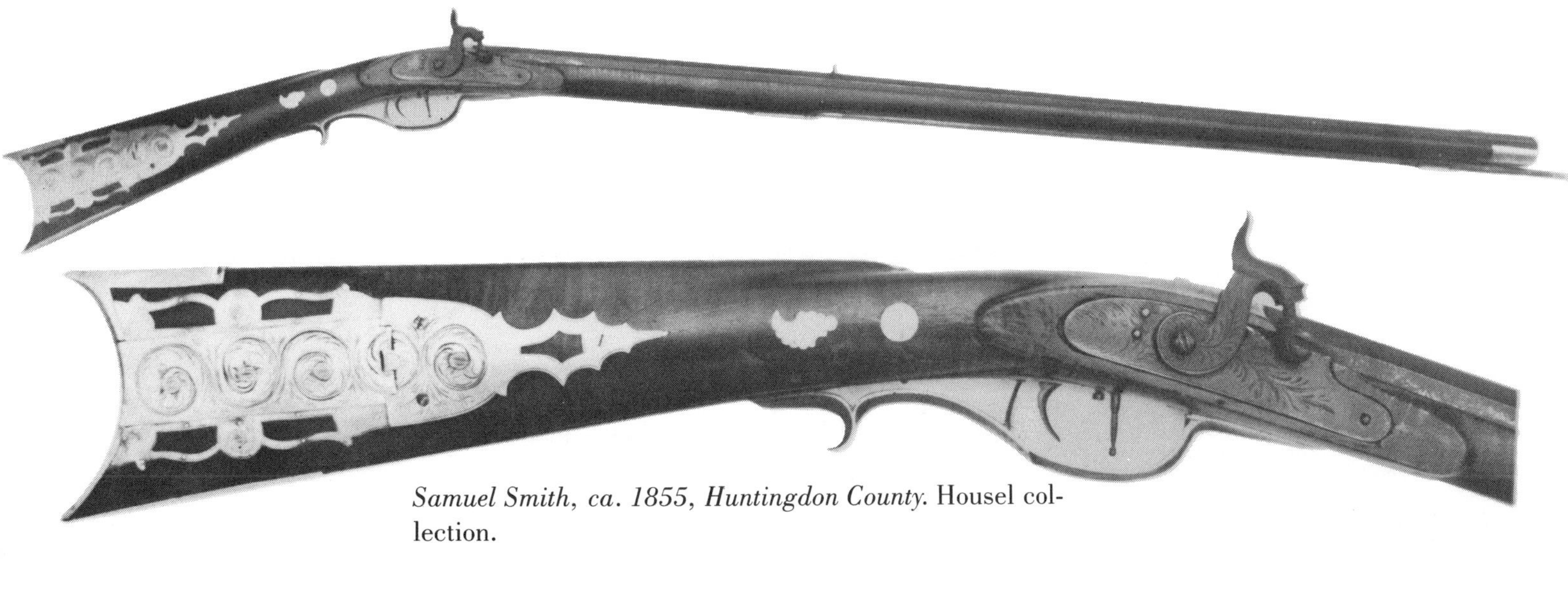

Samuel Smith, ca. 1855, Huntingdon County. Housel collection.

him only with the Pa. state identification and noted a plain but very good gun with a silver inlay with Stophel Smith's name on the barrel. Dillin dated him "before and after 1794." Gardner noted that a known over/under rifle by Stoeffel Smith was dated 1812, and assigned 1790–1812 as dates (RAG, 182).

Smith, Thomas. Gunsmith, 1827–28, Smith Twp., Washington Co. (t).

Smith, William. Arrived in Philadelphia, 1701, a brass founder from Bristol in England (3 Pa. Arch. 3 at 341).

Smith, William. Lock maker, 1853, York (t).

Smith, William. Armorer, 1776, on armed boat *Congress*, Pa. Navy (2 Pa. Arch. 1 at 268).

Smutz (Smuts), Jacob. 1820–24, Fayette Co. (t); 1830 and later, Tuscarawas Co. Ohio (DH, 271). Gunsmith.

Snedden, William Tait. Patents 116,363 and 116,364 of 6/27/1871 for a breech-loading rifle, Johnstown.

Sneider, Frank H. 1876, showed a needle-fire gun at the 1876 Philadelphia Centennial Exhibition (RAG, 182).

Snevely, Jacob. 1810–20, Harrisburg; 1824–35, Miami Co., Ohio (DH, 271). Gunsmith.

Snider, Jacob. Smith, 1769, Northern Liberties, Philadelphia Co. (t).

Snider, Jacob (1821–75). Gunsmith, 1842–43, Hopewell Twp., Bedford Co. (t); 1846–55, Liberty Twp., Bedford Co. (t; c); 1875, land speculator and banker, murdered by a claims jumper at Georgetown, Colorado (f).

Snider, Lewis. Gunsmith, 1803, Donegal Twp., Lancaster Co. (SED, 60).

Snider, Leonard (d. 1823). 1816–23, Warren Twp., Franklin Co. (t; WSB, 27; 31; AMM, 257). Gun-barrel maker.

Snider, Peter and John. Gun-barrel boring mill, 1798, Falling Spring Rd., Chambersburg. Peter, who owned the shop, was a blacksmith, while John was the gunsmith (WSB, 49).

Snider, Tobias (b. 1820). 1848–49, Hollidaysburg, Blair Co. (t); 1850–62, Woodcock Valley, Liberty Twp., Bedford Co. (t); 1867–84, Saxton boro., Bedford Co. (t; c). Gunsmith.

Snoddy, James. Gunsmith, 1757, Derry Twp., Lancaster Co. (SED, 60) (WBB-325, so he was active before 1757).

Snyder, Henry. 1780–83, Lynn Twp., Northampton Co. (t) Gunsmith.

Snyder, Ira. Gunsmith, Haines Twp., Centre Co. (c, 1860).

Snyder, Jacob, Jr. Inventor, before 1866, Philadelphia. His patent 69,941 was assigned to his son, John Vaughn Snyder.

Snyder, J. E. Gunsmith, Haines Twp., Centre Co. (BRD, 1878).

Snyder, John. Gunsmith, 1848, Heidelberg Twp., Lehigh Co. (t).

Snyder, John Ludwick. Gunsmith, 1820–55, Bell Twp., Clearfield Co. (t).

Snyder, Samuel. Gunsmith, 1820, Bushkill Twp., Northampton Co. (t).

Snyder, William. Gunsmith, 14 Green St., Philadelphia (DeSD, 1823).

Snyder, William. Gunsmith, Flatwood, Fayette Co. (BPaD, 1861).

Snyder, William M. (1852–1905). The only mention of his having been a gunsmith was in his obituary in the Bedford *Gazette*, 8/22/1905. Monroe Twp., Bedford Co.

Solliday, Frederick (d. 1804). In 10/1751, he emigrated from Basel, Switzerland, and on 5/6/1762 he purchased 143 acres in Bedminster Twp., Bucks Co. 1762–1803, clock maker and armorer, Bedminster Twp., Bucks Co. (t).

Solliday, Samuel. In Bucks Co. *Intelligencer*, 2/13/1849, he offered knives, razors, and the like for sale in Solebury Twp., Bucks Co. Cutler and silversmith.

Solomon, Theodore. Gunsmith, Market St. and 31st St., Philadelphia (BPaD, 1861). Washington St. West (MD, 1846).

Southart, Charles. Artillery artificer, 1779. (2 Pa. Arch. 15 at 383).

Jacob Snider, Bedford County (Huntingdon school).

Tobias Snider, Bedford County (Huntingdon school).

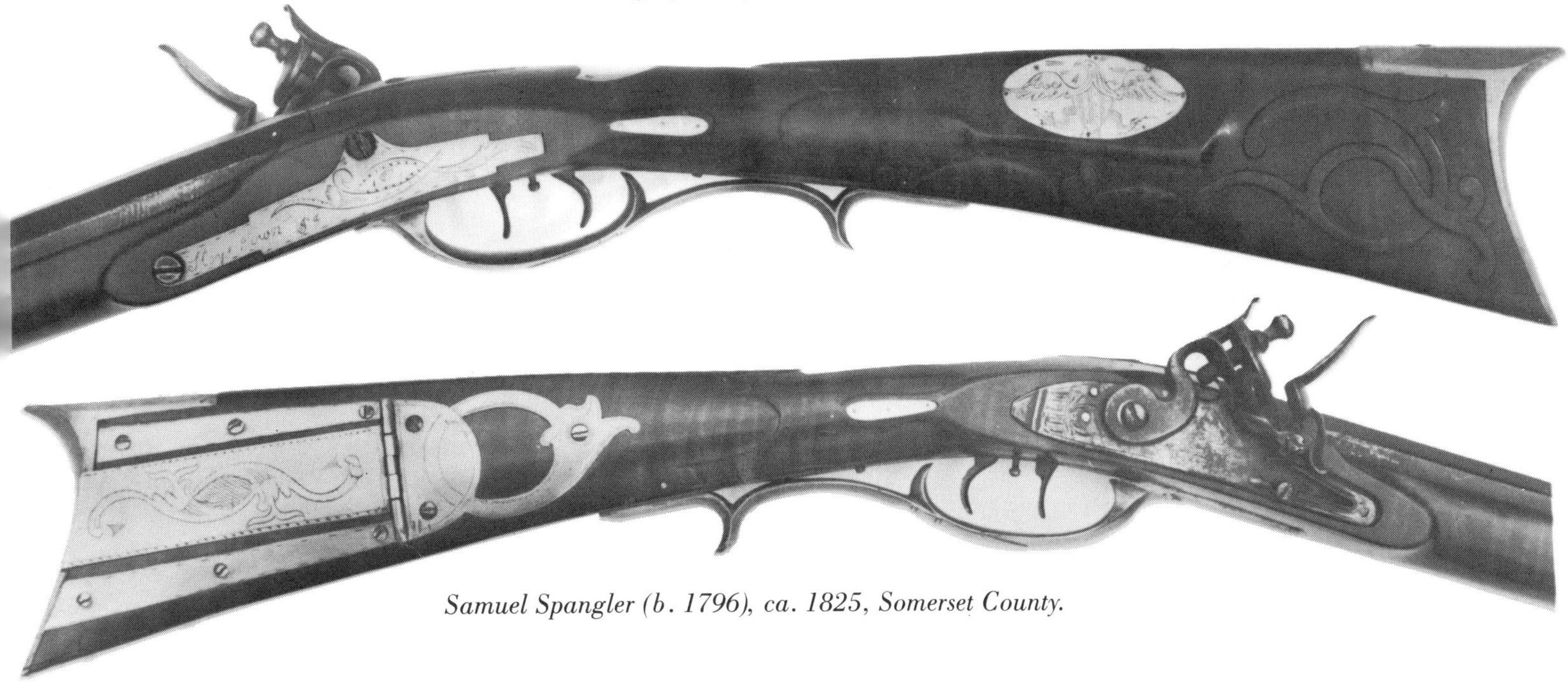

Samuel Spangler (b. 1796), ca. 1825, Somerset County.

Sowers & Smith. Gunlock makers and importers, Philadelphia (MD, 1855).

Spang & Wallace. Gunsmiths, 94 N. 3d St., above Cherry St., Philadelphia (MD, 1830, 1855, 1847, 1841).

Spang, Samuel. 1830–67, 148 N. 3d St., Philadelphia (MD, 1830, 1855, 1859, 1863, 1867). Gunsmith.

Spangel, George. 1825–55, Liverpool, Perry Co. Gunsmith (CHAN, 22).

Spangler, George (1827–1913). In 1827 & 43, Stoystown, Stoneycreek Twp., Somerset Co. Made guns and percussion locks, including the sidehammer "mule ear" style (t). 1843–1900, Madison, Wis. (FSP, 289).

Spangler, Jonas. "before 1812 until 1816" he had a gun-barrel mill in "the short narrows of Jack's mountain, on Kishacoquilla's Creek," in association with George James. Brown and Derby Twps. (Ellis and Hungerford, *Susquehanna and Juniata River Valleys*, 639).

Spangler, Rudolph (1738–1811). Silversmith. Son of Balthasar Spangler. 1760–1811, York. (t).

Spangler, Samuel (b. 1796). Gun and gunlock maker, 1821–43, Stoystown, Stoney Creek Twp., Somerset Co. (t); 1846–50, Monroe, Wis. (FSP, 289).

Sparks, Thomas. Shot tower, 476 S. Front St., Philadelphia (PXD, 1819).

Sparrow, Thomas (b. 1746). Son of Thomas Sparrow of Annapolis, Md. Apprenticed in Philadelphia, 1763–65. Worked for Md. Committee of Safety. Returned to Annapolis (AMM, 257).

Specht, Adam (1768–1872). Beavertown, Snyder Co. Potter and gunsmith. In 1791 he married Catherine, daughter of gunsmith Peter Smith of Union Co. By 1804 he had left New Berlin and come to Snyder Co. where he spent the remainder of his life. He fought in the War of 1812 under Capt. Henry Miller. He died 11/6/1872 at age 84 and was buried at Beavertown Cemetery (t; c; f; Ellis and Hungerford, *Susquehanna and Juniata River Valleys*, 2:1512).

Specht, Arthur. Gunsmith and foundryman, Beavertown, Snyder Co. (f; Ellis and Hungerford, *Susquehanna and Juniata River Valleys*, 2:1512).

Specht, Elias (Ely), (1820–90). Gunsmith and foundryman, Beavertown, Snyder Co. (PD, 1882; Ellis and Hungerford, *Susquehanna and Juniata River Valleys*, 2:1512).

Specht, Moses (b. 1818). Gunsmith and politician, Beavertown, Snyder Co. It was Moses Specht as a major political leader of his day who prompted the entry in *Susquehanna and Juniata River Valleys*, 2:1512.

Speil, Henry. Gunsmith, Susquehanna, Susquehanna Co. (PD, 1882).

Spellerberg, Anton. Patent for a breech-loading firearm of 7/30/1861, and patent 180,803 (1876) for another breech-loader. Philadelphia.

Spellier, August. Patent 30,260 (1860) for a self-loading gun, Philadelphia.

Spere, Henry. Gunsmith, Susquehanna Depot, Susquehanna Co. (BPaD, 1873).

Sperry, I. B. Spring Twp., Crawford Co. (CCD, 1874). Gunsmith.

Springer, Frank D. Gunpowder maker, 842 Drexel, Philadelphia (BPD, 1890).

Springer, Isaac (b. 1808). 1842–56, children of his born in Ohio (he is noted, but not located, in DH, 411); 1860–61, Potter, Raccoon Twp., Beaver Co. (c; BPaD, 1861). Gunsmith..

Springer, J. W. Gunsmith, Raccoon Twp., Beaver Co. (PD, 1882).

Sprouls, John. Artillery artificer, 1780, Capt. Thomas Wylie's Co. at Carlisle (PIR, 2:248).

Spry, Richard S. (1809–82). Born in England. 1840, Honesdale, Wayne Co.; 1844, Springfield, Ohio; 1846, Portsmouth, Ohio (DH, 349).

Squibb, William. On 9/20/1773 Squibb was sold for £13/12/0 to Edward Milne, silversmith of Philadelphia, to be employed as a silversmith's apprentice. Squibb was from Ireland (PIDS).

Stack, John. Gun-barrel smith, 1807, Annville Twp., Dauphin Co. (KKR, 348).

Stackpole, William. Cutler, 1818–19, Penn St. and Cecil's Alley, Pittsburgh. In 1826 his business was operated by R. Bowen (SHIS, 228).

Stahl, John George (b. 1800). Born in Germany, came to Baltimore in 1834; by 1837 was in Quemahoning Twp., Somerset Co. Cutler. Active through 1870 (c; t).

Stahl, William. Gunsmith, 1411 Ridge Ave., Philadelphia (BPaD, 1861); 307 Ridge Ave. (MD, 1855).

Stake, Christian. Silversmith, 1779, York (t).

Stall, Casper. Artillery artificer, 1779, at the laboratory, Philadelphia (2 Pa. Arch. 15 at 386).

Standley, Peter, Jr. On 9/19/1772, apprenticed to Samuel Wheeler, cutler of Philadelphia, to learn the cutler's art, for 10 years, 13 days, and 9 months (PIDS).

Stansberger, John. Artillery artificer and drummer, 1777, Capt. Nathaniel Irish's Co. In 1811, lived in Montgomery Co. (PIR, 2:247).

Stape, Daniel. Gunsmith, 22 N. 2d St., Columbia, Lancaster (BLD, 1874, 1884).

Stapleton, James (1833–1900). Gunsmith, 1854–90, Todd Twp., Huntingdon Co. (c; t).

Stapleton, Joseph. Gunsmith, 1855, Orbisonia, Huntingdon Co. (t).

Stapleton, Solomon (1869–1950). Gunsmith, Todd Twp., Huntingdon Co. (t; c).

Stapleton, William B. (b. 1864). Gunsmith, 1880–90, Todd Twp., Huntingdon Co. (t; c); 1903, Saxton, Bedford Co. (t).

Stare, John. Gunsmith, 1792, Harrisburg (KKR, 349).

Starn, Samuel. Artillery artificer, 1779 (2 Pa. Arch. 15 at 384).

States, Samuel (b. 1797). Gunsmith, born in Bucks Co.; 1821–50, Lackawanna Co. (t; c, 1850).

Statler, John. Gunsmith, 1820, Moore Twp., Northampton Co. (t).

Steck (Stake), John (b. 1795). 1817–42, Hempfield Twp., Westmoreland Co. (t); Franklin Twp., Westmoreland Co. (t; c). Gunsmith and blacksmith.

Steel, James. Artillery artificer, 1777–81, Capt. Jesse Rowe's Co. at the cannon factory, Philadelphia (2 Pa. Arch. 11 at 252).

Steel, Peter. Smith, 1765–70, Chester, Chester Co. (t).

Steel, William. Gunsmith, Philadelphia (MD, 1858).

Steever, Adam. Gunsmith, 1820–44, Union Twp., Huntingdon Co. (t).

Stehle (Staley), Thomas, Sr. (1813–83). 1832–83, S. Main St., Butler, gunsmith. (*History of Butler County* [Chicago: Waterman-Watkins, 1883] 155; James A. McKee, ed., *Twentieth Centry History of Butler and Butler County* [Chicago: Richmond Arnold, 1909] 440).

Stehle, Thomas, Jr. Gunsmith, Butler boro (PD, 1882; see above).

Stein, Frederick. Artillery artificer, 1778, Capt. John Jordan's Co. Brass founder at Carlisle (2 Pa. Arch. 11 at 260).

Steiner, A. B. Lock maker, Bushkill Center, Northampton Co. (c, 1860).

Steiner, Peter. Gunsmith, 1820–28, Bushkill Twp., Northampton Co. (t).

Steinman, Frederick. Gunsmith, 31 Green St., Philadelphia (DeSD, 1825, MD, 1833); Elizabeth St. (DeSD, 1835–36).

Steinman (Stineman), John. Gunsmith, rear of 442 N. 3d St., Philadelphia (PXD, 1808, 1809, 1810); 22 Green St. (DeSD, 1828, 1829, 1830, 1831); 17 Green St. (MD, 1846).

Stekenius, Jacob (1810–83). Born in Germany; 1859, Stoney Creek Twp., Somerset Co. (t); 1860, Harrison Twp., Bedford Co. (c); 1861, Northampton Twp., Somerset Co. (t); 1863, Juniata Twp., Bedford Co. (t); 1870–83, Harrison Twp., Bedford Co. (t; c). Gunsmith.

Stenger, Thomas. Gunsmith, 1840–50, Lycoming Co. (c).

Step, Martin. Artillery artificer and blacksmith, 1779, at the cannon factory, Philadelphia (2 Pa. Arch. 15 at 384).

Stephens, James. Gunsmith, 1810–12, Uniontown (t).

Stephens, John & Co. Committee of Safety arms contractor; was paid £160 on account for 27 guns delivered 4/9/1777 (2 Pa. Arch. 1 at 28).

Stephenson, Joseph. Artillery artificer, 1778, Capt. Thomas Wylie's Co. at Carlisle (PIR, 2:249).

Stevenson, George. 1785–1815, Pine Grove, Middletown Twp., Cumberland Co. Gun-barrel boring mill (WSB, 47).

Stevenson, James. Silversmith, Washington boro (t).

Stewart, Abraham. Partner of M. Baker, 1804–8, in arms manufactory at Haydentown, Fayette Co. after Albert Gallatin withdrew (Gallatin Papers).

Stewart, George W. Mifflin Co. "respectuflly informs the public that he has purchased out the gunshop of Leonard Shuler, late of this place, situated on the corner of Main and Dorcas Sts., and he intends carrying on the gunsmithing business in all its various branches. New guns will be manufactured to order and old ones repaired in a neat and workmanlike manner." (Juniata *Gazette,* 7/26/1832) Gunsmith, Lewistown (c, 1850).

Stiles, Henry. Silversmith, 1784, Reading (3 Pa. Arch. 18 at 657).

Stiles, Joseph. Gun repairman for Committee of Safety. On 11/21/1782, was "paid by order of Council for repairing of arms, £100"; also was paid £100 on 8/26/1783, and £25 on 3/8/1783 for arms repairs. (3 Pa. Arch. 5 at 340–41) Stiles was a commissioner of commissary, but the records are very clear that these payments are separate.

Stille, John. Artillery artificer, 1777, Capt. Nathaniel Irish's Co., Carlisle (PIR, 2:247).

Stilley, Peter (1777–1819). Gunsmith, 1800–1819, Mifflin Twp., Allegheny Co. (t).

Stillwell, C. Gunsmith, Oil City, Venango Co. (t; c; PD, 1882; BRD, 1855).

Stinger, Thomas. Gunsmith, 1835–50, Jersey Shore, Lycoming Co. (t; c).

Stite, Casper. Gunsmith, 1759, Cocalico Twp., Lancaster Co. (SED, 60).

Stober, George. Gunsmith, 1813, Haines Twp., Centre Co. (t).

Stoelzing, Dietrich. Manor Twp., Arstrong Co. Made leather goods for the Union army during the Civil War, including cartridge boxes and gun slings. In 1863 he employed 15 men and used 10,000 hides a year (Smith, *History of Armstrong County* 342).

Stokes, Charles. Gun-barrel maker, Penn St. near Smith St., Pittsburgh (PAD, 1860–61).

Stokes, Enoch. Gunsmith, Mulberry St. and Walnut St., Lancaster boro (BLD, 1857).

Stone, John. Gunsmith, 1847, Rostraver Twp., Westmoreland Co. (t); Polk, Venango Co. (PD, 1882, 1887, 1890).

Stonemitz, John. 5/2/1777: "pay John Stonemitz £24 for 6 muskets and bayonets" (2 Pa. Arch. 1 at 39).

Stoner, Jacob. Artillery artificer, 1779–82, at the cannon factory, Philadelphia (2 Pa. Arch. 11 at 236). 11/1/1782, Capt. John Jordan's Co., stationed at Ft. Pitt (2 Pa. Arch. 11 at 262).

Stotz, Christian. Gunsmith, Williamsport, Lycoming Co. (PD, 1882).

Stoudenour, Jacob (1795–1863). 1819–63, Colerain Twp., Bedford Co. Major gunsmith (t; c).

Stoudenour, John, Sr. Gunsmith, 1814–55, Colerain Twp., Bedford Co. Farmer and part-time gunsmith (t; c).

Stoudenour, John, Jr. Gunsmith, 1838–44, North Wood-

Jacob Stoudenour, Bedford County.

berry Twp., Bedford Co. (t; c).

Stoudt, Joseph. Gunsmith, 1835, Bushkill Twp., Northampton Co. (t).

Stoup, Adam (1792–1863). 1835–63, Unity Twp., Westmoreland Co. (t; c; vendue list 55). Gunsmith.

Stoup, George (1819–90). 1850–90, Unity Twp., Westmoreland Co. (t; c) Gunsmith. His "lott of gunsmith's tools" was valued at only 10¢; boring machine for gun barrels, $2.80; vises, 15¢; powder horn, 25¢.

Stover, Frederick. Major gunsmith, 1781–1824, Paradise Twp., York Co.; 1833-42, Franklin Twp., Adams Co. (WSB, 95).

Straub, Andrew. Silversmith, 1833–34, Brookville, Jefferson Co. (t).

Straub, John (1847–1923). 1870–1923, Snyder Co. Made heavy-barrel target rifles.

Strauss, Reuben (b. 1811). 1847–48, Allentown (t); 1849–60, Saegertown, Crawford Co. (t, c; CCD, 1855). Gunsmith.

Stratton, E. Gunsmith, Green's Ct., Philadelphia (MD, 1841).

Strawick, Andrew (d. 1832). 1810, West Jefferson, Butler Co. (t; c); 1812–32, Clearfield Twp., Butler Co. (t; c). Gunsmith.

Strebey, George. Swordsmith, 1777, Northern Liberties, Philadelphia Co., in employ of Lewis Prahl, gunsmith (2 Pa. Arch. 3 at 104).

Streeley, Joseph. Gunsmith, 61 Germantown, Philadelphia (MD, 1841).

Strieby, William. Gunstock maker, 522 N. 2d St., Philadelphia (MD, 1837).

Strohweg, Andrew. Gunsmith, Water St., Butler (BPaD, 1861).

Strouch, Charles. Gunsmith, Northeast Ward, Pottsville, Schuylkill Co. (c, 1850).

Stuebgin, C. A. Gunsmith, 1861, Saxonburg, Butler Co. (BPaD, 1861); 118 3d St., Pittsburgh (PAD, 1864–65).

Stumpf, Michael (d. 1848). 1826–27, 1832–33, Mahoning Twp., Indiana Co. (t); 1834–40, Washington Twp., Indiana Co. (t); 1841–48, South Mahoning Twp., Indiana Co. Gunsmith.

Stunkard, Robert F. (1827–1902). 1850–81, Wells Twp., Fulton Co. (c; t); 1900, Taylor Twp., Fulton Co. (c; WB 3-269), Gunsmith.

Sturgis, Julius F. Gunsmith, N. Queen St. at James St., Lancaster boro (BLD, 1857).

Subers, James. Gunsmith, Philadelphia (MD, 1855).

Subers, Theodore. Gunsmith, 306 St. John's, Philadelphia (MD, 1846, 1855).

Subreinck, William. Gunsmith, Sawmill Alley near Pine St., Duquesne, Allegheny Co. (PAD, 1864–65).

Sullenberger, Samuel. Inventor, Harrisburg. Patents: 330,354 of 11/10/1885 for a breech-loading gun; 356,338 of 1/18/1887 for a magazine gun.

Sullivan, James, Artillery artificer, 1777, Capt. Jesse Rowe's Co. at the Philadelphia cannon factory; deserted 4/10/1779. (PIR, 2:245)

Suter, Theodore. Gunsmith, Philadelphia. (MD, 1847)

Sutter, John J. (1823–1902). 1840–1902, Ruffsdale, Westmoreland Co. Son of Solomon Sutter (1789–1863). Gunsmith, apprenticed to J. H. Johnston, Pittsburgh. (c; f; t)

Sutter, Worthe G. (b. 1896). 1930–, Ruffsdale, Westmoreland Co. Made gunlocks, especially of the Bedford Co. school, of the highest quality, long after cartridge guns had become standard. (t; c; f) He had a daughter who

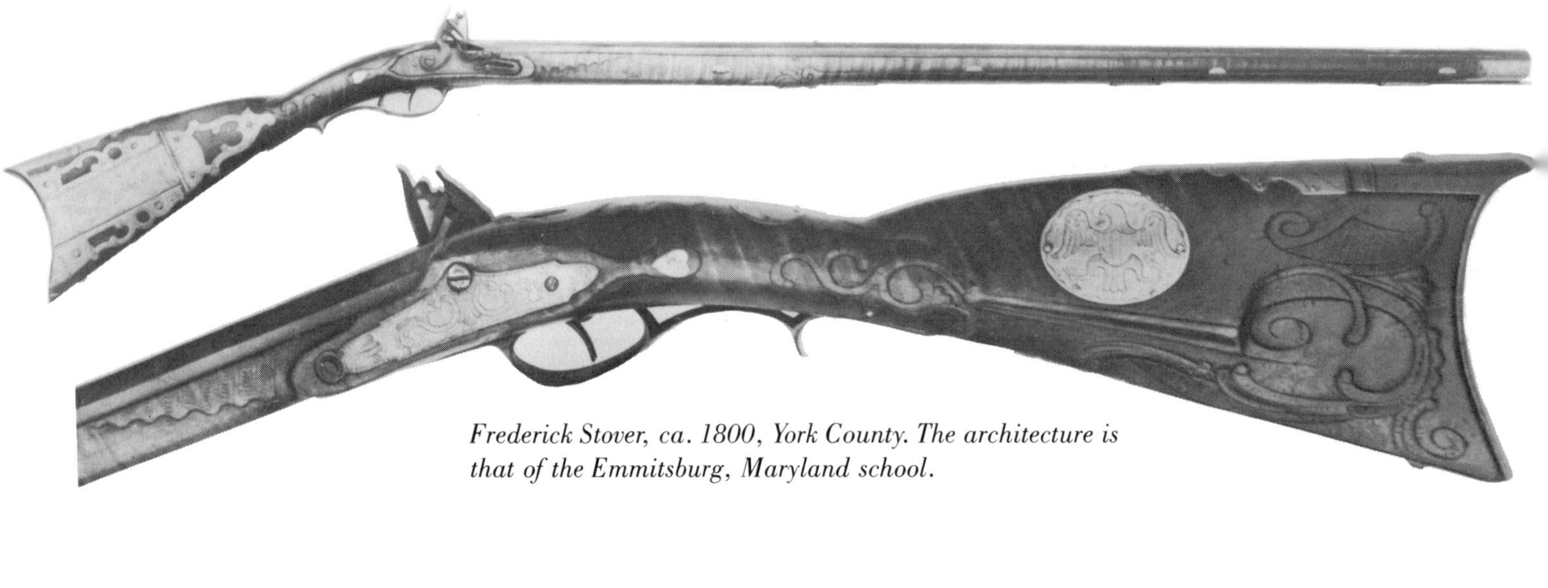

Frederick Stover, ca. 1800, York County. The architecture is that of the Emmitsburg, Maryland school.

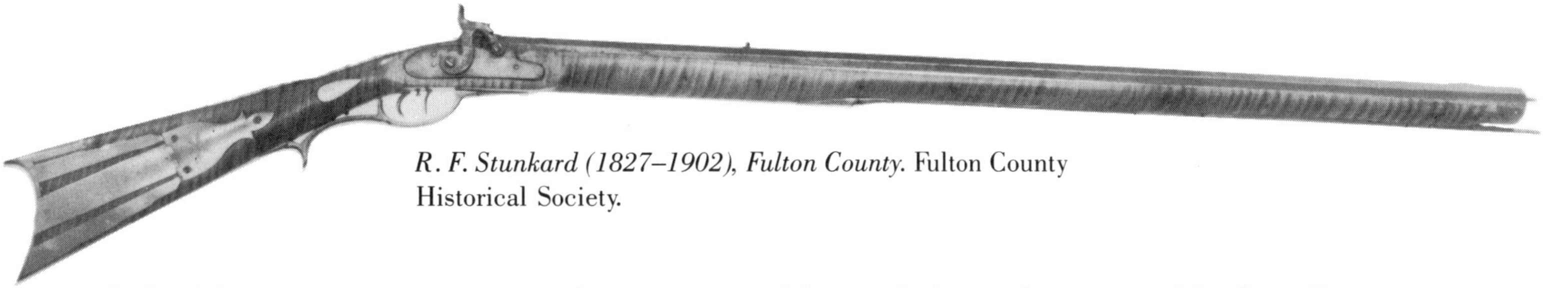

R. F. Stunkard (1827–1902), Fulton County. Fulton County Historical Society.

worked with him as a journeyperson gunsmith.

Sutton & Dunseth. Gunsmiths, Pittsburgh, 1797–1800. (KKR, 350; Pittsburgh *Gazette*, 2/2/1799) George Sutton was active in Pittsburgh from 1796 to 1810 (t; c).

Sutton, George. Pittsburgh *Gazette*, 7/9/1796: "George Sutton, gunsmith from Lancaster, informs his friends and the public in general, that he has commenced business, in the borough of Pittsburgh, in the house lately occupied by Mr Mulholm, in Third St. He has some well finished rifles of Mr John Graeff's make of Lancaster for sale."

Sutton, John. Gunsmith, 55 Duke St., Philadelphia (DeSD, 1819).

Sutvan, Isaac. Inventor, Bridesburg, Philadelphia Co. Patent 46,866 (1865) assigned to B. H. Jenks.

Swain, T. C. Powder-horn maker, 1825, Philadelphia (DeSD, 1825).

Swartz (Swarts), Peter. Gunsmith, York Co. 1/10/1787: "Peter Swarts, for repairing arms, £18/5/4" (3 Pa. Arch. 7 at 102).

Sweger, Billy (1815–92). Gunsmith, near Elliotsburg, Perry Co. (BPaD, 1861; CHAN, 68–83).

Sweigart, Adam. Gunsmith, 1815–17, Halifax, Dauphin Co. (KKR, 352).

Sweitzer, Charles. 1860–70, Mauch Chunk, Carbon Co. (c) Gunsmith.

Sweitzer, Daniel. Gunlock maker, "announced the opening

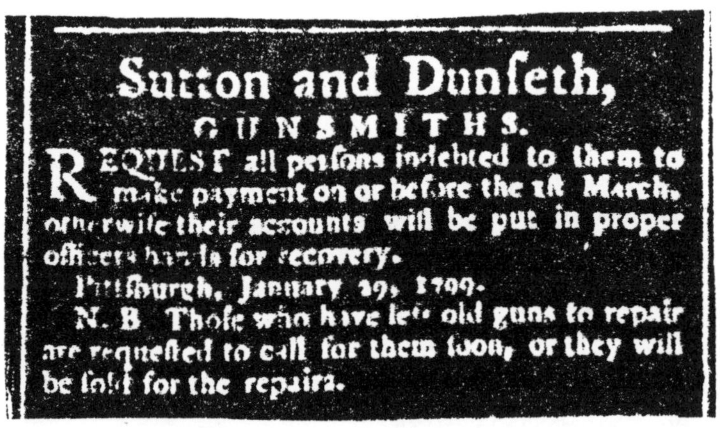

Sutton and Dunſeth,

GUNSMITHS.

REQUEST all perſons indebted to them to make payment on or before the 1ſt March, otherwiſe their accounts will be put in proper officers hands for recovery.

Pittſburgh, January 29, 1799.

N. B Thoſe who have left old guns to repair are requeſted to call for them ſoon, or they will be ſold for the repairs.

Pittsburgh Gazette, February 2, 1799

of their gunlock manufactory west of the Court House on the road to Millerstown." Operated from 1808 until 1829. (RAG, 190; SED, 60; t) Subcontractor for gunlocks on M-1808 muskets. Locks marked "D. Sweitzer" or "Sweitzer & Co." Contract for M-1808 pistols.

Swinehart, Peter. Gunsmith, 1788–1828, West Bethlehem Twp., Washington Co. (t).

Swindles, John. Blacksmith and artillery artificer, 1779, Capt. John Jordan's Co.; made cannon at Carlisle (2 Pa. Arch. 11 at 260).

Swing, Gottfried. Gunsmith, 1779, Philadelphia (t).

Swisher, John. Gunsmith, 1813–15, Mifflin Co. (t).

Switsor, Conrad. Gunstocker, 1777, French's Creek Arsenal (1 Pa. Arch. 15 at 384).

Switzer, A. Gunsmith, Bellefonte (BPaD, 1861).

Swobe, Adam. Gunsmith, 1785–1802, Leacock Twp., Lancaster Co. (SED, 60).

Syfert, John. Gun-barrel maker, Centre Twp., Union Co. (c, 1850).

Sylvis, Jacob (1801–92). Gunsmith, died at 9:00 A.M., 8/23/1892. 1826–31, Hempfield Twp., Westmoreland Co. (t); 1832–91, Washington Twp., Westmoreland Co. (t; c; BPaD, 1861).

Sylvis, Samuel (1833–1904). 1860–1904, gunsmith and farmer, Hempfield Twp., Westmoreland Co. (t; c).

Syng, Philip. Silversmith, Philadelphia. Advertisements in *Pennsylvania Packet:* "Negro named Cato, age about 20, ran away from Philip Syng of Philadelphia, silversmith" (5/5/1748); "Engraving is done by Lawrence Herbert from London at Philip Syng's, Goldsmith" (5/19/1748).

Syphers, John. Gunsmith, Waynesburg, Greene Co. (PD, 1882; BRD, 1876).

T

Taylor, Alexander, Sr. Gunsmith, 1808–9, Peters Twp., Franklin Co. (t); 1823, same (t); 1826, Montgomery Twp., Franklin Co. (t); 1830–32, Peters Twp., Franklin Co. (t; WSB, 22).

Taylor, Alexander Jr. Gunsmith, 1826–28, McConnellsburg, Bedford Co. (t); 1829, name crossed out; single freeman, gunsmith (t).

Taylor, George (d. 1781). Owner and operator, Durham Iron Works, Northampton Co. Made musket barrels and brass cannon during the Revolution (GAAAM, 191).

Taylor, Joseph. Artillery artificer, 1777–1780, from Providence, Chester Co. (PIR, 2:244) Capt. Jesse Rowe's Co. and Capt. Noah Nicholas's Co. (2 Pa. Arch. 11 at 245).

Taylor, Travis. Gunsmith, Chester, Delaware Co. (PD, 1887; BSED, 1900).

Taylor, William. 1771, Philadelphia; had an apprentice to the goldsmith's trade (PIDS).

Teaff, James, Sr. (b. 1796). In 8/1814, James Teaff, apprentice to the gunsmith's trade, ran away from Asa McClelland of Greene Co. In 1816 James Teaff bought James Ross's gunshop, Steubenville, Ohio, and was active through 1850 (DH, 220).

Teague, Joseph. Artillery artificer, 1779, grinding ramrods, files, and the like (2 Pa. Arch. 15 at 386).

Teague, Roger. Artillery artificer, 1779, grinding ramrods, files, and the like (2 Pa. Arch. 15 at 386).

Teel, Joseph. Gunsmith, Eldred Twp., Monroe Co. (PD, 1882).

Tennison, John. 1660: "John Tennison, wife and child, whose property [Sgt. Andres Laurens] bought. . . . John Tennison refuses to go to Manhattan unless his tools . . . are delivered up, but would rather return to the English as he there had the use of the tools" (Hazard's *Annals*, 1:308). Smith, Whorekill on the Delaware.

Terick, I., & Co. Gunsmiths, Mohn's Store, Cumru Twp., Berks Co. (PD, 1882).

Terrill, Ephraim, Jr., Gunsmith, West Fallowfield Twp., Crawford Co. (CCD, 1868).

Tetley, John. Partner in gun manufacturing firm of Bown & Tetley. 1848–62, 37 Fayette St., Pittsburgh (t; c).

Theunissen, Cornelius. Smith, 1659, Dutch Colony on the Delaware. Apparently, he died Ca. 11/25/1659 (2 Pa. Arch. 5 at 441).

Thomas, J. M. Gunsmith, Plum Creek Twp., Armstrong Co. (PD, 1882).

Thomas, John. Artillery artificer, 1779–82. On 11/1/1782 he was stationed at Ft. Pitt. in Capt. John Jordan's Co. (2 Pa. Arch. 11 at 236, 262).

Thomas, Richard. 12/13/1776: "pay Richard Thomas £2/11/6 for repairing rifles" (Col. Rec., vol. 11); brass founder and artillery artificer, 1777–81, U.S. Ordnance Factory at Philadelphia (2 Pa. Arch. 11 at 258). Buried at the Baptist Cemetery, Boswell, Somerset Co.

Thomas, William. Gunsmith, 1800–1801, Upper Milford Twp., Bucks Co.; rear of 198 N. 2d St., Philadelphia (PXD, 1813). Gunsmith.

Thomlinson, Joshua. In 1776 the Committee of Safety bought a lot of gun barrels from Thomlinson (Col. Rec., 10:602); in 1777, he proposed, "for carrying on the business of making gun barrels, first that a sum of £50 be given me as a premium for my superiority and service in the above art; second, that the Committee provide a boreing and grinding barrels and put the same in good order for doing business with every useful machine to complete the work and build a smith's shop for the forging of barrels . . . third, I do agree to provide all manner of working tools. . . . I will deliver musket barrels to the Committee at 20 shillings each . . . fourth, that the Committee advance £200 . . . [which] I will repay . . . out of every 100 barrels I will pay £15 until all is repaid" (1 Pa. Arch. 4 at 772–73).

Thompson, David. West Springfield, Erie Co. (BRD, 1873–74); broom maker, gunsmith, and farmer.

Thompson, George W. Gunsmith, 1870–82, Washington boro (PD, 1882; t; c).

Thompson, John. Artillery artificer and blacksmith, 1779–80, at the cannon factory, Philadelphia (2 Pa. Arch. 15 at 384).

Thompson, Harry. Gunsmith, Erie boro (BRD, 1890).

Thompson, William. On 2/7/1776, proposed erecting a gunpowder mill at Nashameny, Bucks Co.; asked council for an advance of £100, saying that the mill would be producing within 5 weeks (1 Pa. Arch. 4 at 710).

Thopp, J. J. Gunsmith, rear of 214 Green St., Philadelphia (DeSD, 1823).

Thornell, Israel. Artillery artificer, 1778, Capt. Nathaniel Irish's Co. (PIR, 2:247).

Thornell, John. Artillery artificer, 1778, Capt. Nathaniel Irish's Co. (PIR, 2:247).

Thornell, Joseph. 1778, Artillery artificer, Capt. Nathaniel Irish's Co. (PIR, 2:247).

Thorsen, Thomas M. Inventor, Philadelphia. The following patents were issued for automatic guns: 696,118 (3/25/1902); 815,899 (3/20/1906); 867,685 (10/8/1907).

Throckmorton, William A. Gunsmith, Galion, Crawford Co. (CCD, 1858; c, 1870).

Tillgham, Francis. The Committee of Safety bought the

following items from Tillgham on 10/14/1775: 5 cutlasses at 3 shillings each; 7 pairs of pistols (Col. Rec., 10:366, 375).

Tillman, William. 1857–61, Milton, Northumberland Co. (RAG, 193). Gunsmith.

Tilton, John (1760–1849). 1800–1812, Washington Co. (t); 1812, Stark Co., Ohio (t); 1814–31, Wayne Co., Ohio; 1831–49, Orange Twp., Hocking Co. Ohio. Died 8/12/1849. He was a 5-year veteran of the Revolution (DH, 207–08).

Timblin, C. O. Gunsmith, Red Bank Twp., Armstrong Co. (PD, 1882).

Tipton, John. Gunpowder maker, 1830, Antes Twp., Huntingdon Co. (AFR, 40).

Tisdel, Luther W. (d. 1891). 120 Franklin Ave., Scranton, 1848–91, gunsmith (PD, 1882; BRD, 1848; c, 1850, 1860, 1870, 1880).

Tod, James. Artificer, 1789, Pittsburgh; when he died his tools were left to his brother-in-law, George Forbes (Chalfont, *A Goodly Heritage*, 26).

Tommerup, Mathias. Brazier, 1761, Moravian, arrived on ship *Hope* at New York, on his way to the Bethlehem settlement (PaE, 53).

Tonze, John. Gunsmith, 101 S. 2d St., Philadelphia (PXD, 1816; DeSD, 1819, 1824).

Topper, Henry (d. 1840). Gunsmith, 1802, Cumberland Twp., Adams Co. (t); 1810–13, Gettysburg (t); 1815–30, near Romney, Hampshire Co., W. Va. (t; c); 1833, Schellsburg, Bedford Co. (t); 1834–40, Youngstown, Westmoreland Co. (t; WSB, 99).

Topper, Josiah (b. 1813). Gunsmith, son of Henry Topper. 1833, Schellsburg, Bedford Co.; 1834–51, Youngstown, Westmoreland Co. (t; WSB, 99).

Toppin, Johnson. (b. 1808). Born in Md. Gunsmith, 1860, Carmichaels, Greene Co. (c); 1870–80, Cumberland Twp., Greene Co. (c).

Totten, W. T. Gunsmith, Sergeant Twp., McKean Co. (PD, 1887, 1900).

Town, Benjamin. On 12/5/1775, "Benjamin Town, gunsmith, having agreed to make 200 firelocks for this Province at the rate of 55 shillings each" (Col. Rec., 10:422).

Tranks, Casper. Artillery artificer, 1778, Capt. Thomas Wylie's Co. (PIR 2:249). Stationed at Carlisle.

Trayer, John. Gunsmith, 1802–3, Georges Twp., Fayette Co. (t).

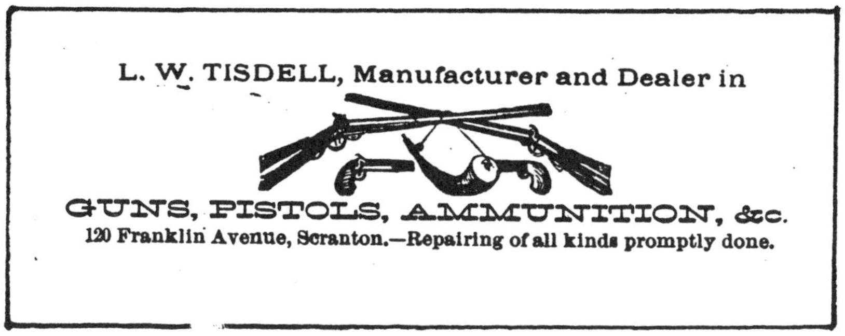

Scranton City Directory of 1879

TAKE NOTICE, Those Indebted.

ALL persons who know themselves in any way indebted to the subscriber for Leather or otherwise, are hereby notified to call on or before the *first day of March next*, and make settlement, by paying off, or giving their notes for the same. Circumstances render it necessary that this be punctually attended to; and however, unpleasant it may be to the subscriber, he will have to place the accounts of those neglecting this notice, without respect to persons, in the hands of an officer for collection after the time above specified. Should any do so, they will have themselves to blame, if costs and trouble follow.

HENRY TOPPER.

Jan. 25, 1834.

Westmoreland Republican, January 25, 1834

Treasted, Richard. On 6/11/1773, an indentured servant, sold for £14/2/7, to be employed at the gunsmith business by John Nicholson. (PIDS); 1778–79, armorer, U.S. Ordnance Yard, Philadelphia (2 Pa. Arch. 11 at 258).

Treibel, Henry. Gunsmith, boarded at Washington Hotel, Lancaster (BLD, 1857).

Treish, John. Gun-barrel maker, 1847–50, Elizabeth Twp., Lancaster Co. (SED, 60; c; t).

Trissler, John (d. 1807). Gunsmith, Lancaster (Burial Book of the Moravian Church, 1744–1821, 35).

Trissler, John, Jr. Gunsmith, 1807–40, Lancaster (SED, 60).

Troad, G. Gunsmith, 158 N. 2d St., Philadelphia (MD, 1855).

Troad, H. Martin. Smith, arrived at Philadelphia on 8/27/1733 (PaE, 1:765).

Trout, Baltzer (d. ca. 1782). Gunsmith, 1767–81, Carbondale Twp., Berks Co. In 1784, there was a "widow Trout."

Trout, John. Gunsmith, 1855–70, 13 W. 4th St., Williamsport, Lycoming Co. (c, 1860; BPaD, 1861; t).

Troutman, Benjamin (1780–1856). 1805–6, Southampton

Twp., Somerset Co. (t); 1807–8, Cumberland, Md., adult apprentice of George Rizer; 1810–56, Southampton Twp., Somerset Co. (t; WSB, 116–17) Gunsmith.

Troutman, Daniel B. (b. 1821). Gunsmith, 1844–47, Southampton Twp., Somerset Co. (t); 1855–64, Londonderry Twp., Bedford Co., gunsmith and justice of the peace (t); 1864–67, Wellersburg, Somerset Co. (t); 1870, Ottawa, Kansas (c, 1870). Son and apprentice of Benjamin (f).

Troutman, Thomas. Gunsmith, 1800, St. John's, Philadelphia (KKR, 353).

Troyer, William. Gunsmith, W. Orange St. near Mulberry, Lancaster boro. (BLD, 1857, 1875–76, 1871–72).

Truax, Abner. Apprentice gunsmith, 1849, working with Jacob Fishel, Thompson Twp., Fulton Co. (t) Fishel owned the land in Fulton Co. and lived just to the west in Hopewell Twp., Bedford Co. (t).

Truby, Jacob (1795–1850). Gunsmith 1820–50, Kittanning, Armstrong Co. (t; c).

Trudelind, William. Gunsmith, Philadelphia (OeSD, 1824).

Truett Brothers. Lock makers and importers, 528 Market St., Philadelphia (MD, 1847, 1855, 1861); Truitt & Co. (MD, 1863).

Trump, Daniel. Artillery artificer, 1779. (2 Pa. Arch. 15 at 383).

Trump, Henry. Gunsmith, 1799–1808, Bullskin Twp., Fayette Co. (t).

Tryon, Edward K. (d. 1892). Prop., Tryon gun works, see above. Medium-size manufactory, frequent government contractor. Philadelphia.

Tryon, Edward K., Jr., (d. 1905). Son of Edward K. Tryon, Sr. Took over the Tryon gun works, Philadelphia, ca. 1868.

Tryon family, gunsmiths: George W. Tryon & Co. (1830–37); Edward K. Tryon (1837–57); Edward K. Tryon & Co. (1857–64); Tryon Brothers (1866–68); Edward K. Tryon & Co. (1868–1905); Edward K. Tryon Co. (after 1905), Philadelphia (FSP, 309).

Tryon, George, & Son. 134 N. 2d St., Philadelphia (MD, 1841). George W. Tryon and his son, Edward K. Tryon.

Tryon, George W. (1791–1878). 1811, operated Tryon & Getz, government contractor, small-arms manufacturer, Philadelphia. The firm became Edward K. Tryon manufactory.

Tucker, Jonathan. Armorer, Gray's Rd., Philadelphia (DeSD, 1814).

Tulle, T. Gunsmith, Philadelphia (DeSD, 1824).

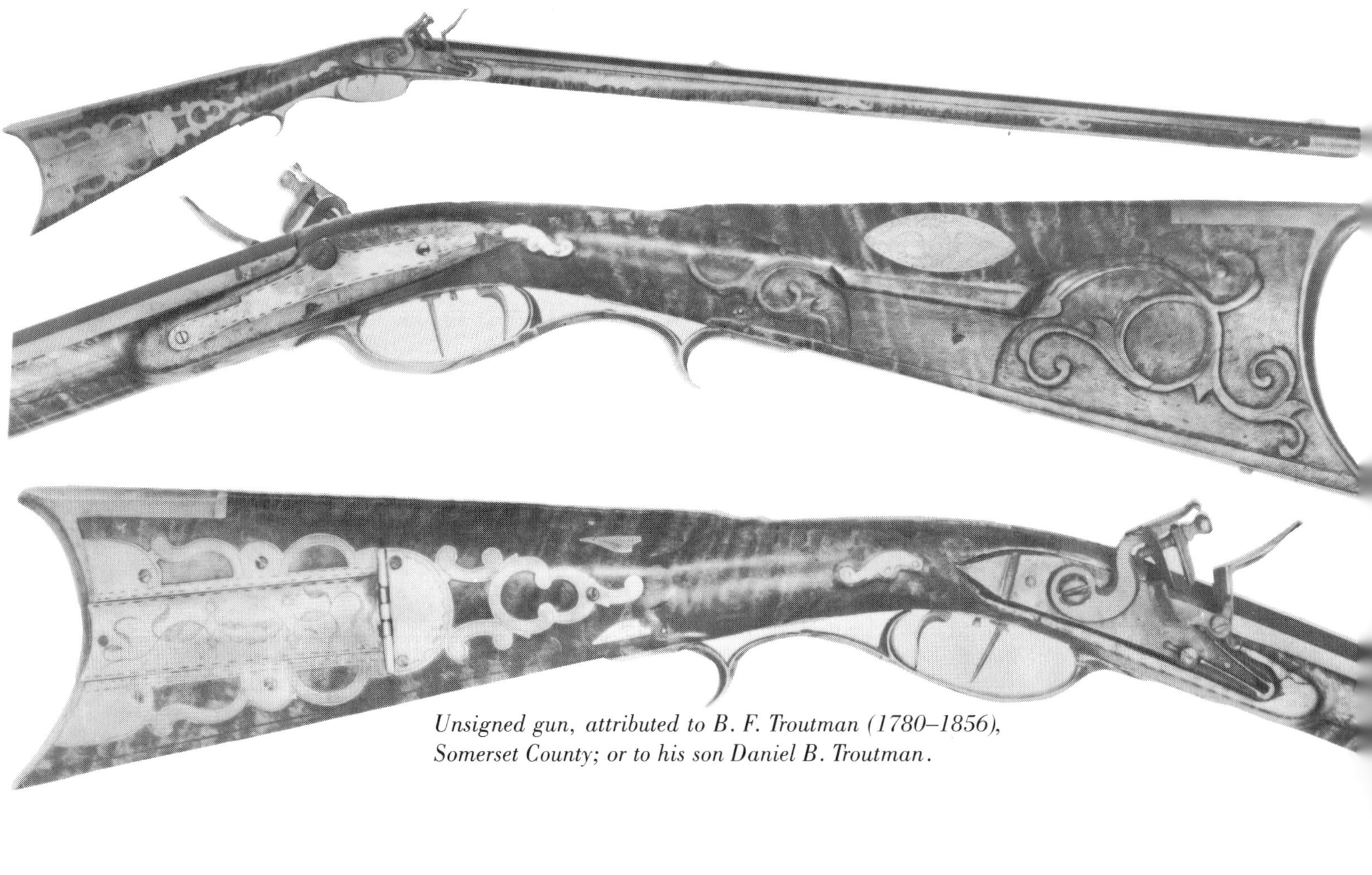

Unsigned gun, attributed to B. F. Troutman (1780–1856), Somerset County; or to his son Daniel B. Troutman.

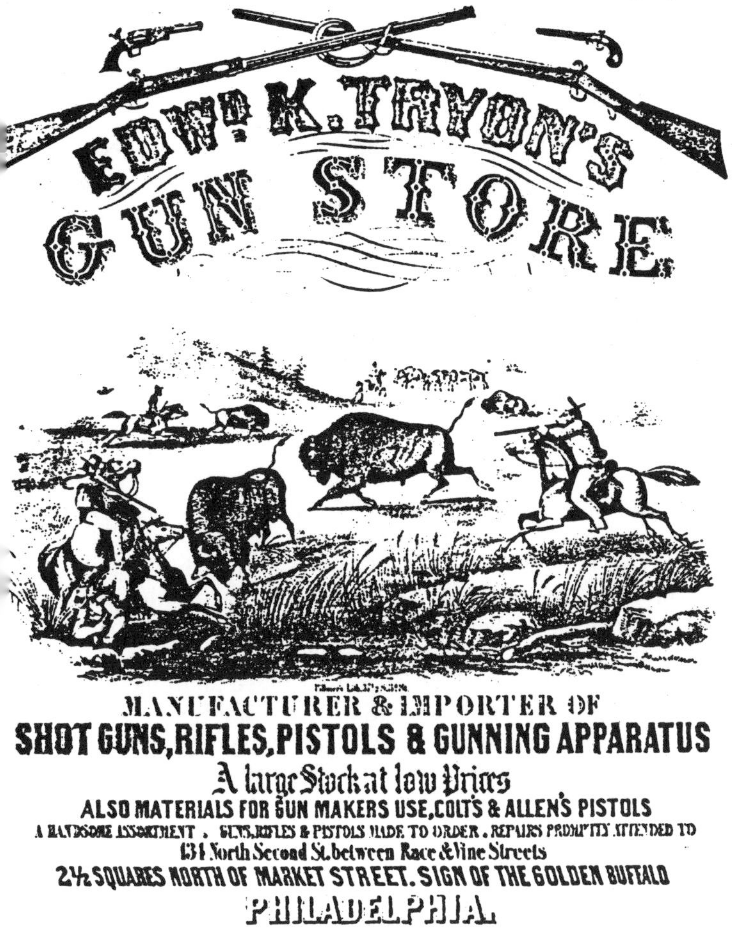

McElroy's Philadelphia Directory, 1844

Tullebahn, Henry. Gun-barrel maker, Elizabeth Twp., Lancaster Co. He had a child born in 1852 (Birth Registration no. 60, 1852) and his occupation was listed as above.

Tully, James. Artillery artificer, 1778, Capt. Thomas's Wylie's Co. at Carlisle (PIR, 2:249).

Turnbull & Marmie Furnace. 1789–1802, on Jacob's Creek, Fayette Co. On 1/12/1792, Major Isaac Craig, military stores keeper at Ft. Pitt, wrote, "As there is no #6 shot here I have taken the liberty to engage Turnbull & Marmie Furnace. . . ." It also cast cannon and made other supplies for Gen. Anthony Wayne's Army (SHIS, 214).

Tustin, J. Tomahawk and accoutrement maker, Soho district, Pittsburgh, before and after 1830 (t).

Tustman, William. Gunsmith, St. John's between Brown St. and Poplar Lane, Philadelphia (NTD, 1799).

Twaddle, James. Gunpowder manufacturer, 105 Spruce St. (office) and 97 South, Wharves (magazine) (PXD, 1808).

Tyler, John. Armorer, 1775, Northampton Twp., Northampton Co. (t); 1777–83, 1 Arch St., Philadelphia (t). Apparently, he had offices both in Philadelphia and in Allentown, for his many contracts were let at one or the other location.

To his Exellency Thomas Wharton, jun[r] Esq., President of the Executive Council for the state of Pennsylv[a] in Council;

May it Please your Excellency—

The Indispensible Duty I owe to your Excellency and the council, Induces me to inform you of the state of my Department at this place; at my first coming here all was hurry and Confusion no place could be procured for carrying on the Repair of the musquits, &c., belonging to this state or even for the People to put their Heads in; but thro' the kind Interference of the worthy Col. Wetzell and other Lieutenents of this County, A place was soon purchased and I have sixteen hands Employ'd and in a few days expect to have three hundred stand of arms repaired and Ready to deliver to such place as your Excellency shall order; (besides a Number already delivered by Col. Wetzells order) Nothing shall be wanting in me to promote with the assistance of the Lieutenants, the utmost decorum and dispatch, some uneasiness it is true prevails among the workmen that the Artificers and Labatary men in the Continental Service draw Liquor, and it is not in their power to draw any, altho' they think they are entitled to it as the allowance from the state, but the difficulty is there is no state Commissary here, and it is the opinion of Col. Hagner and Col. Wetzell that if their was a state Commissary here the people would be better supplied and if they Could then draw their Liquor, everything would go on smooth and it would tend to further the Business of the state. It is therefore requested that your Excellency & the Council would Consider the matter, Col. Hagner & Col. Wetzell are willing to undertake the purchase & Issues for the preasent if they have the Councils Authority for that purpose.

I am
your Excellency's
most Hu'ble Serv[t]
JNO. TYLER, Arm[r]

Northampton, Oct'r 31, 1777.

P. S. As we have no very large Quantity of arms here I hope we may receive such a a supply from time to time as will keep the people in Employ.

Directed,
To His Excellency Tho[s] Wharton, jun[r] Esq., President of the Executive Council of the state of Pennsylvania.

[1 Pa. Arch. 5 at 731–32]

LT. FRED. HAGNER TO PRES. WHARTON, 1778.

Allentown, the 11th May, 1778.

May it please Your Excellency,
According to Your Excellency and Council last Request, dated the 28th April, I have diligently examined and inquired into, but have not been able untill This Day, to give a true and Exact Account what Arms, &c., are now in my possession, & how many shall & can be properly Repaired by the 20th of this month: Which is as follows:

In my possession in Store.

800 Muskets & Bayonets, Scabbords.
550 Bayonet Belts.
750 Cartouch Boxes.
45 Shott Pouches & 18 Powder Horns.
400 Knap & Havre Sacks, new.
75 Blanketts.
25 Tent Cloths.
140 Camp Kettles.

In John Tyler's possession, in good Repair.

31 Rifles.

Will be ready the 20th May.

150 Musketts & Bayonets, from J. Tyler.
150 ditto. ditto. E. Cowell.

This number of Arms, &c., may depended upon, all in good Repair, and shall endeavour, if possible, to exceed the above number, as both Mr. Tyler & Cowell have promised to deliver more muskets, as I have mentioned, because it is difficult to gett the Scabbords for the Bayonets finished. I have Three Saddlers now Employ'd and will not leave a stone unturned, to serve my Country and Your Excellency & Council.

I am with the utmost esteem,
Your Excellency most obedient
and most Humble Serv[t].
FRIEDERICK HAGNER, Sub Lieut.

Directed.
To Thomas Wharton, junr., Esqr., President of the Executive Council, at Lancaster.

[1 Pa. Arch. 6 at 486–87]

Tyrer, James. Inventor, patent of 6/3/1825, expired 6/3/1839, for a firearm. Petersburg, Huntingdon Co.

U

Uhlinger, W. L., & Co. 1861–1890, Philadelphia. Sewing machine manufacturer that bought out William Hankins. Made revolvers that infringed on the Rollin White patent. Made cheap "Saturday Night Special" pistols. From 1861–65, converted old flintlock muskets to percussion for the Union army (FSP, 311).

Uhroe, Glode. Artillery artificer, 1779 (2 Pa. Arch. 15 at 383).

Ulrich, Andrew, & Co. Williamsport, Lycoming Co. Gunsmiths (BRD, 1875, 1880; PD, 1882).

Umholtz, Jacob. 1825, Lykens Twp., Dauphin Co. (KKR, 354). Gunsmith.

Umstead, Abraham. Artillery artificer, 1777–80, Capt. Jesse Rowe's Co. at Philadelphia (PIR, 2:245).

Unangst, Isaac. Gunsmith, West Beaver Twp., Union Co. (c, 1850).

Union Rifle Works. 1856–61, gunlock and gun-barrel makers, N. 2d St., above Dauphin St., Philadelphia (MD, 1857, 1861; BPaD, 1861).

Unkafere (Uncafer), George. In 1760 Frederick Sheetz died in York Co., Pa. His son, Philip, was apprenticed to a local gun smith named George Unkafare to learn the arts and mysteries of the trade. Sheetz was then age 17 and was to be bound until he reached age 21.

> Came into Court, Michael Damer, Esquire, Guardian of Philip Sheetz, a minor orphan son of Frederick Shitz, late of this County, yeoman, deceased., aged 17 years sometime in September last, and prayed that he, the said Philip Schitz, may be bound an apprentice to George Unkafare of this County, Gunsmith. It is considered by the Court and the said Philip Schitz is hereby bound an apprentice to the said George Unkafare until he shall be of the age of twenty-one years. In Consideration whereof the said George Unkafare doeth Covenant and agree to teach or cause to be taught the said Apprentice the Arts or Mysteries of a gunsmith which he now Practiseth, to read the Bible, to write and Arthemick as far as the Rule of Three, direct and to furnish and allow the said Apprentice sufficient Meat, drink, Apparel, Washing and Lodging during the said Term and at the Expiration thereof to pay unto him two suits of Apparel, one whereof shall be new, and of the Value of six Pounds or one new Smiths Bellows and Vize, which the said Apprentice shall then chuse.

George Unkafare was a gunsmith in Little Conewago (present day Hanover), York Co. He died in 1764, just about the time that Sheetz was completing his apprenticeship. Since records do not show Sheetz in York Co., we might surmise that shortly after Unkafare's death Sheetz left for Shepherdstown (t; *Orphan's Court of York Co.*, 1:219).

Unverzagt, Johannes. Gunsmith, Germantown, Philadelphia Co. (1/29/1762, *Pennsylvanishce Geschichte-Schreiber* [Sower's newspaper]).

Updegraff, Abner. Cutler, 1814–15, SW corner 5th St. and Smithfield St., Pittsburgh. Made scalping knives (PAD, 1815).

Updegraff, William. Cutler, 1834–42, Bolivar, Fairfield Twp., Westmoreland Co. (t).

Updegraph, Jacob. 1810, Schuylkill Co. (t). Gunsmith.

Urie, Solomon. Gunsmith. First noted as an adult, in the white campaigns against the Moravian Indian villages. Before 1814, Washington Co.; 1815-46, Orange Twp., Richland Co., Ohio (DH, 37).

Usher, Robert. Gunsmith, Poplar Lane near 5th St., Philadelphia (PXD, 1816).

V

Vache, Francis A. Gunsmith, 1823–37, Philadelphia (DeSD, 1823; MD, 1835–36).

Vail, J. W. Gunsmith, Erie (c, 1850).

Vallee, Prosper. 101 S. 6th St., Philadelphia, gunsmith (DeSD, 1829, 1830, 1831, 1835–36; MD, 1837, 1841, 1846).

Valley Forge Armory (1742–1842). (See Mount Joy Forge.) Established in 1742 as a forge by Stephen Evans, Daniel Welker, and Joseph Williams. On 3/8/1797, received an arms contract for M-1797 pattern muskets; 10/7/1812, O. & E. Evans, delivered 1,960 M-1808 muskets; 1831, made M-1827 pistols (RAG, 198).

VanDam, Jan Jansen. 10/4/1661: "for musket balls delivered" Florins 87/150/00. Whorekill, Dutch colony on the Delaware (2 Pa. Arch. 5 at 455).

Vandergrift, Jeremiah. Gunsmith, 153 St. John's, Philadelphia. (PXD, 1816) He and Isaac Vandergrift were employed by William Henry in 1814 (MPHS 8:149 [25th meeting]).

Vandergrift, John. Committee of Safety contractor. Bucks Co. (RAG, 199).

Vanderslice, Jacob. Gunsmith, Rose St. and William St., Philadelphia (MD, 1847, 1855).

VanEtten & Staples. Gunsmiths, Erie (PD, 1882).

VanLier, Dr. [?]. 2/7/1776, Gloucester Co. Offered to convert his grist mill into a gun powder mill in the summers (1 Pa. Arch. 4 at 710).

VanScriver, Benjamin. Gunsmith, 67 Virgin Alley, Pittsburgh (PAD, 1859–60).

VanScriver, J. B. Gunsmith, Lancaster boro (c, 1850).

VanVoorhis, Abraham. Gunsmith, 1791–1834, Morris Twp., Washington Co. (t).

Varner, John. Gunsmith, St. Clair Twp., Allegheny Co. (c; t, 1800).

Vaughan, John. Agent for DuPont gunpowders, SE corner Walnut St. and Front St., Philadelphia, 1819 (t).

Vaughan, Patrick. Swordsmith, 1777, in the employ of Lewis Prahl, gunsmith, Northern Liberties, Philadelphia Co. (2 Pa. Arch. 3 at 104).

Vaughn, Aaron C. A dentist who patented a pistol, 35,404 of 5/27/1862. Bedford boro. It was never manufactured.

Verner, Andrew. Major gunsmith. Worked on the corner of Bucks and Montgomery Cos., ca. 1770–1800. 1787, Allen Twp., Bucks Co. (t); 1789, Richland Twp., Bucks Co.; 1787, Upper Hanover Twp., Montgomery Co. (t); 1792, Marlborough Twp., Montgomery Co. (t) (GAS-1, 264–69; JKJ, 156–57).

Vester, William. Artillery artificer, 1778, Capt. Thomas Wylie's Co. (PIR, 2:249).

Villa, Isaac. 1693, Gunsmith Philadelphia (research by Dr. George Shumway).

Vogeler (Fogaler), George M. (b. 1769). Goldsmith, arrived at Philadelphia in 1805: age 36, Protestant, "tall and dark;" sailed on the *Margaret* (PaGP, 2:167).

Vogeler, Johann Friedrich. Smith, arrived at Philadelphia 9/14/1753 (PaGP, 1:524).

Voight, Henry. Committee of Safety gunlock maker (RAG, 201).

Volckersten, Abraham. Gunsmith, 11/23/1661, at the Dutch Colony at Whorekill on the Delaware River; was paid 9/60/00 florins for flints (2 Pa. Arch. 5 at 457).

VonShirer, Benjamin Henry. Gunsmith, 1852, Lancaster (A child of his was born, cert. no. 25, in 1852; father's occupation: gunsmith).

Vore, Benjamin Franklin (1814–86). 1840–48 New Paris, Bedford Co.; 1850–78, Oakland, Md., ran a shook and stave factory; 1879–86, New Paris. His early guns had patchboxes; his later guns usually had the date of manufacture on the lock bolt plates (t; c; f).

Vosburg, Orvin. Gunsmith, Port Allegheny, McKean Co. (BPaD, 1861).

W

Wagg, John (b. 1756). On 3/6/1773 Wagg was apprenticed to Michael Brothers, silversmith of Philadelphia, for 4 years, 2 months, and 24 days. (PIDS) 1777–79, brass founder at U.S. Ordnance Yard, Philadelphia; 1834, age 78, pensioner, Ash Co., N.C. (2 Pa. Arch. 11 at 259).

Wagonhurst (Wagenhurst), Johann (1787–1861). Gun-bar-

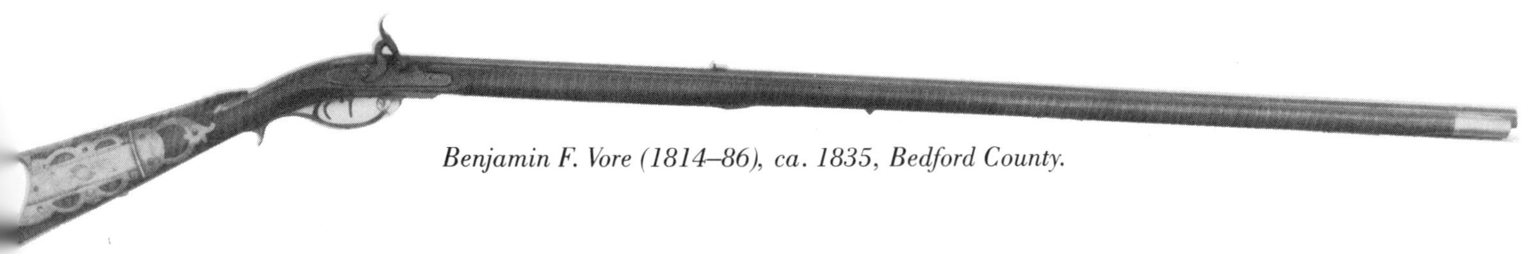

Benjamin F. Vore (1814–86), ca. 1835, Bedford County.

rel maker and gunsmith, Jammer Taht ("veil of misery"), Maxatawny Twp., Berks Co. (t; f; c).

Wagner, David (1852–96). Gunsmith, near Kistler, Madison Twp., Perry Co. (CHAN, 22; c, t).

Walker, G. Patent 20,776 (1858) for a new type of firearm, with E. Brooks, Philadelphia.

Walker, James. Gunsmith, 1802–3 Georges Twp., Fayette Co. (t); after 1804, Morris Twp., Knox Co., Ohio (DH, 226).

Walker, John. Artillery arificer, 1778–79, at the cannon factory at Carlisle, serving under Capt. Thomas Wylie. (PIR, 2:249); 1780–1807, gunsmith, Lancaster boro (SED, 60; t).

Walker, Joseph, Sr. (d. 1816). 1800–1803, Georges Twp., Fayette Co. (t; c); 1804–6, Mt. Vernon, Knox Co., Ohio (DH, 226).

Walker, Joseph. Gunsmith, 1842–43, Shippensburg (t).

Walker, Samuel. Gunsmith, Hublersburg, Centre Co. (BPaD, 1861).

Wall, Isaac. Artillery artificer, 1777, Capt. Nathaniel Irish's Co. (PIR, 2:247) Resided in Centre Co. after the Revolution.

Wallace & Spang. Gunsmiths and supplies. 94 N. 3d St., Philadelphia (MD, 1855).

Wallace, David (d. 1870). Gunsmith, 1834, Franklin Twp., Westmoreland Co. (DB 21, 77–78); Sardis, Franklin Co., Westmoreland Co. (BPaD, 1861); 1860–70, Franklin Township (WB 5, 363).

Wallace, James. On 10/20/1775, Wallace supplied 49 cutlasses at 5 shillings each to the Committee of Safety (Col. Rec., 10:375).

Wallace, John. (b. 1816). Gunsmith, 1839, Pittsburgh; 1850–68, Freeport, Armstrong Co. (t; c); 1870–79, Franklin, Venango Co.; 1880–82, Freeport (c; t; BPaD, 1861; PD, 1882).

Wallace, William H. Gunsmith, Philadelphia (MD, 1841, 1855; BPD, 1860).

Wallis & Birch. Made cheap revolvers, Philadelphia (c. 1850).

Wallis, Thomas M. Inventor, Philadelphia. Patents: 194,489 (1877) for a revolver, and 301,021 (1884) for a breech-loading firearm.

Wallis, William. Artillery artificer, 1777, Capt. Nathaniel Irish's Co. at the cannon factory, Carlisle (PIR, 2:247).

Walsh, James. Armorer, 1775–76, Philadelphia, contractor and subcontractor to the Committee of Safety; 1877–79, Pennsylvania Gunlock Factory, Allentown (2 Pa. Arch. 2 at 698; 2 Pa. Arch. 15 at 384).

Walsh, John. 1775–76, contractor and subcontractor to the Committee of Safety, Philadelphia; 1777–79, gunlock factory at Allentown. (2 Pa. Arch. 1 at 672–73; t) A pair of pistols with brass barrels and marked "J. Walsh" is known.

Walt (Watt), James. Gunsmith, Miller's Ct., Philadelphia (PXD, 1810; DeSD, 1819).

Walter, George (b. 1770). Smith, arrived on the brig *Mary* at Philadelphia, 7/30/1796 (2 Pa. Arch. 17 at 602).

Walter, Henry. 1777, Supplied "powder and lead for the [home] guard" in York Co. at £8/11/0 (3 Pa. Arch. 7 at 61).

Walter, Henry. Gunsmith, 1820–30, Cocalico Twp., Lancaster Co. (SED, 60; t).

Walter, Henry (1821–1903). Gunsmith, Unity Twp., Westmoreland, active 1850–70. His family were wagon makers also (t; c).

Walter, John. Gunsmith and lock maker, 1845–61, Shartleysville, Berks Co. (c; t; BPaD, 1861; BRD, 1860).

Waltner, Conrad. Swordsmith, 1777, Northern Liberties, Philadelphia Co., in the employ of Lewis Prahl (3 Pa. Arch. 2 at 104).

Walton, William. On 5/14/1758, a smith in the employ of the quartermaster general, Ft. Ligonier. (5 Pa. Arch. 1 at 350).

Ward, John. Gunsmith, Otter near Frankford Rd., Philadelphia (PXD, 1813).

Ware, William. Gunsmith, Philadelphia (MD, 1855, 1857).

Warke, John. Smith, 1730, Philadelphia; "settled on a branch of Octoraro" (2 Pa. Arch. 19 at 762).

Warner, George. Gunsmith, Rose Alley, Philadelphia (DeSD, 1819); N. Queen St. near James St., Lancaster (GLD, 1863–64; BLD, 1857); 315 Cherry St., Lancaster (BLD, 1873–74).

Warner, Hieronimus. Brass founder and artillery artificer, 1779, at the cannon factory, Philadelphia (2 Pa. Arch. 1 at 385).

Warner, Hiram (1833–62). Born in Springwater Twp., Livingston Co., N.Y., Hiram Warner moved to Elk Co. by 1844. In 1857 he was listed as a gunsmith working with his brother, Horace. He was a member of the 2d Regt., U.S. Sharpshooters and was killed in action at Antietam, Md. (REH, 228–30).

Warner, Horace (1826–93). Born in Springwater Twp., Livingston Co., N.Y., he came to Elk Co. in 1844. He

Horace Warner (1826–93), Elk County.

enlisted from Elk Co. in the 2d Regt., U.S. Sharpshooters, rising to the rank of 1st lt., and served 1861–65. From 1865 to 1865 he was a farmer near Lake City, Minn. From 1868 to 1879 he was a gunsmith in Elk Co. In 1879 he worked as a gunsmith in Syracuse, N.Y., but returned in late 1879 and lived in Elk Co. until 1885. From 1885 to 1888 Warner was in Syracuse, and in the winter of 1887–88 he was doing business as Warner & Lowe. From 1888 to 1891 Warner was a gunsmith in Williamsport, Lycoming Co. He died in 1893 at Ridgway (REH, 228–30).

Warner, Jacob. Artillery artificer, 1778–80, Capt. Nathaniel Irish's Co. In 1812 he was living in Beaver Co. (PIR, 2:247).

Warner, Joseph. Gunsmith, Rose Alley, Philadelphia (DeSD, 1819, 1820, 1827, 1830).

Warwick, Alexander. Artillery artificer, 1778, Capt. Thomas Wylie's Co. at the cannon factory, Carlisle (PIR, 249).

Warwick Furnace (1738–1867), Berks Co. Associated with Reading Furnace and Coventry Forge. "During the Revolution, it was very active in casting cannon for the Continental Army, some of which were buried upon the approach of the British in 1777" (SHIS, 172).

Warwick, James. Gunsmith, 36 Coates Alley, Philadelphia (MD, 1837).

Washburn, F. C. Gunsmith, Wellsburgh, Tioga Co. (PD, 1882).

Washburn, Ira. Gunsmith, South Gibson, Susquehanna Co. (BPaD, 1861).

Waters, John. Gunsmith, 1779–87, Carlisle, Cumberland Co. (WSB, 55; t). After 1791, he sold his gunsmithing business and did not return to the trade.

Watson, J. M. Gunsmith, 1880–87, Altoona, Blair Co. (t; c; PD, 1882).

Watt, George. Gunsmith, 455 N. 3d St., Philadelphia (PXD, 1810); West Philadelphia (MD, 1841, 1846).

Watt, John (1815–85). Gunsmith, Licking Creek, Juniata Co. (t; c; f) He was the father-in-law and teacher of David Wagner. His shop was at the junction of Willow Run and Tuscarora Creek, near Peru Mills (FSP, 320).

Watt, J. Wesley. Gunsmith, Schuylkill St. and High St., Philadelphia (MD, 1846, 1855); McVeytown, Mifflin Co. (c, 1860, 1870).

Watts, James. Gunsmith, 1860–87, Madison Twp., Perry Co. (CHAN, 26ff.).

Weatherby, Joseph. Gunsmith, 449 N. Front St., Philadelphia (DeSD, 1819).

Weaver, Adam. Gunsmith, Jonestown, Lancaster Co. (SED, 60; t, 1771–79).

Weaver, Anthony. (d. 1702). Smith, 1781–1802, Philadelphia Co. He was from Charlbury, County Oxon (2 Pa. Arch. 19 at 363). His will was dated 2/1702 and probated shortly thereafter (Philadelphia County).

Weaver, Henry. Gunsmith, 1819–34, Southampton Twp., Bedford Co. (t); 1835-37, Hopewell Twp., Bedford Co. (t); 1838, Frankstown Twp., Huntingdon Co. Also a justice of the peace (t).

Weaver, John. Gunstocker, 1777, at French's Creek Arsenal (1 Pa. Arch. 15 at 384).

Weaver, Mathias. Gunsmith, Philadelphia (c, 1850).

Webb, George. Blacksmith, 1778, 10th Pa. Regt. (2 Pa. Arch. 15 at 504).

Webb, William. On 6/3/1777, Webb was paid "£66/17/6 for 2 petards and taking a bullet out of an 18 pounder" (2 Pa. Arch. 1 at 166).

Weber, Mathias. Gunsmith, 1512 N. 2d St., Philadelphia (MD, 1855; BPaD, 1861).

Weddell, Peter M. (1780–1845). Gunsmith, born and trained in Westmoreland Co. Before 1812, Bourbon Co., Ky.; 1820–45, Cleveland, Ohio (f; t; c).

J. Watson, Altoona, ca. 1880, Blair County.

Guns, Revolvers, Powder, Shot, Caps,
SHELLS, LOADING TOOLS, GAME BAGS,
And all kinds of Hunters' Supplies; Also FISHING TACKLE, Base Ball Goods, Pocket Knives, Boxing Gloves and other Sporting Goods. Send for prices.
J. M. WATSON,
1113 11TH. AVE., ALTOONA, PA.

Altoon Mirror, April 11, 1882

Weeks, Daniel (1818–76). Born in Orange Co., N.Y. 1824, Monroe Co., N.Y.; 1848–76, 139 State St., Erie. (t; c; BRD, 1861, 1867) He was also sealer of weights and measures for Erie Co. (PD, 1873–74).

Weeks, E. C. Gunsmith, 1239 State St., Erie (PD, 1882, 1890).

Weidman, Solomon. Gunsmith, Lemon St. near Queen St., Lancaster (BLD, 1857, 1873, 1874–75).

Weiker, George, Sr. (d. 1800). 1780–1880, Richland Twp., Bucks Co. It is not definitely established that he was a gunsmith.

Weiker, George, Jr. (1769–1853). 1790–1800, Richland Twp., Bucks Co. (t; c); 1806–39, Union Co. (t; c); 1801–6, Lower Saucon Twp., Northampton Co.; 1806–39, Union Co. (t; c); 1839–53, Wayne Co., Ohio (GAS-1, 260–63).

Weikusat, August, Sr. Gunsmith, 1774–1816, Kutztown, Lebanon Co. (t).

Weikusat, August. Gunsmith, 1874–1916, Kutztown, Lebanon Co. (t; c; PD, 1882; f).

Weiman, Henry. Gun-barrel forger, Elizabeth Twp., Lancaster Co. (c, 1850; SED, 60).

Weinberger, John L. Gunsmith, Johnstown, Cambria Co. (BPaD, 1861).

Weir (Wier), James. Church Alley at 2d St., Philadelphia. 11/8/1775: "wire & brushes for firearms" (Col. Rec., 10:398).

Weis, Charles (1831–1900). Born in Philadelphia, Weis was a minor inventor and gunsmith, St. Mary's, Elk Co. (REH, 253; PD, 1882).

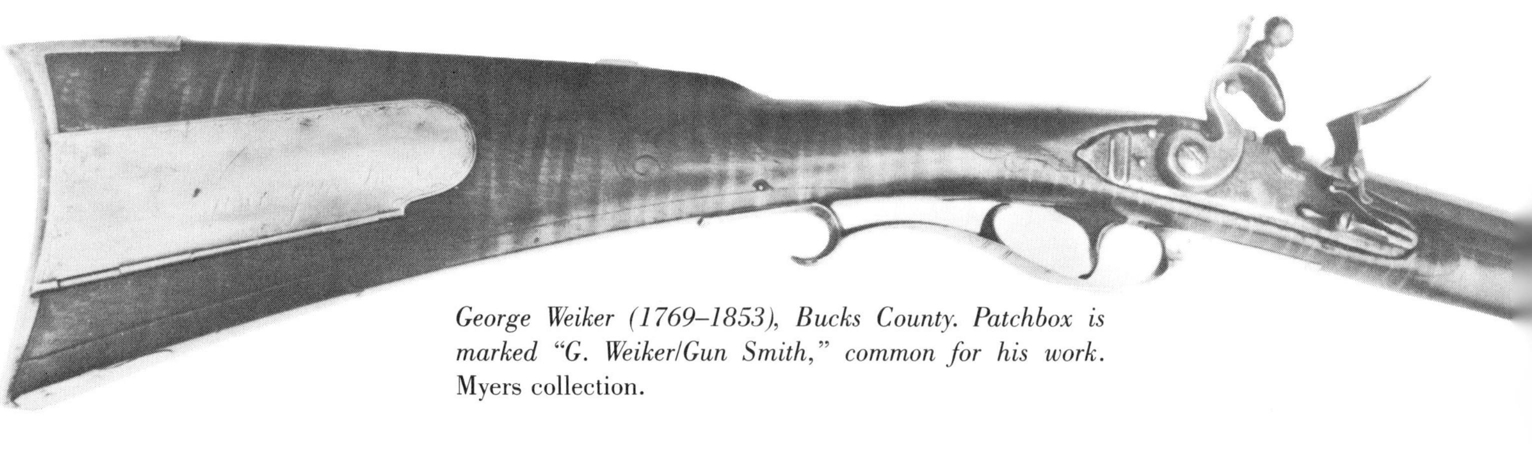

George Weiker (1769–1853), Bucks County. Patchbox is marked "G. Weiker/Gun Smith," common for his work. Myers collection.

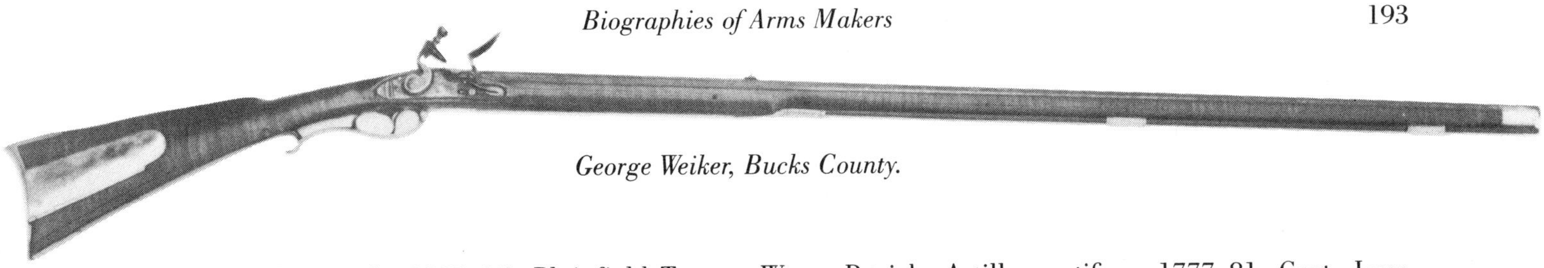

George Weiker, Bucks County.

Weiss, Charles. Gunsmith, 1825–35, Plainfield Twp., Northampton Co. (t).

Weiss, George. Gunsmith, 1785–89, Christian's Spring, near Nazareth, Northampton Co. (GAS-1, 173).

Weiss, Jedediah (1795–1873). Silversmith, Bethlehem, after 1815. (t; c; f). Took over J. S. Krauss's shop after Krass died. (James W. Gibbs, *Pennsylvania Clocks and Watches* [State College: Pennsylvania State University Press, 1984] 182, 184, 185, 190, 191).

Weiss, John George (1758–1811). Lock maker, gunsmith, clock maker, 1800–1811, Bethlehem, Northampton Co. (t; c; f).

Weiss, William. Gunsmith and armorer, 1804–07, Richmond, Va.; 1808–21, Lancaster (SED, 60; FSP, 322).

Welch, Jno. Smith, 1704, Philadelphia Co. Acquired 530 acres by patent (2 Pa. Arch. 19 at 449).

Wells, Richard. Gunsmith, ironmonger, small gun manufactory, Mulberry Ward, Philadelphia. DBA DeHaven & Wells (c, 1790).

Welsh (Welch), James. 1779–90, Dock Ward, Philadelphia (t; c). Gunsmith, armorer, artificer.

Welshans, Abraham. 1780–1804, York boro. Contractor with Jacob Leather for U.S. M-1795 muskets; also worked on militia arms contracts with Jacob Moll and Henry Pickel. Major gunsmith (JKJ, 341).

Welshans, Conrad. Gunsmith, 1781–99, York boro. In 1783 he had 5 men employed in his shop. (HYC, 519; JKJ, 343; t) On 4/11/1798 he was a cocontractor for U.S. M-1795 liberty cap muskets.

Welshans, Jacob. 1778–1800, major gunsmith, York boro. Welshans was paid £9/7/6 for militia arms repairs, 1777–79. (3 Pa. Arch. 7 at 108) In 1783 5 persons were employed in his shop (HYC, 519; JKJ, 345; t). In 1787 he was a subscriber to the York German Reformed Church (f).

Welshans, Joseph, Sr. Gunsmith, 1779–83, York boro (JKJ, 345; t).

Welshans, Joseph, Jr. Gunsmith, 1779–83, York (t; JKJ, 345).

Welshantz, David. Gunsmith 1780–1800, York (t; c).

Wendle (Windle), Isaac (1756–1859). Gunsmith and farmer, 1790–1856, Stoystown, Somerset Co. (t; c; f).

Wenn, Patrick. Artillery artificer, 1777–81, Capt. Jesse Rowe's Co. at the cannon factory, Philadelphia (PIR, 2:245).

Wentzel, Daniel. Gunsmith, Hereford, Berks Co. (BPaD, 1861).

Wenzel, Herman. Gunsmith, 82 VanBraam St., Pittsburgh (PAD, 1872–73; PD, 1882).

Wenzel, Hiram. Gunsmith, 327 North Ave., Pittsburgh (PAD, 1878–79).

Wenzel, J. H. Gunsmith, Smithfield district, Pittsburgh (PAD, 1870–71).

Werner, C. L. Gunsmith, Pittsburgh (PAD, 1873–74).

Werner, George J. Gunsmith, 620 N. Queen St., Lancaster (BLD, 1884).

Werner, J. G. Gunsmith, Lancaster (PD, 1882).

West, Jonathan. Gunsmith, 1809–11, East Bethlehem Twp., Washington Co. (t).

West, Samuel. Gunsmith, 1807–9, Fallowfield Twp., Washington Co.; 1810–30, Millsboro, Washington Co. (t; c).

West, William. Gunsmith, 1812–18, Millsboro, Washington Co. (t).

Westphal, Charles. Arms contractor, 1807–15, Philadelphia. Made M-1808 muskets on contract, doing business as Goetz & Westphall (RAG, 207).

Westward, Edward. Advertisement in the *Pennsylvania Packet*, 3/13/1753: "Edward Westward, an English servant, blacksmith or lockmaker by trade, ran away from James Few and Nicholas Wilson of Wilmington."

Wetzel, Andrew. Gunsmith, 1800–01, Plainfield Twp., Northampton Co. (t).

Wetzel, Henry (1799–1879). Gun-barrel, sickle, and edged-tool maker, north side of Middle Creek, Kreamer, Snyder Co. (c; t; f).

Wetzel, John. Gunsmith, 1781–88, Macungie Twp., Northampton Co. (t).

Wetzel, John. Gunsmith, Sheimersville, Lebanon Co. (BPaD, 1861).

Wharton, Carpenter. Gun suppliers to Sir William Johnson, 1771. Merchants at Philadelphia (Wester White).

Wharton, Frederick. Stockmaker, 1777, at French's Creek Arsenal; Artillery artificer, 1779 (2 Pa. Arch. 15 at 383).

Wheeler & Wilie. Probably, Samuel Wheeler and Thomas Wylie (Wilie), Philadelphia, revolutionary war contractors. 3/24/1777; "£19/0/6 for smith's work on brass cannon" (2 Pa. Arch. 1 at 19); 8/18/1777: "£84/19/6 for 8 gramplins and 200 feet of chain, weight 566½ lbs., . . . for the use of continental fire vessels" (2 Pa. Arch. 1 at 207).

Wheeler, Charles Yandes (1843–99). Inventor, operated major large arms manufactory, McKeesport, Allegheny Co. Developed the Wheeler-Sterling armor piercing projectile, manufactured at Firth-Sterling Steel Co. at McKeesport and, under license, at Thomas Firth & Sons, Sheffield, England.

Wheeler, Samuel. 8/4/1775: "Resolved, that Mr Samuel Wheeler make 100 pikes . . . agreeable to the pattern produced to this Board" (Col. Rec., 10:298); 11/8/1775: "30 pikes rec's from Sam'l Wheeler" (Col. Rec, 10:398). "Pay Samuel Wheeler £141/15/5½ for . . . work done for the brass cannon factory and boring mill." Wheeler was a cutler from at least 1769 on, and during the 1770s had at least 3 apprentices in the cutlery trade: Godfrey Hanible, Joseph Wood, and Peter Standley, Jr. (PIDS).

Whetstone, David (1780–1824). Gunsmith, apprenticed by his father to John Gonter, gunsmith, Hagerstown, Md., 1/1/1796, along with his brother John. 1807–1824, Ayr Twp., Bedford Co. (t).

Whetstone, John. Gunsmith, apprenticed to John Gonter, Hagerstown, Md., 1/1/1796. Gunsmith, 1801, Hagerstown, Md.; 1804, Ayr Twp., Bedford Co; 1806–8, Berkeley Co., W. Va. (DH, 269; WSB, 132).

White, James. Gunsmith, 1816–18, Pittsburgh (t).

White, Job. Gunsmith, Little Britain Twp., Lancaster Co., 1840 (SED, 60).

White, John (1800–1869). Born in Emmitsburg, Md., a son of Peter White. 1818–29, Lisbon, Ohio, working with Samuel Small; 1830–43, Uniontown; 1844-58, 46 St. Clair St., Pittsburgh; 1859–61, Richester, Beaver Co.; 1864–69, Pittsburgh. (t; c; PAD, 1864–65; BPaD, 1861; T. W. Pike in MB 3 [1953] 6) Gunsmith.

White, John. Gunsmith, Philadelphia (MD, 1837).

White, John R. Gun-barrel maker, Reading, Berks Co. (BPaD, 1861)

White, Levi. Gunsmith, Isabella St., Pittsburgh (c, 1850).

White, Peter (1778–1834). Major gunsmith. 1800, Emmitsburg, Md., working with John Armstrong (c); 1803–16, Cumberland Valley Twp., Bedford Co.; 1817–18, Bedford Twp., Bedford Co.; 1818–34, Uniontown, Fayette Co. (t; c) Buried at Old Presbyterian Churchyard, Uniontown. In 1982 his stone was stolen.

White, Samuel. Gunsmith, 1828–29, Uniontown (t).

White, William. Artillery artificer, 1778, Capt. Thomas Wylie's Co. at the cannon factory, Carlisle (PIR, 2:249).

Whitehead, Matthew. Arms repairman for the Committee of Safety (RAG, 209).

Whiting, Ruggles. Cutler, 1816–19, Penn St. and Cecil's Alley, Pittsburgh, bought out C. Cowan's cutlery, which, in turn, was operated in 1826 by R. Bowen (SHIS, 228).

Whitmore & Wolff. Gunlock and hardware distributors and importers, 50 Wood St., Pittsburgh. 1836–52, Whitmore & Wolff; 1852–58, Whitmore, Wolff & Co.; 1858–72, Whitmore, Wolff, Duff & Co. (HJK, 361–62).

Wice, France. Smith, arrived in Philadelphia on 8/27/1733 (PaGP, 1:765).

Wickery, George. Gunsmith, 1813–23, Morgan Twp., Greene Co. (t); 1824–27, Buffalo Twp., Washington Co. (t).

Wickham, Marine Tiler (1781–1833). Gunsmith, armorer, arms contractor. 1802, apprenticed with John Armstrong,

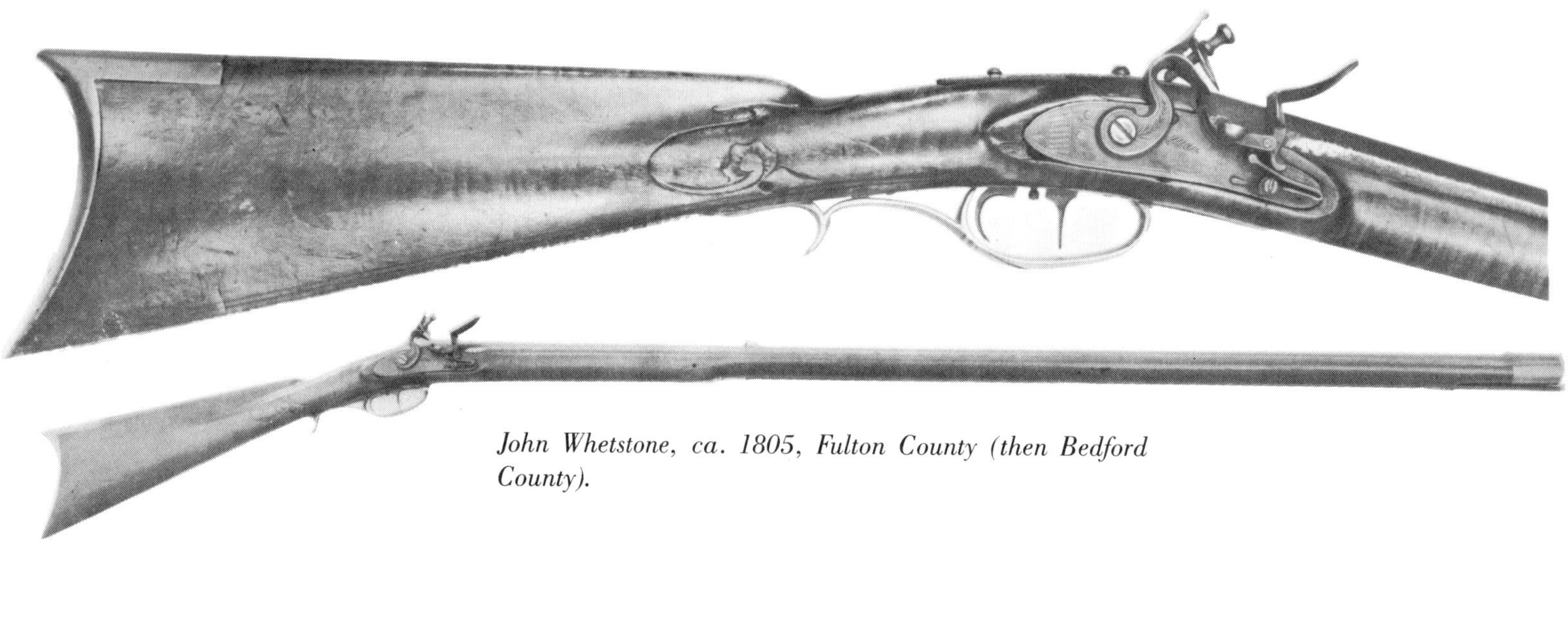

John Whetstone, ca. 1805, Fulton County (then Bedford County).

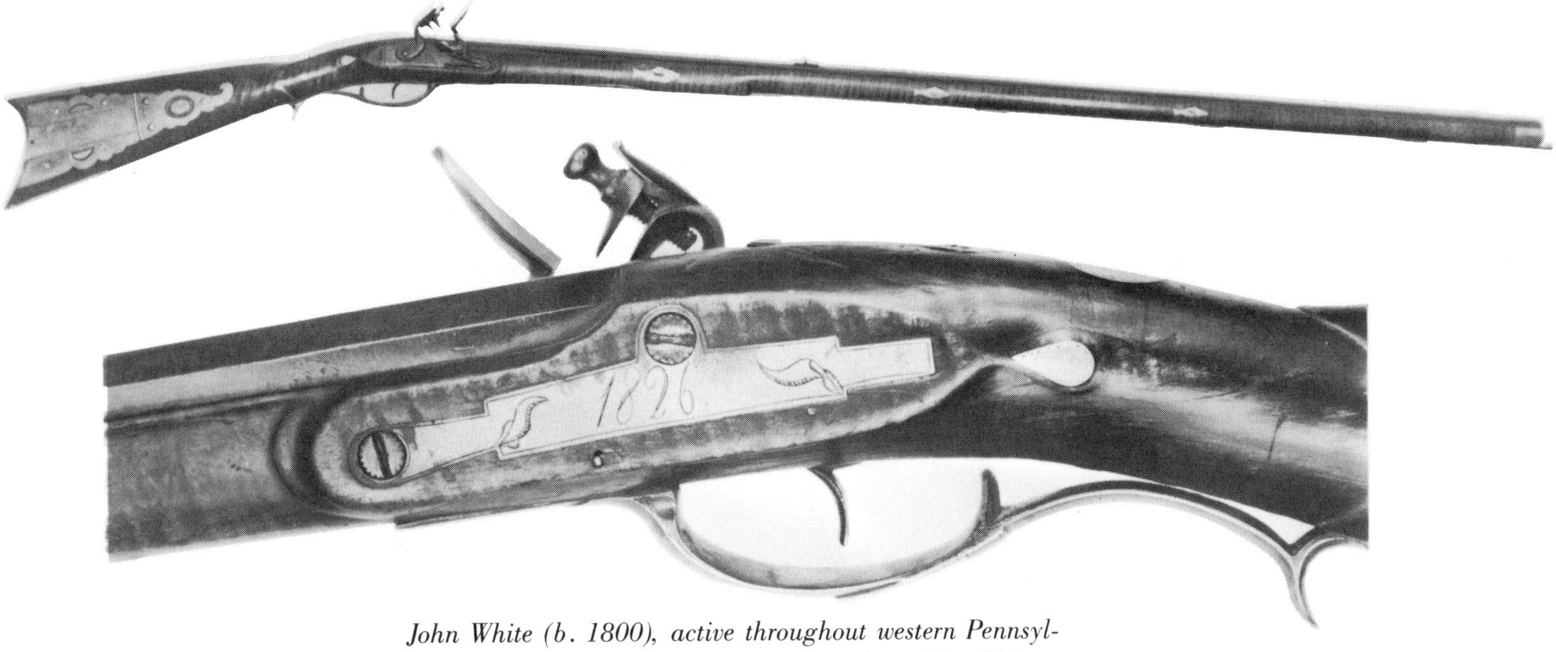

John White (b. 1800), active throughout western Pennsylvania. Gun is dated 1826 and was made in New Lisbon, Ohio (marked "N. Lisbon" on patchbox lid).

GUN MAKING.

THE business of making **Rifle Guns** is still carried on by the undersigned, a few doors West of the Court House, Main street, where he has excellent gun barrels on hand, and is prepared to furnish **New Rifles**, of any pattern or size, on the most reasonable terms and shortest notice, and warranted to perform well, if reasonably used, for a year.

Having two experienced journeymen, of undisputed workmanship, in addition to his other hands, he is prepared to promptly attend to all

Jobs of Repairing,

That those from a distance can have their work to take with them on their return home.

Scissors, Shears and Coffee Mills, repaired in good order and at the shortest notice.

All work not taken away from the shop within three months after it is repaired will invariably be sold to pay the charges for repairing.

A few good Horses,—Curled Maple & Black Walnut, well seasoned for Gun Stocks,—old BRASS and Copper Kettles & Country Produce generally, will be taken in exchange for New Rifles and all kinds of work.

JOHN WHITE.

Uniontown, July 26, 1841.

Uniontown Genius of Liberty, September 13, 1841

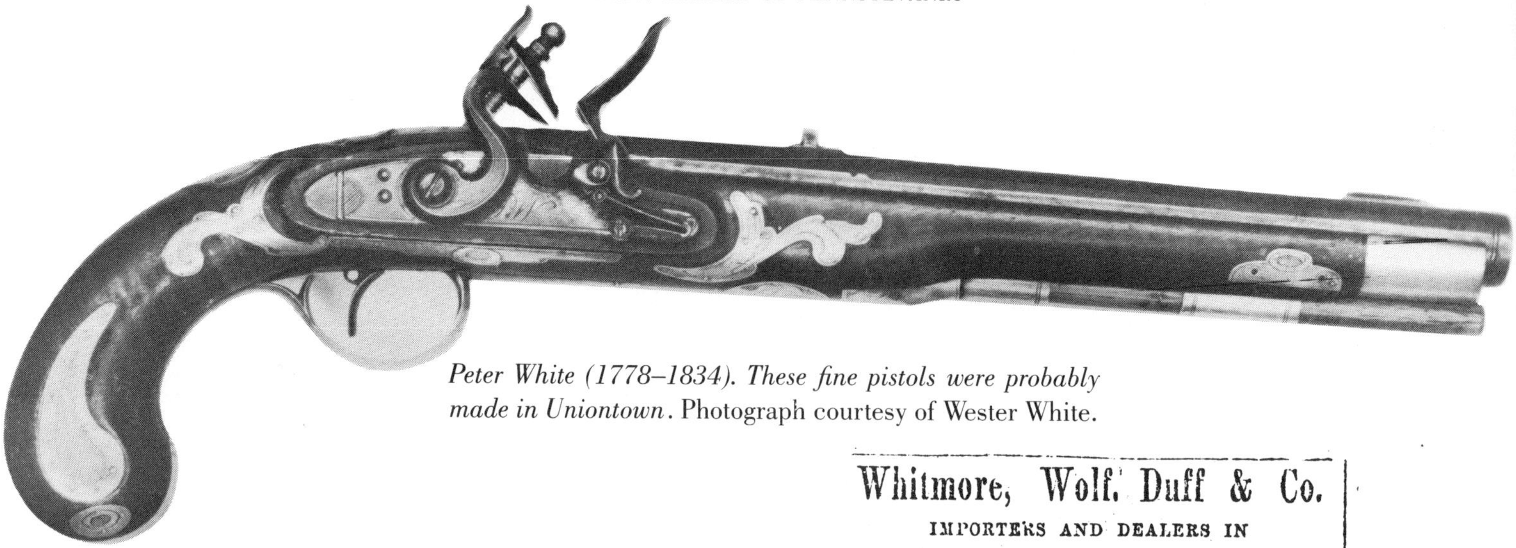

Peter White (1778–1834). These fine pistols were probably made in Uniontown. Photograph courtesy of Wester White.

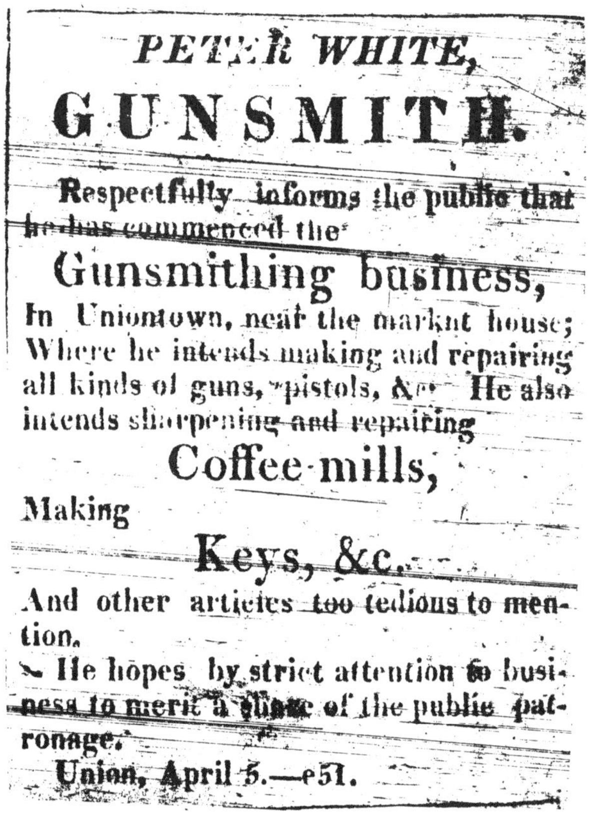

PETER WHITE,

GUNSMITH.

Respectfully informs the public that he has commenced the

Gunsmithing business,

In Uniontown, near the market house; Where he intends making and repairing all kinds of guns, pistols, &c. He also intends sharpening and repairing

Coffee-mills,

Making

Keys, &c.

And other articles too tedious to mention.

He hopes by strict attention to business to merit a share of the public patronage.

Union, April 5.—e51.

Uniontown Genius of Liberty, April 3, 1819

Emmitsburg, Md.; 1805, U.S. Arsenal at Harper's Ferry; 1808–17, Schuylkill Arsenal; 1817–33, NW Noble at 3d St., Philadelphia (DeSD, 1829, 1830, 1831; AAM, 271–72).

Wigfal, Samuel. Committee of Safety arms contractor. On 11/30/1775, Philadelphia, Committee of Safety advanced £100 to him to make gunlocks at £0/22/6 each, to be delivered within 6 months; cocontractor with M. Blackwood. On 12/5/1775, he agreed to deliver 200 gunlocks at the same price, within 5 months (4 Amer. Arch. 6 at 495).

Wigel (Wigle), Peter. York, 1777: "for repairing arms,

Whitmore, Wolf, Duff & Co.

IMPORTERS AND DEALERS IN

HARDWARE,

SIGN OF THE ANVIL,

No. 50 Wood st. three doors above St. Charles Hotel,

PITTSBURG, PA.

HAVE now in stock a full assortment of English, German and American Hardware, of every descrition, which is unrivaled in extent and variety by any other establishment in the country.

Having direct communication with the manufacturers, our facilities for obtaining goods are such as to enable us to sell at prices as LOW as can be found in any of the Eastern Markets.

Buyers will promote their own interest by examining our stock and prices before purchasing.

We are agents for the sale of Wilson, Hawksworth, Ellison & Co's. Celebrated Steel; Watrouss, Shipman & Co's Adjustible Handle Drawing Knives, Troemner's Philadelphia Scales, Kingsland's Patent Box Irons, and sell at manufncturers prices "Novelty" and "Variety" Locks and Latches; Lippincott & Co. and Newmyre & Graff's Axes, Shovels, Saws, &c.; Dodge's Kentucky Cow Bells, Patent and common Boreing machines, Brass Kettles, Apple Pearers, Butcher's, Spencer's and Horner's Files and Rasps, Witherby's Chisels and Drawing Knives, Coe's Wrenches, Wostenholm's Knives; Table and Pocket Cutlery of best English and American brands; Sleigh Bells, Anvils, Bellows, Chains, &c.

Pittsburg, Sept. 26, 1860.

Pittsburgh Gazette, September 26, 1860

£7/5/9" (3 Pa. Arch. 7 at 107).

Wigle (Wigel), Jacob. Gunsmith, 1811–12, South Huntingdon Twp., Westmoreland Co.; 1813–18, Rostraver Twp., Westmoreland Co.; 1818–22, Franklin Twp., Fayette Co.; 1822–26, Rostraver Twp., Westmoreland Co.; 1827–

Two gunlocks by Whitmore, Wolff, Duff & Co., Pittsburgh. These were probably imported from England and sold by the company, a large wholesale hardware company.

44, South Huntingdon Twp., Westmoreland Co. (t; c)

Wigstead, Francis. Artillery artificer, 1779, taken prisoner (2 Pa. Arch. 15 at 384).

Wilbur, Robert (b. 1836). Gunsmith, civil engineer, blacksmith, 1860–80, Fox Twp., Elk Co. (REH, 240) There was a gunsmith by this name, a contemporary of this Robert Wilbur, active in Galesburg, Knox Co., Ill., a different man (FSP, 331).

Wilcocks (Wilcox), John. 1776, placed in charge of the reorganization of the Pennsylvania Gun Lock Factory at Allentown. (3 Pa. Arch. 5 at 7). On 2/18/1833, he received a revolutionary war pension in Allegheny Co. (3 Pa. Arch. 23 at 476).

Wiley, Theodore. Bayonet maker, 1775–77, Philadelphia (RAG, 225).

Wilfong, Vincent W. Gunsmith, Leopard Alley, Lancaster (BLD, 1871–72).

Wilhelm, Jacob. Gunsmith, Middle St. near Factory St., Lancaster (BLD, 1857).

Wilkins, George. Gunsmith and blacksmith, 1804–10, Pike Run Twp., Washington Co. (t).

Wilkes, James. Gunsmith, 134 E. Walnut St., Lancaster (BLD, 1871–72).

Wilkinson, Angus. Artillery artificer, 1778–79, Capt. Nathaniel Irish's Co. (PIR, 2:247).

Will, Wilhelm, Jr. Gunlock maker, 1798–1800, Philadelphia; 1800–2, Zanesville, Ohio (DH, 304).

Williams, Abraham (b. 1789). Gunsmith, 1816–22, Greene Twp., Greene Co.; 1824–26, Monongahela Twp., Greene Co.; 1827–30, Clarksville, Greene Co.; 1833–56, Morris Twp., Washington Co. He had a water-powered factory on Craft Creek (t; c).

Williams, James. Gunsmith, 1843–50, Lancaster (SED, 60; t).

Williams, Jesse. Gunsmith, Rockland Twp. Venango Co. (PD, 1890).

Williams, Joseph. Arms maker, 1846, Upper Saucon Twp., Lehigh Co. (t).

Williams, Levi. Gunsmith, 1820–22, Greene Twp., Greene Co.; 1824–27, Monongahela Twp., Greene Co.; 1831, Morgan Twp., Greene Co. (t).

Williams, Lewis (b. 1800). Gunsmith, 1817–23, Greene Twp., Greene Co.; 1824–33, Monongahela Twp., Greene Co.; 1834–38, Georges Twp., Fayette Co.; 1850–60, Alliance, Ohio (t; c).

Williams, Richard. Gunsmith, 1839–41, Morris Twp., Washington Co. (t).

Williamson, William. "an old man, a good gunsmith, to go to sea." Williamson was a prisoner in the Philadelphia Jail; he was incarcerated on 10/21/1780, and tried on felony charges on the complaint of Benj. J. Paschall, Esq. on 7/3/1781 (1 Pa. Arch. 9 at 249).

Willis, Henry. In his will dated 1757, John Galbreath ordered that his son, Robert, "as soon as he is fit, shall be put to learn to be a gunsmith, either to Henry Willis, or to some other of that trade." (Cumberland Co. WB A, 40) In 1764, Willis was a gunsmith in York Co. (FSP, 333).

Willis, J. E. (d. 1879). Gunsmith, Water St., Kittanning, Armstrong Co. (BRD, 1878) There are over 200 pages of materials related to Willis's death on 8/9/1879 in the Armstrong Co. Court House. There were many items related to the gunsmithing trade in his estate inventory: "2 pistols 45¢, lot of gunlocks 25¢, gunlock springs 50¢, box of hardware and locks 40¢, 1 rifle barrel 65¢, 1 sabre 45¢ 1 lot gunsmith's tools 55¢, 1 new gun barrel 70¢, 1 good rifle $4, 1 good pistol 75¢, 1 air gun 25¢, 1 breech-loading musket 60¢."

Willis, John. 12/5/1775: "John Willis and Benjamin Town, gunsmiths, having agreed to make 200 firelocks for this Province at the rate of £5/4/0 each. . . ." (Col. Rec., 10:422) On 12/6/1775 he agreed to make 200 muskets at £4/05/0 each (4 Amer. Arch. 6).

Willis, Richard. In 1777, Willis took the Oath of Allegiance to the Republic at Lancaster (2 Pa. Arch. 13 at 395), but he was listed in 1778–81 as a gunsmith attainted by treason (1 Pa. Arch. 10 at 250ff.).

Wilson, Andrew (b. 1760). Artillery artificer, 1778, Capt. John Jordan's Co. at Carlisle. (2 Pa. 15 at 263) In 1823 Wilson lived in Lycoming Co., age 63.

Wilson, James. Smith, 1759, employed by quartermaster general at Ft. Loudon (5 Pa. Arch. 1 at 348); artillery artificer, 10/15/1778, Capt. Thomas Wylie's Co. at Carlisle (PIR, 2:249).

Wilson, John. Artillery artificer, 1778, at U.S. Armory, Philadelphia (2 Pa. Arch. 11 at 259).

Wilson, John. Gunsmith, 1828, Mauch Chunk, Lehigh Co. (t).

Wilson, Joseph. On 11/14/1776, Wilson was paid for repairing arms for the Pa. Navy (Col. Rec., vol. 11).

Wilson, Nathaniel. Silversmith, East Pennsboro Twp., Cumberland Co. (t) "freeman at William McChesney's," 1750 (t).

Wilson, Philip, & Co. Gun-barrel makers, 432 Chestnut St., Philadelphia (MD, 1859).

Wilson, William. Artillery artificer and blacksmith, 1778, Capt. John Jordan's Co. at Carlisle (2 Pa. Arch. 11 at 260).

Wilson, W. T. Bridesburg, Philadelphia Co. Patent 49,463 of 8/15/1865 for a breech-loading gun. On 9/3/1874, "John S. Cochennour, maker of rifles and pistols" had in his employ "W. T. Wilson, practical gunsmith" (Olney, Illinois, *Weekly Times*).

Wilt, Jeremiah (1808–84). Gunsmith, 1829–32, Chambersburg: 1837–84, near Dayton, Ohio (DH, 288).

Wineberger, Leonard. 1876–77, south end of Bedford St., Johnstown, Cambria Co., gunsmith (*Johnston Directory*, 1876–77).

Winey, Jacob. Philadelphia. 3/18/1776: "Near 7 [P.M.] went by appointment to meet 11 of the sub-committee who, with me, had before valued 649 guns, bayonets, gun powder, etc., for which said Jacob Winey now objected to, when, after hearing his objections, we confirmed our former settlement, to wit, that our Committee of Secrecy, pay the sum of £3/15/0 per piece, when they are properly put in order by the Commissary and buy the cannon powder at £17 per cwt, provided, upon trial, it be good. The charge of repairing was what Winey objected to, being £80/5/0" (Marshall's *Remembrancer*) [1839]).

Winger, Christian. Militia arms collector, 1777, Lancaster Co.; 1780–1810, gunsmith, Leacock Twp., Lancaster Co. (SED, 60).

Winger, Richard. 1777–78, Lancaster Co. "for repairing muskets," £2/05/0 (2 Pa. Arch. 6 at 384).

Wingerholter, Martin. Gunsmith, 274 Penn St., Pittsburgh (PAD, 1860–61).

Winner & Nippes. Daniel Nippes and James Winner. 1808–12, contractors for M-1808 muskets; 1808–9, Winner, Nippes & Steinman, with John Steinman (RAG, 214). Philadelphia.

Winner, James. Gunsmith, 463 N. 3d St., Philadelphia (PXD, 1810); 104 Walnut St., Philadelphia (PXD, 1813).

Wintable, Abraham. 437 N. 3d St., Philadelphia (PXD, 1813; DeSD, 1813, 1823; MD, 1837). Gunsmith.

Wintafeld, Abel. Gunsmith, 113 N. 2d St., Philadelphia (DeSD, 1819, 1833).

Wirth, John. Gunsmith and inventor, 1869–90, Philadelphia. Patent 490,129 of 1/17/1893 for a recoil check firearm (BPD, 1869; PD, 1882).

Wirth, Louisa. Gunsmith, 3220 Lancaster Ave. and 1431 Market St., Philadelphia (GPD, 1892).

Wirth, Wilhelmina. Gunsmith, 1431 Market St., Philadelphia (BPD, 1890).

Wirth, William P. Gunsmith, 527 N. Christian St., Lancaster (BLD, 1875–76).

Wirtz, Charles. On 5/4/1775, Lancaster Co., council "shall have his powder, being 5 quarter casks and some pounds loose, at the rate of £15/ cwt . . . and his lead, being about 150 lbs at 45 pence/ cwt." (2 Pa. Arch. 13 at 293).

Wise, Nathaniel. Gunsmith, Union Twp., Mifflin Co. (c, 1850).

Wispert, Jacob. Gunsmith, 120 Apple St., Philadelphia (MD, 1837).

Withers, John. Gunsmith, 1771–90, Strasburg Twp., Lancaster Co. (RAG, 214).

Withers, Michael (d. 1821). Gunsmith, 1758–1805, Strasburg Twp., Lancaster Co. (SED, 60) "Michael Withers appeared in Committee and agreed to set to work as soon as he hath completed a few guns which he hath now in hand, and make muskets for this county (part of the number required from this county by the honourable House of Assembly) at the Philadelphia prices; that he will confine himself and his workmen to that work and carry on the same as expeditiously as he can, and that he will deliver in to the Commissioners and Assessors of this county or to this Committee as many muskets" (2 Pa. Arch. 13 at 299; Harris, *History of Lancaster County,* 623).

Withers, William. Gunsmith, 5 Baker Ct., Philadelphia (DeSD, 1819, 1833).

Witman, Solomon. Gunsmith, Lemon St. at Queen St., Lancaster (BLD, 1855, 1857, 1873).

Wittman, James M. Inventor, Bridesburg, Philadelphia Co. Patent 206,991 (1878) for a gunlock.

Witz, Charles. Locksmith, 1767, Douglas Twp., Berks Co.; locksmith and whitesmith, 1779–84, Reading (t).

Woerner, Christian. Gunsmith, Meadville, Crawford Co. (CCD, 1868).

Wolf, Emile. Gunsmith, Lancaster (c, 1850).

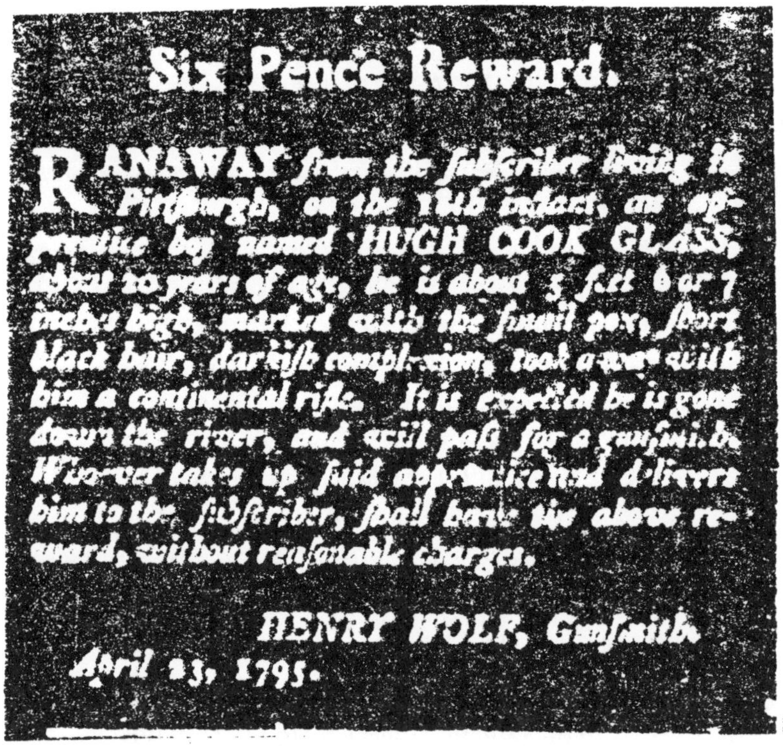

Six Pence Reward.

RANAWAY from the subscriber living in Pittsburgh, on the 18th instant, an apprentice boy named HUGH COOK GLASS, about 20 years of age, he is about 5 feet 6 or 7 inches high, marked with the small pox, short black hair, darkish complexion, took away with him a continental rifle. It is expected he is gone down the river, and will pass for a gunsmith. Whoever takes up said apprentice and delivers him to the subscriber, shall have the above reward, without reasonable charges.

HENRY WOLF, Gunsmith.

April 23, 1795.

Pittsburgh Gazette, April 23, 1795

Wolf, Henry. Gunsmith, 1791, Moon Twp., Allegheny Co.; probably active in the Pittsburgh area, 1787–1808. (t; KKR, 363–65) Wolf had several apprentices, including Hugh Glass. Wolf was evidently a tough master, for at least 3 of the apprentices ran away. Based on the rewards offered, Wolf only wanted one of them to be returned. Although no gun made by him is known, as Professor Kauffman pointed out, he was the principal gunsmith of Pittsburgh in the federal period.

Wolf, Jacob. Silversmith, 1790, Waynesboro, Franklin Co. (t; c).

Wolf, Otto. Gunsmith, 147 Third St., Pittsburgh (PAD, 1852).

Wolfe & Lane. Gunlock makers, 1850–58, Pittsburgh (t; c).

Wolffinger, Frank. Gun-barrel makers, Gouglersville, Berks Co. (c, 1850).

Wolfgang, Michael. Gunsmith, Hegins, Schuylkill Co. (BPaD, 1861).

Wolfheimer, Philip. Apprentice gunsmith, 1776, in Capt. Andrew Graff's Co., Lancaster Militia (5 Pa. Arch. 7 at 1073; 2 Pa. Arch. 13 at 347).

Wood, Jacob. Gunsmith and blacksmith, 1778, Mt. Bethel Twp., Northampton Co. Wood "hath carried on correspondence with the enemies against the State of America and also hath encouraged several to enlist in their service." His goods and land were seized and auctioned off as the property of the Commonwealth. These included: smith's anvil £0/5/0; pair of Handirons, £0/10/0; tongs & grid iron, £0/7/0 (6 Pa. Arch. 12 at 395–96).

Wood, Joseph. On 2/29/1772, Wood was apprenticed to Samuel Wheeler to learn the arts and mystery of the cutler's trade (PIDS).

Wood, Josiah. Cutler, 1775, Norrington Twp., Philadelphia Co. Made pikes and edged weapons on contract for the Committee of Safety (RAG, 215; t).

Wood, Robert. Smith and artillery artificer, 1777, Capt. George Esterly's Co., Philadelphia, "in the smith's shop" (2 Pa. Arch. 13 at 615). In 1828 there was a gunsmith, Robert Wood, in Pocono Twp., Northampton Co. (t).

Woodcock, Bancroft (d. 1811). Goldsmith, Wilmington. (*Pennsylvania Packet*, 5/30/1754) Died in Hopewell Twp., Bedford Co. (t).

Woods, Absalom. Gunsmith, 1816–18, German Twp., Fayette Co. (t).

Woodward, Edward. Gunsmith, 1855–60, Baltimore, Md; 1868, Gettysburg (FSP, 338; AMM, 274–75).

Workman, John. Gunsmith, Lower Mahoney Twp., Northumberland Co., 1838 (t; MB 2 [1968]).

Worley, David, Sr. 1768–71, Gunsmith, plantation owner, and tavern keeper, East Nottingham Twp., Chester Co. (3 Pa. Arch. 11 at 741; t); 11/7/1770, David Worley, gunsmith, and his wife Anah, conveyed land in Martic Twp., Lancaster Co. to Joseph Hains (DB M-197).

Worley, David, Jr. (b. 1797). 1822–32, Donegal Twp., Washington Co. (t); 1833–35, Buffalo Twp., Washington Co. (t); 1836–42, Donegal Twp., Washington Co. (t); 1843–52, East Finley Twp., Washington Co. (t; c). Gunsmith.

Worley, Francis. Cutler, 1779–80, York (t).

Worley, Henry. Gunsmith, Mohn's Store, Berks Co. (BPaD, 1861).

Worley, Jacob. Gunsmith and gun-barrel maker, 1811–20, on the Wyomissing Creek, Berks Co. (RAG, 105, 216).

Wright, Aaron B. (b. 1797). 1832–51, Rich Hill Twp., Greene Co. (t; c). Gunsmith.

Wright, Anthony. 10/1789, supplied gunpowder for the Philadelphia artillery which to celebrate George Washington's birthday (Hazard's *Annals* 2:60).

Wright, Elijah. Artillery artificer, 1778, at the cannon factory at Carlisle, Capt. Thomas Wylie's Co. (PIR, 2:249).

Wright, Moses (1792–1854). Born 10/5/1792 in Cumberland Valley Twp., Bedford Co. Probably an apprentice to Peter White. 1814–16, sickle smith, Cumberland Valley; 1818–54, Freeport, Ohio. Died 6/2/1854 (DH, 201). Major gunsmith.

Wurfflein, Andrew. 1835–71, 208 N. 2d St., Philadelphia. The firm began by making fine long rifles and continued to make target guns of the highest quality, but also offered cheap pistols after the Civil War. Patent 6964 of 12/18/184 for a gunlock (DeSD, 1835–36; MD, 1846, 1859, 1861; BPD, 1868–69).

Wurfflein, John. 1848–53 98 S. 3d St., Philadelphia; 1854–67; 210 S. 3d St. Patent 7334 of 4/30/1850 for a needle-fire gun (MD, 1855, 1859; BPD, 1861, 1865). Gunsmith.

Wurfflein, William. Gunsmiths, 1869–1915, 208 N. 2d St., Philadelphia. Patent of 11/20/1877 for a gallery rifle (t; c; GPD, 1892; BSED, 1880).

Wurtz, Samuel. Gunsmith, 1837, Lancaster (SED, 60; t).

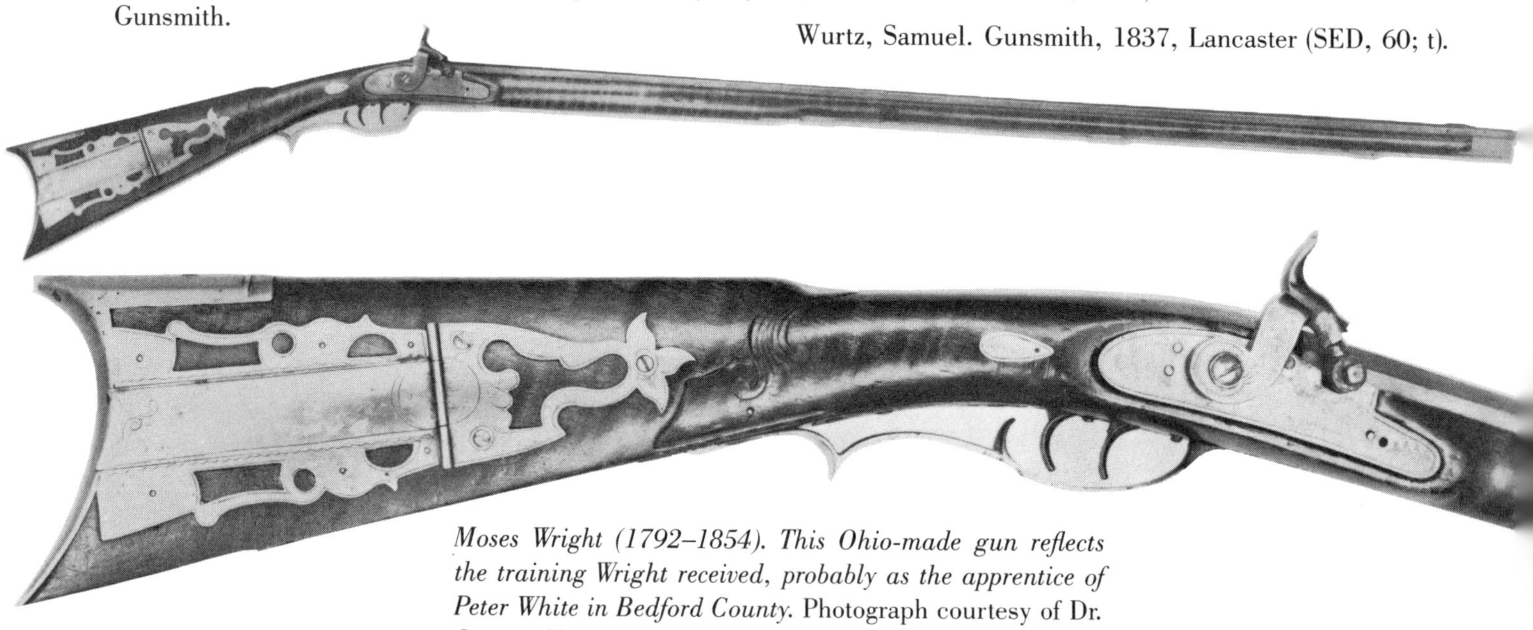

Moses Wright (1792–1854). This Ohio-made gun reflects the training Wright received, probably as the apprentice of Peter White in Bedford County. Photograph courtesy of Dr. George Shumway.

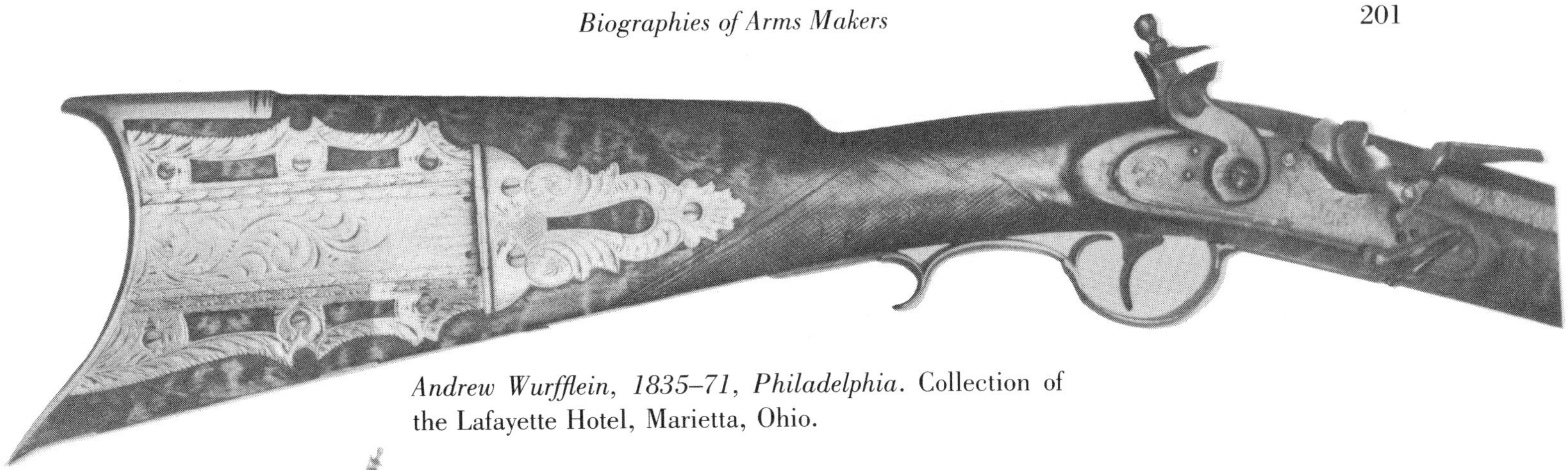

Andrew Wurfflein, 1835–71, Philadelphia. Collection of the Lafayette Hotel, Marietta, Ohio.

A. Wurfflein, Philadelphia.

Wylie, Thomas, Capt. 1777–81, commanded a company of artillery artificers at Carlisle (PIR, 2:248).

Wynkoop, Henry. On 3/26/1776, Wynkoop was authorized to buy and collect arms in Bucks Co. for the militia (1 Pa. Arch. 4 at 725); on 1/24/1776, he was involved in gunpowder manufacture (1 Pa. Arch. 4 at 702); on 11/29/1776, "pay Henry Wynkoop £44/18/8 for repairing arms" for Bucks Co. Militia (Col. Rec., vol. 11).

Wysong, Samuel (1832–98). 1870–90, Rock Island school district, Union Twp., Bedford Co. (t; c). Gunsmith.

XY

Yahner, Henry F. (b. 1880). 1900–30, Loretto, Cambria Co., gunsmith and blacksmith (t; f).

Yeager, Charles F. Gunsmith, E. Main St., Elizabethtown, Lancaster Co. (BLD, 1875–76).

Yeager, Jacob. Gunsmith, Birmingham, Warriors Mark Twp., Huntingdon Co. (c, 1850).

Yeager, John. Pike maker, 1814, Philadelphia (MPHS, 8:143).

Yearhouse, George. Artillery artificer, 1777–81, Capt. Jesse Rowe's Co. at the cannon factory, Philadelphia. Initially he was charged with desertion, but it was later concluded that his enlistment had expired and he was fully entitled to leave (2 Pa. Arch. 11 at 253; PIR, 2:245, 246).

Yocum, J. J. Gunsmith, Orangeville, Orange Twp., Columbia Co. (PD, 1882).

Yoder, Stephen (d. 1863) Cutler, Summit Twp., Somerset Co., 1799–1863 (t; c; f).

Yohn, Philip. Gunsmith, Providence Twp., Montgomery Co. (c, 1800).

Yokum (Yocum), Nicholas, & Son. Wyomissing Creek area, Berks Co. In 1810 he was a gun-barrel maker for Joseph Henry; 1820–39, skelp maker and gun-barrel smith (RAG, 105, 170).

York & Pringle. The Committee of Safety purchased many supplies from York & Pringle, including, 11/22/1775, 48 muskets, 12 cutlasses and 540 lbs of gunpowder (Col. Rec., 10:407). Philadelphia.

Yost, Casper. 1777: "to Casper Yost for repairing arms, £7/15/9" Lancaster Co. (3 Pa. Arch. 6 at 398) He was also paid £13/7/0 for additional repairs, and in 1781, "for repairing arms, £25/13/6" (3 Pa. Arch. 6 at 465). 1779–85, Lebanon Twp., Lancaster Co. (SED, 60; t). Gunsmith.

Young, Alfred. Inventor. Patent 104,394 (1870) for a gunlock, Philadelphia.

Young, D. Gunsmith, Middleburg, Snyder Co. (c, 1850); Boalsburg, Centre Co. (BPaD, 1861).

Young, D. Gunsmith, 1816–31, Franklin Twp., Snyder Co. (t).

Young, Henry. Gunsmith, 1779–86, Easton, Northampton Co. (t) In 1850 there was a Henry Young, gunsmith, age 80, born in Pa., in Union Twp., LaPorte Co., Ind. (c).

Young, J. K. Gunsmith, Easton, Northampton Co. (BRD, 1868).

Young, Jacob. Gunsmith, 1805–19, Wilkes-Barre. There was "a gunsmithery on the north side of the square with Jacob Young as the chief artisan" (t; H. and E. G. Smith, *History of Wilkes-Barre* [1849], 4); 1835, Easton, Northampton Co. (t).

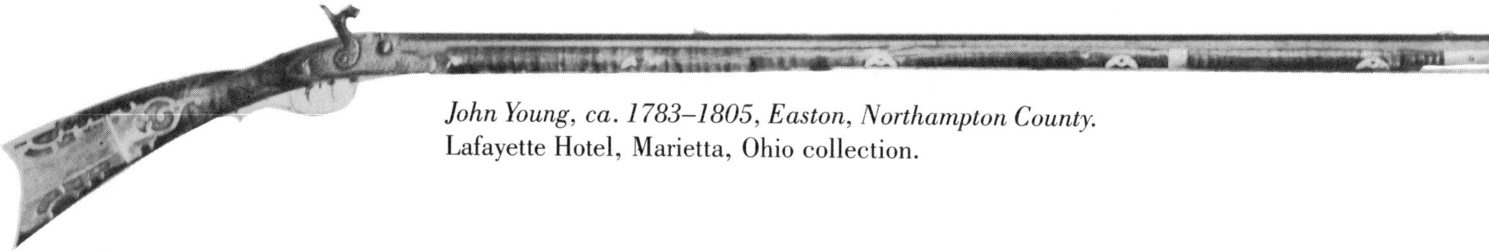

John Young, ca. 1783–1805, Easton, Northampton County.
Lafayette Hotel, Marietta, Ohio collection.

Young, James. Artillery artificer, 1778, Capt. Thomas Wylie's Co. at the cannon factory, Carlisle (PIR, 2:249).

Young, John. On 9/15/1779, John Young, gunsmith, was declared a traitor according to the laws of Pa., and his estate and property were seized and auctioned off. (6 Pa. Arch. 13 at 226) He was then in Forks Twp., Northampton Co. In 1783–1805 he was a gunsmith in Easton. (t) He was an associate of Johnston Smith, making 1,000 muskets, completed 2/1776. On 4/17/1777 he reported having completed 700 firelocks and that he had 130 more on hand in the process of manufacture. On 4/1777 he and Adam Foulkes contracted to make 130 more muskets (GAS-2, 7).

Young, John, Jr. Gunsmith, 1780–1804, Easton, Northampton Co. (t).

Young, Michael. Gunsmith, 1828, Bushkill Twp., Northampton Co. (t).

Young, Peter. Gunsmith, 371 N. 2d St., Philadelphia (t); 1807–35, Easton, Northampton Co. (t).

Young, T. P. Sold guns, ammunition, and the like, Easton (PD, 1882).

Young, William, & Bros. Gunsmiths, 131 Northampton St., Easton (BPaD, 1861).

Young, William. Gunsmith, 1804–35, Easton. (t).

Yous, Joshua (1827–1905). 1850–62, Antrim Twp., Franklin Co. (t); 1865–92, Greencastle, Franklin Co. (WSB, 28–29). Gunsmith and merchant.

Z

Zahm, Matthias. Gunsmith, 1813–19, Lancaster (SED, 60).

Zahringer, Eugene & Egan. Gunsmiths, Corner of Mulberry Alley and O'Hara St., Pittsburgh (PAD, 1860–61).

Zantzinger, Adam. Saltpeter and gunpowder makers and distributors, 1776, DBA Zantzinger & Kuhn, Lancaster. (1 Pa. Arch. 4 at 742) "Inquire into a Report that a Mr Smith of Va. is purchasing the arms of Associators, to carry them out of the Province, and that Mr Zantsinger is buying up all the arms he can procure and selling them out again, by

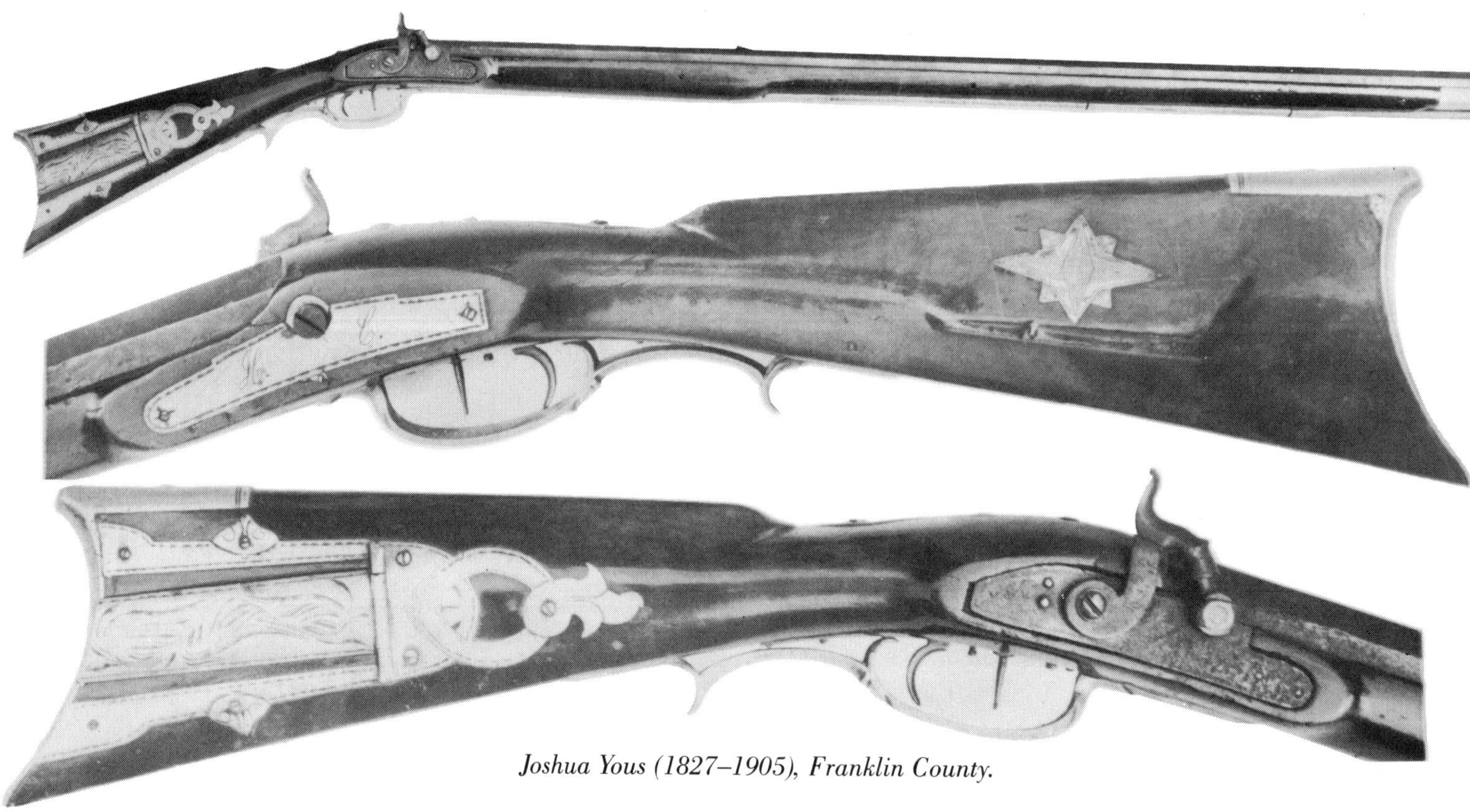

Joshua Yous (1827–1905), Franklin County.

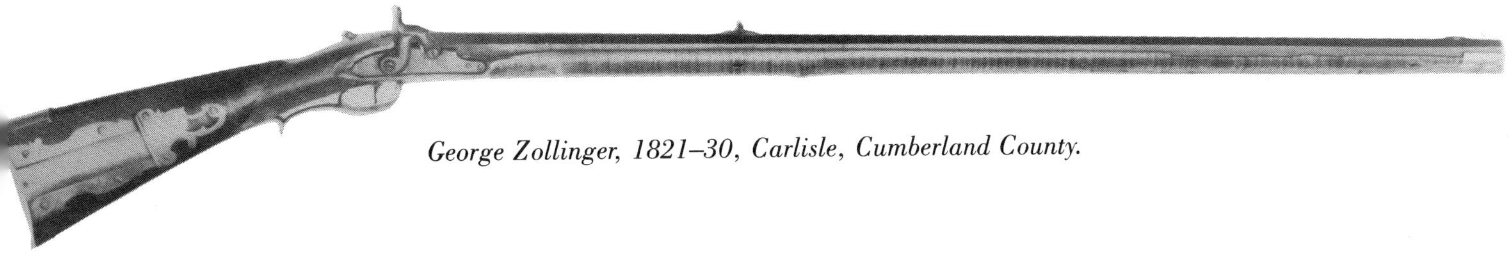

George Zollinger, 1821–30, Carlisle, Cumberland County.

which the price of Fire arms is greatly enhanced" (2 Pa. Arch. 1 at 473).

Zeigler, Amos. Gunsmith, Fraley's Ct., Philadelphia (MD, 1837).

Ziegler, George. Locksmith, 1780, Heidelberg Twp., Berks Co. (t).

Zimmerman, John. Gunsmith, Sunbury, Northumberland Co. (c, 1870).

Zimmerman, Samuel (1803–89). Gunsmith and tinner, 1827–89, Hempfield Twp., Westmoreland Co. (t; c; f).

Zink, Albert (b. 1837). Gunsmith, Port Carbon, Schuylkill Co. (c, 1850, 1860).

Zollinger, George. Gunsmith, 1811, Harrisburg; 1821–30, Carlisle, Cumberland Co. (WSB, 57).

Zorger, C. York Co., ca. 1800 (JKJ, 285). Gunsmith.

Zorger, Frederick (1734–1815). Major gunsmith, lived all his life in York (JKJ, 285–91).

Zorger, George. Gunsmith, Ca. 1800–1810, York (JKJ, 285–86).

Zuin & Clark. Gunsmiths, 12th Ward, Philadelphia. In 1860 the firm made 1,000 guns, but little additional information is available (IC).

Appendix 1
The Pennsylvania Gun Manufactory

"In February 1776, the Committee of Safety deemed it expedient to establish a Provincial gun manufactory, and requested a conference with Mr Benjamin Rittenhouse of Norrington, whom, they thought, a suitable person to superintend it. In March, commissioners were appointed to erect, superintend and conduct the manufactory, and contract for the manufacture of arms. The factory . . . was established on Cherry Street, with Mr Peter DeHaven as principal gunsmith. In April 1778 the gun manufactory, with Mr DeHaven as contracting superintendent, was in operation in Hummelstown, eight miles from Harrisburg. In July, after the tragic events at Wyoming, he recommended its removal to French's Creek or Philadelphia. The Council in 1776 fixed the prices to be paid to gunsmiths for good gun barrels, delivered at the lock factory at 24 shillings apiece. Brass mountings were then made by Lewis Prahl. The Committee also advanced £300 to Lawrence Birnie to enable him to erect an air furnace and mills for the business of file cutting in connection with the gun lock manufactory. Contracts were also made in York and Lancaster for a quantity of arms for the State. The price of a musket and a bayonet and steel ramrod, made according to pattern was £4/5/0, Pennslyvania currency. But it was found difficult to make contracts at less than £4/10/0 or £4/15/0, as materials were scarce, and workmen were unwilling to quit the manufacture of rifles for which the demand was great.

"Muskets were also made at Carlisle and in Bedford and other counties. The latter had but one regular gunsmith [Jacob Saylor, ed.]. There were two boring mills for gun barrels in the vicinity of Lancaster in 1786, and the borough contained 7 gunsmiths, 7 nail makers, 2 brass founders, 3 coppersmiths, 5 silversmiths, 4 tinners, 6 clock and watch makers and 25 black and white smiths.

"John Kerlin [of Philadelphia, ed.] contracted to make muskets and bayonets at 85 shillings each. . . .

"In recommending to the several assemblies in November 1775 the manufacture of firearms and bayonets, Congress directed that each firelock will be made 'with a good bridle lock, ¾ inch bore, and of good substance at the breech, the barrel to be 3′ 8″ in length, the bayonet to be 18″ in the blade, with a steel ramrod . . . [which was to be] trumpet-mouthed, and the price to be fixed by the Assembly or Convention of each Colony; and that until a sufficient quantity of good arms be manufactured they import as many as are wanted by all men in their Province.'" (J. Leander Bishop, *A History of American Manufactures from 1608–1860* [Philadelphia: Young, 1864], 1:572f.)

Appendix 2
The Federal Procession of 4 July 1788

"On Friday, the 4th day of July, 1788, the citizens of Philadelphia celebrated the Declaration of Independence made by the 13 United States of America on the 4th of July 1776, and the establishment of the Constitution or frame of government proposed by the late general convention, and now solemnly adopted and ratified by ten of those states.

"The rising sun was saluted with a full peal from Christ Church steeple, and a discharge of cannon from the ship Rising Sun. . . . Ten vessels in honor of the ten states of the Union, were dressed and arranged through the whole length of the harbor, each bearing a large white flag at the masthead inscribed with the name of the states. . . . Nine gentlemen, distinguished by white plumes in their hats . . . were superintendents of the parade. . . . The different companies of military, trades, and professions had previously met at different places in the city. . . . About half after nine o'clock the grand procession began to move; of which the following is as correct in detail as could be procured. . . .

"I. Twelve axemen dressed in white frocks. . . . .

"VIII. Col. John Shee on horseback, carrying a flag, blue field, with a laurel and olive wreath over the words,
'Washington, the Friend of His Country'. . . .

"XXXIX. Clock and Watch Makers. The Company's arms painted neatly on a silk flag—Motto, 'time rules all things.' . . . .

"XLVI. Plane Makers. Mr William Martin in front, bearing the standard white field a smoothing plane on top, device, a pair of spring dividers three planes, a brace, a square, and gauge. . . . Motto 'Truth.' . . . .

"XLVIII. Black-smiths, White-smiths and Nailers. A machine drawn by 9 horses, representing the blacksmiths, whitesmiths and nailers' manufactory, being a frame ten by fifteen feet, and nine feet high, with a real chimney extending three feet above the roof, and furnished for use. In front of the building, three master black smiths, Nathaniel Brown, Nicholas Hess and William Perkins, supporting the standard, elegantly ornamented with the smiths' arms. Motto 'by hammer in hand, do all arts stand.' . . . .

"LXI. Gunsmiths. A stage erected upon a four wheel carriage, drawn by four horses, being in length fourteen feet, and in breadth eight feet, with a motto in large letters on either side 'armoury' with a number of hands thereon at work, employed in different branches of the trade, conducted by two senior masters, John Nicholson and Joseph Perkins; Abraham Morrow bearing a standard at the head of the company, in the rear of the carriage, the standard decorated with sundry devices representing the arms belonging to the trade. The standard, a large white silk flag, with cross guns in the middle, at the top of the cross guns, the cap of liberty, with the letters 'C P (city proof)' underneath the guns, the cross pistols, with the letter 'V (viewed)' at the end nearest the staff a powder cask at the opposite end, the representation of three balls. The uniform of the company, green baize aprons with green strings. . . . .

"LXII. Goldsmiths, Silversmiths and Jewellers. William Ball, esq., with an urn. Standard bearers, Joseph Gee and John Germon, carrying a silk flag with the silver smiths' arms on one side of it. Motto: 'justita virtutm regina' [justice is the queen of virtues]. And on the reverse the genius of America holding in her hand a silver urn with the following motto 'the purity, brighness and solidity of this metal are emblematical of that liberty which we expect from the new Constitution.' . . . .

(after the parade was over, a banquet was held, and the following were among the toasts offered)

"BLACKSMITHS. May the 13 states be welded into one united empire, by the hammer of conciliation on the anvil of peace; and may the man who attempts to blow the coals of discord be burned by the sparks.

"GUNSMITHS. When the instruments of war are requisite to defend our country's rights, or resent her wrongs, may coolness take the sight and courage draw the trigger" (Samuel Hazard (ed.) *The Register of Pennsylvania,* 5 July 1828, 415–25; 6 September 1828, 136–37).

Appendix 3
The Wyomissing Creek Barrel Makers

"To the south from Reading is Wyomissing Creek, formerly known popularly as the 'Schmutz Deich' [dirty dam] because there were at one time nine dams in a distance of three miles along this creek, each operating a rifle barrel shop. The first of these began work in 1775 and the last one in operation went out in 1906. . . . James F. Schnader and J. T. Huyett were identified with these barrel making shops.

"The Wyomissing Creek has its source in the southern end of Cumru Township; it flows northward and empties into the canal along the Schuylkill River, opposite Reading. There are many mills and factories along this stream, but the rifle barrel shops lay between Mohn's Store and Gouglersville. Among the operatives were Frank Schnader and his son Nathaniel, who made finished barrels and sold them to dealers, who, in turn, sold them to gunsmiths. Daniel Shoeman was a welder; Bennieval Rufner, a grinder; Hiram Hilde welded and finished barrels. The following also made and finished barrels: John Ruth, John Gougler, Isaac Krick, Henry Walley, Cyrus Hornberger, William Pennepacker, [and] Henry Deeds. . . . Frank Wolffinger of Mohntown made repairs. Frank Schnader made from 2500 to 3000 barrels a year, while Frank Miller made 4000.

"Generally, four men would make a dozen barrels a day, including straightening, welding [and] grinding. . . . The days were long, from 12 to 14 hours, and the wages were piece work. The welder received 20¢ a barrel, the straightener 25¢, the grinder (usually a youth) 10¢ a barrel, and a laborer 75¢ to $1.00 a day. Since each shop had its own water power by which the heavy work was accomplished, such as forging under a trip hammer, blast for the fire, etc., the labor was not particuliarly onerous.

"Each shop had its own dam and water wheels, the latter having a diameter of fourteen feet for a hammer forge, and another six feet in diameter for providing air blast. The water wheels were made of wood. They were four feet wide and of the overshot type. Grinding was done by water power, the grindstone being five feet in diameter and sixteen inches wide, running fifty revolutions per minute.

"The main shop usually had a forty foot square floor plan, and there was another shop twenty by sixteen feet. The shop of Frank Miller had a floor plan of thirty by sixty feet, and required three water wheels. The best barrel maker was said to have been Pennepacker. Some barrel makers stamped their names on the underside of the barrel and others had no mark. . . .

"The finished barrel, ready for the gunsmith, usually sold from $2.50 to $3.00 each, for the sport gun size. The ends were squared and ground. . . . The barrel was made of charcoal iron, at first locally made, but subsequently imported from Norway. . . . [This steel] was purchased from Reading hardware stores in strips fifteen feet long, three inches wide, and three-eighths of an inch thick. Barrels for the heavy target guns (fifteen to fifty pounds in weight) were forty-eight inches long.

"The iron cost two to four cents a pound, and the coal

for the forge cost four cents a bushel. This bituminous coal was used to heat the rifle barrel while being forged and welded, with the aid of yellow flux; for the welding of iron or steel, borax was used as a flux. . . . The width of the skelp bar was regulated by a gauge, depending on the diameter desired. . . ." (Commonwealth of Pennsylvania, Department of Internal Affairs, Booklet 12, Volume 15, November 1947.)

"Gun barrel factories along Bennett's Run [in York County were run by] Samuel Grove, George Blymire, John Foster [and] in 1760, Aaron Frazier. . . . The flint lock gun only was made until 1830 . . . after that date the present invention came in vogue. Dr. [Webster] Lewis and Isaac Lloyd first made the new patent here. . . . The first important article of manufacture in this vicinity was the flint lock gun, for the making of which the town [of York] became famous. The business was begun here as early as 1760, and many were made here during the Revolution for the army, by the order of the Committee of Safety of York County" (John W. Gibson, ed., *History of York County* [Chicago: Battey, 1886] 632, 634).

Appendix 4
Memorial of Sundry Gun Manufacturers of the Borough of Lancaster, in the State of Pennsylvania

The memorial was addressed "to the Honorable the Senate and House of Representatives in Congress of the United States."

"The Memorial of the Subscribers, Gun-Manufacturers in the borough of Lancaster, in the commonwealth of Pennsylvania,—

"RESPECTFULLY SHEWETH—

"THAT your memorialists have seen with deep and affecting concern, a resolution to exempt from impost duties, arms manufactured in foreign countries, pass in the house of representatives of the United States. To extend the hand of power and crush the manufatures of our common country in their most infant state, your memorialists would, in time more pressing than the present, consider impolitic: But when, in the full enjoyment of order and peace, the various resources of our country excite improvement and accumulate wealth, the manufacture of arms by ourselves and for our own use, instead of receiving the fostering protection of our government, is defeated after a successful experiment of its utility—your memorialists can behold no point of improvement upon which they can rest with stability,—no manufacture that promises to be permanent,—no experiment that will flatter them with patronage and encouragement.

"Encouraged by the government of the state in which your memorialists live, they have at very great expense established manufactures of arms, and in conjunction with others in the state, have nearly completed twenty thousand stand for the use of the commonwealth of Pennsylvania. Allured by this encouragement they have encreased their establishments, taken in and instructed apprentices, and excited by their undertakings a competition—a spirit of enterprize among their fellow citizens in this manufacture, so essential to national safety, national independence, and national reputation. If the independence and liberty of a nation depend upon the correspondence of its resources to its wants; then is there no want so imminent now which should be more necessitously supplied from the resources of our own country, than the manufacturing of arms. But by giving a loose to the facility of importing arms, the government of the United States will crush this manufacture in its infant establishment which your memorialists hoped to see cherished.—if not for the maintenance of the artist, at least for the safety of the country. Arms may be imported, but who will keep them in repair, after the dispersion of our journeymen and apprentices, who must engage in other pursuits, when the one they have been trained up in will cease to afford them subsistence. Will the day of importation last forever? or will there never be a day when the manufacture of arms will be thought useful and merit the protection of government. A given stock of arms will be exhaustible in war, should it happen, and from whence will it be supplied, if arms are to be procured in another country, when the very cause of the consumption will preclude the possibility of their importation; but the encouragement of such a domestic manufacture will establish an inexhaustible stock of arms in time of need,—artists will be numerous and manufactures convenient; the means will be easy, and their attainment certain. At considerable expense have your memorialists undertaken and with some difficulty progressed in the establishment of this manufacture. Mills for the making of gun-barrels have been erected—gun locks and every other article in a gun have been made in the best manner, and of the most substantial kind. The workmen, the execution, the machinery, and the demand have all progressed apace with each other.—A few years more protection from government, and this manufacture in this country, we believe, will be too firmly established to be destroyed by the importation of foreign arms—break it up by the resolution you have adopted and it will not revive for ages; for no security can afterwards be given to its re-establishment, in which a prudentman will confide. But your memorialists do not consider the manufacture of arms exclusively affected by this resolution which passed your honorable house; they humbly conceive the principle of that resolution strikes deep at all our domestic manufactures; for with this example before them, few manufactures will be disposed to place reliance on the improvement and permanency of any manufacture. Your memorialists beg leave further to represent, that the manufacture of arms in this state is not merely circumscribed within the limits of the county of Lancaster; it is diffused over the state, equally progressing in extent and improvement throughout,—and they confidently assert that twenty thousand stand of arms can be annually manufactured in this state.

"Your memorialists beg leave to call the attention of your honorable body to this extensive manufacture in this state, and they would, with deference, ask if it be prudent or politic to reduce it by a single decision,—to destroy it with one blow. If our sister states to the southward have not established, nor can now practically establish, such extensive manufactures of arms, they can here be supplied on the same terms we have supplied our own government: and though they may purchase arms cheaper in Europe than they can from the American manufacturer, yet as they will suffer no grievance when supplied on the same terms with other states, they should obtain no preference by being supplied by importation, and that, too, at the expense of destroying a useful domestic manufacture.

"If these considerations can command the attention of the honorable the congress of the United States, we trust and pray that the impost duty on arms will not be taken off,—and that the encouragement of manufacturing them among ourselves will be considered not only expedient, but necessary.

"And your memorialists as in duty bound will ever pray.

"*January 28th, 1803*

JACOB DICKERT,
PETER GONTER,
ATRAM HENRY,
JOHN GRAEFF,
HENRY DEHULF,
JACOB & JOHN HAEFFER,
BENJAMIN HUTZ,
ABRAHAM PIEPER,
CHRISTOPHER GUMPP."

Appendix 5
Gunsmith's Chemicals

Dr. Chase's *Recipes* was a standard reference in the United States in the second half of the nineteenth century, and well into the twentieth. He gave the following recommendations in his Gunsmith's Department.

"GUN-BARRELS—Browning Process—Spirits of nitre 1 lb.; alcohol 1 lb.; corrosive sublimate 1 oz.; mix in a bottle and keep corked for use.

"Directions.—Plug both ends of the barrel, and let the plugs stick out three or four inches to handle by, and also to prevent the fluid from entering the barrel, causing it to rust; polish the barrel perfectly; then rub it well with quick-lime by means of a cloth, which removes oil or grease; now apply the browning fluid with a clean white cloth, apply one coat and set in a warm, dark place, until a red rust is formed over the whole surface, which will require, in warm weather, from ten to twelve hours, and in cold weather, from fifteen to twenty hours, or until the rust becomes red; then card it down with a gun-maker's card and rub off with a clean cloth; repeat the process until the color suits, as each coat gives a darker shade.

"2. Quicker and less Laborious Process.—While in Evansville, Ind., I sold one of my books to C. Keller, a man who carries on gunsmithing extensively. He gave me the following, which he was using, and says it makes a dark brown, with but little labor compared with the first.

"Soft water 1 qt., and dissolve in it blue vitriol 2 ozs.; corrosive sublimate 1 oz.; and add 1 oz. of spirits of nitre. Have the barrel bright and put on one coat of the mixture; and in 1 hour after, put on another, and let the barrel stand 12 hours; then oil it and rub it with a cloth, of course having the ends of the barrel tightly plugged, as in the first case.

"But Mr. Sutherland, the gunsmith of this city, says the brown from this recipe will soon rub off; none being permanent unless carded down properly, as directed with the first recipe, that mixture being also superior.

"8. Browning for Twist Barrels.—Take spirits of nitre 1 oz.; tincture of steel ¾ oz.; (if the tincture of steel cannot be obtained, the unmedicated tincture of iron may be used, but it is not so good) black brimstone ¼ oz; blue vitriol ½ oz.; corrosive sublimate ¼ oz.; nitric acid 1 dr. or 60 drops; copperas ¼ oz.; mix with 1½ pts. of rain water, keep corked, also, as the other, and the process of applying is also the same.

"You will understand this is not to make an *imitation* of twist barrels, but to be used upon the real twist barrels, which brings out the twist so as to show; but if you use the first upon the real twist barrels, it will make the whole surface brown like the common barrel.

*"CASE-HARDENING—For Lock-work.—Take old boots and shoes and lay them on a fire, and burn them until charred; now put them into a clean kettle and pulverize them coarsely, while hot; be careful not to get any wood coals mixed with them.

"Directions.—Take the pulverized leather and place in a sheet-iron box, placing the articles to be hardened in the centre of the box, or amongst the pulverized leather, and cover with a sheet-iron cover; or make the box so as to shut up; now blow up a fire of *very dry* charcoal; the coarser the charcoal the better; then open the fire and place the closed box in the centre, cover it up and let stand from forty to sixty minutes, not blowing; but if the coals burn off and leave the box exposed, you will put on more; at the expiration of the time, take the box and pour its contents into clean, moderately cool or cold water—never use warm water; these articles will now be found very hard, and will easily break; so you will draw the temper to suit.

"TINNING—Superior to the Old Process.—Take first, the same as the old way; that is, muriatic acid 1 pt., and as much pure block or sheet zinc as it will cut, in an open dish, a bowl, or something of that character, as much heat is set free and bottles are often broken by it; now take sal-ammoniac 4 ozs.; pulverize it and add to the other, and boil 10 minutes in a copper kettle—bear in mind, only copper is to be used to boil in.

"You will find this will cause the solder to flow right along without difficulty. Keep corked tight when not in use.

"GALVANIZING—Without a Battery.—Dissolve cyanuret of potassium 1 oz., in pure rain or snow water 1 pt., to which add a 1 dr. bottle of the chloride of gold, and it is ready to use. Scour the article to be plated, from all dirt and grease, with whiting, chalk, or rotten stone, pulverized, and put in alcohol, using a good brush—or the "Polishing Compound," No. 3; if there are cracks, it may be necessary to put the article in a solution of caustic potash—at all events, every particle of grease and dirt must be removed; then suspend the article to be plated in the cyanuret of gold solution, with a small strip of zinc cut about the width of a common knitting-needle, hooking the top over a stick which will reach across the top of the jar holding the solution.

"Every five to ten minutes, the article should be taken out and brushed over with the scouring preparation; or on smooth surfaces it may be rinsed off and wiped with a piece of cotton cloth, and return until the coating is sufficiently heavy to suit.

"When the plating fluid is not in use, bottle it, keeping it corked, and it is always ready for use, bearing in mind that it is as poison as arsenic, and must be put high, out of the way of children, and labeled—*Poison,* although you will have no fears in using it; yet accidents might arise, if its nature were not known. The zinc strip, as far as it reaches into the fluid, will need to be rubbed occasionally, until it is bright.

"VARNISH AND POLISH FOR STOCKS—German.—Gum shellac 10-ozs.; gum sandarach 1 oz.; Venice turpentine 1 drachm; alcohol 95 to 98 proof 1 gal.; shake the jug occasionally for a day or two, and it is ready for use.

"After using a few coats of this, you can have a German-polish, by simply leaving out 8 ozs. of the shellac; and a coat or two of the polish makes an improvement on the varnish, and does not require the rubbing, that it would if the full amount of shellac was used, in the last coat or two. It is recommended also to put upon cuts, sores, &c. burns excepted."

Appendix 6
Names for Gunsmith's Chemicals

Common Names	Proper Chemical Name
Blue Stone or Blue Vitroil	Copper sulphate
Butter of Antimony	Antimony trichloride
Butter of Tin	Anhydrous stannic
Butter of Zinc (killed Spirits	Chloride
Copperas (Iron Sulfate)	Zinc chloride
Green Vitriol	Ferrous sulfate
Corrosive Sublimate	Mercuric chloride
Slats of Hartshorn	Ammonium carbonate
Spirits of Hartshorn	Ammonia
Salammoniac	Ammonium chloride
Sugar of lead	Lead acetate
Spirits of Nitre (Sweet)	Ethyl nitrate
Tincture of Steel	Tincture of ferric chloride or iron chloride in pure grain alcohol
Brimstone	Sulphur
Acid of Sugar	Oxalic acid
Aqua fortis	Nitric acid
Blue copperas	Copper sulfate
Calomel	Mercurous chloride
Chili niter	Sodium nitrate
Ferro prussiate	Potassium ferricyanide
Lunar caustic	Silver nitrate
Muriatic acid	Hydrochloric acid
Niter	Potassium nitrate
Red prussiate of potash	Potassium ferricyanide
Spirit of salt	Hydrochloric acid
Vitriol	Sulfuric acid
Yellow prusiate of potash	Potassium ferrocyanide
Water Glass	Sodium silicate
Alkanet root	Alkanna tinctoria
Asphaltum	Mixture of natural bitumins & asphalt
Fustic	Pulverized dried sumac leaves
Logwood	Haematoxylin
Verdigris	Cupric acetate
Madder	Acid alizarine blue
Pearlash	Potassium carbonate anhydrous
Soot	Carbon
Common soda	Bicarbonate
Dragon's blood	Resin extracted from the fruit of the malayan rattan palm
Purple crystals	Potassium permanganate

Appendix 7
George Schalk's Gauge Measurements

These are the micrometer measurements used by George Schalk of Pottsville (1824–92), one of America's best riflemen and the maker of fine precision rifles. His gauge measurements are substantially correct.

No. of balls per lb.	Measurement (in.)	No. of balls per lb.	Measurement (in.)
5	1.15	38	.491
6	.954	40	.485
7	.900	44	.474
8	.861	48	.463
9	.823	52	.453
10	.790	56	.442
11	.760	64	.416
12	.748	70	.400
13	.727	80	.388
14	.708	90	.383
15	.695	100	.364
16	.682	110	.350

17	.665	120	.340
18	.647	130	.332
19	.637	140	.324
20	.625	150	.318
21	.618	160	.310
22	.609	170	.305
23	.601	180	.298
24	.588	190	.292
25	.585	200	.290
26	.575	210	.288
27	.572	220	.284
28	.559	230	.282
29	.549	240	.280
30	.533	250	.278
31	.531	260	.274
32	.530	270	.2715
34	.519	280	.269
36	.506		

Bibliography

Africa, J. S. *History of Huntingdon and Blair Counties*. Chicago: Waterman, 1883.

Albright, R. W. *Two Centuries of Reading*. Reading, Pa.: Historical Society of Berks County, 1948.

Anderson. W. H. "The American Flintlock Pistol." *Newsletter of the Kentucky Rifle Association* 7, no. 3.

Arnold, R. E. "U.S. Model 1841 Rifle: The Famous Mississippi," *Gun Report*, September 1973, 14–21.

Bailey, Dewitt, and Nie, D. A. *English Gunsmiths: The Birmingham and Provincial Gun Trade in the 18th and 19th Centuries*. New York: Arco Publishing Co., 1978.

Bailey, Kenneth P., ed. *Ohio Company Papers*. Ann Arbor, Mich.: Edwards Brothers, 1947.

Balderson, M. R. *Passengers and Ships prior to 1684*. Baltimore: Genealogical Publishing Co., 1970.

Barnhart, Harold A. "A History of the Barnhart Rifle." *Newsletter of the Association of Ohio Long Rifle Collections* 2.

Bean, Theodore Weber. *History of Mongtomgery County*. Philadelphia: Everts & Peck, 1884.

Beck, Herbert L. "Henry Leman, Riflemaker." *Papers of the Lancaster County Historical Society* 40, no. 3.

———. "Martin Meylin." *Papers of the Lancaster County Historical Society* 53, no. 2.

———. "William Henry." *Papers of the Lancaster County Historical Society* 54, no. 4.

Bender, Clarence. *Long Rifle Makers of Fulton County*, 1981.

Berky, A. S. "Elias Brey." *Newsletter of the Kentucky Rifle Association* 2, no. 4.

Bining, A. C. *Pennsylvania Iron Manufacture in the 18th Century*. Vol. 4. Harrisburg: Pennsylvania Historical Commission, 1938.

Bishop, J. L. *History of American Manufactures from 1608 to 1860*. 2 vols. Philadelphia: Young, 1864.

Bivins, John. *Long Rifles of North Carolina*. York, Pa.: Shumway, 1968.

Bowers, William S. *Gunsmiths of Pen-Mar-Va*. Mercersburg, Pa: Irwinton, 1979.

Brandt, F. B. and Gummere, H. V., *Byways and Boulevards in and about Historic Philadelphia*. Philadelphia: Corn Exchange National Bank, 1925.

"Breitenherd," *Bulletin of the Pennsylvania German Folklore Society* 12, no. 202 (1947).

Brooks, J. S. "Wurfflein Gallery Rifle." *Gun Report*, November 1958.

Brown, M. L. *Firearms in Colonial America*. New York: Scribner's, 1940.

Brown, R. C. *History of Crawford County*. Chicago: Warner, Beers, and Co., 1885.

Browne, W. H. *Browne's Archives of Maryland* (1883–1912).

Burnside, Graham. "Anton Spellerberg, Philadelphia." *Gun Report*, December 1959.

Carey, A. M. *American Firearms Makers*. New York: Crowell Publishing, 1953.

Chandler, Roy, et al. *A History of Early Perry County Guns and Gunsmiths*. Duncannon, Pa.: Chandler, 1969.

Chandler, Roy. *Kentucky Rifle Patchboxes and Barrel Markings*. Duncannon, Pa: Chandler, 1969.

Cherepy, Louis. "Nicholas Haw, Monroe County Gunsmith." *Newsletter of the Kentucky Rifle Association* 5, no. 3.

Clegg, M. "French Creek in American History." *Newsletter of the Kentucky Rifle Association* 7, no. 1.

Clippinger, Amanda. "Joseph Clippinger. *Newsletter of the Association of Ohio Long Rifle Collectors* 7, nos. 1, 2.

Conduit, V. W. *History of Easton, 1739–1885*. Easton, Pa.: Centennial Committee, 1888.

Coxe, Tench. *A Statement of the Arts and Manufactures of the*

United States. Philadelphia: A. Cornman, Jr., 1814.

Daskam, Faith Stoek. *Fondersmith-Vondersmith Families of Lancaster and Philadelphia Counties, Pennsylvania*. Columbus, Ohio: privately printed, 1944.

Davis, W. W. H. *History of Bucks County.* Doylestown, Pa.: Democrat Book and Job Office Print, 1876.

Dillin, J. G. W. *The Kentucky Rifle*. 1924. 4th ed. Washington D.C.: National Rifle Association, 1968.

Dubbs, J. H. "Jacob Dubbs of Milford." *In Genealogies of Pennsylvania Families*. 1981, 272–283.

Dunham, L. "Government Kentuckies." *Muzzle Blasts*, May 1943, 8.

Dyker, S. E. "The Baker Family Gunsmiths in Lancaster County, 1717–1754." *Newsletter of the Kentucky Rifle Association* 4, no. 1.

———. "Cornelius Atherton." *American Arms Collector*, April 1958.

———. "The Gumpf Family of Gunsmiths of Lancaster Boro." *Newsletter of the Kentucky Rifle Association* 11, no. 2.

———. "Gunsmiths of Northampton and Lehigh Counties, 1734–1847." *Newsletter of the Kentucky Rifle Association* 9, no. 3.

———. "Henry Mauger." *Newsletter of the Kentucky Rifle Association* 1, no. 4; 9, no. 2.

———. "The Kuntz Brothers." *Newsletter of the Kentucky Rifle Association* 3, no. 4.

———. "Lancaster Gunsmiths and Their Apprentices." *Newsletter of the Kentucky Rifle Association* 11, no. 3.

———. "Moll Family of Gunsmiths." *American Society of Arms Collectors* 25.

———. *The Pennsylvania Rifle*. Lititz, Pa.: Lancaster Bicentennial Committee, 1974.

———. "Peter Smith." *Newsletter of the Kentucky Rifle Association* 3, no. 1.

———. "The Rupp Family." *Newsletter of the Kentucky Rifle Association* 4, no. 2.

———. "William Antes." *American Arms Collector Journal*, October 1957.

Eberhard, D. and Wilson, R. L. "Derringers, Durringers and Diringers." *Arms Gazette*, May 1978.

Ellis, F. and Hungerford, A. N., eds. *History of the Susquehanna and Juniata River Valleys Embraced in the Counties of Mifflin, Juniata, Perry, Union and Snyder. 2 vols. Philadelphia: Everts, Peck and Richards, 1886*.

Farrow, E. S. American Small Arms. New York: The Bradford Co., 1904.

Fellner, Felix. *Loyalhanna Valley in History.* Latrobe, Pa.: The Latrobe Bulletin, 1954.

Feree, G. B. *Ferree Family Gunsmiths*. Indianapolis: privately published, 1967.

Flemish, Joseph, Jr. and Wendling, Dalton. "The Hellertown Molls." *Newsletter of the Kentucky Rifle Association* 5, no. 2.

Foreman, Harry E. *History of the Little Cove*. Mercersburg, Pa.: privately published, 1967.

Fuller, C. E. *The Breech Loader in Service*. Topeka, Kans.: The Arms Reference Club of America, 1933.

Gabel, Ronald. "The Henry Family." *Newsletter of the Kentucky Rifle Association* 5, no. 2.

———. "Hess Family." *Newsletter of the Kentucky Rifle Association* 5, no. 1.

———. "Hess Family Supplement." *Newsletter of the Kentucky Rifle Association* 10, no. 3.

———. "Lehigh Kentuckies." *American Society of Arms Collectors* 35.

———. "Peter Neihardt." *Newsletter of the Kentucky Rifle Association* 2, no. 1.

Gaines, A. J. "Flintlock Swivel-breech Kentucky Rifle." *Gun Report*, September 1972, 60–61. The article deals with the work of Jon DeMuth.

Gardner, R. A. *American Arms and Arms Makers*. Columbus, Ohio: privately printed, 1938.

———. *Five Centuries of Gunsmiths, Swordsmiths and Armourers*. Columbus, Ohio: privately printed, 1948.

———. *Small Arms Makers*. New York: Crown, 1963.

Gibson, John. *Gibson's History of York County.* F. A. Tattey, 1886.

Gill, H. R., Jr. *The Gunsmith in Colonial Virginia*. Richmond: University of Virginia Press, 1974.

Gleichman, R. W. "The Hunters of Kentucky." *Newsletter of the Kentucky Rifle Association* 7, no. 1.

Gluckman, A. and Satterlee, L. D. *American Gun Makers*. 1940.

Goodling, E. C. "Peter Goodling." *Newsletter of the Kentucky Rifle Association* 3, no. 3.

———. "Some Notes on the Pannebecker (Pennypacker) Family of Gunsmiths." *Newsletter of the Kentucky Rifle Association* 3, no. 1.

Goodwin, V. K. "The Hawkens Family." *Newsletter of the Kentucky Rifle Association* 7, no. 3.

"Gun Makers along Wyomissing Creek." *Hobbies*, May 1937.

Guthman, William. "Arms of the Revolutionary Period." *American Society of Arms Collectors*.35.

Halsey, A., Jr. "The Air Guns of Lewis and Clark." *American Rifleman*, August 1984.

Hamilton, D. F. "Long Arms of the Colonists." *Gun Report*, December 1970, 8–13.

Hanger, George. "A Redcoat's View of the Awesome Kentucky Rifle." *Newsletter of the Kentucky Rifle Association* 5, no. 3. The material is a reprint from an English soldier's memoirs.

Hanna, Charles F. *The Wilderness Trail*. 2 vols. New York: Scribner's, 1912.

Harriger, Russell E. *Long Rifles of Pennsylvania: Jefferson, Elk and Clarion Counties*. York, Pa.: Shumway, 1984. This is the first volume of publisher George Alfred Shumway's history of long rifle making in the Commonwealth.

Hartzler, Daniel. Arms Makers of Maryland. York, Pa.: Shumway, 1977.

Hayden, Ronald. "Kopp Family." *Newsletter of the Kentucky Rifle Association* 7, no. 2.

Hazard, Samuel. *Register of Pennsylvania*. Philadelphia: W. F. Geddes, January 1828–December 1835.

Headley, Richard. "Martin Meillen, Gunsmith." *Newsletter of the Kentucky Rifle Association* 4, no. 2.

Heller, W. J. "Gunmakers of Old Northampton." Vol. 17, *Proceedings of the Pennsylvania German Society*. Lancaster, Pa., 1906.

Hench, Stephen. "William Antes." *Newsletter of the Kentucky Rifle Association* 6, no. 3.

"Henry Leman, Riflemaker." *Muzzle Blasts*, August 1960, 4.

History of Bedford, Somerset and Fulton Counties. Chicago: Waterman, 1884.

History of Lodge 43, F. & A. M., Lancaster, 1785–1935. 1935.

"The Hummel Family." *Papers of the Snyder County Historical Society*. 11, no. 243.

"Hunters of Kentucky." *Newsletter of the Kentucky Rifle Association* 3, no. 1.

Hutslar, Donald. *Biographical Information*. Vol. 1; *Gunsmiths of Ohio*. York, Pa.: Shumway, 1973.

Johnson, Curtis. *Illinois Gunsmiths*. York, Pa.: Shumway, 1983. The author has been privileged to correspond and to exchange information with Mr. Johnson.

Jordan, Francis. *The Life of William Henry*. Lancaster Pa.: Press of the New Era, 1910.

Kauffman, Henry, Jr. *Early American Gunsmiths*. Harrisburg: Stackpole, 1952.

———. *The Pennsylvania-Kentucky Rifle*. Harrisburg: Stackpole, 1960.

Kaufmann, Charles. "David Morton." *Newsletter of the Kentucky Rifle Association* 10, no. 1.

Kentucky Rifle Association. *The Kentucky Rifle: A True American Heritage*. Washington, D.C.: Kentucky Rifle Association, 1986. Commonly known as the first KRA book, its purpose is the illustration of Kentucky rifles.

———, *Kentucky Rifles and Pistols, 1750–1850*. Columbus, Ohio: Kentucky Rifle Association, 1976. Commonly known as the second KRA book, its purpose is the same as the first. Some information in both books is incorrect and outdated.

Killikelly, S. *History of Pittsburgh*. Pittsburgh: B. C. & Gordon Montgomery Co., 1906.

Kindig, Joe, Jr. *Thoughts on the Kentucky Rifle in Its Golden Age*. Annotated 2d ed. York, Pa: Shumway, 1984. A monumental study, beautifully illustrated with guns from the author's unrivaled collection of fine, high-art Kentuckies, mostly from eastern Pennslyvania.

King, James C. "The Frontier Gunsmith and Indian Relations." *Western Pennsylvania History Magazine*, 50 (1967): 23–32.

Kleppinger, S. J. *Kleppinger-Clippinger-Kelpinger Family History*. Privately printed, 1956.

Lindert, Albert. *Gunsmiths of Indiana*. Homewood, Ill: Lindert, 1968.

Linn, John Blair. *Annals of the Buffalo Valley*. Harrisburg: L. S. Hart, 1877.

Logan, H. C. "J. E. Evans of Philadelphia." *Arms Gazette*, June 1978.

Logan, James. Account Book, 1712–19 and Ledger, 1720–27. Pennsylvania Historical Society collection, Philadelphia.

Lyon, E. L. "House of Tryon." *Gun Report*, April 1967, 18–23.

Lyon, R. T. "Conrad Horn." *Gun Report*, August 1962, 9–10.

McKnight, W. J. *Pioneer Outline History of Northwestern Pennsylvania*. Philadelphia: J. B. Lippincott Co., 1950.

MacMinn, Edwin. *On the Frontier with Col. Antes*. Camden, N.J.: S. Chew & Sons, 1900.

Mager, D. F. "The So-Called Kentucky Rifle as Made in Lancaster County." *Papers of the Lancaster County Historical Society* 9, no. 3 (1904).

"Memorial of Sundry Gun Manufacturers of the Borough of Lancaster in the State of Pennsylvania, to the Honorable, the Senate and the Representatives in the Congress of the United States." *Newsletter of the Kentucky Rifle Association* 2, no. 1. The gunsmiths who signed the petition, dated 1/28/1803, were Jacob Dickert, Peter Gonter, Abraham Henry, John Graeff, Henry DeHuff, Benjamin Hutz, John Haeffer, Jacob Haeffer, Abraham Pieper, and Christopher Gumpf.

Montgomery, M. L. *History of Berks County*. Reading: Berks County Historical Society, 1886.

———. *Berks County in the Revolution*. Reading: Berks County Historical Society, 1894.

Moyer, Del-Louise. "Upper Places and Nazareth Agreements and Inventories, 1758–1770." Moravian Archives, Moravian College. Contains inventories related to the gunshop at Christian's Spring.

Musser, John M. "Peter Gonter." *Newsletter of the Kentucky Rifle Association* 6, no. 2.

Nazareth Historical Society. *History of Nazareth*. Nazareth: Nazareth Historical Society, 1914.

Patterson, C. M. "Aaron Vaughan's Double-barreled Percussion Revolver." *Gun Report*, December 1959.

"Philadelphia Gunsmiths in 1829." *Gun Collector* 24.

Pippert, Carl. "Christian Hawken." *Newsletter of the Kentucky Rifle Association* 7, no. 3.

G. R. Prowell. *History of York County*. 1883.

Pugh, James. "Charles Roth: Wilkes-Barre Gunsmith." *Muzzle Blasts*, September 1960, 4.

Roberts, Ned H. *The Muzzle Loading Cap Lock Rifle*. York, Pa.: Shumway, 1940.

Rowan, D. C. "Charles Roth, Wilkes-Barre Gunsmith." *Newsletter of the Kentucky Rifle Association* 8, no. 3.

Rupp, I. D. *History of Dauphin, Cumberland, Franklin, Bedford, Adams and Perry Counties*. Lancaster City, Pa.: G. Hills, 1846.

———. *History of Lancaster County.* Lancaster Pa.: G. Hills, 1844.

———. *History of Northampton, Lehigh, Monroe, Carbon and Schuyler Counties.* Harrisburg: Hickok & Cantine, Printers, 1845.

Satterlee, L. D. "Owens Evans and Valley Forge." *Hobbies,* October 1938.

Sellers, Frank. *American Gunsmiths.* 1983. A major compilation drawing on most, if not all, listings and area studies up to Sellers's press time, but not critical or well edited.

Serven, James E. "American Flintlock Pistols." *Gun Report,* April 1969, 8–16.

Shelton, Lawrence. *California Gunsmiths.* Sacramento: Far West Publishers, 1977.

———. "Joseph Allen Craig." *Gun Report,* January 1976, 60–61.

———. "Smith Family Gunsmiths." *Muzzle Blasts,* May 1985, 1.

———. "William Golcher."*Gun Report,* November 1977, 56–58.

Shriver, Samuel S. *History of the Shriver Family and Their Connections.* Privately printed, 1888.

Shumway, George Alfred. "Notes on the Hunters of Kentucky." *Newsletter of the Kentucky Rifle Association* 3, no. 3.

———. *Rifle in Colonia America.* 2 vol. York, Pa.: Shumway, 1980. A beautifully photographed, high-art book, definitive on its subject.

———. *Pennsylvania Long Rifles of Note.* 2d ed. York, Pa.: Shumway, 1977. A study of the major schools of gunsmithing in the Commonwealth.

Smith, Ray. "The Kentucky Rifle," *Bulletin of the Snyder County Historical Society* 5, no. 2 (1946).

"Snyder County Gunsmiths." *Bulletin of the Snyder County Historical Society* 1 (no. 243; 2, no. 14; 11, no. 243.

Stockbridge, V. D. *Digest of U.S. Patents Relating to Breech-loading and Magazine Small Arms.* 1963.

Strassburger, R. B., ed. *Pennsylvania German Pioneers,* 2 vols. Norristown, Pa.: Pennsylvania German Society, 1934.

Swank, James M. *History of Iron and Steel.* Philadelphia: The American Iron and Steel Association, 1892.

J. T. Teesdale. "Gun Making Industry in Wisconsin." *Wisconsin Magazine of History* 32, no. 3 (1949).

———. "Wisconsin Gun Makers." *Gun Collector* 1.

Tepper, Michael, ed. *Emigrants to Pennsylvania.* Baltimore: Genealogical Publishers, 1979.

Tilton, F. G. "Joseph Mills." *Newsletter of the Association of Ohio Long Rifle Collectors* 5, no. 1, 15–16.

U.S. Government. *Specifications and Drawings of Patents for Machine Guns Issued from the U.S. Patent Office Prior to January 1, 1882.* Washington, D.C.: U.S. Printing Office, 1882.

"William Henry, Riflemaker." *Muzzle Blasts,* July 1960, 4.

Williams, Henry W., Jr. "An 18th Century Gunsmith's Pattern Book." *Connoisseur,* July 1943.

Wittlinger, Carlton O. "The Small Arms Industry of Lancaster Co., 1740–1840." *Pennsylvania History* 24, no. 2, 121–136.

Woods, Michael D. "Armourer John Miles at Moose Fort." *The Beaver,* April 1986, 19–26.

Woodside, Sam. "John Watts, Gunsmith of Pennsylvania." *Muzzle Blasts,* December 1954, 9–10.

Yates, W. R. *Bethelem of Pennsylvania: The First 100 Years, 1741–1841.* Bethlehem, Pa.: privately printed, 1868.